Population Genetics
and Ecology

Academic Press Rapid Manuscript Reproduction

Proceedings of the Conference held in Israel
March 1975

Population Genetics and Ecology

EDITED BY

Samuel Karlin
The Weizmann Institute of Science
Rehovot, Israel
and
Stanford University
Stanford, California

Eviatar Nevo
University of Haifa
Haifa, Israel

Academic Press, Inc.   New York   San Francisco   London   1976
A Subsidiary of Harcourt Brace Jovanovich, Publishers

WILLIAM MADISON RANDALL LIBRARY UNC AT WILMINGTON

COPYRIGHT © 1976, BY ACADEMIC PRESS, INC.
ALL RIGHTS RESERVED.
NO PART OF THIS PUBLICATION MAY BE REPRODUCED OR
TRANSMITTED IN ANY FORM OR BY ANY MEANS, ELECTRONIC
OR MECHANICAL, INCLUDING PHOTOCOPY, RECORDING, OR ANY
INFORMATION STORAGE AND RETRIEVAL SYSTEM, WITHOUT
PERMISSION IN WRITING FROM THE PUBLISHER.

ACADEMIC PRESS, INC.
111 Fifth Avenue, New York, New York 10003

United Kingdom Edition published by
ACADEMIC PRESS, INC. (LONDON) LTD.
24/28 Oval Road, London NW1

Library of Congress Cataloging in Publication Data

Main entry under title:

Population genetics and ecology.

    Bibliography: p.
    Includes index.
    1.   Population genetics—Congresses.   2.   Ecology—Con-
gresses.   I.   Karlin, Samuel,   (date)   II.   Nevo,
Eviatar.   [DNLM: 1.   Genetics, Population—Congresses.
2.   Ecology—Congresses.   QH455 P831 1975]
QH455.P66   575.1   75-13075
ISBN 0–12–398560–9

PRINTED IN THE UNITED STATES OF AMERICA

QH455
, P66

Contents

168957

CONTENTS

Part II Models and Evidence

Part III Theoretical Studies

CONTENTS

Contributors

C.M. Ann Baker, Department of Zoology, University of Adelaide, South Australia 5001

Ze'ev Bar, Department of Biology, University of Haifa, Haifa, Israel

John A. Beardmore, Department of Genetics, University College of Swansea, University of Wales, Swansea SA 2, 8 PP, U.K.

Avigdor Beilis, Department of Genetics, The Hebrew University of Jerusalem, Jerusalem, Israel

Robert J. Berry, Royal Free Hospital School of Medicine, University of London, London WC1, England

Julio Bodmer, Genetics Laboratory, Department of Biochemistry, University of Oxford, Oxford OX1 3QU, England

Walter F. Bodmer, Genetics Laboratory, Department of Biochemistry, University of Oxford, Oxford, OX 1 3QU, England

Freddy B. Christiansen, Department of Genetics, University of Aarhus, DK 8000 Aarhus C, Denmark

C. Clark Cockerham, Department of Statistics, North Carolina State University, Raleigh, North Carolina 27607

Dan Cohen, Department of Botany, The Hebrew University, Jerusalem, Israel

R. Dennis Cook, School of Statistics, University of Minnesota, Saint Paul, Minnesota 55101

Walter F. Eanes, Department of Ecology and Evolution, State University of New York at Stony Brook, Stony Brook, New York 11794

Ilan Eshel, Department of Statistics, Tel-Aviv University, Ramat-Aviv, Israel

Warren J. Ewens, Department of Biology, University of Pennsylvania, Philadelphia, Pennsylvania 19174

Marc W. Feldman, Department of Biological Sciences, Stanford University, Stanford, California 94305

Gert Forkmann, Institute für Biologie II, Lehrstuhl für Genetik, Universität Tübingen, D 7400 Tübingen, W. Germany

*Ove Frydenberg,** Department of Genetics, University of Aarhus, DK 8000 Aarhus, C, Denmark

Karl P. Hadeler, Lehrstuhl für Biomathematik, Universität Tübingen, 7400 Tübingen, W. Germany

Daniel L. Hartl, Department of Biological Sciences, Purdue University, West Lafayette, Indiana 47907

William G. Hill, Department of Genetics, University of Edinburgh, Institute of Animal Genetics, Edinburgh EH9 3JN, U.K.

Subodh K. Jain, Department of Agronomy and Range Science, University of California at Davis, Davis, California 95616

Suresh D. Jayakar, Laboratorio Di Genetica Biochimica ed Evoluzionistica, 27100 Pavia, Italy

Samuel Karlin, Department of Mathematics, The Weizmann Institute of Science, Rehovot, Israel, and Department of Mathematics, Stanford University, Stanford, California 94305

Joseph Katznelson, Division of Range and Pasture, Neve Ya'ar Regional Experiment Station, Post Haifa, Israel

Jacob Koach, Genetics Unit, George C. Wise Center for Life Sciences, Tel-Aviv University, Ramat-Aviv, Israel

Richard K. Koehn, Department of Ecology and Evolution, State University of New York at Stony Brook, Stony Brook, New York 11794

Jesse Krakauer, Department of Biological Sciences, Stanford University, Stanford, California 94305

Peter Lange, Institut für Biologie II, Lehrstuhl für Genetik, Universität Tübingen, D 7400 Tübingen, W. Germany

Barrie D.H. Latter, School of Biological Sciences, University of Sydney, N.S.W. Australia

Clyde Manwell, Department of Zoology, University of Adelaide, South Australia 5001

Carlo Matessi, Laboratorio Di Genetica Biochimica ed Evoluzionistica, 27100 Pavia, Italy

---

*The editors express their deep sadness at the passing away of Professor O. Frydenberg some weeks after the Conference. Our heartfelt condolences to the family.

Masatoshi Nei, Center for Demographic and Population Genetics, The University of Texas at Houston, Houston, Texas 77025

Eviatar Nevo, Department of Biology, The University of Haifa, Haifa, Israel

Peter O'Donald, Department of Genetics, University of Cambridge, Cambridge CB4 1XH, England

Josephine Peters, Royal Free Hospital School of Medicine, University of London, London WC1, England

Nira Richter-Dyn, Mathematics Department, Tel-Aviv University, Ramat-Aviv, Israel

Allan Robertson, Institute of Animal Genetics, University of Edinburgh EH9 3JN, U.K.

Wilhelm Seyffert, Institut für Biologie II, Lehrstuhl für Genetik, Universität Tübingen, D 7400 Tübingen, W. Germany

Sajjad A. Shami, Department of Genetics, University College of Swansea, University of Wales, Swansea SA2, 8PP, U.K.

Montgomery Slatkin, Department of Biophysics and Theoretical Biology, The University of Chicago, Chicago, Illinois 60637

Curtis Strobeck, School of Biological Sciences, University of Sussex, Sussex BN1 9QG, England

Renate Strobel, Institut für Biologie II, Lehrstuhl für Genetik, Universität Tübingen, D 7400 Tübingen, W. Germany

John A. Sved, School of Biological Sciences, University of Sydney, N.S.W. Australia

Glenys Thomson, Genetics Laboratory, University of Oxford, Department of Biochemistry, Oxford OX1 3QU, England

John R.G. Turner, Department of Ecology and Evolution, Division of Biological Sciences, State University of New York, Stony Brook, New York 11794

Bruce Wallace, Division of Biological Sciences, Cornell University, Ithaca, New York 14850

Klaus Wöhrmann, Institut für Biologie II, Lehrstuhl für Genetik, Universität Tübingen, D 7400 Tübingen, W. Germany

David Wool, Genetics Unit, George S. Wise Center for Life Sciences, Tel-Aviv University, Ramat Aviv, Israel

Preface

These proceedings derive from a conference-workshop devoted to topics in Population Genetics and Ecology, which took place in Israel, March 13-23, 1975. Lectures and discussion sessions focused on three main objectives: (1) The elaboration and explication of determinants of genetic variation in natural populations; (2) experimental design and analysis of field and laboratory data; (3) theory and applications of mathematical models in population genetics. The papers of this volume report a number of field and laboratory projects describing a variety of spatial and temporal character and enzyme frequency patterns in natural populations, and suggest possible associations of these patterns with ecological parameters. Some papers discuss the origin of race formation, domestication, and the interaction between sexual selection and natural selection. Also, a number of models are proposed for estimating fitness coefficients, testing selective neutrality, the nature of coadapted gene complexes, and relevant evidence provided. Among the theoretical studies presented are facets of selection migration interaction, stochastic selection effects, properties of density and frequency dependent selection, concepts and measures of genetic distance and speciation, aspects of altruism, and kin selection.

The proceedings divide naturally into three parts. Part I includes 10 papers concentrating mostly on field and laboratory studies. Part II consists of 10 papers offering a combination of data analyses and interpretations of related models. The last part, comprising 11 papers, presents mostly theoretical results. Each part has an introduction including brief abstracts of the individual papers. The volume ends with an edited version of the final taped discussion of the conference:

The participants, representing many countries, comprised a mix of naturalists, experimentalists, theoreticians, statisticians, and mathematicians. The sessions constantly exhibited substantial hybrid vigor meshing the intuition and experience of the naturalists with the analytic insights and approaches of the theoreticians and mathematicians.

The conference was conducted at five locations in Israel (of the character of a peripatetic seminar perhaps in the best Socratic tradition). The opening sessions at the Weizmann Institute of Science in Rehovot were devoted to topics in linkage and selection, aspects of multilocus systems, and a discussion of some mixed mating patterns. The conference convened next at the University of Haifa on the northern coast of Israel. The theme of these meetings revolved about the causes and influences of gene frequency variability in natural populations. Afterward the participants moved on to a location near the Sea of Galilee in the northeast of Israel adjacent to the putative site of the origin of wild wheat. Controversies and lively discussions centered on the subject of testing selective neutrality and statistical methodology. We then journeyed down the Jordan Valley and up the Judean Hills to Jerusalem. The talks here concentrated on problems of human genetics and polygenic inheritance. Returning to the Weizmann Institute via the Southern Israeli Desert (Negev) the last sessions dealt with problems of speciation, mixed ecological-genetic systems, altruism and kin selection. A final free-for-all discussion culminated the conference workshop.

We are indebted to The Volkswagon-Stiffung Foundation, The Weizmann Institute of Science, The University of Haifa, The Israel National Academy of Science, The Israel Scientific-Development Council, and The Israel Ministry of Agriculture, for providing financial support of the conference.

Samuel Karlin
Eviatar Nevo

PART I FIELD AND LABORATORY STUDIES

The studies in Part I are mainly descriptive, that impinge upon a variety of problems important for evolutionary theory. Among the topics are gene frequency patterns in space and time of animal and plant populations, the origin of new taxa, animal and plant domestication, variation in heritability related to parental age and problems in the genetics of certain haplo-diploid populations.

Brief abstracts of the papers now follow:

Beardmore and Shami discuss the problem of heritability as a function of parental age. They demonstrate the existence of large age-dependent variation in the heritability of meristic characters in the fruit fly Drosophila and the guppy fish Poecilia. Berry and Peters analyze patterns of genetic polymorphisms in the recently established Skokholm house mouse population. They concluded that the evidence, based particularly on the Hbb locus and skeletal traits, favors a founder effect influence on the genetic structure of the population. Observed cyclic seasonal changes are probably due to selection, with random drift an unlikely agent.

Jain surveys patterns of survival and microevolution in plant populations. Field studies provide evidence for large variation, presumably both genetic and environmental, in each of the variables studied in detail. Some species' specific patterns suggest tests of optimal strategy arguments.

Katznelson describes genetic changes during the domestication processes in Trifolium berytheum. Quantitative changes involve significant increases in seed weight, size and shape of plant, and growth rate. As well, qualitative changes involving seed color and hardness and breakdown of self-compatibility in some lines can be documented. Manwell and Baker review the unusual nature of certain variants in protein polymorphisms in populations of domesticated species as possible evidence for the theory of hybrid origin of these species. Their survey in cattle, sheep, goat, rabbit and duck shows that a number of the major protein polymorphisms in domesticated species involve differences of two or more non-adjacent amino acid substitutions and that the hypothetical intermediate forms are missing.

Nevo relates environmental heterogeneity and genic variation in an attempt to test a niche-variation model. He compares and contrasts genic variation in four species of Israeli toads and frogs occupying increasingly varying and unpredictable environments. Nevo and Bar discuss spatial allozyme variation encoded by 18 loci and the shell banding polymorphism in the land-snail Theba pisana in Israel. These appear to be clines corresponding to a climatic gradient in allelic frequencies at 7 out of the 10 polymorphic loci examined.

Turner discusses the role of ecological islands in adaptive race formation for Heliconius butterflies. He suggests that adaptive radiation, involving alterations in pattern, apparently precedes the evolution of isolating mechanisms.

Wöhrmann, Strobel and Lange are concerned with the characteristics of yeast populations involving vegetative and sexual phases of reproduction. They report both experimental and simulation results pertinent to the variation of frequencies at a mating type locus under two different conditions of medium and relative length of the haploid and diploid phases.

Wool and Koach relate morphological and environmental variations of the gall forming aphid, Geoica utricularia, in Israel. Morphological variation in 19 characters was investigated by factor and multiple regression analyses.

2

Parental Age, Genetic Variation and Selection

J.A. BEARDMORE and S.A. SHAMI

INTRODUCTION

The relation between the heterogeneity in time and space of environments and the genetic heterogeneity of populations of organisms in such environments has attracted both theoretical and experimental attention from a number of workers: Da Cunha and Dobzhansky (1954), Thoday (1953), Mayr (1963), Beardmore and Levine (1963), Long (1969), Powell (1971), Nevo (these proceedings). In general these studies have indicated, either on theoretical or experimental grounds, that a positive relationship between the two types of heterogeneity exists. Perhaps this should be qualified by saying that within any one species such a relationship is expected. As Karlin (these proceedings) has indicated, it is unlikely that absolute measures of environmental heterogeneity can, in fact, be made and comparisons should be on a relative basis which may make comparison between species difficult. One of the problems implicit in Karlin's argument is that it is very difficult, if not impossible, to specify and/or measure all of the relevant variables of the environment. In this paper we wish to discuss some effects of a quantifiable environmental factor which has not received much attention from population geneticists in the past probably because it is a variable of the

<u>internal</u> environment of organisms, namely parental age. At
first sight this may seem to have little to do with considerations of genetic variability in population, but, as the
data presented show, parental age may be of considerable relevance to the study of selection for quantitative characters.
Furthermore, in view of the considerable significance attributed to genetic factors in ageing processes (McFarlane
Burnet, 1974) such studies are worthwhile for other reasons.

In a recent paper (Beardmore, Lints and Al-Baldawi, 1975)
results were presented of a series of experiments in which
the heritability (based on offspring-mid parent regression and
on half sibs) of sternopleural bristle number of <u>Drosophila
melanogaster</u> was studied in relation to parental age. Table
1 shows the mean estimates of offspring mid-parent h^2
obtained in a number of experiments.

The 14, 21 and 28 day estimates do not differ significantly
from each other but the 3 day estimate is significantly
different (P = < 0.01) from those derived from older parents.
A similar result is also found when half-sib estimates of h^2
from 3 day old and older parents are compared. These findings,
which have been supported by results obtained by Lopez-Fanjul
(personal communication), led us to look at the inheritance
of a meristic character in another organism, the guppy <u>Poecilia
reticulata</u>. The character used was the number of rays in the
caudal fin (CFR) which, in the wild type stock used, in large
tank populations maintained at 25°C has a distribution as
shown in Figure 1.

In the first of these experiments a group of 34 pairs of
fish of uniform age was used to produce successive progenies
from single pairs mated assortatively. The fish become
sexually mature at approximately 3.5 months of age, under
laboratory conditions will survive for up to two years and

are capable of producing broods at intervals of about 4-5 weeks.

At intervals of four weeks, estimates of h^2 of CFR number based on offspring mid-parent regression were made using data obtained from the broods produced in the preceding four week period. The estimates so obtained are plotted in Figure 2 against the mean age of the parental fish producing the broods. Figure 2 also gives comparable h^2 estimates derived from a group of 42 single pair matings (Experiment III) set up approximately one year after the start of the first experiment. The agreement between h^2 estimates for comparable ages from the two experiments is strikingly good. Consideration of the data for Experiment I shows that the h^2 values increase, with advancing age, from an initial value of about 0.5 to plateau at approximately 11 months at a value of heritability of effectively 1 . Figure 3 shows the calculated regression lines for the two segments of the data 4.5 - 10.5 months and 10.5 - 22 months in Experiment I and for the period 5 - 11 months in Experiment III, together with the values of b and the attached P values.

It is clear that the increase of h^2 in segment I of Experiment I and in Experiment III is highly significant while for segment II of Experiment I a non-significant value of b is obtained with a mean h^2 of 0.93.

The findings with sternopleural bristle number in <u>Drosophila</u> are thus paralleled by these data from <u>Poecilia</u> and it therefore seems likely that the age dependent effect on heritability may be of some general significance. Extension of these studies to other organisms including mouse, man, and higher plants is being undertaken by Professor Lints and his group in Louvain and by workers in our laboratory. Very sparse data available on the heritability of finger ridge count in man point in the same direction as the fruit fly and fish results but not significantly so.

In the Poecilia studies we are not yet able to say whether the age-h^2 effect is produced by both or only one of the two sexes. In the Drosophila study it is clear that under some conditions the age of a parent of either sex can influence h^2 while in other conditions only the maternal age seems to be an effective variable.

Beardmore et al. (1975) suggested that the age-dependent effect on h^2 could be due to the operation of one or more of three factors.

1. Meiotic drive or an effectively similar process such as gametic selection in the female system.
2. Change in the quantity of genetic material passed into gametes.
3. Change in the state of gene activity of loci influencing bristle number.

The data presently available do not yet enable us to establish which of these hypotheses, if any, is correct. There are cases of age-dependent changes in segregation ratios reported for mutant genes in Drosophila, for example Heuts (1956). In man Hiraizumi et al. (1973) have reported birth order effects on segregation at the ABO locus in man which may have a parental age component.

The latter case suggests that genes which are components of polymorphic variation can display an age-polarized segregation. However, the relevance of this example to the loci which may be segregating for bristle genes (probably ten or more, Mather and Jinks, 1971) is not immediately clear. If non-random segregation at meiosis or post-meiotic gametic selection is involved it would need to work in such a way that gametes containing particular genomes (e.g., largely

6

derived from the paternal set of chromosomes of the individual producing gametes) would be more likely to take part in ultimate zygote formation as parents aged. In view of what is known of the genetic architecture of bristle number in Drosophila, such a process would be exceedingly complicated.

An initial attempt to assay nuclear DNA content in cells from parents of different ages has been made using a micro-densitometer technique (Rees and Walters, 1965; Verma and Rees, 1974) on secondary spermatocytes in Poecilia. The data in Table 2 show the mean values and standard errors for this character in old and young males. The table also gives corresponding values for nuclear area which is, of course, related to nuclear volume. These data show clearly that there is no evidence for the view that the amount of DNA contained in gametes from young and old parents differ (P = > 0.5). The nuclear area, on the other hand, differs significantly in the two classes (P = < 0.001). This difference in average nuclear volume can most plausibly be related to age-dependent changes in nuclear protein. A decrease in the volume of nuclear protein with age has been reported by Herrman et al. (1969) working with mouse liver and Pyhtiliä and Sherman (1969) reported that rat liver nucleo-protein decreased with advancing age.

In view of the regulatory role such proteins are thought to play in the organization of DNA and in the control of the activity of genes in eucaryotic cells (Lewin, 1974) it seems probable that hypotheses 3 is the one that can most profitably be pursued (Hahn and Fritz, 1966).

Brackenridge and Teltscher (1975) report data on the age of onset of Huntington's Chorea in man which indicate that, for a given age of onset in the affected parent, there is a significant inverse relationship between the age at reproduction of that parent and the onset-age in the child. Their interpretation is that this effect is due to an accumulation of somatic mutations but it seems equally likely that the effect is relatable to changes in regulation of gene activity in much the same way as in our data. In a formal sense too the "heritability" of the Huntington's Chorea gene appears to be negatively correlated with parental age though not significantly so.

It is worth pointing out that both of the characters referred to in this paper are not trivial with respect to natural selection. For sternopleural chaeta number in <u>Droso-phila</u>, Scossiroli (1959), Barnes (1968), Kearsey and Barnes (1971) and Linney <u>et al</u>. (1972) have shown that stabilizing selection acts upon laboratory populations with appreciable pressure.

In <u>Poecilia</u> we have examined a number of major components of Darwinian fitness in relation to CFR number. Data on the percentage survival to fifty-two weeks (under moderately competitive conditions at 25°C), growth rate, and total fertility to fourteen weeks and fifty-two weeks of age are given in Figures 4, 5 and 6.

Some measure of ability to adapt to, or at least survive in, a biologically meaningful new environment is indicated in Figure 7 where the survival over a period of eight days in water at a salinity of 1% is plotted in relation to CFR number.

In all of the characters except growth rate it is clear that the 27 CFR class is the optimum phenotype. In growth rate the 25 class is non-significantly higher than the 27 class but when all characters are considered together the superiority of the 27 CFR class is strikingly evident. Despite the fact that we are still unable in any critical way to analyze or predict the genetic consequences of stabilizing selection upon a quantitative or meristic character (Robertson, 1963) it is evident that studies of heritability changes in these characters have some relevance to considerations of the effects of various selection modes on natural populations.

Effects of grand-parental age

The effects of parental age upon heritability so far described in this paper are relatively clear, if unexpected. An additional complexity is introduced when the variable of grandparental age is introduced. An experiment (Experiment II) in which the progenies of old and young parents were to be grown at the same time was set up by taking progeny from the first experiment to use as parents. Fish collected from progenies produced in the first experiment at the time marked approximately by arrow A in Figure 2 represented the old parents while fish collected in a similar way at point B formed the initial group of young parents. To our surprise the estimates of h^2 derived from these two groups of parents were the opposite of what would be predicted from the results obtained in Experiments I and III. The old parents gave an estimate of h^2 of 0.16 while the estimate from the young parents was 0.86. This difference, however, is not significant ($P = < 0.2$).

Further broods were raised from both sets of parents and the resulting estimates of h^2 are shown in Figure 8a. When the two sets of successive broods are compared using an analysis of variance it is seen (Table 3) that they differ significantly both as to mean and to the slope of the regression of h^2 (Figure 8b) on time after the start of the experiment. Although it may be that estimates from later broods will indicate that the appropriate regression for these data is non-linear we can nevertheless be confident that the pattern of h^2 estimates is affected significantly by the age differences in the parents in a way totally unlike that found in Experiments I and III. The explanation appears to lie in the fact that in Experiment II a difference in the age of the parents in the two groups is confounded with an age difference of their parents. Reference to the arrows in Figure 2 makes it evident that the young parents are themselves derived from old parents while the parents of the old parents group were young. In view of the excellent agreement of the results in Experiments I and III in which grand-parental age was random-ized it seems highly probable that the age of the grandparents can affect h^2 estimates in Poecilia. No relevant data for Drosophila are yet available and the effect will be looked for in other organisms. It seems unlikely that the effect has any considerable significance for studies of populations but in special circumstances it is conceivable that it might have.

Age effects and selection

Any general consideration of age effects and selection must necessarily wait upon evidence of the ubiquity of such effects in a range of life forms. It will also be necessary to look at characters which, unlike the two meristic characters so far studied, are liable to vary from time to time during the lifetime of the individual. With these qualifications it is nevertheless worth while to speculate briefly on the potential significance of the results reported in this paper. The implications for characters such as I.Q. in human populations are obvious and need no further comment at this stage. In other organisms exposed to directional selection the age structure of the adults concerned in repro-

10

duction at any point in time could clearly exert a marked effect upon the rate and magnitude of the response to any given selection pressure on a quantitative character.

REFERENCES

Barnes, B.W. (1968). Heredity 23: 433-442.

Beardmore, J.A. and L. Levine. (1963). Evolution 17: 121-129.

Beardmore, J.A., F.A. Lints and A.L.F. Al-Baldawi. (1975). Heredity 34 (1), 71-22.

Brackenridge, C.J. and B. Teltschev. (1975). J. Med. Genet. 12: 64-69.

Burnet, M., Sir. (1974). Intrinsic mutagenesis - a genetic approach to ageing. Medical and Technical Publishing Co. Ltd., London.

Da Cunha, A.B. and Th. Dobzhansky. (1954). Evolution 8: 119-134.

Herrmann, R.L., A.R. O'Meara, A.P. Russel and L.E. Dowling. (1969). Proc. 8th Int. Cong. Geront. 1: 141-142.

Heuts, M.J. (1956). Agricultura 4: 343-352.

Hiraizumi, Y., C.T. Spradlin, R. Ito and S.A. Anderson. (1973). Amer. J. Hum. Genet. 25: 277-286.

Kearsey, M.J. and B.W. Barnes. (1970). Heredity 25: 123-125.

Karlin, S. (1975). These proceedings.

Lewin, B. (1974). Gene expression Vol. 2 Eucaryotic chromosomes. John Wiley & Sons, London.

Linney, R., B.W. Barnes and M.J. Kearsey. (1971). Heredity 27: 163-174.

Long, T.C. (1970). Genetics 66: 401-416.

Lopez-Fanjul, C. (1975). Personal communication.

Mather, K. and J.L. Jinks. (1971). Biometrical Genetics. Chapman and Hall, London.

Mayr, E. (1963). Animal species and evolution. The Belknap Press of Harvard University Press, Cambridge, Mass.

Nevo, E. (1975). These proceedings.

Powell, J.R. (1971). Science 147: 1035-1036.

Pyhtilä, M.J. and F.G. Sherman. (1969). Gerontologia 15: 321-327.

Rees, H. and M.R. Walters. (1965). Heredity 20: 73-82.

Robertson, A. (1963). Proc. XIth Int. Congr. Genetics 3: 527-532.

Scossiroli, R.E. (1959). Comitato Nazionale per le Richerche Nucleari, Divisione Biologica (CNB-4).

Thoday, J.M. (1953). Symp. Soc. Exp. Biol. 7: 96-113.

Verma, S.C. and H. Rees. (1974). Heredity 33: 61-68.

Von Hahn, H.P. and Elvira Fritz. (1966). Gerontologia 12: 237-250.

TABLE 1. *Mean estimates of OP h^2 of sternopleural bristle number in* Drosophila *obtained from randomly mated parents in a number of experiments (from Beardmore* et al. *1975).*

Age	3	14	21	28
h^2	0.281	0.483	0.385	0.408
no. of experiments	6	2	3	7

TABLE 2. *Mean values and standard errors for nuclear DNA content and nuclear area in secondary spermatocytes of* Poecilia reticulata.

Age of Male	DNA (arbitrary units)	Nuclear area (arbitrary units)
Young $\bar{X}$	13.0676	5.2436
s.e.	0.3855	0.1253
Old $\bar{X}$	13.3771	4.7346
s.e.	0.3497	0.0837

13

TABLE 3. *Analysis of variance of regression of heritability on age of parents in crosses started with old and young parents (Experiment II).*

Item	d.f.	ss	MS	F	P
Joint regression	1	683.0955	683.0955	183.5883	< 0.001
Differences between regressions	1	582.8423	582.8423	156.6443	< 0.001
Difference between means	1	4260.3727	4260.3727	1145.0152	< 0.001
Error	135	502.3116	3.7208	—	—

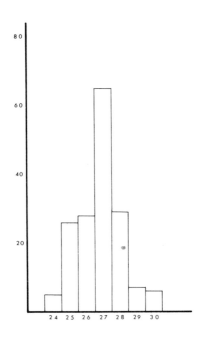

Figure 1. *Distribution of caudal fin ray number (CFR) in Poecilia reticulata*

X - CFR number

Y - number of fish

14

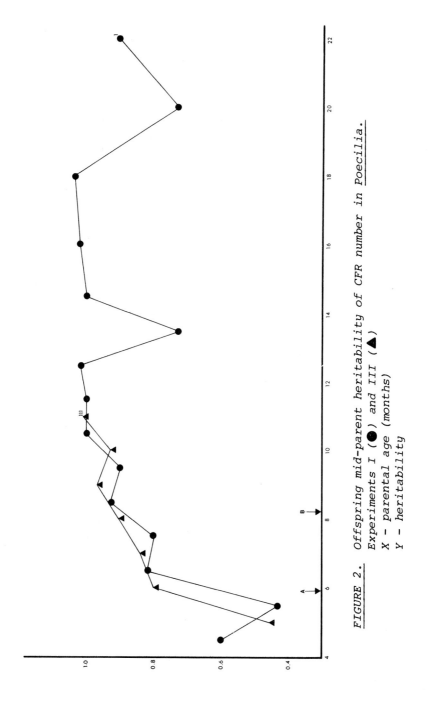

FIGURE 2. *Offspring mid-parent heritability of CFR number in* Poecilia.
Experiments I (●) and III (▲)
X – parental age (months)
Y – heritability

15

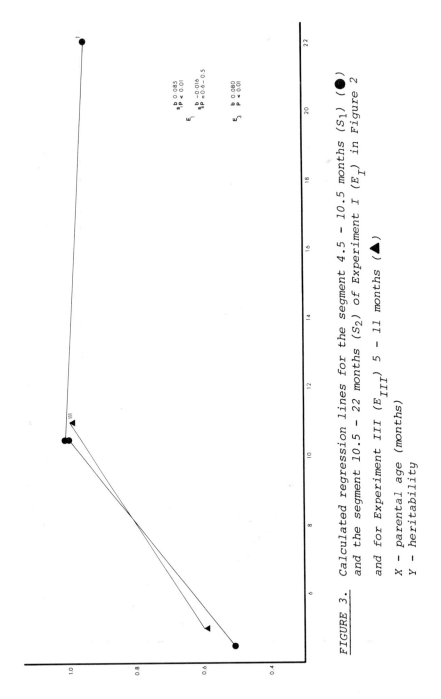

FIGURE 3.   *Calculated regression lines for the segment 4.5 – 10.5 months (S_1)* (●)
and the segment 10.5 – 22 months (S_2) of Experiment I (E_I) in Figure 2
and for Experiment III (E_{III}) 5 – 11 months (▲)

X – parental age (months)
Y – heritability

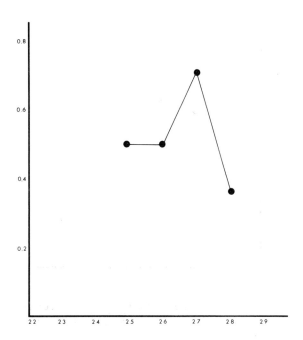

FIGURE 4. *Stabilizing selection in* Poecilia.*-Survival at*
25°C in relation to CFR number (0 - 52 weeks)
X - CFR number
Y - percentage survival

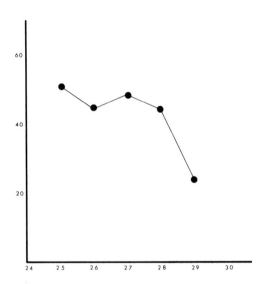

FIGURE 5. *Stabilizing selection Poecilia.-Growth rate (mg/month) in relation to CFR number (0 - 14 weeks).*
X - CFR number
Y - mean growth rate

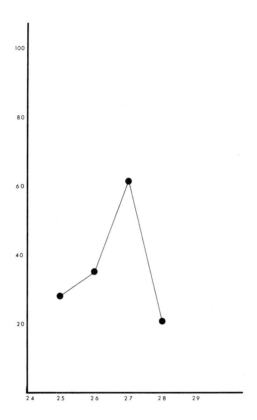

FIGURE 6. *Stabilizing selection in* <u>Poecilia</u>.*-Total fertility in relation to CFR number (to 52 weeks of age)*
X - CFR number
Y - Total progeny

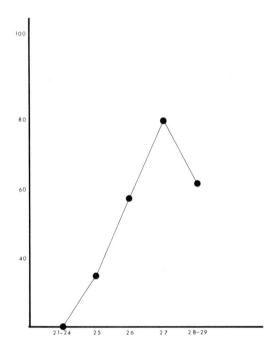

<u>FIGURE 7.</u> *Stabilizing selection in <u>Poecilia</u>.-Survival in*
water at 1% salinity over a period of eight days
in relation to CFR number.
X - CFR number
Y - percentage survival

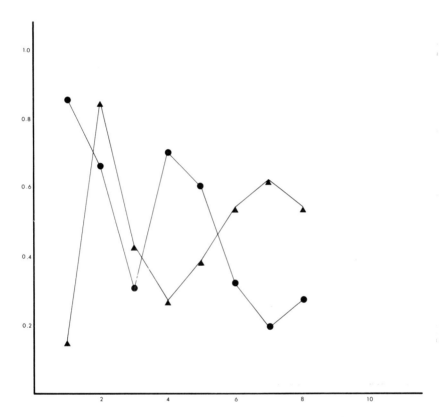

FIGURE 8a. *Heritability (OP of CFR number in* Poecilia *in young parents (●) and old parents (▲). Experiment II.*
X - months after start of experiment
Y - heritability

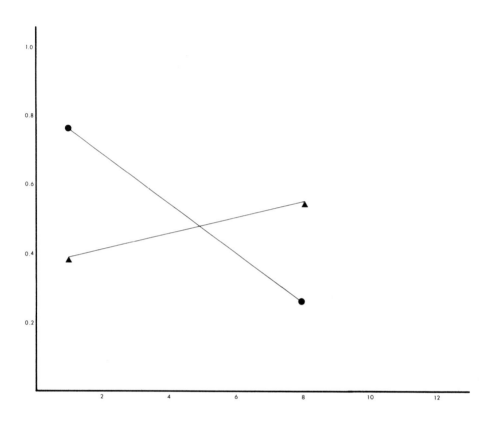

FIGURE 8b. *Calculated regression line of heritability on parental age with young parents (●) and old parents (▲). Experiment II.*
X – months after start of experiment
Y – heritability

Genes, Survival, and Adjustment in an Island Population of the House Mouse

R. J. BERRY and J. PETERS

INTRODUCTION

No amount of intelligent introspection or sophisticated
model building can ever substitute for detailed observation
and experiment on real life populations. This paper is an
account of a study on house mice (Mus musculus L.) living on
an island where they were isolated from immigration and subjected to harsh environmental conditions. We have made sundry
mistakes in a 10-year investigation, and we want to point
these out in the context of presenting some aspects of the
ecological genetics of the island population.

In the 1950s, Grüneberg (reviewed 1963) described the inheritance of a range of minor skeletal variants in inbred
strains of house mice. Weber (1950) found the same characters
in small samples of wild-living mice trapped in London buil-
dings. It seemed possible that these traits could be used as
markers to investigate genetical relationships, and in 1960
one of us (R.J.B.) began collecting mice from a range of localities with the intention of testing their value for this
purpose, using animals caught from corn-ricks when they were
broken down for threshing in early spring (Southern & Laurie,
1946; Rowe, Taylor & Chudley, 1963). The colonizers of each

23

rick can be regarded as a random sample of the local population, and, since every animal is caught, no sampling errors
are introduced.

The most informative results came from 20 ricks on a single
farm at Odiham in Hampshire, central southern England.   No
obvious environmental factors affected the incidence of any
variant, but the frequencies of many variants differed
greatly, even between adjacent ricks (Berry, 1963). Experimental work has shown subsequently that maternal health can
influence the occurrence of particular variants, but the overall spectrum of non-metrical variants possessed by a mouse is
an apparently valid indicator of a substantial part of its
genome (Howe and Parsons, 1967, following Searle, 1954; Deol
and Truslove, 1957).

When the 20 Odiham samples were combined, the pooled sample fitted well into the general South English picture, suggesting that it was the process of sub-division that had produced
the inter-rick differences, but there was no way of proving
this. Pest control workers have described the basic ecology
of mouse rick populations (e.g. Southern, 1954; Southwick,
1958; Newsome & Crowcroft, 1971), but many important genetical questions remain, such as the degree of genetical heterogeneity in agricultural house mice; the amount of individual
movement (and gene-flow); genetical differences between colonizers and refuge-seekers; the numbers of mice that invade
different ricks; intraspecific selection in the closed community of a rick; and similar tantalizers. The answers to these
questions seemed most likely to be obtainable from a genetically closed population where environmental, demographic and
genetical pressures could be estimated.

SKOKHOLM AND ITS MICE

House mice live and thrive on the small Welsh island of
Skokholm (Davis, 1958; Berry, 1968). This island is a rem-
nant of the neighbouring coastal plateau, cut off from the
mainland by marine erosion. It is 3 kms from the mainland,
and is 2 kms long by 0.8 kms wide, having an area of 100 ha
(Figure 1).

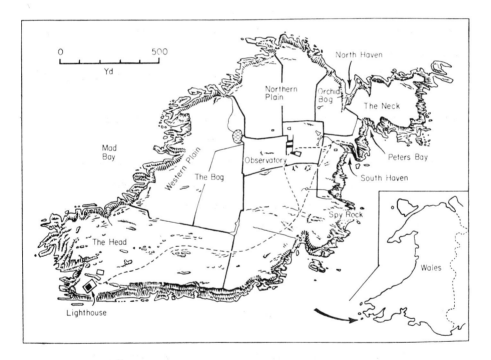

FIGURE 1. *Map of Skokholm (Crown copyright reserved).*

The surface of the island is relatively flat (apart from a
number of rocky outcrops), sloping from about 50m above sea
level at the Lighthouse (south west) end of the island, to

less than 15m at the Neck end. The coastline is composed of
steep rugged cliffs, at the foot of which are immense boul-
ders. The less exposed cliffs have growths of grass and
Armeria maritima (Mill.), while the top of the island is
covered with maritime grass heath (Goodman & Gillham, 1954).
During the winter, surface water collects in the Bog, and
slowly drains away during spring. The only permanently inhabited human dwelling is a lighthouse, but an old farmhouse is
used as a bird observatory during the summer.

One of the advantages of the Skokholm population was that
the origins of the mice were apparently known. There were no
mice there up to 1881, when the island was last farmed intensively. No one lived on the island from 1881 to 1907 and
Lockley (1943, 1947) recorded an introduction of mice in 1907
when a new tenant took over the farm. He noted also that
mouse plagues occurred in 1913-15. Now the time between the
arrival of a few animals and the great numbers a few years
later is very short. Since mice had failed to colonize the
island successfully during the years it was farmed (for at
least 150 years before 1881), it seems that the conditions
for establishing a mouse population are stringent. In particular, the island mice are highly susceptible to low
temperatures, and there were a succession of cold winters in
1907-09. Thus there are grounds for believing that there
were already mice on Skokholm prior to 1907.

CHARACTERISTICS AND RELATIONSHIPS OF THE SKOKHOLM MICE

The original aim of studying the Skokholm mouse population
was to discover whether changes occurred in non-metrical
variant incidences from year to year, and to relate these to
environmental factors. However the first sample trapped on

26

Skokholm in May, 1960 proved to have such a markedly different
array of frequencies to southern British mice that the distinctiveness of the island population seemed worth investigating in its own right (Figure 2).

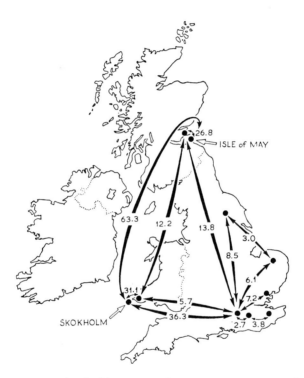

FIGURE 2.   Genetical distances between mouse populations. The
             figures are multi-variate measures of divergence
             based on differences between 35 non-metrical ske-
             letal frequencies - the higher the value, the more
             different are the populations compared (from Berry,
             1967).

Now if the divergence between Skokholm and its mainland
neighbours (and closest relatives) had been produced by adaptation to feral island life, similar changes might be expected

27

in other populations living under similar ecological conditions to Skokholm. This has not been observed. A comparison between the Skokholm population and the ecologically closest population that could be found (the Isle of May, 600 km to the north) showed they were very different to each other*, although both islands have the large mice characteristic of small islands, and the same degree of distinctiveness from their nearest population (Berry, 1964). Explanation of this difference between the island populations became clear when the finer details of Skokholm with its nearby mainland are considered.

According to Lockley, the original Skokholm mice came from Martinshaven, at the end of a short peninsula. However, modern mice from Martinshaven are significantly more unlike the island population than samples from elsewhere in the neighbourhood - even from ones caught some distance away in the main farming area of north Pembrokeshire (Figure 3). In contrast, mice from near Dale are much more like the Skokholm population than any others. All the mainland samples are similar to one another and to mice from other parts of southern Britain.

The historical links of Skokholm were usually via Martinshaven, because this was the best local landing place, but for a time, in the 1890s the island was rented by a farmer from St. Ishmaels.

---

* The value of comparing two populations as far apart as those on Skokholm and the May is debatable. For this reason, a similar study of Apodemus sylvaticus (L.) island populations was made, in which mice living on geographically close islands could be compared. The results and conclusions were the same as for the Skokholm-May house mouse investigation (Berry, 1969, 1973).

28

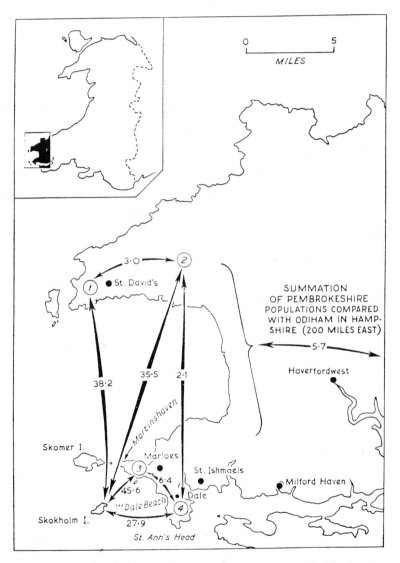

FIGURE 3. Measures of divergence between the Skokholm population and populations on the neighbouring mainland (from Berry, 1965).

29

Hence it seemed possible that mice may have originally got to Skokholm during his tenancy, particularly since a successful introduction at that time would have allowed the mice to be well established before the plague(s) of 1913-15. Another factor was a series of very mild winters from 1895-1900 which would have allowed immigrants time for adjustment to the demands of island life and hence able to survive the cold conditions of the 1900s. It was gratifying that a story of mouse introduction during the tenure of the St. Ishmaels farmer was subsequently reported (Howells, 1968).

Skokholm mice differ from British wild mice mainly in high incidences of three skeletal variants: interfrontal bone, dyssymphyses of thoracic vertebral arches, and foramina transversaria imperfecta of the sixth cervical vertebra. Interfrontal presence is a very rare variant in wild mice but both it and the other variants occur in Dale mice. Presumably the original colonizers happened to carry these variants, and subsequent genetical processes have established them in the genome (Berry, 1967). Mice taken from Skokholm and sib-mated for a number of generations in the laboratory maintain these characteristics (Berry & Jakobson, 1975a). Unusual alleles are common in the genetically closed population because the presumed selection acting after isolation has had to use the available variation. In other words, the Skokholm (and, by inference, the Isle of May) mice are examples of the operation of the founder effect (Berry, 1975).

Whether the Skokholm and May populations would converge if both were isolated for much longer periods is unknown: all one can say in the short-term is that they have achieved sufficient efficiency for survival - and that is the acid test of evolution.

INITIAL AND SUBSEQUENT DIVERGENCE

The founder effect is probably the most powerful way there
is of changing gene frequencies, and may produce enough
difference from the parent population to give instant sub-
speciation. A more important question is the extent of
change (and its causes) following the initial isolating event.
For this reason the Skokholm mice were monitored fairly
intensively for a 10-year period (1960-69).

The 1962 sample from Skokholm was skeletally (and therefore genetically) different from 1960 and 1961 samples trapped
at the same time of year. The obvious possibility was that
these changes resulted from drift, since negligible breeding
occurs on the island during the winter months (October to
March), and the number of survivors to spring each year are
sometimes few. Furthermore, some authors have suggested that
the breeding unit in house mice is very small indeed - perhaps
as few as four (Philip, 1938; Lewontin & Dunn, 1960; Anderson,
1964; Reimer & Petras, 1967; Levin, Petras & Rasmussen, 1969;
Selander, 1970; DeFries & McClearn, 1972). Consequently it
became important to discover the range of population sizes on
Skolkhom - and hopefully to make estimates from them of the
effective breeding size of the population.

Mark-release-recapture experiments were begun in 1964, and
over 3000 mice were marked and released during the next 5
years. Combining Lincoln Index and "trap-out" estimates
(q.v. Hayne, 1949; Berry, 1968, 1970), it was possible to
deduce the number of individuals on the island at all times
of the year. In general, population size increased eight to
ten fold in most years. The lowest number observed was about
75 pairs in Spring, 1963 (Figure 4). Since virtually all the
females roamed only within the territory of individual males

31

and about two-thirds of female winter survivors in any year became pregnant, this meant that the smallest breeding size of the population between 1960 and 1970 was about 100 individuals.

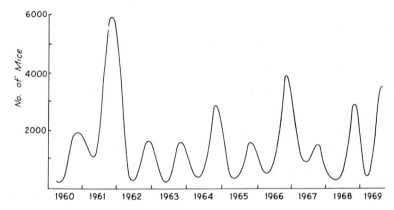

FIGURE 4. *Numbers of mice in the Skokholm population, 1960-69 (methods of estimation given by Berry, 1968) - (from Berry & Jakobson, 1971).*

Breeding mice on Skokholm are highly territorial (Figure 5). Individually marked mice were used to determine how often mice changed their residence, and whether there was any sub-division of the island population. Since virtually all animals in a particular area could be caught (as judged by varying the density of trapping; by "trapping-out" an area and monitoring the movement of mice into it; and by conclusions from recaptures), it was possible to get good estimates of population mixing (cf. Crowcroft, 1955; Crowcroft & Jeffers, 1961).

More than a quarter of the mice on Skokholm breed at a site other than the one at which they were born ("site" being defined as a trapping area, usually at least 50-100m from another) (Table I). Confirmatory evidence for this conclusion comes from a study of rare alleles at allozyme loci (Berry & Jakobson, 1974). There can be no doubt that individual territories form only temporary barriers to gene flow.

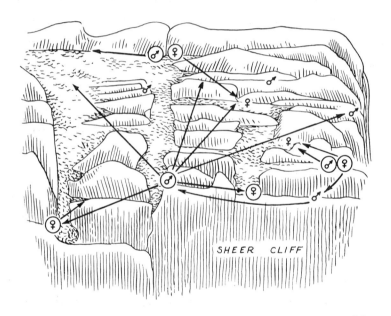

FIGURE 5.   Results of grid-trapping with 40 traps for 11 nights on an area of 50 m^2 on Skokholm in July. If a mouse was caught in the same trap for three or more nights (indicated by ♂ or ♀), this was regarded as its "home" trap; movement from this point (in the sense of being caught in another trap) is shown as a line radiating from the "home" trap. Animals trapped only once or twice are shown unringed (from Berry, 1970).

TABLE I. Change in residence of marked mice which survive
the winter (% of total population).

	Between autumn and the following spring	Between spring and the following autumn
Males	33.6	25.7
Females	29.7	23.6

Population increase and decrease

Estimates of population size also permitted the measurement of individual survival, the distribution of animals on
the island, and the construction of a life-table (Berry &
Jakobson, 1971).

Reproduction on Skokholm begins in mid-March and has
usually finished by the end of September (Figure 6). In most
years the number of mice on Skokholm increases ten-fold
during the breeding season. Since litter size and pregnancy
rates are known, this means there is about a 50% mortality
between birth and weaning (i.e. when animals become trapp-
able). There is no way of knowing how much of this is the
result of whole litters dying, and how much is differential
survival within individual litters.

Population growth and individual survival were the same in
the five summers this was studied (χ^2_{19} for homogeneity =
10.6 ; $P \simeq 0.9$). With the exception of first-born (March-
April) males, all animals have an average 60% chance of surviving two months. In contrast, mortality rates in different
winters are highly heterogeneous ($\chi^2_5 = 43.9$; $P < 0.001$).
The proportion of deaths is much higher in colder winters
(Berry, 1968), and high or "plague" numbers have always been
related to a mild preceding winter (Table II).

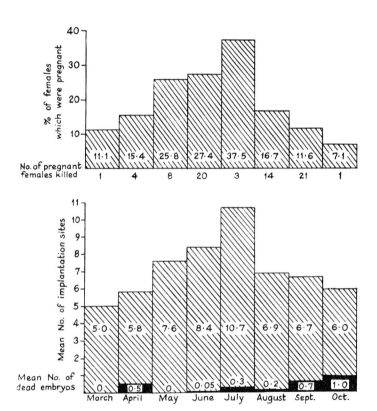

FIGURE 6. *Litter size and breeding intensity in mice on Skokholm (modified from Batten & Berry, 1967).*

TABLE II. *Climate and fluctuations in mouse population number on Skokholm.*

A. Records of mice from Skokholm archives.

		Difference from mean temperature* (°C)		
		January	February	March
1895	"The mice were originally	-2.7	-5.6	-0.9
1896	shipped to Skokholm	+0.3	+0.7	+1.3
1897	.... in the mid to	-2.1	+1.3	+0.3
1898	late 1890s"	+2.2	+0.8	-0.8
1899				
1907	"Mr Jack Edwards rented the	-0.3	-1.1	+0.5
1908	island in 1907....... A few	-1.3	+1.3	-0.9
1909	years later (the mice) were	+0.1	-0.5	-1.7
1910	abundant and caused an influx	-0.3	+0.7	-0.5
	of owls in the winter"			
1913	"The lighthouse was rendered	+0.9	+0.2	+0.4
1914	proof against the entry of mice;	-0.9	+1.6	+0.3
1915	whilst it was being built a plague	-0.2	-0.6	+0.8
	of these mice caused special			
	precautions to be taken"			
1936	In 1938..... "Continues to increase,	-0.3	-0.8	+1.1
1937	apparently reaching a high peak	+0.9	+1.4	-1.6
1938	of population after its low	+1.3	+0.3	+1.9
	numerical status 3 years ago"			
1947	'After human reoccupation, they were	-1.6	-6.6	-1.3
	not observed until late in August'			
1948	"Very common throughout the year"	+1.5	+0.2	+4.2

B. Direct population size determinations

	Spring	Autumn	January	February	March
1960	150-400		+0.2	-0.1	+0.2
1961	500-1000		+0.2	+2.6	+1.4
1962	150-400	1500-3000	+0.5	+0.4	-2.5
1963	100-300		-6.5	-4.2	-1.1
1964	200-500	2000-4000	-0.2	+0.4	-1.3
1965	150-400	1000-3000	-0.2	-1.1	-0.7
1966		2500-5000	-1.2	+0.9	+0.7
1967	300-600	1000-2000	+0.7	+1.1	+0.9
1968	250-400	2000-3000	+1.1	-0.9	+0.2
1969	200-300	2000-4000	+1.3	-1.9	-1.2

*10 year means: January   5.9°C
                  February 5.2°C
                  March   7.0°C

In the years we worked on Skokholm the proportion of the autumn population surviving to the following spring varied between 6% and 20%. During an "average" winter, the 2-monthly death rate of mice is 55% for first year animals, but higher for second year ones. No mouse has ever been found to survive two winters on Skokholm, although in equable commensal

situations more than one in five live over two years, and
pampered laboratory mice often survive into a fourth year
(Varshavskii, 1949; Russell, 1966). The highest winter survival rate on Skokholm occurs on the cliffs, where littoral
animals form an important part of the diet (particularly the
amphipod Talitrus saltator (Montagu)). The interior of the
island becomes very wet in the winter; few animals survive
there, and the area is recolonized by young animals every
spring. Their territorial organization never becomes as complete as on the cliffs. Only a very few animals make a
reverse migration from the centre to the periphery.

The main causes of death seem to be "physiological" :
there is no interspecific competition, no obvious sign of
disease, and little predation. Thus we can summarize the
life-cycle as two distinct phases (an increase and a decrease
phase), with temperature-related factors playing an important
part in mortality during the winter decline.

GENETICAL CHANGES

The genetical situation in the Skokholm mice became
clearer when electrophoretic techniques made it practicable
to detect a number of segregating genes in blood (Berry &
Murphy, 1970). The animals from which blood was taken
suffered no harm, as shown by their subsequent survival.

We classified the genotypes of individual mice at six
(possibly seven) loci: haemoglobin-β (Hbb - linkage group I),
serum transferrin (Trf - linkage group II), two peptidases
(Trip-1 and Dip-1 - linkage group XIII), and two esterases
(Es-2 and Es-5 - linkage group XVIII). Fourteen alleles were
segregating at these loci. This was surprising, since only
one further allele is commonly found in British mainland

37

populations, and a population founded by only a few individuals would be expected to have an acute poverty of inherited variation. Wheeler and Selander (1972) have commented that this implies "the original propagule (if, indeed, only a single introduction occurred) was large enough to carry most of the variability present in the parental mainland population". Unfortunately it is impossible to know whether they are right, or if the extremes of environmental change which a Skokholm mouse undergoes during its life and the genetical responses this engenders (see below) have led to a high rate of incorporation of alleles arising since isolation.

There were also changes in allele and genotype frequencies between the beginning and end of single breeding seasons, changes which were repeated in two consecutive years. Only rare alleles occurred at 3 loci (Trf, Trip-1, Es-5), but the other loci all showed remarkable volatility (Table III).

TABLE III. *Biochemical loci showing selective change.*

		Hbb				Es-2			Dip-1
	N	Freq. Hbb^S	% excess of heterozygotes from expectn. Cliffs	Interior	Freq. $Es-2^b$	% excess of heterozygotes from expectn. ♂♂	♀♀	Cliffs	Freq. $Dip-1^b$
Spring 1968	47	0.343	2.4	37.1	0.409	50.4	23.0	28.0	0.925
Autumn 1968	76	0.438	43.3	16.5	0.242	34.5	30.8	27.6	0.815
Spring 1969	32	0.386	27.0	13.0	0.312	56.5	26.5	30.8	0.807
Autumn 1969	67	0.464	72.8	45.4	0.296	42.9	17.6	30.0	0.663
1971: Older mice	56	0.260	-29.8	-10.0	0.216	33.3	20.7	30.0	0.989
Younger mice	61	0.281	9.2	14.6	0.271	35.8	42.9	47.5	0.981
1972: Older mice	27	0.593	-11.0	28.6	0.289	62.8	33.3	62.8	0.981
Younger mice	49	0.510	31.0	-43.4	0.135	12.4	7.1	16.3	0.980

The most regular changes took place at the Hbb locus: frequencies of heterozygotes increased on the cliffs during the breeding season and decreased during the winter. Genotypic fluctuations in the island interior were irregular, and to some extent mirrored the excess of young produced by the coastal breeders. In the autumn of any year, the old mice in the population will be the survivors of the previous winter or their first born young of the spring, and the proportion of heterozygotes among them would be expected to be less than that in younger mice. This expectation is fulfilled.

Previous workers (e.g. Petras, 1967) have reported a deficiency of heterozygotes in populations segregating for Hbb, but some selection is suggested by the widespread occurrence of a polymorphism at the locus (Wheeler and Selander, 1972). Differences in reproduction and survival between Hbb genotypes have been found in field-living mice in California (Myers, 1974), and Berry & Peters (1975) have recorded a significant difference in allele frequency between young and old mice on the sub-Antarctic Macquarie Island where the animals breed the whole year round.

Nothing is known of the Hbb alleles which helps to understand any effect on fitness (Russell & Bernstein, 1966). One allele (Hbb^d) is a duplication, with the two polypeptide chains differing from each other at six positions; the other allele (Hbb^s) has three out of 196 amino acids different from that in the duplicated chain (Popp & Bailiff, 1973). Gilman (1974) has suggested a mechanism for heterozygous advantage at the locus based on the functional capabilities of the haemoglobin molecule, but this is still speculative. A histocompatibility locus (H-1) is only two map units from Hbb (Russell & McFarland, 1974), and it is conceivable that the haemoglobin polymorphism is the result of "hitch-hiking" on an adaptive chromosome segment.

In the years for which biochemical data are available for Skokholm, the frequency of Hbb^S (which is the commoner allele in most mainland populations) ranged from 26% to 59%. Changes at other loci were almost as marked, but completely uncorrelated: the commonest Dip-1 allele decreased from 92% to 66% during two breeding seasons, but then increased again to 98%. At the Es-2 locus, an excess of heterozygotes in males increased in the 1968-9 winter, decreased in the summers in 1968 and 1969; in 1971 there was an increase in female heterozygous frequency most marked on the cliffs: and in 1972 there was a decrease in heterozygotes during the summer in both males and females, particularly on the cliffs.

Cyclical variation was also detected in the frequencies of skeletal variants (Berry & Jakobson, 1975a). During the period of study, autumn samples were more homogeneous than spring ones - indicating that stabilizing selection acts during the summer months. One of the peculiarities of this summer selection was that the midline dorsal weakness characteristic of Skokholm mice (gaps between the frontal bones of the skull and a failure of the neural arches of some vertebrae to fuse) actually increased. The incidences of these variants then decreased during the winter.

The genetical difference between Skokholm and Isle of May mice has already been noted. In contrast to Skokholm, the Isle of May mice are monomorphic for Hbb^S - indeed they showed no allozymic variation at 17 loci tested. This highlights the difficulties of relating genetical changes at a locus with the action per se of the locus: the May mice have similar ecological problems to the Skokholm ones, yet their genetical diversity is very low. Herein lies an advantage of morphological investigation: inherited traits are affected

40

by many loci, and multivariate comparisons between samples are
likely to be more sensitive detectors of genetical differences
and pressures than ones based on a limited number of bio-
chemically scored loci (cf. Soulé & Yang, 1974).

The pattern of skeletal variants on Skokholm changed pro-
gressively during the 1960s (Figure 7).

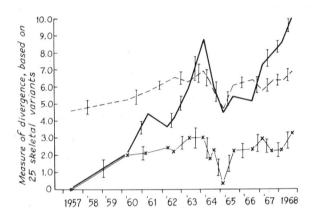

FIGURE 7. *Genetical change in the Skokholm mouse population*
based on non-metrical skeletal frequencies. The
continuous line shows genetical distances, where
each step is the change from the previous sample;
- X - X - X - X plots the genetical distances from
the first sample collected; the dotted line plots
successive Skokholm samples from the neighbouring
mainland.

This change was neither large nor continuous, but was enough
to show that the genome was subject to adjustment over a number of years. It is impossible to predict whether an isolate
like Skokholm will ever come into a state of balance with its
environment that such genetical change will stop. Clearly
both cyclical and long-term genetical adjustment form part of
the present survival strategy of the population.

SURVIVAL AND SELECTION

Mortality in Skokholm mice is not random with respect to genetical constitution. An obvious practical question is 'what causes the mice to die?'. Since the death rate in the winter is related to external temperature, this has been investigated by measuring the responses of individuals to cold in temperature-controlled metabolism chambers, and then following their subsequent survival when released back into the population.

At first it seemed that particular traits (such as high basic metabolic rate) were highly correlated with winter survival. This proved too simplistic. Different physiological measures were more important in different years, and older animals were less able to cope with environmental pressures than younger ones. The conclusion from direct physiological testing of individuals is in fact the same as that from statistical comparison of survival in different winters, that the characteristics which lead to a mouse surviving or dying in a particular set of circumstances will vary according to those circumstances.

It is possible to generalize more confidently about the effect of age. Second year animals always succumb to winter conditions (see above), although their survival in the less harsh summer weather does not differ from that of younger mice. Consequently any measurement of the reaction of an individual of any genotype to its environment has to take age into account. We have been able to show that a whole range of phenotypic measures (organ weights, ionic contents of bones, haematological values, overall body sizes) are significantly correlated with age, even when the data are "corrected" to eliminate the effects of size (by adjusting

42

them to a standard body size). In a discriminant function analysis based on these measures it was possible to separate young (under 3 months), adult, and over-wintered mice from each other with only 15% error (Bellamy, Berry, Jakobson, Lidicker, Morgan & Murphy, 1973). The different age classes in this analysis were genetically alike. However, it was also possible to produce a similar degree of separation between biochemical phenotypes using only the same quantitative data (Table IV). In other words, there are differences between phenotypes which are subject to different selection pressures, and which include characters which change with age - and age affects survival.

TABLE IV. *Amount of mis-classification between biochemical phenotypes shown by a discriminant analysis using 17 quantitative measures (after Bellamy et al., 1973).*

	Males		Females	
Locus	Number	Percentage incorrect	Number	Percentage incorrect
Hbb	40	17.5	38	21.0
Es-2	40	12.5	37	10.8

This led us to propose a model of survival (Figure 8) which emphasizes the dynamic nature of the problems an animal faces at different stages through life (Berry, Jakobson & Triggs, 1973; Berry & Jakobson, 1975b). The aim of the model is primarily conceptual rather than predictive: it is only too easy to characterize a population sample in terms of its genes, physiology, or morphology, and then invent an explanation for the characteristics. This may work sometimes, but

43

it will tend to naivety and may obscure important happenings.

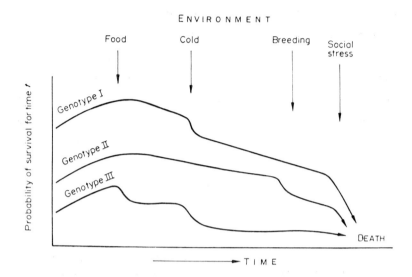

FIGURE 8. A model to describe the interplay of individual, environment and genotype: different phenotypes (i.e. the result of genotype plus age) react differently to environmental factors (from Berry, Jakobson & Triggs, 1973).

For example, the Hbb genotype frequencies in the September 1971 sample from Skokholm closely fitted the binomial expectation. Only because previous samples collected at different times of the year had shown that seasonal selection occurs at this locus was the data divided by age and habitat. This showed a change from a 30% deficiency of heterozygotes in older mice on the cliffs to a 9% excess in younger ones.

The Skokholm study began as an apparently simple genetical investigation and ramified into morphology, physiology, ecology, behaviour, reproductive and ageing studies. At times it seemed that the obvious problems had been solved and further work would give ever diminishing returns, while at

44

others that wild-mammal work produced more problems than it answered. Nevertheless useful results were obtained and the work has seemed worth describing in some detail because of the light it has thrown on many genetical problems - even if the light has often done no more than show the surrounding umbra of ignorance.

ACKNOWLEDGEMENTS

Many people have helped us with field work on Skokholm. Our thanks are due to them all, especially to Dr. M.E. Jakobson who was responsible for the physiological studies, and to the West Wales Naturalists' Trust and the Field Studies Council for making it possible to work on the island; also to Professor C.A.B. Smith for statistical help, particularly his derivation of the simply-computed measure of divergence which raised the original questions about heterogeneity in wild mouse populations, and to Mr. A.J. Lee for drawing the figures. The studies have been supported at various stages by the Rockefeller Foundation, the Royal Society, and the Medical Research Council, and this is gratefully acknowledged.

REFERENCES

Anderson, P.K. (1964). Science 145: 177-178.

Batten, C.A. and R.J. Berry. (1967). J. Anim. Ecol. 36: 453-463.

Bellamy, D., R.J. Berry, M.E. Jakobson, W.Z. Lidicker, J. Morgan and H.M. Murphy. (1973). Age and Ageing 2: 235-250.

Berry, R.J. (1963). Genet. Res. 4: 193-220.

Berry, R.J. (1964). Evolution 18: 468-483.

Berry, R.J. (1965). Nature Wales 9: 110-115.

45

Berry, R.J. (1967). Eugen. Rev. 59: 78-96.

Berry, R.J. (1968). J. Anim. Ecol. 37: 445-470.

Berry, R.J. (1969). J. Zool. 159: 311-328.

Berry, R.J. (1970). Field Stud. 3: 219-262.

Berry, R.J. (1973). J. Zool. 170: 351-366.

Berry, R.J. (1975). J. Zool. 176: 292-295.

Berry, R.J. and M.E. Jakobson. (1971). Exp. Geront. 6: 187-197.

Berry, R.J. and M.E. Jakobson. (1974). J. Zool. 173: 341-354.

Berry, R.J. and M.E. Jakobson. (1975a). J. Zool. 175: 523-540.

Berry, R.J. and M.E. Jakobson. (1975b). J. Zool. 176: 391-402.

Berry, R.J., M.E. Jakobson and G.S. Triggs. (1973). Mamm. Rev. 3: 46-57.

Berry, R.J. and H.M. Murphy. (1970). Proc. Roy. Soc. Lond. B 175: 255-267.

Berry, R.J. and J. Peters. (1975). J. Zool. 176: 375-389.

Crowcroft, P. (1955). J. Mammal. 36: 299-301.

Crowcroft, P. and J.N.R. Jeffers. (1961). Proc. Zool. Soc., Lond. 137: 573-582.

Davis, R.A. (1958). Skokholm Bird Observatory Report for 1957, 26.

DeFries, J.C. and G.E. McClearn. (1972). In Evolutionary Biology, 5: 279-291. Th. Dobzhansky, M.R. Hecht and W.C. Steere (Eds.). New York: Appleton-Century-Crofts.

Deol, M.S. and G.M. Truslove. (1957). J. Genet. 55: 288-312.

Gilman, J.G. (1974). Ann. N.Y. Acad. Sci. 241: 416-433.

Goodman, G.T. and M.E. Gillham. (1954). J. Ecol. 42: 296-327.

Grüneberg, H. (1963). The Pathology of Development. Oxford: Blackwell.

Hayne, D.W. (1949). J. Mammal. 30: 399-411.

Howe, W.L. and P.A. Parsons. (1967). J. Embryol. Exp. Morph. 17: 283-292.

Howells, R. (1968). The Sounds Between. Llandysul: Gomerian.

Levin, B.R., M.L. Petras and D.I. Rasmussen. (1969). Am. Natur. 103: 647-661.

Lewontin, R.C. and L.C. Dunn. (1960). Genetics 45: 705-722.

Lockley, R.M. (1943). Dream Island Days : A Record of the Simple Life. London: Witherby.

Lockley, R.M. (1947). Letters from Skokholm. London: Dent.

Mayr, E. (1954). In Evolution as a Process: 157-180. J. Huxley, A.C. Hardy and E.B. Ford (Eds.). London: Allen & Unwin.

Myers, J. (1974). Ecology 55: 747-759.

Newsome, A.E. and W.P. Crowcroft. (1971). CSIRO Wildl. Res. 16: 41-47.

Petras, M.L. (1967). Evolution 21: 259-274.

Popp, R.A. and E.G. Bailiff. (1973). Biochim. Biophys. Acta 303: 61-67.

Rowe, F.P., E.J. Taylor and A.H.J. Chudley. (1963). J. Anim. Ecol. 32: 87-97.

Russell, E.S. (1966). In Biology of the Laboratory Mouse, 2nd ed.: 511-519. E.L. Green (Ed.). New York: McGraw-Hill.

Russell, E.S. and S.E. Bernstein. (1966). In Biology of the Laboratory Mouse, 2nd ed.: 351-372. E.L. Green (Ed.). New York: McGraw-Hill.

Russell, E.S. and E.C. McFarland. (1974). Ann. N.Y. Acad. Sci. 241: 25-38.

Searle, A.G. (1954). J. Genet. 52: 68-102.

Selander, R.K. (1970). Amer. Zool. 10: 53-66.

Soulé, M. and S.Y. Yang. (1974). Evolution 27: 593-600.

Southern, H.N. (Ed.) (1954). Control of Rats and Mice, Vol. 3. House Mice. Oxford: University Press.

Southern, H.N. and E.M.O. Laurie. (1946). J. Anim. Ecol. 15: 135-149.

Southwick, C.H. (1958). Proc. Zool. Soc., Lond., 131: 163-175.

Varshavskii, S.N. (1949). Zool. Zh. 28: 361-371. (In Russian).

Weber, W. (1950). J. Genet. 50: 174-178.

Wheeler, L.L. and R.K. Selander. (1972). Univ. Texas Publs, No. 7213: 269-296.

Patterns of Survival and Microevolution in Plant Populations

S.K. JAIN

Plant population studies are cited in the literature on evolutionary ecology primarily in relation to the variations in breeding systems, ecotypic variation, or the role of phenotypic plasticity. The patterns of geographical variation, life cycle components, species coexistence, and other population features are often deduced from largely qualitative arguments. For example, Darlington (1958) has predicted a large number of correlations among such features of the genetic system as recombination rate, chromosome number, genetic variation, level of sexual/asexual reproduction and inbreeding vs. outbreeding. Similar deductions were advanced by Salisbury (1942), Dansereau (1957), and others about the correlated features of life form, reproductive capacity, and distribution within certain plant communities. Recently, many of these generalizations have stimulated quantitative field studies as well as some theoretical developments. Among the recent reviews are those by Barber (1965), Allard, Jain and Workman (1968), Baker and Stebbins (1965), Janzen (1971), Stebbins (1971, 1974), and Stern and Roche (1974). The description of genetic variation has improved in terms of the multilocus polymorphisms as well as quantitative genetic studies in both

49

wild and domesticated species. There is now a great deal of interest in the evolution of strategies of long term survival versus reproduction by annual seed production under varying environments (e.g., Giesel (1974); Schaffer and Gadgil (1975)). Plant demography and the dynamics of population regulation have become attractive fields of research, mainly due to the elegant work of J.L. Harper, G.R. Sagar and their associates (Harper (1967), Harper et al. (1961); Harper and White (1970, 1974); Sagar and Mortimer (in press)).

Thus, it would appear timely that we begin to examine jointly the genetic and ecological aspects of evolutionary processes in terms of the following issues:

I.   Comparative studies of coexisting related species for their patterns of distribution and niche breadth;

II.   Parametrization of population structure in terms of subdivisions in space and gene exchange among the subdivisions (islands or neighborhoods);

III.   Consequences of different population structures on the patterns of geographic variation (genetic as well as phenotypic);

IV.   Evaluation of the relative roles of drift and selection in genetic changes using information on population numbers, life cycle components of population regulation, etc.; and finally,

V.   The evolution of life history patterns, given a genetic variety of adaptive responses to varying environments described in terms of certain optimizing properties.

All too often we find descriptions of genetic variation surveys in the literature without any direct link with the variance in fitness or the dynamic role in adaptive processes,

50

or we find ecological models of optimal strategies assumed to
have evolved in most organisms as we find them today but no
real effort to establish the genetic variation needed to provide for such evolutionary changes. For the sake of generality, we are confronted with the issues of definition,
detection, causal meaning and predictability of patterns
emerging from such studies. Hutchinson (1953) defined the
concept of pattern in ecology as "the structure which results
from the distribution of organisms in or from, their interactions with their environments." He discussed the abiotic
(vectorial), reproductive, social, coactive and stochastic
components of a pattern. Evolutionists are naturally concerned further with the patterns of genetic variation and of
newly arising adaptations under changing environments. In
this paper, a summary of our studies in California grasslands
will be presented to initiate a synthesis along these lines.
The detailed analyses of data are presented elsewhere (IBP
synthesis volume on Grassland Biome Study in California, eds.
Duncan and Woodmansea).

Species distributions and population structure

The six species of primary interest in our studies are:
Avena fatua and A. barbata (wild oats), Bromus mollis and B.
rubens (brome grasses), Trifolium hirtum (rose clover), and
Medicago polymorpha (bur clover). All of them were introduced into the California range and roadwise weedy flora from
the Mediterranean region, and resemble each other in being a
highly successful group of predominantly self-pollinated
colonizing species. Several features of their geographical
variation and breeding systems were discussed by Jain (1975).
All six species have large areas of overlap in their distribution. A. barbata and B. mollis are very widespread, forming

51

almost continuous stands on the grazed lands; B. rubens has
characteristically patchy distribution in the northern part
of its range, A. fatua is widespread more as a ruderal weedy
species with small, scattered but almost pure stands; M.
polymorpha is also often highly restricted to rather well
defined discontinuous stands of varying sizes, whereas T.
hirtum was selected as an example of a species actively expanding its distribution by the colonization of roadside areas
in several regions. M. polymorpha and B. rubens represent a
wide spectrum of mesic to xeric adaptations and rarely occur
together. Several community analyses have recently begun to
describe the patterns of coexistence among these six species
(e.g., Wu (1974), Jain (unpublished data)). Within many
grassland sites, there occur small depressions (vernal pools)
characterized by endemic floristics. Several Limnanthes spp.
(meadowfoam) have been sampled in recent years that provide
comparisons of their population features with those of the
introduced range species. All of these annuals offer excellent opportunity for genetic and demographic studies.

Population structure and species distributions were mapped
in the following way: Nearly five hierarchial levels were
used for sampling populations, and observed for the occurrence
of these species, namely, (i) climatic regions (such as Central Valley, North Coast range, San Francisco Bay area, South
Coast range and the Sierra Nevada foothills on the eastern
side of the Valley), (ii) hillsides within regions (designated as sites), (iii) transects within each site over 30-
50 meters scale, (iv) patches within each subsite (several
meters square in area), and finally, (v) highly localized
description of plant stands (runs a´ la Pielou (1969)) within
these patches. The distance scales roughly correspond to a
few hundred miles between the extreme regions, one to a few

miles between neighboring sites, and down to patches and mean
run lengths within patches to the order of say 10 to 100 cms.
Thus, on a suitable logarithmic scale we obtain a linear
scale of sampling range. Due to numerous practical considerations, these scales are often used only as approximate
guidelines. _Avena_ _fatua_ and _B._ _mollis_, as noted earlier,
provide area continuity over large areas, with some exceptions
such as in the small clearings at the edges of a forest, or
as understory in the fruit orchards. In contrast, _M._ _poly-_
morpha distribution varies from large to very small colony
sizes, and likewise _T._ _hirtum_ occurs both as large continuous
stands in some sown pastures to a highly patchy local distribution along many roadsides which it has colonized only
recently. As noted above, _Limnanthes_ spp. often occur in
small well-defined micro-habitats (vernal pools and edges
along temporary streams) with distinct patterns of subdivided
isolates. Linear habitat is approached by many roadside
populations of _A._ _fatua_ that adjoin many cultivated fields,
townships, vineyards, or similar man-made habitat boundaries,
or by _Limnanthes_ populations occurring in a narrow band
around the vernal pools.

Plant ecologists use various forms of pattern analysis of
species distributions based on numerous schemes of sampling
including the use of contiguous quadrats. Kershaw (1973)
reviewed in great detail the results of such studies of
species distributions in relation to the role of vegetative
propagation, plant morphology, seed dispersal, allelochemics,
physical environment, herbivory, and random events of local
population extinction and recolonization. For our purposes
here, we recognize the dynamic nature of local patches in
terms of the so-called "vegetation cycles". Thus subdivisions,
changes in population numbers and local events of colonization

are features well recognized by plant ecologists. Levin
(1974) developed a model for dispersal and population interactions that predicts the species distribution in a patchy
habitat. Patchiness or mosaic nature of ecological maps can
be described in a variety of ways which provide a measure of
niche occupancy in subdivided habitats, or discrete neighborhoods. Population growth, migration, competition, and random events were shown to determine the species coexistence
and turnover. Furthermore, dispersal through pollen within
the cross-pollinating fraction of breeding system and through
seed or vegetative propagules needs to be quantified in order
to define the size of neighborhoods under an isolation-by-
distance model. Such studies of dispersal and population
density repeated over sites and successive generations provide a description of population structure in terms of neighborhood dimension, density and turnover rates.

Different species provided data on changes in the patterns
of their patchy distribution within a local area over two to
four years. Except for highly disturbed sites, patchiness
seemed to be stable with no local extinction of any such
neighborhood (a breeding colony defined as unit of evolutionary genetic change, after Wright (1969)). The measures
of niche breadth derived by autecological studies as well as
competition experiments, will be reported elsewhere.

Local dispersal of pollen and seed was studied in the
following way: Using dominant allele class as pollen source,
placed as a colony in the center, and colonies of recessives
placed in various geometric designs, the frequency of outcrosses is scored as heterozygotes among the offspring of
sampled recessives. Such outcrosses directly estimate the
effective rate of gene flow through pollen as a function of
distance from central colony. Likewise, gene flow through

54

seed is estimated by suitable census of seed morphs planted
in various relative arrangements. Besides founded colonies
in a controlled experimental area, certain natural stands can
also be used if morphs occur as large monomorphic patches.
Examples of pollen and seed flow studies for plotting dispersal rates as functions of distance are given in Figures 1
and 2. Note that there is a significant leptokurtosis or
over-dispersion as measured by the fourth moment (kurtosis
parameter β) such that $(\beta-3) \gg 0$, or variance/mean
ratio $\gg 1$ represent leptokurtosis, an excess in the long
range component of gene flow. As shown by Wright (1969) and
others, many compound distributions with two or more components having unequal means or variances involve kurtosis.
Highly autogamous species do not allow as large a role of
pollen flow as that of seed dispersal, although occasional
long range cross-pollination events might have a large influence on the measures of differentiation (e.g. Jain and
Bradshaw (1966)). Seed flow has two distinct phases:
initially, seeds are placed onto the soil surface through
aerial dispersion around each mother plant, and subsequent
phase of movement by wind, water, animal activity, etc. along
the soil surface. Most distance curves suggest strong leptokurtosis due to either seed polymorphisms such that different
morphs (e.g. hairy vs. nonhairy lemma in Avena) have unequal
dispersal rates, or due to large ecological heterogeneity in
space (litter, soil cracks, etc.). Most distance curves of
seed or pollen dispersal in our study showed a striking similarity, with rather small differences in the range for
different species. The long range component, however, is
perhaps underestimated and deserves further experimental work.

The pollen flow component in highly inbreeding species
relates to the random outcrossing rates (t) which were

estimated to be as follows:

Estimates of outcrossing parameter (t)
and standard deviation (s_t)

Species	t	s_t	number of estimates
A. barbata	.024	.028	37
A. fatua	.028	.022	21
B. mollis	.035	.033	26
B. rubens	0	0	6
T. hirtum	.045	.024	10
M. polymorpha	.082	.031	6

Note that occasionally these highly selfed species may show as much as 16-18% outcrossing in a locality, due to partial chasmogamy (opening of glumes during pollen dehiscence) in grasses, insect pollinators' activity, presence of male steriles, etc. Following Wright (1969), we may estimate the size of effective neighborhood under linear continuity as:

$N_e = \frac{2\sigma d\sqrt{\pi}}{(1+F)}$; where $d = n/2\sigma$, density of n individuals in a strip 2σ units long, F as fixation index related to selfing levels, σ = standard deviation of pollen and seed dispersal curves. For area continuity, $N_e = 4\pi\sigma^2 d/(1+F)$. Lepokurtosis further reduces N_e in proportion to β such that for β in the range of 15-25 these N_e estimates are nearly halved. (See Pielou (1969), for further discussions of a random walk model which would yield leptokurtosis if different seed classes disperse by unequal step lengths.) Note that based on all these data together, the order of N_e was estimated to be roughly as follows: M. polymorpha 30-50; T. hirtum 30-100; B. mollis 100-150; A. barbata 100-150;

A. fatua 50-100; and B. rubens 50-200. These differences and
wide range are largely related to the variation in plant density, seed output per plant, prevailing wind, seed size, seed
weight, appendages such as glume parts and awn, and also the
role of animals.

It should be emphasized that the estimates of N_e for any
of these species vary even within wider limits than shown
here, depending on site location, wind direction, etc. and
of course pose large statistical problems of estimation. In
particular, note that in all our studies, the sampling transect was truncated around 15-20 feet from the central colony
due to a low recovery of seed by a tedious screening process,
but long range gene flow, albeit small and not measured here,
should be recognized. Yearly fluctuations in plant density,
large variance in progeny number per individual, and the high
rate of population growth in newly founded patches (Nei,
et al. (1975)), as observed particularly in Medicago polymorpha and B. rubens would give lower estimates of N_e,
whereas taking long distance component into account would
yield higher values. Our data suggest strongly that values
of N_e must enter into any predictions of the variation
patterns of individual species. Levin and Kerster (1969,
et seq.) studied extensively a variety of features of gene
flow in plant populations from which they concluded that the
values of N_e were indeed as low as 20-100 and the leptokurtosis was a key feature.

Pattern of geographical variation

Based on leptokurtic dispersal curves, and gene flow rates
as monotonically decreasing functions of distance, some
expectations of variation patterns can be derived. Wright's
F-statistics, for instance, provided a theoretical basis of

correlating expected genetic variance parameters with the
patterns of hierarchial subdivision. If we represent the
leptokurtic migration curves by a combination of long range
component m_1 as represented in the Wright's island model,
and m_2 for stepping-stone model allowing gene exchange only
between adjacent colonies, their relative role can be obtained
in terms of the correlation of gene frequencies between
colonies, i = 1,2,...,K , steps apart, and population genetic predictions can be made with increasing order of complexity in the model with no selection. Further, certain
kinds of selective forces are assumed to become divergent
with increasing distance. Genetic measures of differentiation
could be based on Wright's F-statistics, Nei's D or I statistics (Nei (1973)), some numerical taxonomic measure utilizing
data from both allelic frequencies and quantitative variation,
or simply, a hierarchial partitioning of variances. The
rates of increase and levels of differentiation vary among
species, or species groups, as shown by the studies of Lewontin (1974), Nevo (this conference), Selander (Selander and
Kaufman, (1973); Selander and Hudson (in press)), and others
in animal populations.

Table 1 gives a summary of variation observed in B. mollis
populations sampled from 54 sites in six regions of California; of these, 24 sites were further studied for genetic
variation components by growing individual plant families for
quantitative data. Data from natural stands vs. greenhouse
cultures showed a great amount of phenotypic change. In
general, most quantitative traits in all the species studied
gave no regional or consistent ecotypic patterns but within
a region or group of sites, one finds between-sites, between-
families and within-families sources of variation in a decreasing order. This is not surprising for an inbreeding

species with a great deal of inter-site variability and
rather low levels of heterozygosity. However, principal component analyses of variation in Avena spp. and Bromus spp.
did not show the two main climatic variables, viz. temperature
and rainfall, to account for regional differences. Also, note
that not all regions showed the same amount of genetic variation. This hierarchial pattern increasingly breaks down for
a species like B. rubens having small scattered stands and
greater degree of patchiness within a site.

The pattern of variation in A. barbata is perhaps most
unique and deserves a detailed discussion here. Figure 3
shows four climatic regions in California and allelic frequency distribution at a locus Ls/ℓs (pubescence of basal
leaf sheaths). Note that almost the entire Valley and South
Coast regions are monomorphic for ℓs allele; in fact, monomorphism was found at loci B,H and Np (Jain (1969)) and
for a majority of electrophoretic loci scored (Singh and Jain
(unpublished data). Regions I and II are polymorphic, with
loci B and Ls highly polymorphic, and loci H and Np
polymorphic only at a small fraction of sites. Detailed surveys within an area, at sites located 1/2 - 2 miles apart,
bring out a pattern as shown in Figure 4. Thus, many sites
show a markedly patchy (non-uniform) distribution of genotypic variation. Still more detailed surveys within sites
using transects, 10 - 30 meters long and sampling each quadrat
30×30 cm^2 along these transects, brings out a highly significant local patchiness (Figure 5). This pattern is frequently repeated at a majority of sites and although patches
shift in certain disturbed areas, repeated samples in marked
transects show year-to-year stability of this pattern.

Statistical tests were made using χ^2 for homogeneity of
binomial or multinomial proportions (phenotypic frequencies)

59

and ANOVA following the arc sine transformation of frequency data. Tests for clinal or other geographical networks (Sokal, (personal communication) suggested a new test) gave no evidence for any geographical pattern in these data.

Avena fatua and B. mollis, in contrast show a greater degree of variation among sites and regions, especially between the so-called central and peripheral areas of distribution but much less patchiness within sites. An example of clinal variation in A. fatua is shown in Figure 6 (Marysville site) with marked between-years variation in the slope of cline.

More commonly, both A. fatua and B. mollis populations have allelic combinations distributed homogeneously without marked patchiness or localized clines. A. barbata shows a number of "steep clines" at the boundary of regions of I and III (Figure 3) where polymorphism changes within a distance of say 20-30 meters to monomorphism, apparently correlated with the influence of coastal fog and leeward exposures in region I. Rai (1972) discussed in some detail the possible role of random drift and migration in certain examples of clines within an overall patchwork (Figure 7). Since the role of local founder effects cannot be ruled out, these scattered examples of clines are not to be taken necessarily as "proofs" of selection-migration balance.

However, another line of evidence for population interactions between congeneric species is provided by the analyses of polymorphisms in pure and mixed stands of A. fatua and A. barbata. Polymorphisms index, (defined as $\Sigma\ p_i(1-p_i)$ where p_i are allelic or phenotypic proportions and Σ is summation over loci) based on several morphological markers, and quantitative variation for seed size, was higher within pure stands and lower within each species in mixed stands

60

(Figure 8). This "variance displacement" is correlated with the character mean displacement for seed size and germination times, which further experimental tests have shown to be related to the seed characteristics of niche widths.

B. rubens is highly monomorphic like the valley ecotype of Avena barbata and showed only a small amount of genetic differentiation at the level of regions (Wu, (1974)). The northern valley and south coast populations showed a significant difference in flowering time. Tables 3 and 4 show evidence of genetic differentiation in M. polymorpha and T. hirtum but no consistent pattern between regions or even sites. Thus, these two species again differ from the patterns discussed above. Clearly, we need a more systematic and critical survey of these scales of genetic divergence along the following lines: (1) accumulation of genotypic data at many more loci, (2) repeated frequency data in marked sites over years, and (3) controlled experiments on measures of ecotype differentiation.

These different patterns of geographic variation, along with a similar conclusion reached from a literature survey (Jain (1975)), bring several hypotheses to mind: (1) Are different scales of macrodifferentiation related to the estimates of neighborhood sizes (N_e) as discussed above? The answer seems to be yes, at least at the level of patches and runs in relation to the range of local seed dispersal, albeit there are rather wide variations around any mean values of our estimates and thus at best predict only low correlations. Perhaps N_e estimates will need drastic revision with further work in this area. (2) Could some differences in the long range gene flow through seed or pollen be an important factor? Again, although there are no quantitative estimates available, the rates of pollen flow over long distances might vary between the outcrosses through wind-borne pollen in grasses to the insect-mediated outcrosses in the legumes. T. hirtum has been used to test the rates of pollen flow among distinct colonies in an area, using a correlation between the polymorphism index and the size of colonies as these expand in numbers over time. Figure 9 shows clearly that larger colonies have higher amounts of polymorphism which our observations in the field suggest to be due to the pollinator bees interchanging pollen among colonies as these become larger and attract their attention. Another long range factor in gene flow in these species is the role of grazing animals as they are moved frequently over large areas by man. Seeds of different species survive at different rates the paths of animal's digestive tract, or agricultural practices.

(3) Data on outcrossing rates might suggest that even small differences in this parameter, through the role of heterozygotes in selective regime, may be correlated with the patterns of local polymorphisms. For example, populations of Medicago polymorpha and Trifolium hirtum have an average of 8-10% heterozygotes, Bromus mollis and A. fatua on an average 4-8%, and A. barbata nearly 1-2% per locus, whereas in B. rubens, it is probably even less than 1%. In a simulation study of various selection models in inbreeding populations and with stochastic variations in outcrossing rates, the maintenance of polymorphisms was found to depend critically on the balance among the amounts of heterozygote advantage, asymmetry between homozygote selective values, and variation in the outcrossing component. In fact, a great deal of reported evidence on heterozygote advantage in inbreeders might need reexamination with a better knowledge of the mating system parameters as well as the neighborhood structure of populations. (4) Another important feature of quantitative

variation in plants is the different degree of phenotypic plasticity, such as reported in Avena fatua vs. A. barbata (Marshall and Jain (1968)). In Bromus spp., Moraes (1972) and Wu (1974) analyzed the relative sizes of genetic and non-genetic components of variation by several techniques based on measurement data on experimental cultures grown under different environments. It appears that B. mollis has, like A. fatua, greater genetic variation and less plasticity than B. rubens; and interestingly enough, both A. barbata and B. rubens gave evidence for some ecotypic differentiation.

We close this section by emphasizing the many unknowns about the historicity, ecology of dispersal and neighborhood structure, variation in breeding system, role of polyploidy levels, modifier polymorphisms (e.g. see Feldman and Krakauer), etc., such that the patterns of geographical variation on various distance scales appear to be largely serendipitous and perhaps require further surveys through both descriptive and experimental field studies. Population genetic surveys of variation will in the near future become more meaningful in terms of both adaptive explanation of certain specific allozyme polymorphisms as well as better, long term field assessment of population ecological variables.

Variation in life history and demography

The nature and intensities of selective forces may be examined in detail through what Cavalli-Sforza and Bodmer (1971) christened as genetic demography. Birch (1960), among others, had begun the search for correlations between the parameters of variation and population numbers. Following Mortimer (1971) and Sagar and Mortimer (in press), Figure 10 shows the stages of life history in terms of components a to g for annuals, and two routes for perennial and asexually reproducing species. They reviewed the literature extensively for availability of estimates of these components and provided their own estimates in several species. Harper and White (1974) may be consulted for an excellent review of demographic literature on plants. Clearly, a large variety of life history patterns have been at least qualitatively investigated. The notion of optimal strategies of size, cycle, energetics, and adaptive response in different species would require such developments in correlative studies of variation, survival, reproduction, with age-specific birth and death schedules, etc. A few of Mortimer's estimates are summarized in Table 5. Note that most components vary indivi-dually and must certainly vary as combinations within an overall strategy of replacement in successive generations. Unfortunately, the estimates of h and j components are not yet available for these perennial species. A few of our estimates for annuals are given in Table 6. Three points to note are: (1) variation among years and replicate plots; (2) variation between Avena and Medicago for the seed dor-mancy component (d) ; (3) variations in plant density in these populations being nonrepresentative of variation in disturbed habitats or peripheral populations where local patches might occasionally become extinct. Again, there are unfortunately very few studies of the genetic variance in these components. An outstanding analysis of various selec-tion components by Schaal (1974), and Antonovics (1971, et seq.) showed that population census data provided some estimates of the rates of survival of competing genotypes through seed, seedling and adult stages. Our germination studies in natural stands of Avena fatua and Medicago poly-morpha (unpublished data of Rai and Jain) showed that different seed morphs provided a large selection differential

64

among different genotypes within small patches.

Seed dormancy allows overlap between the successive generations and perhaps accounted for the stability of genetic variation patterns in Linanthus parryai (Epling, Lewis and Ball (1960)) and Stephanomaria exigua (Gottlieb (1974)). Seed carryover patterns vary in terms of the degree of dormancy and seed viability under storage in soil. A computer simulation program has been written to evaluate the role of seed dormancy in the lag of response to changing environments. Tolbert (unpublished results) showed that the seed carryover rates and seed longevity interact with the pattern of environmental variations such that the strategies might even switch from a high to low seed carryover under selection.

A model by Cohen (1966) examined the strategy argument for variations in dormancy to be correlated with the probabilities of "good year" and "bad year" in terms of survival and reproduction. Mountford (1971) showed that if instead of a maximax strategy, we use maximin arguments, the optimal germination rates for avoiding extinction might be even lower. MacArthur (1972) extended Cohen's model to examine not only the result of most probable combination of good and bad years, but an average of all possible combinations. This allowed him to predict that if a species occurs in many local subpopulations, mortality risks will be better covered by high dormancy. These predictions need to be tested with the published data and our own studies in progress.

The genetic component of variation in dormancy has been analyzed in several species. We discovered an interesting pattern by comparing dormancy in freshly collected seeds of several species during the months of summer. Table 7 summarizes the data of these tests which show large interpopulation variation as well as intrapopulation variation. Dormancy breakdown occurred in grasses such that by autumn these are fully ready for germination after the first good rainstorm. The role of dormancy and a mechanism of selection maintaining it are understood when we look at the climatic data of various sites and regions. Summer rainfall is highly erratic but apparently sites with occasional years of one or more summer rainstorms (Table 8) have consistently more dormancy in the July seed batches. Apparently, selection is periodic and requires long-term evolution of such strategies, and there-fore, much broader framework of environmental heterogeneity studies are needed than a single contemporary environment.

Schaffer (1974) examined the optimality of life history patterns in terms of the evolution of perenniality and itero-parity (repeated reproduction within a life time) in plants. Briefly, his arguments are based on the assumption about fecundity (reproductive effort) and survival as seed, seed-ling or plants. The age dependent fertility, $G(E)$ and sur-vival rate, $P(E)$, beyond the first age of breeding, and $V(E)$, growth rate, were defined in terms of varying repro-ductive effort, E. Their shapes determine the evolution of parity, $E = 1\%$ (one time reproduction) or iteroparity $0 < E < 1$, and polymorphisms in some others. Note that although very few studies provide evidence of polymorphism for differing longevity within species, it cannot be ruled out since most demographic studies are based on rather limited species samples. Our studies on several grasses including long term perennials such as Deschampsia and Anthoxanthum, short term perennial like Holcus and annuals like Bromus mollis suggest that there might indeed be large variation in the time, frequency and success of reproduction by seed. A complete genetic and demographic analysis of these poly-

66

morphisms would be extremely interesting. Studies on the use
of percent reproductive effort in plants have also stimulated
interest in r- and K-selection models. Although many issues
remain polemic, we can show that percent reproductive effort
(R) would be one and only partial measure of these strategies
(e.g. see Sarukhan and Gadgil (1974), Solbrig and Simpson
(1974)). Table 9 shows some estimates for several annuals
which shows R values with rather high variances over sites,
season, etc. Many unknown variables underlie the selective
mechanisms for these variations. The point to note is that
reproductive effort as measured here and by numerous other
ecologists is inadequate for definitions of alternative
reproductive strategies. We plan to collect genetic variance
data on each of the components referred to in Table 5 before
we can test the notion of optimality in life histories.

Population regulation

Our studies on population numbers suggest that most local
populations of these grassland annuals are 'regulated' in the
sense of fluctuating within limits and thus rarely show local
extinction. Seed dormancy in legumes (Medicago, Trifolium)
further assures that failures in any one season do not cause
extinction. Occasional long range dispersal certainly allows
many new areas to be explored for colonization which may take
initially many trials through repetitive dispersal and seed
storage in soil but followed by a rapid expansion phase. The
annual fluctuations in numbers were found to be correlated
with the patterns of autumn rainfall and temperature changes
during the critical phenological stages such as flowering and
seed maturation. Detailed studies in Avena (Yazdi-Samadi and
Jain (unpublished data)) and in Bromus (Wu (1974)) showed
that different population genotypes provided a variety of

67

output/input ratios in the competition studies using varied
frequencies and densities. Therefore, the genetic variation
in survival or reproductive traits appears to be important in
the outcome of competition. Genetic polymorphisms, then,
seem to provide a fine tuning factor in the overall population
regulation. In all our tests of germinability under differing
environments, species show a great deal of inter- and inta-
population variation. Reproductive rates at low densities
(r's) and survival rates at high densities (K's) have been
estimated for several genotypes within each of these species,
which show evidence for genetic variation in various test
environments. Whether the observed life histories represent
optimum survival or evolutionary rates and whether genetic
variation in each of these demographic variables represents a
regulatory system, is not known since as Rosen (1967) pointed
out, optimality principles can be applied only when structures
and their relations can be analyzed in terms of some well-
defined benefit-cost measures.

An index of evolutionary opportunity, defined in terms of
survival rates and fecundity, may usefully measure what we
may call the demographic surplus[*], meaning the amount of fitness variance provided by the excess progeny per individual.
Genetic variance of fitness then relates to the opportunity
for selective change. Vickery (1974) and Walley, et al.
(1974), for instance, attempted to predict the potential for
novel adaptive response under stress in relation to the

---

[*] This term, like genetic load, may be interpreted differently
depending on one's viewpoint whether a species is often said
to be gambling or struggling for existence (a topic of great
ecological interest).

presence of certain infrequent genotypes in any population. Our studies suggest that evolutionary opportunity further depends on rare but repetitive trials through long range dispersal, occasional outcrossing, rare genetic recombination, etc. For example, recent colonization by artificially founded colonies in rose clover showed the role of certain specific seed sources which might represent rare gene combinations. The role of weak coupling in complex ecogenetic systems by such events was explored in another context by Karlin and McGregor (1972). They showed the role of low rates of migration in the conditions for maintaining multiniche polymorphisms. Thus, certainly apparently rare events, often ignored by disregarding the range of estimates of a parameter, might be of unusual interest in novel adaptive changes.

Thus, we have a dilemma about the notion of patterns in population biology and the value of detailed quantitative studies in this area. We see different species sharing several common features of their genetic systems and predict their expected common patterns of microevolution. But are different species somehow unique that any search for patterns would be a priori futile? Other things being equal (which they rarely are), we hope to explain similarities and differences on the basis of their shared as well as their divergent features of life cycle, population structure, etc. Patterns result presumably from certain associations among these features into some discontinuous modes and from the assumption of somehow integrated, balanced solutions of various adaptive problems. These studies reviewed here clearly suggest that comparisons among species patterns would have to take into account much more than single factor dichotomies or deductions based on simple models of breeding system, balanced polymorphisms, migration, or life cycle differences.

69

Lewontin (1974) posed the question, "How much genetic variation is there that can be the basis of adaptive evolution?" The answer demands a predictive theory of evolution which is to be built not only with our current knowledge of variations in genome size, karyotype, levels of polymorphism, or regulation of gene action, mutation, second order selection, etc., but as noted by Slobodkin (1968) in his elegant essay, a knowledge of environmental challenges to come and a sound theory of physiological, biochemical as well as genetic responses to these challenges. The criteria of optimality or pay-offs in the evolutionary games would be evolution of homeostasis, reproductive success, as well as origin of phyletic lineages. The genetic and ecological parameters discussed in this paper need to be extended to many test environments on one hand and developmental-biochemical variables on the other.

SUMMARY

In order to determine the patterns of survival and microevolution in colonizing annual plants, a comparative study of six grassland species was undertaken. The following population variables were emphasized: degree of random outcrossing, rate and range of local gene flow through pollen and seed, estimation of neighborhood sizes, genetic and nongenetic components of variation along several scales of geographical and local distances, life cycle components of variation in population numbers, and finally, potential correlation between genetic variation and population size or density. Small differences in outcrossing rates and seed dispersal range were found and neighborhood sizes in these species were roughly estimated to be ordered as follows:

M. polymorpha 30-50; T. hirtum 30-100; B. mollis 100-150;
A. barbata 100-150; A. fatua 50-100; and B. rubens 50-200.
Avena fatua and Bromus mollis showed polymorphisms at several
loci to be ubiquitous along both local and wide geographic
range, whereas A. barbata showed a large region to be monomorphic and another region to be polymorphic with patchiness
and occasional clinal variation at the boundary of two climatic regions with distinct ecotypes. Bromus rubens showed
evidence of a very low rate of outcrossing, and of some local
and ecotypic variation between the regions studied. Both
Medicago polymorpha and Trifolium hirtum showed higher levels
of outbreeding, high levels of local variation in Medicago but
not in rose clover, and very little or no evidence of patchiness even though they often occur as distinct colonies. Evidence was presented to suggest that as colonies grow in rose
clover, interchange of pollen by insect pollinators allows
mixing and homogeneity of colonies over large areas. In
Avena and Bromus, there appears to be a negative relationship
between the high levels of genetic polymorphism and a smaller
degree of phenotypic plasticity in Avena spp. and Bromus spp.
There is also evidence for differing amounts of genetic
variation within pairs of congeneric species in mixed stands
versus pure stands, perhaps resulting from some degree of
niche differentiation and character displacement.

Population census data during the seed, seedling and adult
stages and specific germination tests allowed certain measures
of the relative role of seed dormancy, seedling survival and
fecundity in the regulation of population numbers (locally).
In a few instances, specific polymorphisms could be followed
to assess the sizes of various selection components. Grasses
as a group seem to have much less seed dormancy and seedling
deaths and a greater role of progeny size variation than the

71

legumes which showed a greater component of seedling mortality.

Besides some broad patterns of climatic factors, however, very little is known about the factors of local environmental heterogeneity (e.g. soil texture and nutrients, moisture, herbivores, etc.), only a small part of which could be simulated in many of our greenhouse experiments on inter- and intra-species competition. The choice of few morphological marker loci and quantitative traits related to size, growth habit, fecundity, etc. might also not fully represent the patterns in such large scale surveys. Besides long-term continuation of the description of population properties and environmental changes, certain specific ideas on the potential for new adaptive responses are now being tested using artificially founded colonies with the known genetic makeup, and in relation to stress environments at the periphery of species distributions. It was argued that the patterns of survival and adaptive responses, as derived from the comparative ecogenetic studies on related species are often complex, multifactorial associations among their biogeographical and evolutionary characteristics.

REFERENCES

Allard, R.W., S.K. Jain and P.L. Workman. (1968). Adv. Genet. 14: 55-131.

Antonovics, J. (1971). Am. Sci. 59: 593-599.

Baker, H.G. and G.L. Stebbins. (Eds.) (1965). The Genetics of Colonizing Species. Academic Press, New York.

Barber, H.N. (1965). Heredity 20: 551-572.

Birch, L.C. (1960). Am. Natur. 94: 5-24.

Cavalli-Sforza, L. and W.F. Bodmer. (1973). The Biology of Human Populations. Freeman.

Cohen, D. (1966). J. Theor. Biol. 12: 119-129.

Dansereau, P. (1957). Biogeography: An ecological perspective. Ronald, N.Y.

Darlington, C.D. (1958). Evolution of Genetic Systems. 2nd Ed. Oliver and Boyd.

Endler, J.A. (1973). Science 179: 243-250.

Epling, C., H. Lewis and F.M. Ball. (1960). Evolution 14: 238-255.

Feldman, M.W. and J. Krakauer. (This conference).

Giesel, J.T. (1974). Am. Natur. 108: 321-331.

Gottlieb, L.D. (1974). Genetics 76: 551-556.

Harper, J.L. (1967). J. Ecol. 55: 247-270.

Harper, J.L., J.N. Clatworthy, I.H. McNaughton and G.R. Sagar. (1961). Evolution 15: 209-227.

Harper, J.L. and J. White. (1970). In Proc. Adv. Study Inst., Dynamics of Numbers in Populations, pp.41-63.

Harper, J.L. and J. White. (1974). Ann. Rev. Ecol. Syst. 5: 419-464.

Heady, H.F. (1958). Ecology 39: 402-416.

Hutchinson, G.E. (1953). Proc. Acad. Natur. Sci. 105: 1-12.

Jain, S.K. (1969). Evol. Biol. 3: 73-118.

Jain, S.K. (1975). In Plant Genetic Resources: Today and Tomorrow. O.H. Frankel and J.G. Hawkes (eds.), pp.15-36. Cambridge Univ. Press.

Jain, S.K. and A.D. Bradshaw. (1966). Heredity 21: 407-441.

Janzen, D.H. (1971). Ann. Rev. Ecol. Syst. 2: 465-492.

Karlin, S. (This conference).

Karlin, S. and J. McGregor. (1972). Theor. Pop. Biol. 3: 210-238.

Kershaw, K.A. (1973). Quantitative and Dynamic Plant Ecology. American Elsevier.

Levin, D.A. and H. Kerster. (1969). Evolution 23: 560-571.

Levin, S.A. (1974). Am. Natur. 108: 207-228.

Lewontin, R.C. (1974). The Genetic Basis of Evolutionary Change. Columbia Univ. Press.

Marshall, D.R. and S.K. Jain. (1968). Am. Natur. 102: 457-467.

Mortimer, A.M. (1972). Ph.D. Dissertation, Univ. College of North Wales, Bangor.

Mountford, M.D. (1971). J. Theor. Biol. 32: 75-79.

Nei, M. (1973). Proc. Nat. Acad. Sci. USA 70: 3321-3323.

Nei, M., T. Maruyama and R. Chakraborty. (1975). Evolution 29: 1-10.

Nevo, E. (This conference).

Pielou, E.C. (1969). An Introduction to Mathematical Ecology. Wiley-Interscience.

Rai, K.N. (1972). Ph.D. Dissertation, Univ. of Calif., Davis.

Rosen, R. (1967). Optimality Principles in Biology. Butterworths.

Sagar, G.R. and A.M. Mortimer. Ann. Appl. Biol. (in press).

Salisbury, E.J. (1942). The Reproductive Capacity of Plants. Bell, London.

Sarukhan, J. and M. Gadgil. (1974). J. Ecol. 62: 921-936.

Schaal, B.A. (1974). Ph.D. Thesis, Yale Univ.

Schaffer, W.M. (1974). Ecology 55: 291-303.

Schaffer, W.M. and M.D. Gadgil. (1975). In The Ecology and Evolution of Communities. M. Cody and J. Diamond (eds.). Harvard Univ. Press.

Selander, R.K. and R.O. Hudson. Am. Natur. (In press).

Selander, R.K. and D.W. Kaufman. (1973). Proc. Nat. Acad. Sci. USA 70: 1186-1190.

Slobodkin, L.B. (1968). In Population Biology and Evolution. R.C. Lewontin (ed.). Syracuse Univ. Press.

Solbrig, O.T. and B.B. Simpson. (1974). J. Ecol. 62: 473-486.

Stebbins, G.L. (1971). Ann. Rev. Ecol. Syst. 2: 237-260.

Stebbins, G.L. (1974). Flowering plants: Evolution above the Species Level. Belknap Press, Cambridge.

Stern, K. and L. Roche. (1974). Genetics of Forest Ecosystems. Springer-Verlag.

Vickery, R.K. (1974). Ecology 55: 796-807.

Walley, K.A., M.S.I. Khan and A.D. Bradshaw. (1974). Heredity 32: 309-319.

Wright, S. (1969). Evolution and the Genetics of Populations. Vol. 2. The Theory of Gene Frequencies. Univ. of Chicago Press.

Wu, K.K. (1974). Ph.D. Dissertation, Univ. of Calif., Davis.

TABLE 1. Variation in *Bromus mollis* populations.

Region	No. of Sites	Frequency of Recessive Class bb	Frequency of Recessive Class hh	Days to Flower G.H.	Plant Height Nature	Plant Height G.H.	Panicle Length Nature	Panicle Length G.H.	Seed Weight Nature
South Coast	8	.09	.01	102	27.1	64.1	4.7	7.9	16.4
Monterey	10	.14	.05	111	31.9	58.2	4.7	7.6	19.8
Sierra Foothills	6	.39	.15	107	37.2	60.0	4.9	7.5	20.1
Central Valley	6	.08	.19	103	30.6	62.8	4.5	7.2	18.6
North Coast	18	.05	.27	108	30.7	56.4	3.8	7.2	21.0
Redding-Eureka transect	6	.09	.06	112	28.1	51.5	3.3	6.8	17.1

Locus B/b = presence/absence of a brown spot on chromatogram, H/h = pubescent/smooth glume.
Nature - Data on plants from natural stands.
G.H. - Greenhouse culture.

TABLE 2. Hierarchial variation pattern in *Bromus mollis*.

Source of variation	Plant DF	Height MSS	Panicle Length MSS
Among sites	7	5736.95	39.48
Among families within sites	184	1353.45	9.34
Among replicates within families	384	377.70	3.55
Within families (error)	2880	74.69	1.42

TABLE 3. *Microdifferentiation in Medicago polymorpha populations (Winters site, 1972 data): allelic frequencies at four morphological markers.*

Sample	Number of plants scored	Locus Sp # a	b	Locus Pl # a	b	c	Locus R # a	b	c	Locus W # a	b
H_1	295	.053	.947 (.207)*	.358	.053 (.835)	.589	.126	.284 (.930)	.589	1 (0)	0
H_2	330	.558	.442 (.686)	.069	.012 (.316)	.918	.118	.072 (.613)	.809	.982 (.090)	.018
H_3	328	.791	.209 (.513)	.021	0 (.102)	.979	.073	.006 (.298)	.921	.994 (.037)	.006
H_4	338	.380	.620 (.664)	.018	.041 (.260)	.941	.068	.081 (.073)	.851	.970 (.135)	.030
H_5	351	.487	.513 (.693)	.214	.048 (.700)	.738	.259	.111 (.882)	.634	.359 (.653)	.641
H_6	316	.570	.430 (.683)	.082	.003 (.090)	.905	.038	.028 (.288)	.934	.987 (.069)	.013
S_P^2		.0610		.0018	.00057	.0226	.0062	.0152	.0240	.0658	
F_{ST}		.245		.016	.022	.019	.062	.134	.151	.632	

*Values of $H' = -\Sigma\ p_i lnp_i$ are in parentheses.

#Sp = pink spot on leaflet, Pl = pink leaf petiole.
R = red basal ring on leaflet, and W = white specks on first leaflet.

TABLE 4. Polymorphism in Rose Clover Populations.

Locus Site	W/w #		Sp/sp #		U/u #		R/r #		Diversity index H'
	p	Het	p	Het	p	Het	p	Het	
									(based on 3 loci)
Madera #20	.98	.04	.95	.10	.46	.26	.49	.20	0.986
#26	.95	.10	.95	.10	.51	.32	.46	.30	1.089
MA 1	.87	.10	.96	0	.62	.11	*--	--	1.218
MA 2	.94	.06	.88	.06	.46	.31	*--	--	1.284
MA 3	1.00	0	.74	.23	.87	0	*--	--	0.959
MA 4	.85	.11	.67	.15	.69	.23	*--	--	1.676
						*(this trait did not express well)			
Shasta FPC	.92	.12	.90	.16	.43	.31	.43	.26	1.287
HOR	.95	.08	.92	.16	.44	.24	.41	.25	1.133
SH 1	.92	.04	.92	.08	.38	.08	--	--	1.222
SH 2	.83	.04	.96	0	.31	.38	--	--	1.243
Glenn BHR	.92	.16	.91	.18	.56	.18	.39	.16	1.267
BLP	.95	.10	.93	.13	.52	.38	.52	.31	1.362
GL 1	.76	.17	.41	.14	.36	.53	--	--	1.881
GL 3	1.00	0	.40	.12	.48	.28	--	--	1.365
GL 4	.73	.23	.36	.18	.44	.52	--	--	1.923
Sacramento (BRR)	1	0	.88	.06	.53	.12	--	--	1.058
Temblor Range (TR)	1	0	.26	.07	.22	.02	--	--	1.100

Het = Proportion of heterozygotes, p = freq. of dominant allele; $H' = -\Sigma\, p_i \ln p_i$.

#W/w = presence or absence of white markings on leaflets; Sp/sp = pink spot on leaflets; U/u = position of white markings; R/r = red vs. green veins on inflorescence bracts.

TABLE 5. *Summary of Mortimer's Population Flux Studies (Mortimer, 1974).*

Species	a	b	c	d	e	f	h	j
				Component of Regulation [*]				
Dactylis glomerata	200	.15	.24	.16	.03	.03	?	?
Holcus lanatus	275	.35	.90	.37	.08	.08	?	?
Plantago lanceolata	63	1.0	.90	.11	.02	.02	?	?
Poa annua	70	?	1.0	.07	.01	.01	?	?

[*]
a = No. of seeds/plant.
b = Seed rain on soil surface.
c = Prob. of a seed arriving in "seed bank".
d = Prob. of germination (1-d = dormancy measure).
e,f = Prob. of seedling establishment and becoming mature plant.
h,j = Routes II and III (Fig.9) - rates of survival as plants after the first year and asexual propagation respectively.
? = Indicates that no reliable estimates are available as yet.

79

TABLE 6. *Some components of "population regulation" in two grassland annuals (SE based on eight reps per census).*

Species	Population & year	a ‡	d ‡	(e+f) ‡	g ‡	Plant density per m^2
Avena barbata	#1 1971	31 ± 6	.94	.64 ±.17	0	64 ±10
	1972	20 ± 8	.95	.38 ±.06	.02	39 ± 7
	1973	27 ±10	.97	.73 ±.29	0	40 ± 8
	#2 1971	14 ± 2	1.00	.76 ±.16	0	134 ±21
	1972	10 ± 2	.99	.68 ±.12	.04	111 ±12
	1973	18 ± 7	1.00	.79 ±.22	.01	78 ±16
Medicago polymorphia	#1 1971	26*	.28	.22 ±.17	.07	75 ±32
	1972	20*	.62	.38 ±.24	.11	76 ±30
	1973	19*	.38	.19 ±.08	.16	29 ± 13
	#2 1971	21*	.58	.64 ±.19	.03	49 ± 15
	1972	20*	.31	.70 ±.30	.04	72 ±30
	1973	11*	.51	.42 ±.28	.11	30 ±16

* Seed number per plant is estimated in Medicago by counting number of pods and multiplying with the average number of seeds per mature pod.

‡ a = seed output/plant, d = % germination, (e+f) = seedling survival rate;

g = % nonreproductives (Fig.9).

TABLE 7. *Inter- and intrapopulation variation for dormancy in some grassland annuals.*

Species	MSS		F *	% among population variation total
	Among pop.	Within pop.		
Avena	(July) 211.56	14.88	14.22	93.4
barbata	(Sept) 735.60	14.54	50.59	98.1
A. fatua	(July) 258.53	12.20	21.19	95.5
	(Sept) 786.19	34.83	22.57	95.8
Bromus	(July) 1647.65	11.64	141.55	99.3
mollis	(Sept) 1177.45	20.24	58.17	98.3
B. rubens	(July) 772.19	102.27	7.55	88.3
	(Sept) 1135.14	13.60	83.48	98.8
B. rigidus	(July) 1062.21	72.91	14.57	93.6
	(Sept) 460.27	15.26	30.16	96.8
	Among families	Within families		
A. barbata	MSS	MSS	P	
Prop. #2	163.61	10.32	<.01	
Prop. #8	405.46	24.20	<.01	
A. fatua				
Prop. #4	225.20	30.27	<.01	
Prop. #11	172.80	26.21	<.01	

* *F-values are highly significant (P < .01).*

TABLE 8. *Mean and coefficient of variation for average monthly precipitation (in inches; based on 1957-1969 period).*

Site		July	August	September	October	November
Redding	$\bar{X}$	.06	.32	.98	2.08	5.10
	CV	355.6	191.4	199.3	126.4	62.8
Marysville	$\bar{X}$	.004	.072	.081	1.64	4.05
	CV	210.8	259.2	170.6	170.6	65.0
Healdsburg	$\bar{X}$	.007	.197	.100	2.57	9.78
	CV	213.5	145.3	180.1	123.8	80.5

TABLE 9. *Estimates of percent "reproductive effort" (R) per plant.*

Species	Habitat	Time of Harvest	R̄	C.V.	Correl. with total biomass	No. of plants sampled
Poa annua	Fallowed field	January	.180	.316	.525	50
" "	"	March	.208	.363	.064	50
" "	"	April	.231	.406	.069	50
Capsella bursapastoris	"	January	.490	.283	-.199	30
"	"	March	.254	.377	-.117	30
"	"	April	.331	.288	-.003	60
Senecio vulgaris	Edge of vernal pools	March	.161	.372	-.075	50
"	"	April	.332	.780	-.142	50
"	"	May	.460	.621	-.255	50
Montia perfoliata	"	March	.267	3.072	-.180	30
Calandrinia caulescens	Fallowed field	March	.081	.353	-.512	30
Medicago polymorpha	Greenhouse culture	April	.387	.437	-.009	30
"	"	May	.474	.303	-.005	30
Trifolium grayi	Vernal pool	June	.474	.276	-.612	30 } 2 different sites
"	"	June	.485	.166	-.473	30 } sites
T. dubium	"	June	.481	.179	-.337	18
T. microcephalum	"	June	.261	.400	-.559	30 } 3 different sites
"	"	June	.530	.159	-.336	20 }
"	"	June	.369	.299	-.338	30 } sites

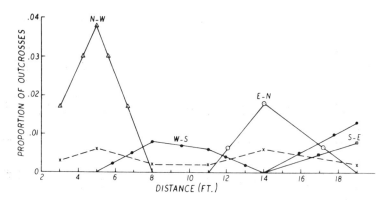

FIGURE 1. *Gene flow through pollen as a random outcrossing component in* A. barbata. *N,S,E,W refer to the directions of transects sampled and distance is given from the central colony. Note the effect of prevailing wind from S-W direction on the outcrossing rate or distance. (After Rai, 1972).*

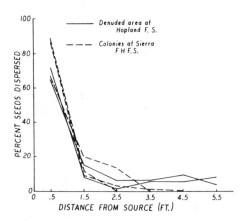

FIGURE 2. *Gene flow through seed dispersal in* A. barbata, *scored in terms of distance seeds dispersed within several weeks after maturing and shattering. Data were gathered in colonies at the UC field stations at Hopland and Sierra Foothills. (After Rai, 1972).*

REGIONAL FREQUENCIES			
I	II	III	IV
0.750	0.373	0.999	1.000

FIGURE 3. Distribution of Ls and ls alleles (dark and clear sectors) in A. barbata populations. Four regions in California are shown along with a summary table of ls allelic frequencies. Note that several other loci including B/b, Np/np, esterases and anodal peroxidases also showed a remarkably similar regional pattern of differentiation. The climatic factors that account for this pattern are mainly mean daily and monthly temperatures and summer fog in the coastal areas.

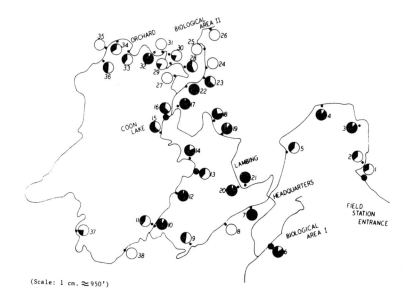

(Scale: 1 cm. ≈ 950')

FIGURE 4. *Distribution of Ls (dark) and ls (clear) alleles in A. barbata at Hopland Field Station. (After Rai, 1972).*

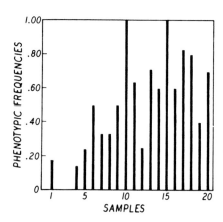

FIGURE 5. *Distribution of Ls allele along a representative transect where samples are drawn in 30 × 30 cm² quadrats. Note very highly localized variation along a 10 meters transect. (After Rai, 1972).*

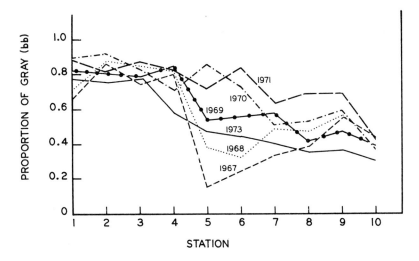

FIGURE 6. Clinal variation at locus B/b in A. fatua population at Marysville site scored for six different generations. Note a wide range of fluctuations in the central area. Observations suggest that burning of grass litter during some years and disturbance due to some erosion might have resulted in these fluctuations. (Detailed analysis in a manuscript in preparation).

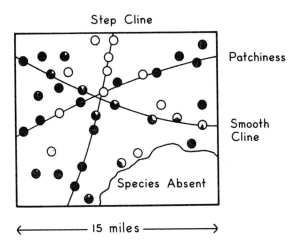

FIGURE 7. A representative sample of patchiness in gene frequency distribution in A. barbata populations at the boundary of regions I and III. Note that within a two-dimensional sampling matrix, different linear transects show steep cline to extreme patchiness.

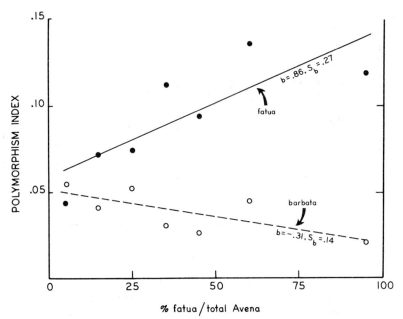

FIGURE 8. *Character displacement in Avena, shown as relative changes in the amounts of polymorphism with the proportion of A. fatua and A. barbata in mixed stands. Note that A. fatua is less polymorphic in predominantly A. barbata stands. Relative abundance of the two species was averaged from three-years' data.*

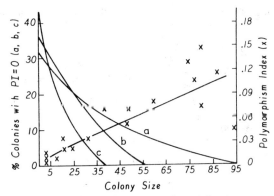

FIGURE 9. *Graph showing the relationship between colony size and the amount of polymorphism scored at four diallelic loci in T. hirtum. Polymorphism index ($PI = p_i(1-p_i)/n$, n = number of loci) is higher in larger colonies. Alternatively, the proportion of colonies with PI = 0 at three locations (a,b,c) was negatively correlated with the colony sizes.*

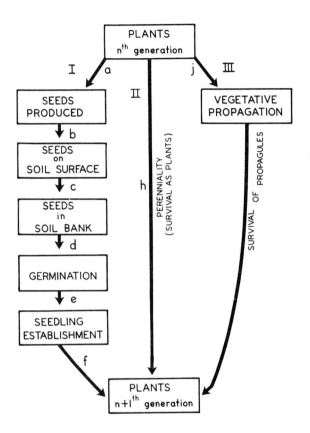

FIGURE 10. Diagram showing the life cycle stages scored in population census studies. Route I represents an annual life cycle (semelparity) and routes II and III refer to perennials reproducing by seed or vegetatively, once or repeatedly (iteroparity) during their life period. (After Sagar and Mortimer, in press).

* g - not all seedlings reproduce.

Domestication Processes in *Trifolium berytheum* Boiss*

J. KATZNELSON

INTRODUCTION

Domestication is a change imposed on a wild species, and
is usually a very long process (Harlan, 1970). It took
thousands of years of man's activity to achieve the current
small grains (Zohary, 1969), and wild and cultivated types
differ in many characters, due to contrasting selective forces (Oka and Chang, 1964). To become suitable for cultivation, the wheat plant has lost traits such as rachis brittle-
ness and hairyness, lemma hardness, and germination-regulating
mechanisms; became accustomed to a dense monoculture; and
lost its great population variability in many physiological
and morphological traits through selection towards self-
fertilization and homozygosity. A similar process of adjustment, but on opposite direction, i.e., to become suitable to
be "wild" plant, is seen in stands of non-harvested, cultivated barley, if they are bordered by wild Hordeum spontaneum

---

*
Contribution from the Agricultural Research Organization,
Bet Dagan, Israel. 1975 series, No. 136-E. The study was
financed in part by Grant F6-IS-222 from the U.S. Department
of Agriculture under P.L. 480.

Koch. This is obtained by introgressive hybridization and
selection for "wild" characteristics (Zohary, 1960).

Forage and pasture plants differ considerably in this respect from other crops, especially in two traits:
a) uniformity of product is not so important as in other crops;
b) in cases of self-seeding annuals, such as Medicago, Trifolium and Lolium, the germination-regulating mechanism
through hard-seededness and dormancy is as important in cultivars as in the wild state (Quinlivan, 1970). After all,
both cultivated and wild types are used by man for the same
purpose, and cultivars of pasture plants naturally invade and
even dominate unsown land (Morley and Katznelson, 1965). This
is why domestication of forage plants may be achieved within
15-30 years, after selection for agronomic traits such as
yield, quality, seedling vigor and production, and disease and
pest resistance. Very little is known on the dynamics of domestication processes even in pasture plants.

T. berytheum Boiss. is an annual wild legume, quite common
in heavy and waterlogged soils along the coastal plain and in-
land valleys in Israel, from the Lebanese border to the
Lakhish region. It is closely related to the cultivated berseem, T. alexandrinum L. (Oppenheimer, 1961; Katznelson, 1971;
Putievsky and Katznelson, 1973), but differs from it by being
more prostrate and hairy, and having much smaller seed (1 mg
vs 3 mg) polymorphic in color, and thinner branches. It is
self-incompatible although some interspecific F_1 hybrid
plants produced self-compatible progenies (Putievsky and
Katznelson, 1970, Putievsky, Katznelson and Zohary, 1975).

Twelve years ago we collected seeds from a population of
Trifolium berytheum Boiss. NYT 529, and grew them in an introduction nursery. The agronomic traits of NYT 529 were quite
impressive, and seeds were collected and multiplied. After

ten generations of bulk multiplication the line (T. berytheum cv. Akhziv) is now at the final stage of testing and is being recommended as a new hay crop (Katznelson, 1974). The changes that were observed in this population during domestication are described herewith.

MATERIALS AND METHODS

The original seeds of Trifolium berytheum Boiss. (NYT 529) were collected along the seashore at Akhziv, Western Galilee, in 1963 (O_{63}); in 1964 they were sown in an introduction nursery along with many other Trifolium populations, including T. berytheum and the closely related, cultivated T. alexandrinum L. The sowing resulted in a very dense stand. Seeds were collected (S_{64}), and since then the line was multiplied further in bulk in every year except 1969. These formed seed sources denoted S_{65}, S_{66}, etc. The original population at Akhziv was sampled again in 1967 (O_{67}) and 1971 (O_{71}). All samples and sowings comprised thousands of seeds from hundreds of plants.

The O and S bulk seeds were used to study germination, seed weight and color, and average weight of each color. Altogether, we could distinguish nine different colors, but as some were very rare, we have a complete set of reliable data for only four of them, viz., yellow, orange, purple and black.

Single O and S plants were studied individually in nurseries in 1967 and 1972. In 1967 a comparison was made between O_{63} and S_{66}, and in 1972 between O_{71} and S_{71}. Data were collected on germination percentage (1967, 1971), plant height and width (1967 and 1972), number of branches, hairyness (scored 1-hairy, 2-pubescent), leaf size, flowering

93

condition, head size, plant weight, and seed size.

The colors of the parent seed were determined in 1972 only. Germination percentages were calculated in November of some years, on hand-threshed seed.

RESULTS

Original population

Seed size and seed color pattern hardly changed between 1963 and 1967 and 1971. Most of the seeds are dark and their weight is 1.1-1.2 mg, as can be seen in Table 1. In the three years, there was 3-7% germination of unscarified, harvest-threshed seed. There was no change in plant habitus between 1967 and 1971, the plants remaining semi-prostrate (height/diameter ratio was 0.30 and 0.29 in the two years, respectively). Plant habitus, hardseededness, and seed size in the wild NYT 529 are more or less the same as those of many other populations of T. berytheum collected in the region between Qadesh and Gezer. The distinct differential characteristic between populations is seed color; a population approximately 10 km south of Akhziv had either yellow or orange seeds, and they were somewhat smaller (weight = 0.82 mg) than those of the Akhziv population. On the other hand, a population from Lod region had mostly black seeds.

Changes in seed size and color with time

Several changes occurred in seed characteristics with time. Seed size almost doubled, and the percentage of bright seed (yellow and orange) increased from 10.8 in 1963 to 72.4 in 1974 (Figure 1). These changes were gradual, and there is a significant positive correlation between seed size and percent bright seed (r = 0.943). There was no intentional selection whatsoever for either of these traits.

94

Germination and seed color

Most of the non-germination can be attributed to hard-seededness, at least within the limits of the temperatures in which germination was tested.

We have three sets of data - original seed, S_{70} and S_{72} (Table 1). Original seed with different colors did not differ in their germination pattern, even though softer seeds were found more often among orange seeds than those of other colors, hardseededness exceeded 90% even in orange seeds. The pattern changed a great deal in the S_{70} seeds, in which most of the orange seeds were soft and most of the dark seeds were hard. Between 1970 and 1972 there was an obvious reduction in hardseededness in the dark seeds. The change toward light-colored seeds was thus highly correlated with softening of seeds, and germination of bulk seed in 1974 exceeded 55%.

Germination of seeds collected from single plants grown in 1972 from O_{71} and S_{71} seeds, separated according to color, was also studied, and the frequency distribution of the results of single plants is presented in Table 2. The data shows that the decrease in hardseededness occurs in a few lines, rather than in the whole population.

Other changes

Some of the morphological parameters based on single plants of O_{71} and S_{71} are summarized in Table 3, which shows that selection caused distinct shifts in population averages. Plants became broader and more erect due to thickening of the branches. They became heavier but had fewer branches, and the size of a single branch thus almost doubled. Another impor-tant change was an increase in growth rate up to the age of 40 days. Selected plants at that age were at least three times larger than those of the original population. Varia-

95

bility in these morphological traits was more or less the
same in S_{71} and O_{71}. On the average, variance values were
about 50% lower in O_{71}, corresponding to the lower mean
values. Thus, values of the coefficient of variability (c.v.)
were more or less the same (Table 3).

Another important change observed was that while many
bagged plants of the original population did not set any seeds,
several S_{71} lines from single plants were completely self-
compatible (D. Globerson, personal communication).

DISCUSSION

Causes of the change

The data presented show that a rapid change occurred in
many traits in NYT 529, changes that in total can be called
the syndrome of domestication. Included were seed softening,
an increase in seedling vigor, a change from prostrate to
erect habitus and a trend toward self-compatibility. These
changes occurred as a result of selection caused by a change
in growth conditions and/or by some introgression from T.
alexandrinum.

There were two changes in growth conditions. (i) In the
wild habitat the stand is quite sparse, despite the dominance
of T. berytheum in the vegetation. This is so as hardseededness is almost complete, and although entire hoads (with
35-50 seed) often serve as dispersal unit, it is very rare
that two plants will emerge from one such unit. On the contrary, seeding and germination in introduction rows were very
dense, and intrapopulation as well as general competition was
much stronger than in natural stands. This may account for
the increase in seedling vigor, thicker stems, and possibly
also the more erect plant habitus. (ii) While a natural

stand is composed of plants derived from seed of various ages, the nursery stand was derived from 6-month-old seed only; only soft seeds will germinate and the rest are lost to the population. Repeating this procedure for several generations will select for softseededness. That not all lines are softseeded is due to the fact that, (i) even in a very hardseeded lot of seed some will soften due to various external factors, and, (ii) a few seeds will be soft anyway. The speed of the softening depends on both these factors. This cannot explain, however, the difference in softening between orange, yellow and black seeds, unless we assume that a certain dark pigment has some fungicidal effect and that fungi are responsible for some seed softening. This is similar to the case of cathecol in onion coats (see Williams, 1964).

Introgression from T. alexandrinum may explain some of these phenomena. With the syndrome of domestication NYT 529 is somewhat intermediate in several traits between the original NYT 529 and T. alexandrinum. This is so in seed color (berseem cultivars have orange-yellow seeds), early vigor, erect habitus, and seed size. Introgression in the last character, however, does not fit our earlier findings, that T. alexandrinum × T. berytheum F_1 hybrids produced seed of two distinct sizes: small (1.0 - 1.2 mg), as in T. berytheum, or large (2.5 - 3.5 mg), as in T. alexandrinum, but no intermediates (Katznelson, 1969). S_{74} seeds were almost twice as large as O_{71}, but individual seed never attained the size of small T. alexandrinum seed.

The domestication syndrome does not include characters of floral structure and branching habit. In local berseem, branching is either basal (in cultivars of the Musgawi group) or upper (in Fahli group). In wild T. berytheum and in cv. Akhziv, secondary and tertiary branches appear along the whole

97

stem.

Although we cannot exclude the possibility of some gene flow from T. alexandrinum to NYT 529, our impression is that this contribution to the changes described is very small compared with that achieved due to selection caused by the change in growth condition. Further studies have shown that other changes in growth conditions (harvest regime, late sowing, etc.) impose considerable changes in genetic parameters such as seed color and size even after one generation. The S_{74} population provides a very interesting tool for the study of domestication processes. It is still as variable as the wild population, in contrast to other species where the variability was reduced with the process of domestication (Harlan, 1970; Robertson, 1965).

The present study points out two important general conclusions:

First, the population pattern of wild annual species that have a germination regulating mechanism is, in fact, perennial. The perenniality of such stands is a potent buffer against changes in population structure and gene frequencies. This is so especially when conditions vary between years, and alleles have different selective preferences in these years. An average generation time of such a stand of annuals may be of several years, i.e. the average age of germinating seeds, or, more precisely, those that get to maturity. As germination pattern will vary between years so will average generation time.

Second, the unintentional selective forces, that operated in the present process in response to change of growth conditions were very powerful. This may be the case in other introduction nurseries and may cause loss of valuable genetic material if it is associated with any of the "wild" syndrome characteristic. This should be considered seriously by plant breeders.

SUMMARY

Changes were noted in genetic parameters during domestication of <u>Trifolium berytheum</u> Boiss. The characters that were affected by the bulk, repeated growth and seed multiplication were: seed weight, almost doubled; seed color pattern, changed from mainly dark to mainly light colored; hardseededness, partial breakdown of that typical to the wild population; plant height and width, increasing such that the plant habitus changed from semi-prostrate to semi-erect; growth rate, increasing at early growth stages; and self-incompatibility, breakdown in a few lines. There was hardly any change in population variability. All these changes, forming the domestication syndrome, occurred within nine generations as a result of "natural" unintentional selection. Connections between the direction of selection and the changes occurring are discussed.

ACKNOWLEDGEMENTS

The help of Miss Nitza Shermann in this project is greatly appreciated.

REFERENCES

Harlan, J.R. (1970). In: <u>Genetic Resources in Plants</u>.
   (O.H. Frankel and E. Bennett, Eds.), pp.7-17. Blackwell, Oxford.

Katznelson, J. (1969). Third Annual Report to U.S. Dept. Agric., Proj. No. Al0-CR-56.

Katznelson, J. (1971). Final Report to U.S. Dept. Agric., Proj. No. Al0-CR-56.

99

Katznelson, J. (1974). Hassadeh 55: 185-190. (In Hebrew).

Morley, F.H.W. and J. Katznelson. (1965). In: The genetics of colonizing species. (H.G. Baker and G.L. Stebbins, Eds.) pp. 269-285. Academic Press, N.Y.

Oka, H. and W. Chang. (1964). Bot. Bull. Acad. Sinica 5: 120-138.

Oppenheimer, H.R. (1959). Bull. Res. Counc. Isr. 7D: 202-221.

Putievsky, E. and J. Katznelson. (1970). Chromosoma (Berl.) 30: 476-482.

Putievsky, E. and J. Katznelson. (1973). Theor. Appl. Genet. 43: 351-358.

Putievsky, E., J. Katznelson and D. Zohary. (1975). Theor. Appl. Genet. 45: 355-362.

Quinlivan, B.J. (1970). In: Proc. XI Int. Grassland Congr. (M.J.T. Norman, Ed.), pp. 583-586. Univ. Queensland, St. Lucia, Australia.

Robertson, F.W. (1965). In: The genetics of colonizing species. (H.G. Baker and G.L. Stebbins, Eds.), pp.95-115. Academic Press, N.Y. and London.

Williams, W. (1964). Genetical principles and plant breeding. Blackwell, Oxford.

Zohary, D. (1960). Bull. Res. Coun. Isr. 9D: 21-42.

Zohary, D. (1969). In: The domestication and exploitation of plants and animals. (P.J. Ucko and G.W. Dimbleby, Eds.), pp. 47-66. Duckworth, London.

TABLE 1. *Seed weight (mg) and color pattern in T. berytheum NYT 529 sampled in 1963, 1967 and 1971, and germination percentages of 0_{67}, S_{70} and S_{72} seeds.*

Source	Avg. seed weight (mg)	Seed color			
		yellow	orange	purple	black
		Seed number (% of total)			
0_{63}	0.985	4.2	6.6	57.2	32.0
0_{67}	1.201	2.7	8.1	61.5	27.7
0_{71}	1.105	2.3	7.2	60.0	30.5
		% Germination			
0_{67}		4.1	9.3	5.4	4.7
S_{70}		25.7	65.1	8.6	9.3
S_{72}		40.2	68.5	36.2	40.1

TABLE 2. *Frequency distribution of hardseededness in single plants derived from seed of different colors of 0_{71} and S_{71} plants.*

Seed color		Hardseededness (%)								
		10	20	30	40	50	60	70	80	90
	Source - 0_{71}									
Yellow									1	7
Orange								1		2
Purple							1		1	3
Black									2	5
	Source - S_{71}									
Yellow		1	1	3		1		4	2	
Orange		3	4		2	2	3	1		1
Purple			1	1	2	1		2	3	
Black		3		2			3	1	2	1

TABLE 3. Some morphological parameters of plants derived from seed of different colors of O_{71} and S_{71} plants.

| Seed color | Source | Number of plants | Plant | | | Number of branches |
			diameter (cm)	height (cm)	weight (g)	
Yellow	O_{71}	15	71	15	119	24
	S_{71}	15	88	35	110	24
Orange	O_{71}	5	95	20	86	20
	S_{71}	13	87	32	210	25
Purple	O_{71}	13	62	28	116	28
	S_{71}	13	87	65	163	22
Black	O_{71}	12	62	19	100	27
	S_{71}	13	120	36	171	21
Average	O_{71}	45	69	20	100	26
	S_{71}	54	96	41	169	23
Variability C.V. (%)						
	O_{71}	45	37.6	71.8	48.2	47.8
	S_{71}	54	33.3	56.2	55.4	55.2

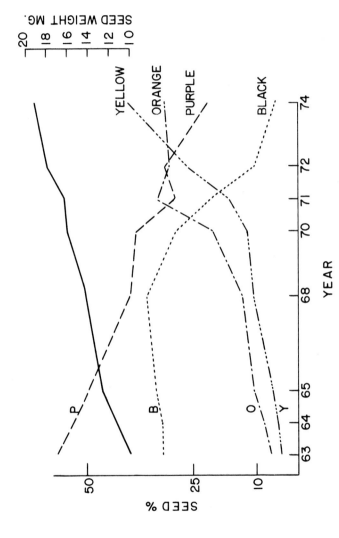

FIGURE 1. Changes in seed color patterns (% of total, left side scale) and seed weight (top right) during domestication process.

103

Protein Polymorphisms in Domesticated Species: Evidence for Hybrid Origin?

C. MANWELL and C.M. ANN BAKER

INTRODUCTION

It has long been recognized that the evolution of domestic
plants and animals is of great importance, both for its
applications to agriculture and for its contributions to
general evolutionary principles (Darwin, 1905). In choosing
this topic as a subject for discussion at this symposium we
were influenced by the recent contributions of Israel to this
field: for example, Zohary and colleagues' (1969) studies on
the progenitors of modern wheats; Epstein's (1969, 1971)
monographs on the domestic animals of China and Africa; and
Moav and Wohlfarth's (1966, 1970) genetic improvement of
domesticated carp.

Since the introduction of high resolution electrophoretic
methods information of value in understanding the evolution
of domesticated species has been obtained (Mourant and Zeuner,
1963; Manwell and Baker, 1970). A survey of the literature
reveals a greater accumulation of population data on the
genetic polymorphism of proteins in various farm animals than
for any other species except man. There are fewer data on
protein polymorphisms in plants, both domesticated and wild,
but some of the recent plant research is especially elegant

105

and one example is given now to illustrate the utility of the approach through protein electrophoresis.

The origin of tobacco: The commercial tobacco plant, Nicotiana tabacum, has twice the chromosome number of related wild species of the genus Nicotiana. It has been suspected that N. tabacum arose by hybridization of N. sylvestris with another species, followed by amphiploidy. Protein electrophoresis has allowed not only the identification of the second species but also the identification of which species contributed the male parent and which species contributed the female parent in the actual hybridization event (Gray, Jung, Wildman and Sheen, 1974). These workers studied the two subunit types, 'large' and 'small', which make up the enzyme ribulose 1,5 diphosphate carboxylase, the enzyme involved in CO_2 fixation as the primary process in C_3 plants and as a secondary process in C_4 plants. The small subunit is coded from nuclear DNA; the large subunit is coded from chloroplast DNA. The latter is inherited from the maternal parent only. It was observed by isoelectric focusing that the large subunit of N. tabacum is identical only to that of N. sylvestris and, therefore, this species must have contributed the female parent to the cross. There are two distinct small subunit types in ribulose 1,5 diphosphate carboxylase from N. tabacum, one electrophoretically identical to that from N. sylvestris and the other identical to that from N. tomentosiformis, which must have been the species contributing the pollen in the original hybridization. The results were checked by forming the N. sylvestris x N. tomentosiformis amphidiploid and observing an identical electrophoretic pattern for the subunits of ribulose 1,5 diphosphate carboxylase as observed for the commercial tobacco plant.

Single versus multiple amino acid differences between
polymorphic protein variants: No such elegant approach to
hybrid origin has been shown for animal domesticates which
compares with the tobacco example just summarized. However,
there are unusual data on certain common protein variants in
cattle and sheep which are most readily explained by the
hypothesis of hybrid origin, perhaps interspecific hybridization and introgression (Manwell and Baker, 1970).

Studies on gene mutation as expressed in the amino acid
sequence phenotype of proteins have shown that nearly all
variants of a particular protein, whether tryptophan syn-
thetase of E. coli or human haemoglobin, differ by single
amino acid substitutions (Dayhoff, 1972). These amino acid
changes can, in turn, be explained by changing one of the
three nucleotide bases in the corresponding codon.

For example, in Dayhoff's (1972) compilation there are
some 145 different human haemoglobin variants listed, nearly
all of which differ from normal by just one amino acid. Only
one of these 145 listed human haemoglobin variants, Hb C
Harlem, differs from normal by two amino acids, each at a
different point, not adjacent to each other, in the polypeptide chain. Hb C Harlem is

$$\alpha_2 \beta_2 \quad 6 \text{ Glu} \rightarrow \text{Val}, \quad 73 \text{ Asp} \rightarrow \text{Asn} \quad .$$

The 6 Glu → Val substitution characterizes the most common
known abnormal human haemoglobin, sickle cell haemoglobin,
Hb S. Therefore, the most likely explanation of Hb C Harlem
is that it represents a 'second site' mutation which took
place somewhere in the relatively large assemblage of Hb S
genes in the human gene pool. Thus, although Hb C Harlem
differs in its β chains from Hb A by two amino acids, this
condition is bridged by a known (and very common) intermediate

variant, Hb S , i.e., the mutational divergence can be represented

$$Hb\ A\ \longrightarrow\ Hb\ S\ \longrightarrow\ Hb\ C\ Harlem$$

In other words, there is no 'missing link'.

An alternative explanation for the origin of Hb C Harlem cannot be ruled out: intracistronic crossing over in an individual heterozygous for Hb S, with the 6 Glu → Val substitution, and Hb Korle-Bu, which has the 73 Asp → Asn substitution. This alternative hypothesis is considered unlikely in view of the rarity of Hb Korle-Bu.

Several rare types of mutation can give rise to multiple amino acid differences between variants, though the result is usually distinct from the accumulation of single amino acid substitutions, in that all of the multiple differences are at one location in the polypeptide chain, not scattered along its length at random. A clean deletion may involve only one amino acid, e.g., Hb Freiberg or Hb Tours, or more than one amino acid, e.g., Hb St. Antoine, lacking two residues, Hb Tochigi, lacking four residues, or Hb Gun Hill, lacking five residues (Wajeman, Labie and Schapira, 1973). A clean insertion mutation of 9 nucleotides has resulted in three extra amino acids, duplicating the three residues just before them on the N-terminal side of the sequence, being intercalated into the middle of the α chain of Hb Grady (Huisman, Wilson, Gravely and Hubbard, 1974).

On rare occasions a single mutational event alters two or more adjacent nucleotides with the possibility of changing adjacent amino acids. Hb J Singapore is α 78 Asn → Asp and α 79 Ala → Gly (Blackwell, Quentin, Boon, Liu and Weng, 1972) and so far no haemoglobin variant has been found with just one or the other of those two changes.

A few chain termination mutations are known, e.g., Hb Constant Spring, where several extra residues occur at the C-terminal end of the molecule (Dayhoff, 1972). A similar addition of residues at the C-terminal end is observed in a frameshift mutation deleting a single nucleotide towards the end of the ∝ chain cistron, specifying Hb Wayne (Seid-Akhavan, Winter, Abramson and Ricknagel, 1972).

There are two types of mutations usually explained by intracistronic crossing over: 1. Mutations, such as the various Hb Lepores, 'anti-Lepores' (Badr, Lorkin and Lehmann, 1973), and Hb Kenya (Huisman, Wrightstone, Wilson, Schroeder and Kendall, 1972), where a new polypeptide chain of the same length as other non-∝ chains is formed which has part of its sequence traceable to one cistron and part of its sequence traceable to another cistron, e.g., Hb Kenya is the result of an apparent fusion of the γ and β chain cistrons so that the resulting polypeptide chain has the γ sequence for the N-terminal half of the molecule and the β sequence for the remainder. 2. Mutations, such as haptoglobin 2, where the polypeptide chain (∝) is not of the same length as that in haptoglobin 1 but is elongated to nearly twice the length -- in the case of the haptoglobin '2∝' chain by a postulated mismatched synapsis followed by crossing over in a heterozygote for the hidden '1F∝' and '1S∝' polymorphism.

Whereas these unusual types of mutation have the potential for creating allelic variants which differ by several amino acids, they are, nevertheless, single mutational events. Furthermore, they are distinguishable from the result of accumulation of single amino acid substitutions in that the latter occur at random along the length of the polypeptide chain, whereas the former occur in juxtaposition. The exception is intracistronic crossing over; however, providing that the original polypeptide chain types are still present, such fusion chains are readily traceable, e.g., the various Hb Lepores are traceable to different combinations of the N-terminal portion of the δ chain and the C-terminal portion of the β chain.

By way of contrast, consider one of the first protein polymorphisms of a domesticated animal to be studied thoroughly at the level of primary structure: bovine haemo-globins A and B (Schroeder, Shelton, Shelton, Robberson and Babin, 1967). The β chains of these haemoglobins differ by not one but by three amino acids, the substitutions being (counting from the N terminus): 15 Ser $\rightarrow$ Gly; 18 His $\rightarrow$ Lys; and, 119 Asn $\rightarrow$ Lys , comparing B with A . [The Atlas of Protein Sequence and Structure (Dayhoff, 1972), p. D-72, gives the substitution at position 18 His $\rightarrow$ Gly with only the Schroeder et al. (1967) paper cited. More recent publication from Schroeder's laboratory still lists the substitution as 18 His $\rightarrow$ Lys (Schroeder, Shelton, Shelton, Apell, Huisman, Smith and Carr, 1972).] In either case, the mutation at position 18 would require not one but two nucleotide changes; for example, the messenger RNA codons for histidine are CAC and CAU, whereas the codons for lysine are AAA and AAG ; thus, the 18 His $\rightarrow$ Lys substitution actually requires changes in the first and third nucleotide. For the three amino acid differ-ences between bovine haemoglobins A and B a minimum of four nucleotide changes occurred.

The questions posed for discussion in our present paper are:

1. Do other common protein variants from domesticated species show multiple amino acid differences combined with the absence of intermediate forms?

2. Are such variants differing by more than one amino acid evidence for hybridization between distinct populations, possibly populations which had already diverged to the species level?

MATERIAL AND METHODS

Our studies for this paper are based upon compilations of data from the literature. There is no general published compilation known to us for the protein polymorphisms of domesticated species and we are assembling such a generalized summary. To reduce the number of references cited here we refer to certain already published compilations on specific proteins and animals; for example, cattle haemoglobin (Lush, 1966); cattle transferrin (Jamieson, 1966); cattle milk proteins (Aschaffenburg, 1968; McKenzie, 1971); sheep haemoglobin (Agar, Evans and Roberts, 1972); sheep transferrins (Rasmusen and Tucker, 1973); chicken egg and serum proteins (Baker, 1968; Baker, Croizier, Stratil and Manwell, 1970). For information on the number of amino acid differences we have used Dayhoff (1972) as a starting point to which more recent material has been added.

For inclusion into our present table on polymorphic proteins of domesticated animals two criteria had to be met:

1. There must be evidence that the polymorphism is the result of two or more alleles segregating at one locus. Care has been taken to avoid examples which are likely to represent non-allelic genes where relatively recent gene duplication has been followed by slight mutational divergence. For an example of the problem posed by certain studies, consider information available on the primary structure of certain chicken proteins. Although Elleman and Williams (1970) report

two amino acid differences between a minor and a major chicken transferrin, there is no information provided in that study to allow one to decide whether or not this represents a protein polymorphism segregating in a Mendelian manner or non-allelic duplicated loci. Similarly, a single ambiguous amino acid position involving either threonine or isoleucine has been reported in the study of the primary structure of chicken egg white avidin (Delange and Huang, 1971). Whether this represents coding ambiguity, an otherwise unknown genetic polymorphism, or a recent gene duplication, comparable to the $\gamma^{136 \text{ Gly}}$ and $\gamma^{136 \text{ Ala}}$ non-allelic foetal haemoglobin loci in man, gorilla and chimpanzee, remains to be established.

2. Sufficient primary structure data must be available to allow accurate determination of the number and position of amino acid substitutions. We cannot include the single peptide difference for chicken ovalbumins A and B (Fothergill and Fothergill, 1970a, 1970b) because only small amounts of the amino acid sequence were determined and the actual number and nature of the differences not firmly established. On the other hand, the kappa-casein data can be used because several workers have independently established that there are two amino acid differences between A and B variants for the kappa-casein macropeptide fraction; also, whether or not additional differences are found elsewhere in the molecule does not affect the fact that the two variants differ by more than one amino acid substitution (e.g., Hill, Naughton and Wake, 1970; Grosclaude, Mahé, Mercier and Ribadeau-Dumas, 1972a).

RESULTS

Data on the number of amino acid substitutions differentiating variants of polymorphic proteins from domesticated
animals are assembled in Table 1. In addition, a brief
comment has been included on the breed distribution and
approximate range of gene frequencies for reasons to be discussed shortly.

For some 25 different comparisons there are 10 different
protein polymorphisms where variants differ by two or more
amino acids and where no bridging intermediates are known.
[The α_{s1} casein variant A is not included in such a
listing of multiple differences, for it can be derived from
α_{s1} B by a clean deletion of 13 amino acids, representing
the effect of a single mutational event which deleted 39
nucleotides from the α_{s1} casein cistron.]

The data available at present certainly suggest an unusual
situation, distinct from the usual understanding of mutant
proteins, in that approximately one-third of the polymorphic
variants of domesticated animals differ by two, three, in one
case, by seven, amino acids -- in no case with any known
bridging intermediates.

To establish definitely that the situation is unusual, two
questions need to be answered:

1. Is there something different in the mutation process
for these proteins in this particular sample of species?
Phrased another way, is it possible for some proteins in some
species that multiple mutations occur more frequently than
they do for human haemoglobin and various bacterial and viral
proteins?

2. Do polymorphic proteins in wild species show such a
high frequency of multiple amino acid differences? Phrased

another way, is the observation of some examples of multiple differences among polymorphic proteins in domesticated species simply a sampling artefact in that most research has been done on the primary structure of proteins from conveniently obtainable domesticated species?

Neither question can be answered with certainty but some useful information is available:

1. The first question can be tentatively answered in the negative by partitioning the data in Table 1 so that rare variants are used as a 'control' for common variants. The logic of this procedure is as follows: Assuming that there is something unusual about mutation for these proteins in domesticated species so that multiple differences arise quite frequently, then one might expect approximately similar frequencies of multiply different variants among those that are rare as among those that are common.

Such partitioning is, obviously, arbitrary. However, most of the variants in Table 1 can be sorted into two categories: common, found in many breeds and often at gene frequencies of 0.1 or more. Such common variants are often the first to be discovered and, thus, bear letters A and B (e.g., sheep haemoglobins, cattle haemoglobins, cattle β-lactoglobulins, cattle kappa-caseins); rare, variants found in only one or a few breeds and usually at gene frequencies below 0.1 .

Five marginal cases exist: for example, α-lactalbumin A is very rare in European cattle breeds, although often found at gene frequencies in the range 0.05 to 0.2 in some Zebu breeds. At present there are insufficient breed and population data to classify the haemoglobin polymorphisms in the goat and the rabbit, and the lysozyme polymorphism in the duck; however, these variants, most of which differ by more than one amino acid, are probably common.

114

Nevertheless, in spite of the inevitable arbitrariness about such a classification of common versus rare variants for what must obviously be a continuous distribution of gene frequencies (and with the added complication that a variant may be in high gene frequency though restricted to only one or a few breeds), the partitioning does reveal an important fact: For cattle, 4 out of 6 of the comparisons for common variants reveal multiple differences (haemoglobins A vs B, kappa-caseins A vs B , β-lactoglobulins A vs B , and carboxypeptidase A 'val' vs 'leu'). In contrast, for 11 comparisons involving variants which are rare or marginal, 10 differ in the conventional manner of single amino acid substitutions; the one exception, Hb D, differs from Hb A by two amino acids, and this variant is classified as marginal because it is restricted in breed distribution although occurring in gene frequencies in the range 0.2 to 0.3 in the Muturu. [An added problem is that it is not known whether or not all the various bovine 'Hb D's' are the same.]

The common variants are likely to represent polymorphisms of some antiquity, especially those such as the sheep haemoglobins A and B which occur in nearly all breeds. Rare variants may represent either 'old' variants which are being displaced (or have never attained high gene frequencies), or rare variants may represent recent mutations. The latter situation is most likely where the rare variant is confined to only one breed, or a few closely related breeds. In Table 1 none of the rare variants and only one of the marginal variants differ from the appropriate 'normal' by more than one amino acid substitution. Therefore, it is likely that protein mutation in these domesticated species is basically similar to that in E. coli and man: the commonest type of mutation results in single amino acid substitutions; multiply different variants are usually the result of accumulation of a number of separate single amino acid substitution mutations.

That over half of the differences among common variants in Table 1 involve multiple differences without bridging intermediates being known in present polymorphisms, whereas none of the rare variants differ by more than a single amino acid substitution [α_{s1} casein A is an exception, though a single deletion mutation], can also be explained in another way, by no means exclusive to the conclusion about mutation. There might be greater selective advantage in a polymorphism where common variants differed by more than one amino acid. The very existence of such a situation argues that these common polymorphisms are not selectively neutral, whatever their mode of mutational origin. An interesting point is whether or not the strong new selection pressures that man places on species which are domesticated tend to favour multiply different allelic variants. Of relevance is the observation of 11 amino acid differences in a short region of common rabbit immunoglobulin G allotypes 1 and 3 (Wilkinson, 1969); multiply different variants without bridging intermediates in a polymorphism might represent a means for decreasing the rate of spread of an infectious disease until the immunoglobulin titre is sufficiently raised. Man has frequently kept his domesticated animals in conditions of crowding -- and infectious disease has been a recurrent problem.

2. The second question, do polymorphic variants in wild species show such a high frequency of multiply different protein variants, is more difficult to answer at present. There are few data on complete amino acid sequences of polymorphic proteins from wild species. Such data as are available, on haemoglobin variants in the Grey Kangaroo and a number of

116

primates, reveal only single amino acid substitutions (Day-hoff, 1972).

If our explanation of multiply different amino acid sub-stitutions in common variants resulting from hybridization in the origin of domesticated species is a correct explanation, then it is predicted that occasional examples, though at a lower percentage of the total polymorphisms, will be found in wild animal populations, where it is becoming increasingly recognized that hybridization plays an important evolutionary role (Scudder, 1974).

Some multiple amino acid differences exist among both α and β chain variants in the White-tailed deer, Odocoileus virginianus (e.g., Kitchen, Putnam and Taylor, 1966; Harris, Huisman and Hayes, 1973; Harris, Wilson and Huisman, 1972). However, the precise interpretation of these data is diffi-cult in view of biochemical evidence for non-allelic loci in some but not all White-tailed deer and in view of the paucity of genetic data. For the $Hb^{I}\alpha$ locus, the three variants are bridged by an intermediate in the same way as human Hb S forms a bridge between Hb A and Hb C Harlem, or human Hb C forms a bridge between Hb A and Hb Arlington Park (Adams and Heller, 1973). Segregation of multiply different haemoglobins in some populations of White-tailed deer would not be sur-prising; the species has a wide distribution, extending over much of North America; because of its popularity for hunting there have been numerous deer 'improvement' schemes, some of which have included importing deer from different parts of the species range and releasing them into wild populations or cross-breeding them in managed reserves.

Though the situation may also be complicated by non-allelic loci, there is evidence that the common α and β chain variants in the mouse Mus musculus differ by several amino

117

acids (Dayhoff, 1972). Only recently has man intentionally
domesticated the house mouse, usually for 'fancy' or for the
laboratory. However, man has had a long and close association
with this pest of stored grain, often accidentally transpor-
ting mice to many parts of the world. Thus, by man's migrations since the neolithic there has been frequent mixture
of formerly long isolated mouse populations.

DISCUSSION

The role of hybridization in the origin of domesticated
animals: That the multiple amino acid differences in allelic
variants for a protein polymorphism in domesticated animals
might be the result of hybridization of distinct populations
was first suggested by S.H. Boyer and colleagues (1966) in
their studies on the amino acid sequences of sheep haemoglobins A and B. The results presented in Table 1 of the
present paper indicate that most of the sufficiently studied
common protein variants in domesticated animals differ by more
than one amino acid substitution and by the absence of the
hypothetical intermediates.

In one case, the β-caseins, some intermediates do exist
among the inter-related variants (diagrammed in Figure 1).
But, even here it appears that two variants characteristic of
Zebu breeds, β casein B_Z , and β casein D , differ by
more than one amino acid from other known variants, although
complete sequence details are not available (M.P. Thompson in
McKenzie, 1971).

The β casein B_Z variant also illustrates the problem
of isoalleles. B_Z is electrophoretically identical to B
in a variety of buffers, yet it has several differences in
total amino acid composition. It is believed that B vs B_Z

118

distinguish B. taurus and B. indicus breeds of cattle, respectively -- a point which should be checked with further
studies. The Jersey breed has a high frequency of β casein
B; the high frequency of haemoglobin B in the Jersey, plus
some other characters, has been used as evidence for some
Bos indicus contributions to that European breed. However,
M.P. Thompson (in McKenzie, 1971) says that the β casein B
of the Jersey is not B_Z .

Even among the caseins, there are no known intermediates
to bridge the gap between the two common kappa-casein variants.
The caseins present another complication in that the loci
specifying α_{s1} , β and kappa caseins are very closely
linked and, thus, a number of possible combinations of
variants do not occur.

For the bovine haemoglobins, at least four nucleotide substitutions separate the common A and B variants. There
are other bovine haemoglobin variants, usually occurring at
low gene frequencies and in only a few breeds, e.g., the
various C's and D's , E_{Muk} , G , X , Y and 'Khillari'. Are
any of these among the hypothetical intermediates between A
and B ? The African variants $C_{Rhodesia}$ and D_{Zambia} are
clearly not (Schroeder, Shelton, Shelton, Apell, Huisman,
Smith and Carr, 1972).

Are any of these other variant haemoglobins indicative of
hybridization and introgression among various bovine species?
That cattle comprise two distinct but readily interbred
species has been recognized for some time, both Bos taurus and
Bos indicus being Linnean names. Several of the less well
characterized haemoglobin variants have been found in cattle
breeds in Northern India. The Simoons (1968, p.24) consider
the Eastern Himalayas "the foremost area of bovine hybridi-

zation in the world" and comment on the numbers of B. indicus x Mithan, B. frontalis, hybrids in Bhutan. Marshall (1973) considers that the Gaur, B. gaurus, the wild progenitor of the Mithan according to some researchers, was common in the Indus Valley at the time of the Mohenjo-Daro civilization, likely to be part of the centre of bovine domestication.

Even such seemingly unlikely contributors as the Yak, Bos grunniens, the Kouprey, Bos sauveli, or the Banteng, Bos sondaicus, cannot be ruled out. Yak-cattle hybrids have been utilized by peoples indigenous to central Asia for hundreds of years. Female hybrids between domestic cattle and the Banteng are fertile, although males are not as a rule. Some workers assume that because a few attempts yield only sterile hybrids the cross has no evolutionary significance. Even if only one cross in a thousand is fertile, the opportunity for introgression of genes exists. Also, it has been possible to select for genetic compositions which are more tolerant of hybridization, e.g., the recent improvement in obtaining fertile descendants from the intergeneric cross of domestic cattle and the American bison ('buffalo'), Bison bison, to form the 'cattalo'.

Quite apart from accidental hybridization as a consequence of neolithic man's bringing formerly isolated species into juxtaposition, there may well have been deliberate attempts at species hybridization quite early in the process of domestication; for example, the mule appeared soon after the horse was domesticated, about 5000 years ago (Brentjes, 1972). Some 'primitive' peoples have elaborate vocabularies and breeding programmes for animal hybridization (e.g., see Simoons and Simoons, 1968). Spurway (1956) has postulated that an important requirement for domestication is relatively promiscuous mating behaviour.

The relationship of Sheep haemoglobins A and B to
Mouflon haemoglobin: The sheep haemoglobins A and B
were the first variants of domesticated animals to be recognized as multiply different -- although the amino acid
sequences have still not been completely specified as several
residues are given only as Glx (glutamic acid or glutamine)
or as Asx (aspartic acid or asparagine) and there are some
minor differences in the results from different laboratories
(Dayhoff, 1972). Although no intermediate sequence between
sheep Hb A and B has been found for other haemoglobin
variants in the domestic sheep, the amino acid sequence of
the β chain from haemoglobin B of the Mouflon, Ovis
musimon, is intermediate in its sequence between domestic
sheep β^A and β^B (see Figures 2 and 3).

The simplest explanation for these results is that the
Mouflon β^B chain has diverged less from a common ancestral
haemoglobin β chain than either sheep β^A or sheep β^B
(Figure 3). One can postulate the divergence of two ancestral
ovine populations, which ultimately evolved sheep A and
sheep B haemoglobins, respectively, followed by hybridization to contribute both A and B genes, but no intermediates, to the present Ovis aries gene pool. With its
higher oxygen affinity, and its linkage in a supergene with
Hb C (discussed later), the ovine A haemoglobin is more
adapted to life at high altitudes. Sheep haemoglobin B is
more adapted to lowland conditions. Migrations of man between low and high altitudes (transhumance) might well have
brought together these differently adapted ovine species in
the early phases of domestication.

[Two minor points should be mentioned here. It is possible,
when all the glutamic and aspartic acids, and their amides,
are precisely located, that the Mouflon β^B chain will not

121

C. MANWELL AND C.M. ANN BAKER

be exactly intermediate between sheep β^A and sheep β^B.
Nevertheless, it is likely that the Mouflon β^B will be
closer to such a hypothetical intermediate than either of the
two common sheep haemoglobin β chains.

Secondly, one could postulate that the origin of Mouflon
β^B is by two intracistronic crossing over steps, analogous
to the formation of a combined 'Lepore' and 'anti-Lepore'
haemoglobin. Mouflon β^B agrees in sequence with sheep β^A
in the vicinity of the N- and C-terminal portions of the
chain, but resembles sheep β^B in the middle of the chain
(Figure 2). It is believed that such intracistronic crossing
over events are rare relative to single amino acid substitution mutations. Furthermore, such a postulate requires yet
additional complications, that both sheep β^A and sheep β^B
were in the ancestral Mouflon gene pool.]

Are any of the species which contributed to the formation
of the domestic sheep still extant? Or are such wild ancestral sheep as may still survive now introgressed with domestic
sheep genes? Results on chromosome numbers and on protein
variants of wild Iranian sheep show promise. These wild
Iranian sheep have some but not all of the domestic sheep
transferrin variants and only sheep haemoglobin B , as
judged by electrophoretic similarity (Lay, Nadler and Hassinger, 1971). The taxonomic position of these wild sheep seems
uncertain and there appear to be intergrading chromosomal
races, the wild Iranian sheep being variously termed Ovis
ammon (Nadler, Lay and Hassinger, 1971), Ovis linnaeus (Lay,
Nadler and Hassinger, 1971), or Ovis orientalis (Clark, 1964),
in a region where the last-mentioned author recognizes other
Urial species, Ovis gmelini and Ovis vignei, as well.

Even allowing for the taxonomic confusion, for the likelihood that some of these ovine groups are semi-species in a

super-species complex, and the possibility that the hypothetical ancestral populations have either become extinct or
contaminated with genes from domestic sheep, there is no
shortage of candidate populations to study which may reveal
the persistence of intermediate variants (such as haemoglobin
B of the Mouflon) and give some information on the origin
and adaptation of different biochemical variants found in
modern sheep.

Information from animal behaviour and natural history may
also reflect the hybrid origin of domestic sheep. Interspecific hybrids sometimes show behavioural abnormalities,
usually maladaptive. Curtain (1971) points out that domestic
sheep have not gone feral in Australia, although a number of
other domesticates have. Sheep as feral domesticates can
only survive for prolonged periods in restricted environments
lacking large mammalian predators. Also, the sheep has a
wide range of grazing and browsing tastes (e.g., see Griffiths,
Barker and MacLean, 1974); domestic sheep, protected by man,
readily overgraze many situations, a condition which suggests
that the species is of recent origin and has not yet come
into adaptive equilibrium with its environment.

Origin of multiply different variants by pseudoallelism:
There is another hypothesis to explain the existence of
allelic variants differing by several single amino acid substitution mutations without bridging intermediates: tandem
gene duplication, followed by mutational divergence, followed
in turn by the loss (or repression of expression) of non-
allelic portions of the duplicated locus, yielding a pair of
pseudoalleles coding for variants differing by several amino
acids.

The first requirement for such a system, tandem gene duplication to form a supergene, is met by the complex haemo-globin loci. In man not only are the loci specifying the β, γ and δ chains very closely linked, but the γ chain locus is itself made up of at least two and possibly four non-allelic structural genes (Huisman, Schroeder, Bannister and Grech, 1972). There is also evidence that the α chain locus is duplicated in a number of mammalian species (e.g., Harris, Wilson and Huisman, 1972, 1973).

Several species of caprine artiodactyls synthesize a different non-α chain type, called β^C, upon being made anaemic, and the genetic locus for this β^C chain is believed to be part of the non-α haemoglobin supergene. The β^C locus has diverged considerably from the other β chain locus, there being approximately 20 amino acid differences when caprine β^C chains are compared with caprine β chains.

Several groups of workers have observed that domestic sheep, Ovis aries, which are homozygous for the β^B variant cannot synthesize β^C upon anaemic stress, although sheep homozygous or heterozygous for β^A can (reviewed: Manwell and Baker, 1970; see also Baldy, Gaskill and Kabat, 1972). One interpretation of these results is that, besides the seven amino acids difference between β^A and β^B, there is also a further difference: the actual polymorphism is of a single locus versus a double locus, as follows:

$$\beta^B \qquad \text{and} \qquad \beta^A - \beta^C \ .$$

This single versus double locus effect further emphasizes the distinctness of the major sheep haemoglobin polymorphism. As several caprine species synthesize very similar β^C chains, the duplicated locus is probably the ancestral condition. The hypothetical ancestral sheep species with the β^B gene also lost (or is unable to express) the β^C gene. A search

of different domestic sheep breeds and wild species of the
genus Ovis for the hypothetical intermediate supergene,

$$\beta^B - \beta^C \quad ,$$

would be of interest. Lay, Nadler and Hassinger (1971) report
one wild Iranian sheep with only haemoglobins B and C ,
perhaps evidence for this intermediate supergene.

The single versus double locus effect can also suggest
another method for the origin of the multiple differences
between domestic sheep β^A and β^B : gene duplication of
the original β chain locus, followed by divergence until
seven amino acid differences had occurred,

$$\beta^A - \beta^B \quad ,$$

followed in turn by the loss of β^A from some of the supergenes and the loss of β^B from other supergenes, so that β^A
and β^B are actually the product of pseudoalleles. The hypothetical intermediate supergene, $\beta^A - \beta^B$, must be rare, if
it exists at all, in present populations of the domestic
sheep because the Hb A , Hb B polymorphism shows normal
Mendelian segregation ratios and populations approximate the
Hardy-Weinberg Equilibrium -- in fact, in one case with a
significant deficit of heterozygotes (Manwell and Baker, 1970)
-- whereas segregation of such a supergene, together with β^A
and β^B , should generate an excess of heterozygotes. However, a search for unusual segregation ratios and different
proportions of Hb A to Hb B in heterozygotes might reveal
evidence for such a crucial test of the pseudoallele hypothesis.

The pseudoallele hypothesis is not incompatible with the
theory of hybrid origin for multiply different polymorphic
variants. Indeed, as the pseudoallele hypothesis requires a
series of steps, isolation of diverging populations, followed

by hybridization, would account for the absence of inter-mediates, such as the missing β^A - β^B supergene in sheep.

Pseudoallelism might explain the remarkable haemoglobin polymorphism in the Barbary Sheep (or Barbary Goat) Ammotragus lervia (Huisman and Miller, 1972). Besides duplicate ∝ chain loci, some individuals of this species have a typical adult caprine β chain (called β^B although it differs in sequence from sheep β^B) and, in addition, synthesize a typical caprine β^C chain upon being made anaemic. Other individuals of the Barbary Sheep have, whether anaemic or not, only a C-type chain, termed $\beta^{C(na)}$, the letters na = non-anaemic. The genetic polymorphism appears to involve the segregation of one single allele,

$$\beta^{C(na)} \qquad ,$$

and one double allele,

$$\beta^B - \beta^C \qquad .$$

The three chain types all differ by several amino acids; $\beta^{C(na)}$ is separated by at least 8 different amino acid sub-stitutions from β^C and 20 substitutions from β^B (Huisman and Miller, 1972). So far the data are on Barbary Sheep in captivity from various zoos. It remains to be seen if this remarkable polymorphism occurs in part or all of the native range of this species in Northern Africa.

Monophyletic versus polyphyletic origin of the domestic fowl: Not all studies on protein polymorphisms suggest hybrid ancestry in domesticated species. The domestic fowl is frequently stated to have had a 'polyphyletic ancestry'; indeed, Williamson and Payne (1959) say that all four species of Jungle Fowl, Gallus gallus, G. sonnerati, G. lafayettei and G. varius, contributed to the ancestry of the domestic fowl. Hashiguchi and colleagues (1970) claim that a diallelic amylase polymorphism which occurs in some breeds of chickens and the Ceylonese (Sri Lanka) Jungle Fowl, G. lafayettei, is evidence for such hybrid ancestry. However, many examples of shared polymorphisms in related species are known. Such shared polymorphisms can be the result not only of hybridization but also of the retention of the polymorphism in evolution from a common ancestor; in addition, claims for the identity of variants surveyed in a single electrophoretic buffer system should be viewed more skeptically (Manwell and Baker, 1970).

Our own studies on egg white protein polymorphisms indicate that the most common variants in the domestic fowl also occur as the major or only variant in the Red Jungle Fowl, Gallus gallus; identity has not been established with amino acid sequence data but does rest upon electrophoresis in a variety of buffers over the pH range 4.7 to 9. Certain ovoglobulin types found in Gallus sonnerati do not occur in the domestic fowl (Baker and Manwell, 1972). Similarly, Stratil (1968) has found that the G_2 ovoglobulin of G. lafayettei is distinct from any of the known G_2 variants in domestic fowl. A fast variant of G_3 ovoglobulin has been found in the Yokohama breed and resembles a variant found in the Red Jungle Fowl (Baker, 1964, 1968).

Although the postulate of hybrid origin is very attractive in explaining the extreme diversity of breeds of domestic fowl, there is as yet no rigorous biochemical evidence for it. The evidence at present suggests that the Red Jungle Fowl, Gallus gallus, is the main progenitor of the domestic fowl, at least for egg white proteins. However, there are several different subspecies of G. gallus; and, of course, other contributing species may now be extinct. In the latter case it may well be that finding common variants in the domestic fowl which differ by more than one amino acid substitution is, if not conclusive evidence, at least the best we can get in the absence of fossil evidence or living representatives of hypothetical hybridizing species.

The high level of polymorphism in some domesticated animals: There is another approach to the study of protein polymorphism in domesticated species which may ultimately be related to hybrid origin of some domesticates. It has been suggested that the amount of polymorphism is relatively constant in certain situations (Manwell and Baker, 1970) and in certain phylogenetic assemblages (Selander and Kaufman, 1973). That the situational aspect may partly overrule the phylogenetic aspect is suggested by the relatively low level of protein polymorphism in burrowing mammals such as the mole rat Spalax ehrenbergi (Nevo and Shaw, 1972), the pocket gopher Thomomys talpoides (Nevo, Kim, Shaw and Thaeler, 1974), and the hairy-nosed wombat Lasiorhinus latifrons (Manwell, Baker and Wells, unpublished); adaptation to a rigorous environment or to a specialized ecological niche may result in a decrease in genetic variability.

The animal domesticates which are most likely to have had a complex hybrid origin, sheep and cattle, are also highly variable for a number of proteins and blood groups, e.g., the sheep transferrins are close to exhausting the alphabet for their nomenclature, and the bovine blood groups include some of the most permultiallelic loci known.

The difficulty is that our speculations are based on data biased by emphasis on polymorphic rather than monomorphic proteins, and on common and economically useful domesticates. However, the impression of greater variability in some domesticates has held in surveys by the same researcher on different species:

Whereas most major milk proteins show polymorphism in small population samples from various breeds of cattle, researchers who have discovered many of these variants have, using the same techniques, observed that 102 out of 105 Indian Water Buffalo, Bubalus bubalis, were identical and the other three individuals were heterozygous for only one milk protein, a β casein (Aschaffenburg, Sen and Thompson, 1968). Although the Water Buffalo has a long history as a domesticated animal, it is notable among bovines for the lack of hybrids (Gray, 1954).

For the domesticated quail Coturnix coturnix 16 out of 28 protein loci are polymorphic, whereas for the wild quail Coturnix pectoralis only 7 out of 36 protein loci are polymorphic, most of the variation in the two species being measured under similar electrophoretic conditions (Baker and Manwell, 1975). The domestic quail has a higher number of alleles per polymorphic locus and a higher heterozygosity. Some of the domestic quail populations are known to have had a hybrid ancestry of different populations of C. coturnix.

The major task will be to obtain measures of polymorphic loci, number of alleles per polymorphic locus, and average heterozygosity, for a number of breeds of different domesticated species and their wild relations. If the higher polymorphism of certain domesticated species holds up under such more rigorous and standardized methods of comparison, and if the amount of polymorphism in related wild species is itself not too variable from species to species, then it may be possible to measure the extent of hybridity as the excess polymorphism.

REFERENCES

Adams, J.G., III and P. Heller. (1973). Am. J. Human Genet. 25: 10A.

Agar, N.S., J.V. Evans and J. Roberts. (1972). Animal Breeding Abstracts 40: 407-436.

Aschaffenburg, R. (1968). J. Dairy Res. 35: 447-460.

Aschaffenburg, R., A. Sen and M.P. Thompson. (1968). Comp. Biochem. Physiol. 27: 621-623.

Badr, F.M., P.A. Lorkin and H. Lehmann. (1973). Nature, New Biology 242: 107-110.

Baker, C.M.A. (1964). Comp. Biochem. Physiol. 12: 389-403.

Baker, C.M.A. (1968). Genetics 58: 211-226.

Baker, C.M.A., G. Croizier, A. Stratil and C. Manwell. (1970). Adv. Genet. 15: 147-174.

Baker, C.M.A. and C. Manwell. (1972). Anim. Blood Grps. Biochem. Genet. 3: 101-107.

Baker, C.M.A. and C. Manwell. (1975). Comp. Biochem. Physiol. 50B: 471-478.

Baldy, M., P. Gaskill and D. Kabat. (1972). J. Biol. Chem. 247: 6665-6670.

Blackwell, R.Q., W.H. Boon, C.-S. Liu and M.-I. Weng. (1972). Biochim. Biophys. Acta 278: 482-490.

Boyer, S.H., P. Hathaway, F. Pascasio, C. Orton, J. Bordley and M.A. Naughton. (1966). Science 153: 1539-1543.

Brentjes, B. (1972). Säugetierkundliche Mitteilungen 20: 325-353.

Clark, J.L. (1964). The Great Arc of the Wild Sheep. Univ. of Oklahoma Press, Norman.

Curtain, C.C. (1971). Antiquity 45: 303-304.

Darwin, C. (1905). The Variation of Animals and Plants Under Domestication. 2 volumes, 2nd edition. John Murray, London.

Dayhoff, M.O. (Editor). (1972). Atlas of Protein Sequence and Structure. National Biomedical Research Foundation, Washington, D.C.

Delange, R.J. and T.-S. Huang. (1971). J. Biol. Chem. 246: 686-709.

Elleman, T.C. and J. Williams. (1970). Biochem. J. 116: 515-532.

Epstein, H. (1969). Domestic Animals of China. Commonwealth Agricultural Bureaux, Farnham Royal, Bucks., England.

Epstein, H. (1971). The Origin of the Domestic Animals of Africa. 2 volumes. Africana Publishing Co., N.Y.

Fothergill, L.A. and J.E. Fothergill. (1970a). Biochem. J. 116: 555-561.

Fothergill, L.A. and J.E. Fothergill. (1970b). Eur. J. Biochem. 17: 529-532.

Gray, A.P. (1954). Mammalian Hybrids. Commonwealth Agricultural Bureaux, Farnham Royal, Bucks., England.

Gray, J.C., S.D. Kung, S.G. Wildman and S.J. Sheen. (1974). Nature 252: 226-227.

Griffiths, M., R. Barker and L. MacLean. (1974). Aust. Wildl. Res. 1: 27-43.

Grosclaude, F., M.-F. Mahé, J.-C. Mercier and B. Ribadeau-Dumas. (1972a). Annales de Génétique et de Sélection animale (INRA) 4: 515-521.

Grosclaude, F., M.-F. Mahé, J.-C. Mercier and B. Ribadeau-Dumas. (1972b). Eur. J. Biochem. 26: 328-337.

Harris, M.J., T.H.J. Huisman and F.A. Hayes. (1973). J. Mammal. 54: 270-274.

Harris, M.J., J.B. Wilson and T.H.J. Huisman. (1972). Arch. Biochem. Biophys. 151: 540-548.

Harris, M.J., J.B. Wilson and T.H.J. Huisman. (1973). Biochem. Genet. 9: 1-11.

Hashiguchi, T., M. Yanagida, Y. Maeda and M. Taketomi. (1970). Japan. J. Genet. 45: 341-349.

Hill, R.J., M.A. Naughton and R.G. Wake. (1970). Biochim. Biophys. Acta 200: 267-274.

Huisman, T.H.J. and A. Miller. (1972). Proc. Soc. Exptl. Biol. Med. 140: 815-819.

Huisman, T.H.J., W.A. Schroeder, W.H. Bannister and J.L. Grech. (1972). Biochem. Genet. 7: 131-139.

Huisman, T.H.J., J.B. Wilson, M. Gravely and M. Hubbard. (1974). Proc. Nat. Acad. Sci. U.S.A. 71: 3270-3273.

Huisman, T.H.J., R.N. Wrightstone, J.B. Wilson, W.A. Schroeder and A.G. Kendall. (1972). Arch. Biochem. Biophys. 153: 850-853.

Jamieson, A. (1966). Heredity 21: 191-218.

Kitchen, H., F.W. Putnam and W.J. Taylor. (1966). Abstract of paper presented at the Symposium on Comparative Haemoglobin Structure, Greece, 1966.

Lay, D.M., C.F. Nadler and J.D. Hassinger. (1971). Comp. Biochem. Physiol. 40B: 521-529.

Lush, I.E. (1966). The Biochemical Genetics of Vertebrates Except Man. North-Holland, Amsterdam.

Manwell, C. and C.M. A. Baker. (1970). Molecular Biology and the Origin of Species. Sidgwick and Jackson, London.

Marshall, J. (1973). Mohenjo-Daro and the Indus Civilization. Vol. 1. Indological Book House, Delhi.

McKenzie, H.A. (1971). Milk Proteins: Chemistry and Molecular Biology. Vol. 2. Academic Press, N.Y.

Moav, R. and W.G. Wohlfarth. (1966). F.A.O. Fisheries Report 44: 12-29.

Moav, R. and W.G. Wohlfarth. (1970). J. Heredity 61: 153-157.

Mourant, A.E. and F.E. Zeuner (Editors). (1963). Man and Cattle. Royal Anthropological Institute of Great Britain and Ireland, Occasional Paper No.18.

Nadler, C.F., D.M. Lay and J.D. Hassinger. (1971). Cytogenetics 10: 137-152.

Nevo, E., Y.J. Kim, C.R. Shaw and C.S. Thaeler. (1974). Evolution 28: 1-23.

Nevo, E. and C.R. Shaw. (1972). Biochem. Genet. 7: 235-241.

Rasmusen, B.A. and D.M. Tucker (1973). Anim. Blood Grps. Biochem. Genet. 4: 207-220.

Schroeder, W.A., J.R. Shelton, J.B. Shelton, G. Apell, T.H.J. Huisman, L.L. Smith and W.R. Carr. (1972). Arch Biochem. Biophys. 152: 222-232.

Schroeder, W.A., J.R. Shelton, J.B. Shelton, B. Robberson and D.R. Babin. (1967). Arch. Biochem. Biophys. 120: 124-135.

Scudder, G.G.E. (1974). Canad. J. Zool. 52: 1121-1134.

Seid-Akhavan, M., W.P. Winter, R.K. Abramson and D.L. Ruck-
nagel. (1972). Blood 40: 927 (Abstract).

Selander, R.K. and D.W. Kaufman. (1973). Proc. Nat. Acad.
Sci. U.S.A. 70: 1875-1877.

Simoons, F.J. and E.S. Simoons. (1968). A Ceremonial Ox of
India. Wisconsin Univ. Press, Madison.

Spurway, H. (1956). J. Genet. 53: 325-362.

Stratil, A. (1968). Thesis, Czechoslovak Acad. of Sciences,
Lab. of Physiol. and Genet. of Animals, Libechov.

Wajeman, H., D. Labie and G. Schapira. (1973). Biochim.
Biophys. Acta 295: 494-505.

Wilkinson, J.M. (1969). Biochem. J. 112: 173-185.

Williamson, G. and W.J.A. Payne. (1959). An Introduction to
Animal Husbandry in the Tropics. Longmans, London.

Wilson, J.B., A. Miller and T.H.J. Huisman. (1970). Biochem.
Genet. 4: 677-688.

Zohary, D., J.R. Harlan and A. Vardi. (1969). Euphytica 18:
58-65.

Page 1 of Table 1.

TABLE 1. *Number of amino acid differences in the primary*
structure of protein variants in domesticated animals.

Protein	Number of amino acid differences	Common or rare: comment on breed distribution and approximate gene frequency of the variants
CATTLE [Bos taurus (European breeds) and B.indicus (Indian and African Zebu)]		
Haemoglobin Hb A vs Hb B (β chain)	 3	Hb B occurs at gene frequencies of 0.3 to 0.6 in the Jersey, around 0.1 in a number of European breeds, and 0.4-0.5 in many Zebu breeds.
Carboxypeptidase A 'val' vs 'leu'	 3	Gene frequency for the two alleles close to 0.5 in the original sample of small numbers of various European cattle breeds.
α - Lactalbumin (specifier protein in the lactose synthetase system)		
A vs B	1	Very rare in Western breeds, but marginal in that values 0.05 to 0.2 occur in several Zebu breeds.
β- Lactoglobulin A vs B	 2	A and B are the common variants, with the gene frequency of A usually between 0.1 and 0.5.
B vs C	1	Rare, largely confined to the Jersey (and not all herds); low frequency (around 0.05) in some Russian breeds.
B vs D	1	Rare, French Montbeliarde and Italian Reggiana.
α- S₁ casein	1 deletion (residues 14-26)	
A vs B		B is the major or sole variant in a number of cattle breeds. A is rare, confined to some Friesian and Danish RDM herds.
B vs C	1	C is usually at frequencies below 0.1 in European breeds but becomes the major variant, gene frequencies in the range 0.6 to 0.95 in African and Indian Zebu breeds.
B vs D	1	D is rare, gene frequency of around 0.04 in the French Flamande and around 0.01 in the Italian Bruna Alpina.
β casein A^1 vs A^2	 1	A^1 and A^2 are the common variants in many breeds of European and African cattle, although A^2 is much commoner than A^1 in Indian Zebus.
A^2 vs A^3	1	A^3 is rare, most reports being from herds of Friesians (Holsteins) at gene frequencies in the range 0.02 - 0.05 but also occurs at a frequency of 0.02 in the Normande breed.

Page 2 of Table 1

A¹ vs B	1	B is marginal; although its gene frequency is usually below 0.1 in European breeds, it is higher in the Jersey. B in the Zebu breeds is a different variant (discussed in text).
A¹ vs C	1	C is relatively rare; it is confined to a few European breeds and occurs at low gene frequencies, although approaching 0.1 in the Simmentaler.
A² vs E	1	E is rare, gene frequency of 0.02 in the Piedmont breed.
kappa - casein		
A vs B	2	The A-B kappa casein polymorphism is present in nearly all cattle breeds, the gene frequency of A usually ranging from 0.2 to 0.8.
growth hormone		
'leu' vs 'val'	1	? Must be relatively common to be detected originally as an apparent ambiguity in amino acid sequencing of pooled material.
SHEEP [Ovis aries] Haemoglobin		
A vs B (β chain)	7	The A-B polymorphism occurs in nearly all breeds, the frequency of A being often in the range 0.1 to 0.5.
A vs D (α chain)	1	D is very rare; native Yugoslavian sheep.
GOAT [Capra hircus] Haemoglobin		
A vs D (β chain)	1	Differences in nomenclature and possible confusion in some cases with non-allelic genes make it difficult to ascertain breed distribution and frequency of these variants; it is suggested that the β chain variant occurs as a polymorphism in English, Sardinian and Algerian goats, whereas the α chain variant occurs in Indian goats. That the latter should differ by 3 substitutions is of interest for it has been suggested that Indian goats have some admixture from the Markhor, Capra falconeri.
A vs E (α chain)	3	
RABBIT [Oryctolagus cuniculus] Haemoglobin		
α-chain variant	3	Breed distribution and gene frequency data are limited because most information on rabbit haemoglobin has come from studies on amino acid sequence of electrophoretically silent polymorphisms.
β-chain variant	3	
β-chain, 112 position	1	
DUCK, [Anas platyrhynchos] Lysozyme		
II vs III	6	Few population data available, but II and III are probably the major variants in lysozyme polymorphisms reported for Japanese, French and American domestic ducks.

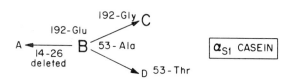

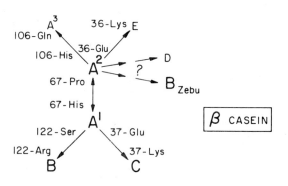

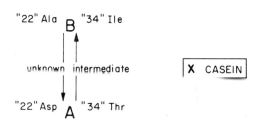

FIGURE 1.   *See legend on following page.*

FIGURE 1. Interrelations of the variants of cattle milk
caseins.

The arrows denoting direction provide only one of several
possible interpretations of the evolutionary divergence. While
it is likely that some of the rare variants have evolved
recently from common variants, it is possible that other rare
variants are really ancestral forms which have largely been
replaced. Common variants are denoted with large letters;
rare variants with small letters.

It can be seen that most casein variants can be linked by a
series of single amino acid substitution mutations through
known variants, e.g., β casein A^3 is separated from β
casein C by three such mutations, the intermediates being
the common β casein variants, A^1 and A^2.

The position of the <u>Bos indicus</u> β casein variants B_z and
D is not clear, although total amino acid composition data
are suggestive of multiple differences (Thompson in McKenzie,
1971). In our diagram these Zebu variants are derived tentatively from β casein A^2, which is the commonest casein
variant in many Zebu breeds; however, Thompson (in McKenzie,
1971) believes that β casein D is most like β casein C
in its total amino acid composition and, thus, might be
derived from that variant.

Although 6 of the variants of β casein can be bridged by
single amino acid substitutions through known intermediates,
the common variants of kappa casein cannot. The hypothetical
intermediate, which would be either "22"-Ala, "34"-Thr, or
"22"-Asp, "34"-Ile, has not been found. (The numbers are
placed in quotation marks because the kappa-casein data are
counted from the C-terminal end of the macropeptide obtained
in partial digests of kappa-casein, not from the N-terminal
end as is the standard practice.)

The diagrams are based on the amino acid sequence data of
Grosclaude, Mahé, Mercier and Ribadeau-Dumas, 1972a, 1972b.

	49	57	74	75	119	128	143
SHEEP A	Ser	Ala	Val	Glx	Ser	Glx	Arg
MOUFLON B	Ser	Ala	Met	Lys	Ser	Glx	Arg
SHEEP B	Asx	Pro	Met	Lys	Asn	Asp	Lys

FIGURE 2. *The seven amino acid differences when the* β *chains of sheep haemoglobin A , Mouflon haemoglobin B and sheep haemoglobin B variants are compared (based on data of Wilson, Miller and Huisman, 1970). Note that so far, no amino acid substitution has been found which separates Mouflon* β^B *from both sheep* β^A *and sheep* β^B ; *thus, the Mouflon* β^B *chain appears to be an exact intermediate between sheep* β^A *and sheep* β^B .

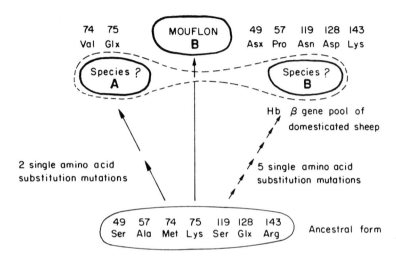

FIGURE 3. A model for the evolutionary divergence for sheep
          haemoglobin β chain variants A and B ,
followed by hybridization to form the present sheep Ovis aries
gene pool. Mouflon β^B has retained a possible ancestral
ovine sequence in the β chain, whereas in one hypothetical
species, A , two amino acid substitutions have occurred, and
in another hypothetical species, B , five amino acid substitutions have occurred. Hybridization of these two hypothetical species brings the two multiply different haemoglobin
cistrons into the present sheep gene pool without any intermediates. An alternative model, based on pseudoallelism, is
discussed in the text.

Adaptive Strategies of Genetic Systems in Constant and Varying Environments

E. NEVO

INTRODUCTION

If heterozygosity of allozyme loci is evolutionarily significant, and is not adaptively neutral, it must vary in natural populations at least partly with environmental heterogeneity. Varying and unpredictable environments should select for more heterozygous species, while relatively constant environments should select for more homozygous ones. Available evidence is not only scanty but ambiguous and indecisive (Lewontin, 1974; Nevo et al. 1974), and critical testing is needed to elucidate this relationship. To test the environmental variation model as a predictor of genetic variation, I contrasted the protein polymorphisms of four anuran species in Israel which range from relative ecological constancy to environments that vary greatly in both space and time.

Anurans represent a successful group of amphibians whose evolutionary origin dates back to Triassic times (Estes and Reig, 1973). They are adapted to a variety of burrowing, aquatic, arboreal and terrestrial habitats. Four anuran species were selected for this study because they live in an inferred gradient of environmental variability: spadefoot-

toads, Pelobates syriacus, edible frogs, Rana ridibunda,
lemon-yellow tree-frogs, Hyla arborea, and green-toads, Bufo
viridis. Beginning with the narrow specialist Pelobates
which lives underground in a relatively constant and predictable environment, the series runs through semi-aquatic and
aquatic Rana and arboreal Hyla, terminating with the terrestrial generalist Bufo which experiences great environmental
variation (Blair, 1972).

The progressive variation in environmental heterogeneity
and predictability exhibited by these 4 anuran species,
elaborated below, make them an appropriate test case for
evaluating the relationship between environmental heterogeneity or predictability and genetic variation. The detailed
studies of each of the 4 species will be published elsewhere.
If environmental heterogeneity increases heterozygosity Pelobates is expected to support minimal, while Bufo is expected
to support maximal heterozygosity. This paper demonstrates a
significantly positive correlation between environmental
heterogeneity and heterozygosity.

EVOLUTIONARY HISTORY, GEOGRAPHICAL AND ECOLOGICAL DESCRIPTION OF THE SPECIES

Spadefoot-toads, Pelobatidae, originated in Cretaceous
times and the pelobatines, including Pelobates, emerged in
the early Oligocene (Estes and Reig, 1973). The evolutionary
history of pelobatines involved progressive adaptations to
increasing aridity in both the Old and the New Worlds, by
developing burrowing habits that made them narrow habitat-
specialists (Bragg, 1961).

Modern Eurasian species of Pelobates (P. cultripes, P.
syriacus and P. fuscus) probably evolved by Miocene times.
During the Pleistocene the advancing ice sheets restricted

P. cultripes and P. syriacus to the Iberian Peninsula and
the Eastern Mediterranean basin respectively (Estes, 1970).
Pelobates syriacus balcanicus ranges from South Yugoslavia
through the Balkans and Asia Minor to Israel where it is
found infrequently in the Coastal Plain, Galilee Mountains
and the Golan Heights. P. syriacus is a thoroughly burrowing
and nocturnal species. It spends the greater part of its
existence underground and sometimes reaches one meter below
the surface of the ground. Spadefoot toads are chiefly seen
above ground at nights during the short breeding season in
December and January in the Coastal Plain but later, in March
and April, in the upper Galilee Mountains (Nevo, in preparation).

The enormous genus Rana involves 50 fossil species, some
dating back to Miocene time (Estes and Reig, 1973), and 250
recent species ranging universally from the Arctic to the
Tropics (Savage, 1973). Rana ridibunda ridibunda ranges from
Central Europe to West Asia and North Africa and is primarily
semi-aquatic to aquatic inhabiting rivers, creeks, lakes and
ponds and preferably, but not invariably, in permanent and
deep water. It abounds in the Mediterranean region of Israel
and in a few desert oases. It is usually restricted to permanent bodies of water which serve as buffers to climatic
perturbations involving both humidity and temperature. These
habitats guarantee moisture almost throughout the year thus
serving as effective safeguards from drought, dessication and
freezing temperatures. Breeding takes place from February to
September, varying with the ecogeographical location. Rana
ridibunda thus represents a habitat-intermediate species,
much less restricted ecologically than Pelobates syriacus
(Nevo, 1975).

The genus Hyla comprises numerous species, is nearly world-

143

wide, and is known from the Oligocene, while Hylidae dates
from the Paleocene (Estes and Reig, 1973). The only hylid in
Israel is the arboreal Hyla arborea savignyi, a biologically
good species (Schneider and Nevo, 1972), which is found in
the East Mediterranean zone where it breeds from December
through August in marshes, ponds, rivers and sometimes in
temporary, but preferably in permanent, bodies of water. In
the fall tree-frogs spend the daytime in trees and bushes in
the vicinity of water sources. Hence they are sheltered most
of the year from climatic extremes of both temperature and
humidity, either being shaded by trees or immersed in water.
The arboreal habitat probably exposes H. arborea to more
terrestrial conditions than those of R. ridibunda, but still
leaves it within the broad definition of a habitat-
intermediate.

The huge toad genus Bufo, comprising more than 200 species
and dating back to the Paleocene (Estes and Reig, 1973), is
nearly cosmopolitan. The terrestrialism of Bufo is one of
the important features of the habitat-generalism of this
group (Blair, 1972). Bufo viridis viridis ranges widely from
south Sweden to North Africa and eastwards to Mongolia and
Tibet. Not only is it a geographically widespread and abundant species, but it is also found in extremely variable
environments. From 400 meters below sea-level in the Dead
Sea of Israel, it reaches 4,672 meters in the Himalayas,
colonizing along the way desert enclaves, brackish water and
areas of high salinity. Within Israel it ranges across sharp
southward (300 km) and eastward (80 km) transects involving
ecological gradients of temperature and humidity. It is an
active colonizer penetrating further south into the northern
Negev and Sinai deserts than any of the above-mentioned
species of Pelobates, Rana and Hyla (Figure 1).

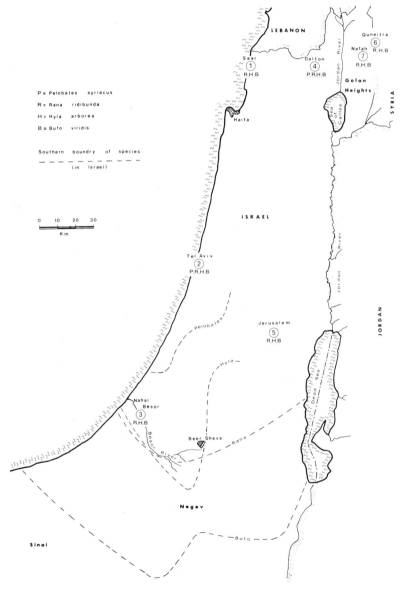

FIGURE 1. *Southern boundaries of the 4 anuran species in Israel, the localities sampled (1-7), and the species collected in each.*

Bufo viridis is active during the hot, dry summer and is exposed to extreme drought periods far from available water
sources. The larvae develop in shallow temporary most uncertain ponds, where temperature variation is extreme, and
some of which dry up causing mass mortalities of young un-
metamorphosed larvae. Green toads are thus exposed both as
larvae and adults to extremely varying and unpredictable
environments in space and time where severe selection
pressures are expected to operate (Nevo, 1976).

MATERIALS AND METHODS

Allozyme variation in proteins encoded by 26-32 loci was
compared in 4 anuran species from 7 central and marginal
localities sampled along a mesic-xeric transect in Israel
(Figure 1). Only in 2 localities (Dalton and Tel-Aviv), were
all 4 species sampled. In the remaining 5 localities (Saar,
Nahal Besor, Jerusalem, Quneitra and Nafah) only 3 species
(Bufo, Hyla, and Rana) were collected in each locality
(Figure 1). The localities sampled classified according to
population structure are: (a) Central populations: Saar,
Tel-Aviv, Dalton, Quneitra and Nafah; (b) Marginal populations: Nahal Besor and Jerusalem. Detailed descriptions
of all localities are given in Dessauer, Nevo and Chuang
(1975).

Blood from each specimen was separated into plasma and
hemolysate by centrifugation. Tissue samples of kidney, liver,
heart and leg-muscle were mixed and homogenized together
after observing identical results in pilot experiments in
Pelobates, Rana and Hyla, but were treated separately in Bufo
(Dessauer, Nevo and Chuang, 1975). Electrophoretic procedures have been described elsewhere (Yang, in Selander et al.,

1971). Studied loci are given in Table 2.

RESULTS

The four species studied vary widely in their mean alleles per locus $(\bar{A})$, mean proportion of loci polymorphic $(\bar{P})$, and mean proportion of loci heterozygous per individual $(\bar{H})$, with genetic variation increasing in the following order: Pelobates < Rana < Hyla < Bufo (Tables 1, 2; Figures 2, 3).

TABLE 1. *Comparison of genetic variation in 4 anuran amphibians from 7 localities in Israel.*

Species	No. of populations	Sample size (N)	No. of loci tested	Mean no. of alleles per locus $(\bar{A})$	Mean proportion of loci	
					Polymorphic per population • $(\bar{P})$	Heterozygous per individual $(\bar{H})$
Bufo viridis	7	294	26	1.69 ±.063 •••	.47 ± .029	.141±.004
			14 ••	1.84 ±.079	.56 ± .029	.169 : .010
Hyla arborea	7	211	27	1.68 ±.067	.43 ± .037	.074 : .007
			14 ••	1.86 ±.083	.50 ± .043	.088 : .008
Rana ridibunda	7	203	27	1.46 ±.023	.38 ± .032	.073 : .004
			14 ••	1.55 ±.020	.44 ± .024	.088 : .005
Pelobates syriacus	2	56	32	1.09 ± .015	.09 ± .013	.023 ±.021
			14 ••	1.11 ± .015	.11 ± .013	.052 ±.027
total	23	764				

• Criterion of polymorphism .01

•• 14 loci shared by the four species: aGpd, Ldh-1, Ldh-2, 6-Pgd, Pgm-2, Pept-2, Est-1, Est-2, Ipo, Got-1, Prot-1, Prot-2, Alb, Tf.

••• $\bar{y}$ ± S.E.

E. NEVO

TABLE 2. Protein loci and their heterozygosity in seven localities of four species of anuran amphibians ($\overline{Y}$ ± S.E.)

Protein locus (& Abbreviation)		Bufo	Hyla	Rana	Pelobates
Malate dehydrogenase 1	(Mdh-1)	.015 ± .012	-	-	.000
Malate dehydrogenase 2	(Mdh-2)	-	.066 ±.014	-	.000
a- glycerophosphate dehydrogenase	(aGpd)*	.000	.056 ±.024	.005 ± .005	.000
Lactate dehydrogenase 1	(Ldh-1)*	.000	.177 ±.048	.401 ± .049	.000
Lactate dehydrogenase 2	(Ldh-2)*	.320 ± .058	.018 ±.014	.019 ± .014	.000
Isocitrate dehydrogenase 1	(Idh-1)	.536 ± .035	.012 ±.008	-	-
Isocitrate dehydrogenase 2	(Idh-2)	.024 ± .010	.005 ±.005	-	.017±.004
6- Phosphogluconate dehydrogenase	(6-Pgd)*	.056 ± .016	.019 ±.019	.185 ± .032	.000
Phosphoglucomutase 1	(Pgm-1)	-	.413 ±.073	.056 ± .027	.000
Phosphoglucomutase 2	(Pgm-2)*	.461 ± .047	.051 ±.024	.014 ± .010	.000
Phosphoglucomutase 3	(Pgm-3)	-	.000	.020 ± .015	.000
Phosphoglucose isomerase	(Pgi)	-	.121 ±.079	.000	.000
Glutamate dehydrogenase	(Gdh)	.000	-	.000	.000
Fumarase	(Fum)	-	.045 ±.025	-	.000
Sorbitol dehydrogenase	(Sdh)	-	.064 ±.030	.025 ± .016	.000
Peptidase 1	(Pept-1)	.000	.000	.000	.000
Peptidase 2	(Pept-2)*	.268± .048	-	.000	.000
Esterase 1	(Est-1)*	.306±.062	.244±.038	.316 ± .059	.660± .027
Esterase 2	(Est-2)*	.438±.059	.155±.039	.212 ± .041	.071± .013
Esterase 3	(Est-3)	-	-	.313 ± .068	-
Esterase 4	(Est-4)	-	-	.056 + .025	-
Indophenol oxidase	(Ipo)*	.007± .007	.004±.004	.000	.000
Glutamate oxaloacetic transaminase 1	(Got-1)*	.048± .015	.462±.094	.005 ± .005	.000
Glutamate oxaloacetic transaminase 2	(Got-2)	.017± .009	-	.019 ± .009	-
Octanol dehydrogenase	(Odh)	-	.000	.009 ± .006	.000
Xanthine dehydrogenase	(Xdh)	-	.014±.009	.189 ± .068	.000
Prolidase	(Prol)	.003± .003	-	-	-
Cyanoperoxidase	(Cnp)	.000	-	-	-
Acid phosphatase 1	(Acph-1)	.029± .016	-	-	.000
Acid phosphatase 2	(Acph-2)	.056± .026	-	-	-
Glutamate pyruvate transaminase 1	(Gpt-1)	.000	-	-	-
Glutamate pyruvate transaminase 2	(Gpt-2)	.600± .050	-	-	-
Aldehyde oxidase	(Ao)	-	-	-	.000
Hemoglobin 1	(Hb-1)	-	.000	-	-
Hemoglobin 2	(Hb-2)	-	.000	-	-
Protein 1	(Prot-1)*	.000	.000	.000	.000
Protein 2	(Prot-2)*	.004± .004	.005±.005	.000	.000
Albumin	(Alb)*	.004± .004	.000	.087 ± .067	.000
Transferrin	(Tf)*	.415±.027	.004±.004	.004 ± .010	.000
Haptoglobin	(Hp)	-	.004±.004	.000	.000

* Loci common to the four species (N = 14)

GENETIC VARIATION IN 7 LOCALITIES OF 4 ANURAN
SPECIES BASED ON 26 - 32 LOCI

BUFO———HYLA ———— RANA —·—·— PELOBATES •

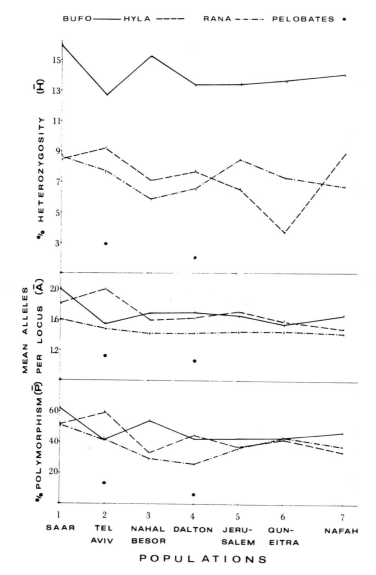

FIGURE 2. Genetic variation of the 4 anuran species, based
on 26-32 loci, in the 7 localities sampled.

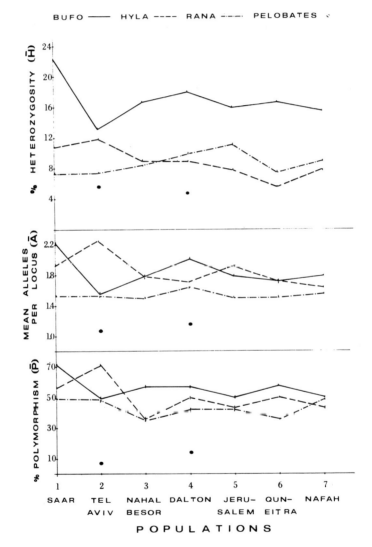

FIGURE 3. Genetic variation of the 4 anuran species, based
on the 14 shared loci, in the 7 localities sampled.

As expected, heterozygosity increases with environmental
heterogeneity. Pelobates is largely homozygous, with only 2
esterases out of 32 loci being polymorphic. In contrast, Bufo
is highly heterozygous; higher than any vertebrate yet studied
(Dessauer, Nevo and Chuang, 1975). Bufo is more heterozygous
than Hyla (Table 1, P < .001) , and has significantly higher
values of $\bar{A}$, $\bar{P}$ and $\bar{H}$ than Rana (P < .005; P < .01;
P < .001; respectively). Hyla is significantly higher in $\bar{A}$
than Rana (P < .01), marginally higher in $\bar{P}$ and equal in
$\bar{H}$. Pelobates is significantly lower in all parameters than
Rana, Hyla and Bufo. The patterns of the three measures of
variation differ between the species. Bufo has high values
in $\bar{A}$, $\bar{P}$ and $\bar{H}$ (i.e., it is highly polymorphic and highly
heterozygous). Hyla is characterized by high $\bar{A}$ and $\bar{P}$ but
by medium $\bar{H}$ (i.e., it is highly polymorphic but only medium
in heterozygosity), and it displays a pattern of many polymorphic loci some of which are weakly polymorphic (Table 2).
Rana has medium $\bar{A}$, $\bar{P}$ and $\bar{H}$. Pelobates is low in all
three measures.

The relative genetic patterns differentiating the four
species are similar for all 26-32 loci and for the 14 shared
loci, with the $\bar{A}$, $\bar{P}$ and $\bar{H}$ estimates higher in the latter
case (Table 1). Likewise, the differentiating pattern is
largely consistent in the 7 localities studied (Figure 2,3).
Heterozygosity ($\bar{H}$) is the best differentiating genetic parameter between Bufo, Hyla-Rana, and Pelobates, displaying high,
medium and low estimates respectively across the ranges of
the 4 species in Israel. Of the 14 shared loci, the following
are either monomorphic or weakly polymorphic in all 4 species:
Gpd, Pept-1, Ipo, Prot-1, Prot-2 and Albumin. In contrast
the following loci are strongly polymorphic in either one or
more of the species: Ldh-1, Ldh-2, Idh-1, 6-Pgd, Pgm-2,

Pept-2, Est-1, Got-1, Tf. The only loci which are strongly
polymorphic in all species are Est-1 and Est-2. Bufo is
characterized by uniquely high Pgm-2 and Tf; Hyla is distinguished by high Got-1; and Rana by a high 6-Pgd. Both
Bufo and Hyla display high Pept-2 while Hyla and Rana exhibit
high Ldh-1.

DISCUSSION

The idea that genetic variation is related to environmental
variation dates back to Dobzhansky (1951) and others, framed
theoretically by Levins (1968) and tested experimentally by
Powell (1971) and McDonald and Ayala (1974). Levins' (1968)
theory of fitness suggests that the amount of genetic
variation may be regarded as an adaptation to environmental
heterogeneity and uncertainty. The latter gives rise to
broad-niche or habitat generalist species richer in heterozygosity in contrast to the homozygosity patterns expected in
narrow-niche or habitat specialist species.

The environmental variability model seems the best predictor of the differential genetic variation between the four
anuran species. Increased environmental heterogeneity and
unpredictability leads to increased heterozygosity. The
narrow habitat-specialist Pelobates syriacus, living in a
relatively constant and predictable underground environment,
and thus insulated from short term environmental changes, has
significantly lower genetic variation than the habitat-
intermediates Rana ridibunda and Hyla arborea. Contrariwise,
habitat-generalist Bufo viridis, living in spatially and
temporally distinctly varying and unpredictable environments
with a much larger range of environmental fluctuations, displays the highest degree of genetic variation not only in the
series studied, but also higher than in any vertebrate yet

studied (Dessauer, Nevo and Chuang, 1975). These results are particularly striking in view of the fact that niche-breadth has not been quantified for the anurans studied and the series was graded chiefly on qualitative grounds.

The environmental variability model is also supported circumstantially by the range-ratio of the 4 anuran species. The range ratio of B. viridis, R. ridibunda, H. arborea savignyi and P. syriacus is roughly 10:6.5:3.5:1 , respectively, and it is positively correlated with heterozygosity ($r = .95$; $P < .05$). Geographically widely ranging anurans show wide ranges of thermal tolerance while anurans with restricted ranges show poor ability to adjust physiologically to thermal acclimation (Brattstrom, 1963, 1968). If so, a larger geographic range implies larger environmental heterogeneity matched by higher heterozygosity coupled with greater ecological tolerance and physiological function. These result in a better colonizing ability as displayed both regionally and locally by Bufo viridis (Figure 1).

Additional support for the environmental variability model derives from the comparison of subterranean and surface environments. The subterranean environment is certainly ecologically more uniform and stable than surface environments in terms of annual and daily fluctuations of temperatures and relative humidity (McNab, 1966, Table 2 and his references), both being crucial factors for anurans. Relatively low heterozygosity is shared by all subterranean and fossorial mammals (Nevo et al., 1974, Table 8), subterranean mole crickets (Nevo, in preparation) and subterranean Pelobates. All these taxonomically unrelated animals ranging in different continents and differing in their history and population structure and dynamics are narrow habitat specialists. It seems therefore plausible to relate their relatively low $\overline{H}$ to selec-

tion for homozygosity as an optimal strategy in the relatively constant and predictable subterranean environment.

The surface, more heterogeneous and unpredictable environments, select for heterozygosity as a general optimal strategy in the 3 remaining anurans, but their genetic patterns
differ. Hyla arborea savignyi displays a unique pattern: it
shares a medium level of heterozygosity with Rana ridibunda
but it also shares a high degree of polymorphism with Bufo
viridis. This is due to the relatively high number of weakly
polymorphic loci characterizing it. Though the adaptive significance of this pattern is unknown, it may be related to
higher environmental uncertainty characterizing the open
arboreal habitats occupied by Hyla. Such a strategy keeps a
potentially ready store of genetic variation which may be
actualized in rapidly changing environments. Evidence derived
from geographic variation in body size of B. viridis in
Israel (Nevo, 1972) reinforces the idea that Bufo is indeed
under strong phenotypic selection across increasingly arid
climates. Its body size increases clinally southwards as a
function of humidity, probably as an adaptation to arid climates. None of the other 3 anurans display similar sharp
size clines along the same ecological transect. The high
genetic variation of Bufo makes it a successful colonizer of
distinctly varying and uncertain environments, far deeper
into the northern Negev and Sinai deserts than the other 3
anurans (Figure 1).

Alternative models for explaining the genetic variation
found in the 4 anurans can be ruled out. The genetic variation found is neither correlated with geological age nor with
population size and structure. First, Pelobates syriacus
dating back to Miocene time (Estes, 1970), has the least
heterozygosity. If heterozygosity was a function of time

Pelobates, which is probably the oldest of the 4 anurans, should have the highest $\overline{H}$. Second, the 3 species B. viridis, R. ridibunda, H. arborea have very large breeding populations making genetic drift an unlikely evolutionary agent. Even less abundant Pelobates, which is elusive owing to its sub-terranean life, displays medium effective breeding population sizes involving hundreds of individuals in local populations (Nevo, unpublished). Hence its homozygosity is unlikely a result of random processes. The genetic patterns of Pelo-bates and Bufo negate neutrality also as an explanatory model. On the average it requires 4N generations for a new neutral mutant to go to fixation or to become prevalent in a popu-lation of size N (Kimura and Ohta, 1969). But if this were the case in Pelobates why were all 30 systems fixed except the two esterases in the disjunct populations of Dalton and Tel-Aviv, as well as in one specimen from the Golan Heights and 3 specimens of P. cultripes from Spain? (Nevo, in pre-paration). Nor can migration explain the esterase poly-morphism since the populations mentioned are either isolated geographically or reproductively. Similarly, desert isolates of Bufo viridis in Israel (Dessauer, Nevo and Chuang, 1975) share the main allele with central populations despite appa-rent lack of interpopulation migration. These genetic patterns are best explained as adaptive patterns to the rela-tively constant and varying environments occupied by P. syriacus and B. viridis, respectively.

Present evidence relating genetic and environmental varia-tions is still inconclusive and needs critical analytical studies. Supportive evidence for the hypothesis was summa-rized in Bryant (1974), who showed that populations exposed to increased environmental variability evolve greater genetic variation than similar populations in constant environments.

Furthermore, geographic variation in heterozygosity and polymorphism is largely predicted by environmental variation
(Bryant, 1974; Nevo et al., 1974; Gorman et al., 1975 and
Nevo and Bar, this volume). Despite some evidence to the
contrary (Gooch and Schopf, 1972) a growing body of evidence
reinforces the idea that genetic polymorphism, including protein polymorphism, is adaptively important rather than neutral. Environmental heterozygosity probably selects for
higher heterozygosity and thereby for higher fitness. The
latter may not only depend on individual polymorphic loci but
also on greater overall enzyme heterozygosity per individual
as suggested by Wills (1973).

SUMMARY

Allozymic variation of 4 species of Israeli toads and
frogs encoded by 26-32 loci was compared electrophoretically
in 769 specimens from 23 populations collected at seven
localities. The 4 species occupy increasingly varying and
unpredictable environments in this order: the spadefoot-toad,
Pelobates syriacus, a subterranean narrow habitat-specialist;
the edible frog, Rana ridibunda, a semi-aquatic and aquatic
habitat-intermediate; the tree-frog, Hyla arborea, an arboreal
habitat-intermediate, and the green-toad, Bufo viridis, a
terrestrial broad habitat-generalist.

Genetic variation increased in the following order:
Pelobates < Rana < Hyla < Bufo. Mean number of alleles per
locus, $\overline{A}$, mean proportion of loci polymorphic per population,
$\overline{P}$, and mean proportion of loci heterozygous per individual,
$\overline{H}$, based on 14 shared loci, for Pelobates, Rana, Hyla and
Bufo, respectively, were: $\overline{A}$ = 1.13, 1.55, 1.86, 1.84; $\overline{P}$ =
.095, .44, .50, .56; and $\overline{H}$ = .029, .088, .088 and .169.

The environmental variability model seems the best predictor of the genetic variation displayed by the 4 anuran species. Heterozygosity is positively correlated with environmental heterogeneity and unpredictability. It is concluded that protein polymorphism is adaptively important and is maintained in the species studied by natural selection.

ACKNOWLEDGEMENTS

Mr. M. Avrahami, Mrs. H. Bar-El and Mr. G. Heth assisted in field and lab work. Dr. S.H. Yang assisted in setting up the laboratory and collaborated in part of the work to appear elsewhere. Professors J.A. Beardmore, W.F. Bodmer, S. Karlin, R.K. Koehn, R.C. Lewontin and Mr. S. Mendlinger commented on an earlier draft of the paper. This study was supported by the U.S.-Israel Binational Science Foundation, Grant No. 330. I am grateful to all.

REFERENCES

Blair, W.F. (Ed.) (1972). In Evolution in the genus Bufo, pp.3-7. Austin Univ. Texas Press.

Bragg, A.N. (1961). Anim. Behav. 9: 178-186.

Brattstrom, B.H. (1963). Ecology 44: 238-255.

Brattstrom, B.H. (1968). Comp. Biochem. Physiol. 24: 93-111.

Bryant, E.H. (1974). Am. Natur. 108: 1-19.

Dessauer, H.C., E. Nevo and K.C. Chuang. (1975). Biochem. Genet. (In press).

Dobzhansky, Th. (1951). Genetics and the origin of species. Columbia Univ. Press, N.Y.

Estes, R. (1970). Bull. Mus. Comp. Zool. 139: 293-340.

Estes, R. and O.A. Reig. (1973). In Evolutionary biology of the anurans, pp.11-63. J.L.Vial (Ed.). Columbia Univ. Missouri Press.

157

Gooch, J.L. and T.J. Schopf. (1972). Evolution 26: 545-552.

Gorman, G.C., M. Soulé, S.Y. Yang and E. Nevo. (1975). Evolution 29: 52-71.

Kimura, M. and T. Ohta. (1969). Genetics 61: 763-771.

Levins, R. (1968). Evolution in changing environments. Princeton Univ. Press.

Lewontin, R.C. (1974). The genetic basis of evolutionary change. Columbia Univ. Press, N.Y.

McDonald, J.F. and F.J. Ayala. (1974). Nature 250: 572-574.

McNab, B.K. (1966). Ecology 47: 712-733.

Nevo, E. (1972). Isr. J. Med. Sci. 8: 1010.

Nevo, E., Y.J. Kim, C.R. Shaw and C.S. Thaeler, Jr. (1974). Evolution 28: 1-23.

Nevo, E. (1975). Ranidae. Encyclopaedia Hebraica. (In Hebrew; in press).

Nevo, E. (1976). Bufonidae. Encyclopaedia Hebraica. (In Hebrew; in press).

Powell, J.R. (1971). Science 174: 1035-1036.

Savage, J.M. (1973). In Evolutionary biology of the anurans, pp. 351-445. J.L. Vial (Ed.). Columbia Univ. Missouri Press.

Schneider, H. and E. Nevo. (1972). Zool. Jb. Physiol. 76: 497-506.

Selander, R.K., M.H. Smith, S.Y. Yang, W.E. Johnson and G.B. Gentry. (1971). Univ. Texas Publ. 7103: 49-90.

Wills, C. (1973). Am. Natur. 107: 23-34.

Natural Selection of Genetic Polymorphisms along Climatic Gradients

E. NEVO and Z. BAR

INTRODUCTION

The nature, dynamics and evolutionary significance of protein polymorphisms is still largely unresolved and remains a major problem of evolutionary genetics (see detailed alternative views in Le Cam et al., 1972). Is protein variation largely adaptive as contended by Darwinians (Dobzhansky, 1970; Lewontin, 1974), or is it largely irrelevant to natural selection as advocated by non-Darwinian evolutionists (King and Jukes, 1969; Kimura and Ohta, 1971)? The first objective of this study is to analyze the genetic differentiation of sedentary populations along climatic gradients in an attempt to test between the two alternative theories.

Clines, or character gradients in space and time, may originate in response to ecological gradients and/or as a result of migration between populations of differing genetic composition. The theory of clines assumes that the genetic structure of populations reflects both local adaptation (ecotypic differentiation) combined with the influence of gene flow (Haldane, 1948; Fisher, 1950; Mayr, 1963; Dobzhansky, 1970). Some recent investigators have concluded on obser-

vational, experimental and theoretical grounds that in many
cases migration may be of only minor consequence to among-
population differentiation (Jain and Bradshaw, 1966; Ehrlich
and Raven, 1969; Thoday and Gibson, 1970; and Endler, 1973),
though estimating the relative contribution of selection and
migration to a specific cline is formidable (Karlin and Dyn,
this volume).

The second objective of this study is an attempt to
evaluate the effects of selection and migration to spatial
variation of allozyme and shell banding polymorphism in the
Mediterranean land snail Theba pisana along two climatic
gradients. T. pisana well suits this purpose because of its
abundance, severely limited migration, and extreme phenotypic
variation. The genetic pattern of protein and visible polymorphisms discovered suggests that climatic selection is the
major mechanism of genetic differentiation of T. pisana populations.

DISTRIBUTION, ECOLOGY AND POPULATION STRUCTURE
OF THEBA PISANA

The common white garden snail, Theba pisana, is predominantly a coastal plain circummediterranean species extending also to Atlantic and sub-Atlantic regions, including a
recent colonization in North America (see details of distribution and ecology in Sacchi, 1971). It is a generalist
species, ranging in dry and arid regions within sea influence,
occupying there remarkably varying habitats involving dry
sandy plains, gardens, fields, roadsides, hedges, hill slopes,
thistles, tree trunks and walls, usually in great numbers and
in places fully exposed to the sun. Its phenotypic variability in color, size, shell thickness and banding patterns
is remarkable (Taylor, 1912).

The distribution, ecology, life cycle, population structure and dynamics of T. pisana in Israel have been extensively
studied by Harpaz and Oseri (1961) and summarized briefly by
Avidov and Harpaz (1969). T. pisana is the most prevalent
land snail along the coastal plain of Israel, outnumbering
14:1 all other 50 species of terrestrial snails. It occupies
a variety of habitats, involving variable substrates and
diverse vegetation types where it feeds on many wild and cultivated species, and is a notorious crop damaging snail. It
is a biannual, hermaphroditic, panmictic outcrossing species.
Colonies are usually very large, frequently continuous along
many kilometers but sometimes disjunct by several kilometers.

The life cycle and activity patterns of T. pisana are
largely determined by climatic factors including temperature,
relative air humidity, and rainfall. Towards summer, mean
temperature above 22°C plays a decisive role in inducing
summer dormancy, or aestivation, during the hot season. The
latter starts in late June and lasts from 95 days in irrigated
groves to 131 days in dry sand dunes. Snails aestivate on
posts, wire fences, vegetation, and walls totally exposed to
solar radiation. Aestivation is terminated by the first winter rains, if sufficient. Winter activity of populations in
irrigated citrus groves and heavy alluvial soil is largely
controlled by relative humidity. The 65-70 per cent level of
mean air humidity seems to be the lower limit of normal activity. During rainy or cloudy days when humidity is high, the
usually nocturnal Theba may also be active during daytime. In
dry habitats, such as uncultivated calcareous sandstones and
sandy dunes, activity appears to be suppressed even at times
when humidity is adequately high. Here, exposure of snails
to the sun seems to be the determining factor. Evidently,
the activity rhythms and population dynamics of T. pisana are

humidity and temperature-dependent.

MATERIALS AND METHODS

A total of 262 specimens representing 8 populations of the
land snail Theba pisana were sampled along the coastal plain
of Israel, 5 populations in July, 1973 (1,4,5,6,8) and   3
(2,3,7) in May, 1975 (Table 1; Figure 1). Sampling was
carried out along two transects of increasing aridity. The
first, a north-south 180 km. transect, runs from Gesher Haziv
(population 1), through Kurdani (2), Nahal Alexander (3),
Rishon Le-Zion (5), Ashdod (7), to Jabaliya (8), the southern-
most limit of the species range in Israel. The second, a
west-east 17 km. transect, runs from Tel-Yona (4), again
through Rishon Le-Zion to Lod (6), the easternmost limit of
the species in Israel. Increasing aridity in the two transects is indicated by decreases in rainfall and rain days and
increases in evaporation and days in which temperatures rise
above $30^{\circ}C$. This is also expressed by Thornthwaite (1948)
moisture index which is essentially based on potential evapo-
transpiration and it summarizes the annual water balance of a
region, thereby defining climatic regions (Table 1).

The soil substrates on which the 8 populations have been
sampled differed so greatly that it seems probable that they
generated different microclimates. The substrates of populations 1 and 4 consisted of calcareous sandstone; population
2 of permanent sand; population 3 of sand dune; population 5
of sandy loam: population 6 of alluvial dark soil; and populations 7,8 on a mixture of permanent sand and sandy loams.

TABLE 1. *Summary of geographical and climatological data for the examined populations of* Theba pisana.

Populations	Gesher Haziv	Kurdani	Nahal Alexander	Tel Yona	Rishon Le-Zion	Lod	Ashdod	Jabaliya
	1	2	3	4	5	6	7	8
Sample size (N)	35	30	29	35	35	35	29	35
Israel Grid 250.000 — Longitude	161	158	138	125	131	142	117	101
Israel Grid 250.000 — Latitude	272	240	198	156	152	151	138	103
Elevation (meter)	20	10	20	5	60	50	5	50
I. Rainfall and Humidity								
Rainfall (mm)	601	546	529	542	532	496	479	371
No. of rain days	66.8	66.0	61.0	64.3	55.5	58.3	46.0	43.5
No. of dew nights	120-160	150-190	180-220	120-160	120-160	160-200	120-160	120-160
Evaporation (cm)	174	176	179	185	190	195	178	162
Relative humidity at 08:00 (%) (Mean Annual)	66	67	70	70	74	67	73	72
Relative humidity at 14:00 (%) (Mean Annual)	60	65	68	63	58	50	60	61
Mean daily relative humidity (%)	67	68	68	69	70	64	70	70
Moisture index Thornthwaite (1948)	C_1-20-0	C_1-20-0	C_1-20-0	D-40 - 20	D-40 - 20	D-40 - 20	D-40 - 20	E-60 - 40
II. Temperature								
Temperature at 14:00 (°C) (Mean Annual)	23.7	22.3	23.5	23.0	23.9	24.7	23.9	24.0
Difference between daily minimum and maximum temp. (Mean Annual)	9.5	7.6	9.9	7.2	10.0	12.5	10.3	10.4
No. of days above 30 C	12	15	30	35	65	85	83	85
Minimum Temp. (°C): August (Mean monthly)	19.4	21.0	19.5	17.9	17.2	16.6	18.4	18.7
: January	4.1	5.0	4.6	3.0	2.3	1.7	3.7	4.3
Maximum Temp. (°C): August (Mean)	32.3	33.0	32.8	31.8	33.5	35.2	33.3	33.8
: January	22.3	22.5	22.9	23.4	23.9	24.4	23.8	24.1
III. Radiation								
Total solar radiation absorbed by horizontal surface (Kcal cm^2) (Israel Atlas)	<182	<182	<182	<182	<182	<182	<185	189-195

· Climatological records were taken from the nearest station to the sampled locality.

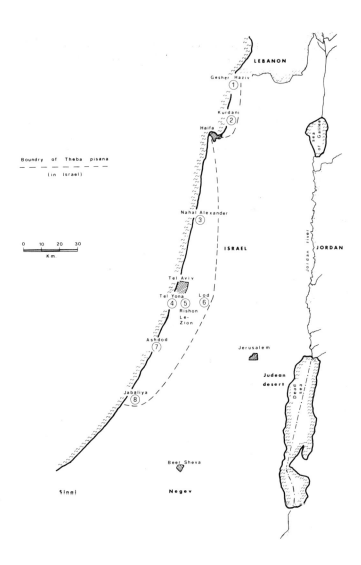

FIGURE 1. Distribution of _Theba pisana_ in Israel and sampled
localities (1-8).

Live specimens were homogenized in the laboratory after recording morphometric data and shell banding patterns. Allozymic variation of enzymes and other proteins encoded by 18 gene loci was studied by standard starch gel electrophoresis with procedures described in Selander et al., (1971). The 18 loci studied coding for soluble proteins, classified according to function, are: Group I. Glucose metabolizing enzymes: malate dehydrogenases, two loci (Mdh-1 and Mdh-2); 6-phosphogluconate dehydrogenase (6-Pgd); phosphoglucomutases, two loci (Pgm-2 and Pgm-3); phosphoglucose isomerase (Pgi); gluconate dehydrogenase (Gdh); glucose-6 phosphate dehydrogenase (G-6-pd); hexokinase (Hk). Group II. Other enzymes: peptidases, two loci (Pept-1 and Pept-2); esterases, four loci (Est-1, Est-2, Est-3 and Est-4); indophenol oxidase (Ipo); glutamic oxaloacetic transaminase (Got). Group III. Nonenzymatic proteins: protein-1 (Prot-1). Alleles were designated alphabetically in order of decreasing mobilities of their allozymes. The banding polymorphism was studied by the procedure described for Cepaea nemoralis where effectively unbanded shells include all shells whose banding is unexposed (Ford, 1971).

RESULTS

Pattern of variation. The allelic frequencies of 18 loci examined electrophoretically are given in Table 2 and are shown graphically in Figures 2, 3 and 4. Of the 18 loci examined, 13 (= .72) were polymorphic in at least one of the 8 populations studied, of which 4 (Mdh-1, Mdh-2, Pgi and Gdh) were locally, and 9 regionally polymorphic. Monomorphic loci included: 6-Pgd, Ipo, Got, Hk, Prot-1. There was no case of alternative allele fixation among populations.

TABLE 2. Allele frequencies at 18 loci of 8 populations of *Theba pisana*.

Locus & Allele		Popu. 1	2	3	4	5	6	7	8
	N°	35	30	29	34	35	35	29	35
1. Mdh-1	a	1.00	1.00	.90	1.00	1.00	1.00	1.00	1.00
	b	-	-	.10	-	-	-	-	-
2. Mdh-2	a	1.00	.62	.94	1.00	1.00	1.00	.93	1.00
	b	-	.38	.06	-	-	-	.07	-
3. 6-Pgd	a	1.00	1.00	1.00	1.00	1.00	1.00	1.00	1.00
4. Pgm-2	a	.64	.64	1.00	.91	.89	1.00	1.00	1.00
	b	.36	.36	-	.09	.11	-	-	-
5. Pgm-3	a	-	.14	-	.02	-	-	-	-
	b	.22	.50	-	.29	.33	.28	.06	.49
	c	.78	.36	1.00	.69	.67	.72	.94	.51
6. Pgi	a	.93	1.00	1.00	1.00	1.00	1.00	1.00	1.00
	b	.07	-	-	-	-	-	-	-
7. Gdh	a	1.00	1.00	1.00	.99	1.00	1.00	1.00	1.00
	b				.01				
8. G-6-Pd	a	.44	.79	.10	.22	.06	.08	-	.03
	b	.56	.21	.83	.78	.94	.89	.66	.84
	c	-	-	.07	-	-	.03	.34	.10
	d	-	-	-	-	-	-	-	.03
9. Pept-1	a	-	-	-	-	-	.09	-	-
	b	-	-	-	.12	-	.21	.10	-
	c	.56	1.00	.31	.53	.57	.29	.48	.31
	d	.44	-	.69	.35	.43	.41	.42	.69
10. Pept-2	a	.33	.05	-	-	-	-	-	.06
	b	.64	.95	1.00	.96	.89	.89	.98	.60
	c	.03	-	-	.04	.11	.11	.02	.34
11. Est-1	a	.20	.10	-	-	-	-	-	-
	b	.14	.01	-	-	.10	.16	-	.03
	c	.14	.51	-	.31	.29	.70	.70	.88
	d	.06	-	-	-	-	-	-	-
	e	.46	.38	1.00	.69	.61	.14	.30	.09
12. Est-2	a	.03	-	-	-	-	.20	.12	.50
	b	.93	1.00	1.00	1.00	1.00	.80	.84	.50
	c	.04	-	-	-	-	-	.04	-
13. Est-3	a	.81	.63	1.00	.94	.85	.37	.73	.37
	b	.19	.37	-	.06	.15	.63	.27	.63
14. Est-4	a	.86	.97	.64	-	.52	.55	.66	.31
	b	.14	.03	.36	1.00	.48	.45	.34	.69
15. Ipo	a	1.00	1.00	1.00	1.00	1.00	1.00	1.00	1.00
16. Got	a	1.00	1.00	1.00	1.00	1.00	1.00	1.00	1.00
17. HK	a	1.00	1.00	1.00	1.00	1.00	1.00	1.00	1.00
18. Prot-1	a	1.00	-	-	1.00	1.00	1.00	-	1.00
Polymorphism 1%		.56	.42	.26	.44	.44	.44	.50	.44
Polymorphism 5%		.56	.37	.26	.33	.44	.44	.50	.44

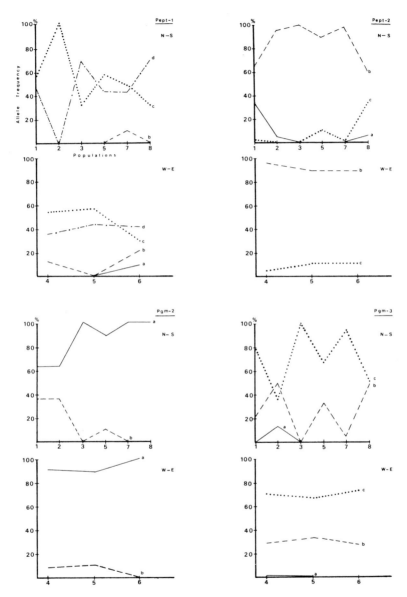

FIGURE 2. *Allozymic clines of Theba pisana showing allele*
frequencies at Pept-1, Pept-2, Pgm-2 and Pgm-3.
Allele frequencies appear on the Y axis and popu-
lations on the X axis. N-S describes the north-
south transect (populations 1,2,3,5,7,8); W-E des-
cribes the west-east transect (populations 4,5,6).

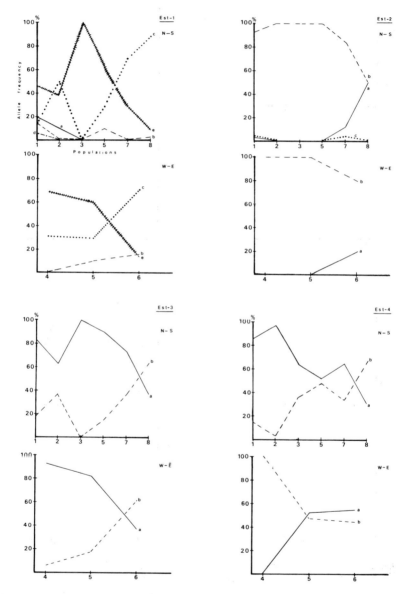

FIGURE 3. Allozymic clines of Theba pisana showing allele
frequencies at Est-1, Est-2, Est-3 and Est-4.
Symbolism of axes as in Figure 2.

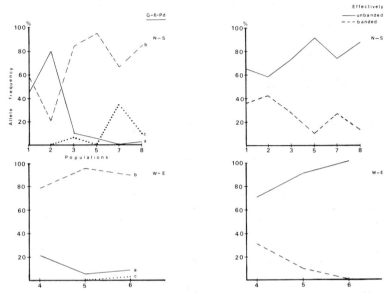

FIGURE 4. *Allozymic clines of* Theba pisana *showing allele frequencies at* G-6pd *and the shell banding polymorphism. Symbolism of axes as in Figure 2.*

The main feature of variation was either the prevalence of one allele across populations, or spatial variation in allelic frequencies. A remarkable feature is the clinal variation in allelic frequencies of 6 out of the 9 regionally polymorphic loci, or 7 out of 10 regionally polymorphic loci when the shell banding polymorphism is included. The non-clinal loci were Pgm-3, Pep-1 and Pep-2. The loci varying clinally along the north-south transect involve: the four esterases Est-1,2,3,4 , Pgm-2, G-6pd, and shell banding.

Similar trends in allelic frequencies to those of the southward clines are expressed in the eastward transect by 5 loci: three esterases Est-1,2,3 , Pgm-2, and the shell banding. The following strongly polymorphic loci show no clinal pattern: Pep-1, Pep-2 and Pgm-3.

A summary of the genetic data on the 8 populations of T. pisana is given in Table 3.

TABLE 3. *Genetic variation based on 18 loci in 8 populations of Theba pisana.*

Populations	Sample size (N)	Mean No. of alleles per locus ($\bar{A}$)	Mean proportion of loci polymorphic per population ($\bar{P}$)	Heterozygous per locus per individual ($\bar{H}$)
1. Gesher Haziv	35	1.83	.56	.165
2. Kurdani	30	1.56	.42	.064
3. Nahal Alexander	29	1.37	.26	.054
4. Tel – Yona	34	1.56	.44	.101
5. Rishon Le – Zion	35	1.50	.44	.114
6. Lod	35	1.67	.44	.131
7. Ashdod	29	1.67	.50	.065
8. Jabaliya	35	1.67	.44	.110
Total	262			
Mean $\pm$ S.E.		1.60$\pm$.049	.44$\pm$.027	.100$\pm$.013
Range		(1.37-1.83)	(.26-.56)	(.054-.165)

On the whole, T. pisana is a genetically variable species even in the marginal populations of Lod and Jabaliya. Noteworthy, populations living on sands are significantly less heterozygous than those living on calcareous sandstones, sandy-loams and alluvial soils ($\bar{H}$ = .124 and .061 respectively; P < .01) . This difference is not apparent in either $\bar{A}$ or $\bar{P}$ values; except in Nahal Alexander. Thus the populations of Kurdani and Ashdod are highly polymorphic but low in heterozygosity.

Values of $\bar{A}$, $\bar{P}$ and $\bar{H}$ (Table 3) have been recalculated separately for three groups of proteins in two ways: according to the method of Gillespie and Kojima (1968) and according to Johnson (1971). The first method involves: glucose metabolizing enzymes (group I), all other enzymes (group II), and nonenzymatic proteins (group III). The second method includes: variable substrate enzymes, regulatory enzymes and non-regulatory enzymes. These results are given in Table 4.

TABLE 4. Genetic analysis of number of alleles per locus, polymorphism and heterozygosity per locus per individual in 8 populations of Theba pisana, for glucose metabolizing enzymes (group I), other enzymes (group II) and nonenzymatic proteins (group III).

Locus	No. of alleles per locus	POPULATIONS — Heterozygosity per individual								Mean proportion of — Polymorphic populations	Mean proportion of — Heterozygosity per individual
		1	2	3	4	5	6	7	8		
Group I (glucose metabolizing enzymes)											
Mdh-1 ...	2	—	—	—	—	—	—	—	—	.000	.000
Mdh-2 ...	2	—	.100	—	—	—	—	—	—	.125	.013
6-Pgd ...	1	—	—	—	—	—	—	—	—	—	.000
Pgm-2 ..	2	.429	.233	—	.125	.171	—	—	—	.500	.125
Pgm-3 ..	3	.250	.200	—	.353	.429	.441	.103	.457	.875	.280
Pgi ..	2	.029	—	—	—	—	—	—	—	.125	.004
Gdh ...	2	—	—	—	.029	—	—	—	—	.125	.004
G-6-Pdh ..	4	.294	.066	.138	.147	.057	.114	.138	.086	1.000	.130
HK ...	1	—	—	—	—	—	—	—	—	—	.000
Mean	2.11	.125	.075	.016	.082	.082	.068	.030	.068	.390	.061
Group II (other enzymes)											
Pept-1 .	4	.314	—	.206	.394	.229	.588	.069	.171	.875	.246
Pept-2 .	3	.429	.100	—	.029	.114	.114	.035	.400	.875	.262
Est-1 .	5	.771	.233	.670	.625	.735	.319	.074	.057	1.000	.497
Est-2 .	3	.029	—	—	—	—	.114	—	.257	.375	.050
Est-3 .	2	.323	.346	—	.125	.294	.514	.435	.457	.875	.312
Est-4 .	2	.191	—	.675	—	—	.107	.522	.088	.625	.276
Ipo .	1	—	—	—	—	—	—	—	—	—	.000
Got ...	1	—	—	—	—	—	—	—	—	—	.000
Mean	2.63	.229	.075	.192	.130	.153	.195	.126	.159	.578	.155
Group III (non enzymatic proteins)											
Prot-1	1	—	—	—	—	—	—	—	—	.000	.000

. variable substrate enzymes
.. regulatory enzymes
... non regulatory enzymes

Summary:	No. loci	Ā	P̄	H̄	Summary:	No. loci	Ā	P̄	H̄
Group I	9	2.11	.390	.061	Variable substrate enzymes	7	2.86	.661	.235
Group II	8	2.63	.578	.155	Regulatory enzymes	4	2.75	.625	.135
Group III	1	1.00	.000	.000	Non regulatory enzymes	6	1.50	.042	.003

Heterozygosity ($\overline{H}$) in group II is significantly ($P < .05$) higher than in group I. Likewise, both variable substrate and regulatory enzymes are significantly ($P < .05$ and

P < .01, respectively) more heterozygous than loci encoding nonregulatory enzymes.

Climatic correlates of genetic variation. A test for the best predictor of each of 10 polymorphic loci, total polymorphism ($\bar{P}$) and heterozygosity ($\bar{H}$) was conducted by a stepwise multiple regression analysis (SPSS regression analysis) employing a series of climatic independent variables listed in Table 1. Allele frequencies were transformed to arctang values for normalization. The analysis was performed on the mean estimates of 8 populations. The results are given in Table 5, first for each of the polymorphic loci, then for $\bar{P}$ and $\bar{H}$. A combination of 2 or 3 humidity and/or temperature variables explains significantly a high proportion of the genetic variation among populations, in most loci. Noteworthy, $\bar{H}$ is explained by only 3 variables in the 5 original populations sampled in 1973, but it needed 6, though correlated variables, to explain variation in all 8 populations. This was probably caused by the low $\bar{H}$ values in the additional 3 populations sampled in 1975. Temperature variables are the best predictors for: Est-3, Est-4, G-6pd, and shell banding polymorphism. A combination of humidity and temperature variables are the best predictors for: Est-1, Est-2, Pgm-2, $\bar{P}$ and $\bar{H}$.

Coefficients of genetic distance and similarity (Rogers, 1972) were calculated for paired combinations of all 8 populations. Mean genetic similarities, $\bar{S}$, among the 8 populations are high $S = .92$ (range, 86-97), and decreased not only with geographic distance but also possibly with local habitats.

Theba pisana is polymorphic for shell banding pattern. The frequencies of individual morphs vary geographically in a southwards clinal pattern paralleling the trends of protein polymorphisms. For population samples of 40 specimens each, the proportion of effectively unbanded shells, increased southwards (Geshel Haziv .64; Kurdani .58; Nahal Alexander .72;

Table 5. Coefficients of multiple determination (R^2) employing allele frequencies at each of 10 polymorphic loci, total polymorphism and heterozygosity at all 18 loci of Theba pisana *as dependent variables. Symbols and significance levels are defined in the footnote.*

Locus	Variable combinations I	II	III
Pgm-2	Th .74••	Th·Ev .87••	Th·Ev·Rd .90••
Pgm-3	TMP .21	TMP·Amax .56	TMP·Amax·M-M .64
G-6-pd	Jmax .64•	Jmax·D30 .82•	Jmax·D30·Amin .96••
Pept-1	P .10	P·Amin .12	P·Amin·Jmin .13
Pept-2	Ev .36	Ev·Th .66	Ev·Th·D30 .91•
Est-1	P .30	P·Jmin .47	P·Jmin·Ev .93••
Est-2	P .49	P·Th .86••	P·Th·Jmax .90•
Est-3	Jmin .26	Jmin·Amax .76•	Jmin·Amax·Ev .88•
Est-4	Amin .30	Amin·Jmin .83•	Amin·Jmin·Ev .91•
Effectively unbanded shells	Jmax .84••	Jmax·Th .92••	Jmax·Th·Amin .98••
Polymorphism 1%	TMP .46	TMP·M-M .79•	TMP·M-M·H14 .83•
Heterozygosity (populations 1,4,5,6,8.)	Jmax .41	Jmax·Amax .99••	Jmax·Amax·Amin .99•
Heterozygosity (8 populations)	H14 .38	H14·Amax .62	H14·Amax·Amin·Ev·D30·Jmin .99••

Significance level: •p<.05; ••p<.01.

Symbolism:
P : Mean annual precipitation.
Ev : " " evaporation.
Rd : " " no. of raindays.
H14 : " " relative humidity of 14:00.
TMP : " " temperaure at 14:00.
M-M : " " difference between daily minimum and maximum temperature.
D30 : " " no. of days above 30°C.
Amin : " monthly minimum temperature for August.
Jmin : " " " " January.
Jmax : " " maximum " " " .
Amax : " " " " " August.
Th : Thornthwaite moisture index.

Tel Yona .70; Ashdod .73 and Jabalyia .865) and eastwards
(Tel Yona .70; Rishon Le Zion .90 and Lod 1.00). Detailed
account of the banding polymorphism will appear elsewhere.

DISCUSSION

I. The cline structure, climatic selection gradients and
   gene flow.

1. Allozymic variation. The cline structure of allelic
frequencies of T. pisana (Figures 2,3,4), corresponds remarkably well with some of the background climatic gradients
(Table 1). Seven polymorphic systems, out of 10 regionally
polymorphic ones, vary clinally in association with climatic
gradients southwards, and in 5 cases they also show similar
trends eastwards. All clinal systems are significantly predicted by humidity and/or temperature variables (Table 5).
Neither genetic drift nor neutrality can probably be invoked
to explain these patterns. The effective breeding population
size of T. pisana is very large, involving many thousands of
individuals, making genetic drift insubstantial. The relationship of the cline structure to climatic gradients negates
neutrality. The spatial variation of allozymes and shell
banding polymorphism seems to be primarily due to climatic
selection. Complementarily, polygenic traits such as shell
size and thickness vary also geographically, and are positively correlated with increasing aridity (Bar, in preparation).

Spatial differentiation of populations without distinct
barriers have been extensively documented in both plants and
animals (Ehrlich and Raven, 1969, and references therein).
Theoretically, the effect of migration on cline structure will
be insubstantial if migration is relatively uniform and

174

limited on a smooth environmental gradient, since the net
effect of gene flow will be small on a deme receiving immigrants on both sides (Wright, 1931). Experimental and theoretical models (Endler, 1973) show that it is possible for
local differentiation to evolve parapatrically in spite of
considerable gene flow if the selection gradients are relatively uniform. Gene flow may be unimportant in population
differentiation along environmental gradients. Slatkin (1973)
concluded that Endler's result that gene flow has no effect
on the cline is strongly dependent on the assumption of
linearity of selection pressures. Karlin and Dyn (this
volume) questioned the general validity of Endler's model
which is based on an anomalous symmetry. They found that the
influence of increased migration flow, in a two-range environmental model, is unambiguous and substantial: the cline does
flatten directly with increasing migration rate. They also
found that the swamping effect of increased gene flow is more
pronounced in a few deme, rather than in a multi deme model.
This was also concluded by Endler (1973) who stated that in a
few deme cline the effect of gene flow will be much greater
than in a multi deme cline, with the greatest possible effect
on gene flow in the two deme model.

Gene flow is presumably unimportant in the cline structure
of T. pisana. First, the background climatic gradients vary
smoothly and continuously across the range of T. pisana in
Israel, rather than existing in a two range geographical
selection gradient, or ecotone (see Karlin and Dyn, this
volume for a comprehensive theoretical analysis of migration-
selection interaction in a cline in a geographical two range
environment). Second, gene flow in T. pisana is presumably of
the stepping stone migration model, but it is severely limited. Sacchi (1971) reported homing behavior in this species

175

following marked individuals. Likewise, distinct sedentari-
ness of both young and adults have been observed in Israel
(Nevo, unpublished observations). Third, T. pisana is not
only distinctly sedentary, but it also involves numerous very
large demes across its range. The gradient climatic background, severely limited migration, and numerous large demes
argue that the cline structures under discussion are largely
the result of climatic selection gradients rather than affected by the attenuating effect of gene flow. This conclusion
may be reinforced by the kinky shape of the clines (Figures
2, 3 and 4).

The slope of a cline between regions is partly indicative
of population differentiation and may possibly aid in estimating the selective pressures maintaining the cline. A
steep cline means sharp differentiation while a gentle one
implies small differentiation between regions. The steepness
of the T. pisana allelic frequency clines differ between and
within loci (Figures 2, 3 and 4). The steepest clines are
those of esterases, particularly Est-1 and Est-2, and the
gentlest cline slopes are those of Pgm-2 and the shell banding polymorphism. Remarkably, the southward and eastward
slopes are very similar in Est-1, Est-3 and Pgm-2. Since all
loci were measured on the same sample of snails, gene flow,
no matter how low, must have been equivalent for all loci.
Hence, the different clinal slopes may be the result of
varying selection pressures on different loci. It is noteworthy that esterases, which are expected to reflect best
environmental selection gradients as members of group II
enzymes (Gillespie and Kojima, 1968) are indeed the steepest
of all clines.

The graphical display of the clines shows kinks and irregularities particularly in populations 2 and 3. These pro-

bably attest to the small attenuating effect of gene flow on the cline structure of T. pisana and may result from one or more combined factors. First, spatial irregularities in selection intensity may be expected due to varying local substrates; populations 2 and 3 are from sandy substrates. Second, temporal changes in selection intensity may be expected due to seasonal and yearly climatic fluctuations. Populations 1, 4, 5, 6, 8 were sampled in July, 1973 whereas populations 2,3,7 in May, 1975. Third, as Karlin and Dyn (this volume) have shown, non-homogeneity of gene flow creates more kinks in the cline shape. The spatial and temporal variations in allelic frequencies are testable and we plan to compare and contrast geographically close populations that differ in substrates and follow them over several years.

Evidence suggesting that allozymic variation is adaptive and may affect fitness is gradually accumulating in both plants and animals (Koehn, 1969; Lewontin, 1974; Johnson, 1974; McNaughton, 1974; Bryant, 1974; Nevo et al., 1974; Gorman et al. 1975 and their references). Climatic selection of differential adaptive enzymatic properties of leaf malate dehydrogenase of Typha latifolia, involving activation energy, thermo-stability and activity levels, have been demonstrated by McNaughton (1974). Likewise, adaptive clinal esterase polymorphism in a freshwater fish mediated by temperature selection appears to be based on differential level of activity of the three genotypes in different temperature optimums (Koehn, 1969). Climatic, primarily humidity-dependent, selection causes genetic differentiation of wild oat populations (Clegg and Allard, 1972). While in most cases, including that of Theba, only ecological correlates of allozymic variation are known at present, they strongly suggest the potential existence of biochemical and/or physiological

correlates. The search for such potential correlates must
certainly be rewarding.

   2. Shell banding polymorphism. The large scale geographical variation in morph frequencies of the shell banding
polymorphism of T. pisana seems likewise the result of climatic selection. Noteworthy, the proportion of effectively
unbanded snails increases clinally both southwards and eastwards, where it reaches fixation, paralleling the trends described for the 6 enzyme polymorphisms. Such clines are most
likely due to regional climatic gradients rather than the
result of selective predation on variable backgrounds (see
Cain and Sheppard, 1954 for the latter case). The variation
of the shell banding pattern is largely and significantly
explained by a temperature variable combination (Table 5).
Effectively unbanded T. pisana are commoner in the southern
and eastern portions of its range in Israel, or alternatively, effectively banded snails are at a higher frequency
in the colder and more humid parts of the species range. The
cline structure of the banding pattern seems to be heat-
dependent and can possibly be explained physiologically.
Effectively banded snails are pigmented, hence they heat up
more rapidly than do the effectively unbanded light snails,
which thereby minimize overheating. The internal temperature
of fully pigmented banded individuals of the snail Cepaea
vindobonensis was indeed shown to be significantly higher than
that of faintly banded individuals when both were exposed to
sunshine (Jones, 1973). Likewise, unbanded C. nemoralis
shells transmit less light, which is associated with heat,
thereby minimizing overheating (Emberton and Bradbury, 1963).
Lamotte (1966) has also shown experimentally that mortality
rises with increasing banding upon heating individuals of

Cepaea for 40 minutes under a 150 watt lamp bulb. Some large scale gene distributions in Cepaea vindobonensis and C. nemoralis are explicable on differential thermal relations with the environment, the light-colored morphs are found preferentially in the warmest parts of their ranges (Jones, 1973). On a microgeographical scale, Arnold (1969) has also shown that arid hillslope populations tend to have high frequencies of unbanded C. nemoralis with banded snails more frequent near the river, and invoked climatic selection to explain this local population differentiation. T. pisana is largely nocturnal but it is exposed to direct solar radiation on the posts and plants where it either rests during daytime or aestivates during summertime. Hence effectively unbanded Theba having light colored shells will be adaptively selected in those portions of the range involving severe heat load to avoid overheating.

II. Heterozygosity of enzyme classes and the theory of selective neutrality.

Significant differential variation among functional classes of enzymes seems incomptaible with the theory of selective neutrality. Gillespie and Kojima (1968) proposed that levels of enzyme polymorphism may reflect environmental variation in substrates, and that far less heterozygosity is expected at loci of enzymes involved in energy production (group I) than at other enzyme loci (group II). Latter, (this volume) has likewise shown that group I and group II enzymes in Drosophila differ significantly from predictions based on the neutral model. He has shown that the glycolytic enzymes in particular have a strikingly different gene frequency from the remainder and only half of the frequency of heterozygotes. On the other hand Johnson (1971) suggested that enzyme polymorphisms are often associated with regulatory reaction in metabolism. In his words "those enzymes which exert acute control over flow through metabolic pathways should be most individually sensitive as sites of action of selective forces". Summarizing overall heterozygosity estimates for (1) variable substrate enzymes, (2) regulatory enzymes and (3) nonregulatory enzymes (see Johnson, 1974 for group definitions and identification) he found in _Drosophila_, small vertebrates, and man, that $\bar{H}$ is by far higher in groups I and II as compared with III. In other words, the degree of heterozygosity at an enzyme locus is significantly nonrandom and it reflects the regulatory role of the enzyme.

The _Theba_ evidence supports the hypotheses that allozymic variation is nonrandom and is correlated with physiological variation (Table 4). Glucose metabolizing enzymes (group I) are significantly less heterozygous than all other enzymes (group II), substantiating the Gillespie-Kojima (1968) hypothesis. Likewise, as indicated in Table 4, both variable substrate and regulatory enzymes are significantly more heterozygous than nonregulatory enzymes, supporting Johnson's (1971) hypothesis. The clinal patterns of enzyme polymorphisms, their climatic correlates, and the differential variation among enzyme classes negate neutrality and suggests natural selection as a major factor in population differentiation of _T. pisana_. The fact that climatic gradients of humidity and temperature explain significantly most of the genetic variation of _T. pisana_ (Table 5) suggests that climatic selection, both regionally and locally, is the prime mechanism of population differentiation. This remarkably corresponds with the life cycle and activity patterns of _T. pisana_ which are strictly humidity and temperature-dependent

(Harpaz and Oseri, 1961).

Theba pisana is truly remarkable in displaying the inter-relations between regional climatic patterns of selection and spatial variation of gene frequency, not only of proteins, but also of a visible polymorphism and polygenic shell traits. This conclusion does not exclude the operation of local selection due to differential microclimates, variable back-grounds, differential population densities, or even the gene-tic background. The evidence presented does suggest, however, that regional geographical trends in allelic frequencies are primarily associated with increasingly arid and hot environ-ments both southwards and eastwards. Natural selection through climatic gradients seems to be the prime mechanism regulating the genetic structure and differentiation of popu-lations in T. pisana.

SUMMARY

Allozymic variation in proteins encoded by 18 loci was analyzed electrophoretically in 262 specimens from 8 popu-lations of the white graden snail, Theba pisana, sampled along two transects of increasing aridity in the coastal plain of Israel. The same specimens were analyzed for the banding polymorphism.

Genetic variation in Theba pisana is high. Mean number of alleles per locus, $\bar{A}$, is 1.60 (range, 1.37 - 1.83); mean proportion of loci polymorphic per population, $\bar{P}$, is .44 (range, .26 - .56); and mean proportion of loci heterozygous per individual, $\bar{H}$, is .100 (range, .054 - .165). Estimates of genetic variation of functionally different classes of enzymes indicate that heterozygosity is significantly non-random. Genetic similarity between populations is high

$(\bar{S} = .92;$ range, .86-.97). Of the 18 loci examined, 13
(= .72) are polymorphic in at least one population, and 9 are
regionally and strongly polymorphic. There is no alternative
fixation; the same allele is either prevalent or it varies
across the range. The predominant feature is clinal variation
in allelic frequencies in 7 out of 10 polymorphic loci,
corresponding to the background climatic gradients.

Genetic variation is significantly predicted in 8 polymorphic loci, polymorphism and heterozygosity by climatic
variables of either temperature or temperature and humidity.

The genetic pattern of proteins and shell banding polymorphism suggest that natural selection through climatic
gradients, rather than drift, neutrality, or gene flow, is
the major mechanisms of genetic structure and differentiation
of populations in T. pisana.

ACKNOWLEDGEMENTS

M. Avrahami, H. Bar-El, R. Broza, M. Fried, S. Friedman
and S. Mendlinger assisted in field and laboratory work.
W.F. Bodmer, J.A. Beardmore, S. Karlin, R.C. Lewontin,
S. Mendlinger and an anonymous reviewer commented on the
paper. This study was supported by the U.S.-Israel Binational
Science Foundation Grant No. 330 to E. Nevo and R.C. Lewontin.
We are grateful to all.

REFERENCES

Arnold, R. (1969). Evolution 23: 370-378.
Avidov, Z. and I. Harpaz. (1969). Plant pests of Israel.
Israel Univ. Press, Jerusalem.
Bryant, E.H. (1974). Am. Natur. 108: 1-19.
Cain, A.J. and P.M. Sheppard. (1954). Genetics 39: 89-116.

Clegg, M.T. and R.W. Allard. (1972). Proc. Nat. Acad. Sci. U.S.A. 69: 1820-1824.

Dobzhansky, Th. (1970). Genetics of the evolutionary process. Columbia Univ. Press, New York and London.

Ehrlich, P.R. and P.H. Raven. (1969). Science 165: 1228-1232.

Emberton, L.R.B. and S. Bradbury. (1963). Proc. Malac. Soc. London 35: 211-219.

Endler, J.A. (1973). Science 179: 243-250.

Fisher, R.A. (1950). Biometrics 6: 353-361.

Ford, E.B. (1971). Ecological Genetics, 3rd Ed., Chapman and Hall, London.

Gillespie, J. and K. Kojima. (1968). Proc. Nat. Acad. Sci. U.S.A. 61: 582-585.

Gorman, G.C., M. Soulé, S.Y. Yang and E. Nevo. (1975). Evolution 29: 52-71.

Haldane, J.B.S. (1948). Genetics 48: 277-284.

Harpaz, I. and Y. Oseri. (1961). Crop damaging snails in Israel and their control. Hebrew Univ. Faculty of Agriculture, Rehovot.

Jain, S.K. and A.D. Bradshaw. (1966). Heredity 21: 407-441.

Johnson, J.B. (1971). Nature 232: 347-349.

Johnson, J.B. (1974). Science 184: 28-37.

Jones, J.S. (1973). Science 182: 546-552.

Karlin, S. and N. Dyn. (This volume).

Kimura, M. and T. Ohta. (1971). Theoretical aspects of population genetics. Princeton Univ Press, Princeton, N.J.

King, J.L. and T.H. Jukes. (1969). Science 164: 788-798.

Koehn, R.K. (1969). Science 163: 943-944.

Lammote, M. (1966). Lavori della societá Malacologica Italiana 3: 33-73.

Latter, B.D.H. (This volume).

Le Cam, L.M., J. Neyman and E.L. Scott (Eds.) (1972). Proc. of the sixth Berkeley Symp. on Math. Stat. and Prob. Vol.5. Univ. of Calif. Press, Berkeley and Los Angeles.

Lewontin, R.C. (1974). The genetic basis of evolutionary change. Columbia Univ. Press, N.Y. and London.

Mayr, E. (1963). Animal species and evolution. Belknap Harvard Univ. Press, Cambridge, Mass.

McNaughton, S.J. (1974). Am. Natur. 108: 616-624.

Nevo, E., Y.J. Kim, C.R. Shaw and C.S. Thaeler, Jr. (1974). Evolution 28: 1-23.

Rogers, J.S. (1972). Univ. of Texas Publ. 7213: 145-153.

Sacchi, C.F. (1971). Nature Soc. It. Sc. Nat. Museo. Civ. St. Nat. e Acquario Civ. Milano 62/3: 277-358.

Selander, R.K., M.H. Smith, S.Y. Yang, W.E. Johnson and G.B. Gentry. (1971) Univ. of Texas Publ. 7103: 49-90.

Slatkin, M. (1973). Genetics 75: 733-756.

Taylor, J.W. (1912). In The Monograph of the land and fresh-water mollusca of the British Isle. Part XIX, London: B.M.N.H. pp. 368-397.

Thoday, J.M. and J.B. Gibson. (1970). Am. Natur. 104: 219-230.

Thornthwaite, C.W. (1948). Geogr. Rev. 38: 55-94.

Wright, S. (1931). Genetics 16: 97-159.

Muellerian Mimicry: Classical 'Beanbag' Evolution and the Role of Ecological Islands in Adaptive Race Formation*†

J.R.G. TURNER

INTRODUCTION

<u>Heliconius</u> are a group of very beautiful South American butterflies which have undergone an extraordinary pattern of evolution, the formation of parallel races. The best studied and most dramatic example of this is shown in Figure 1: with the exception of the western rift valley of Colombia, where only one of the species is present, <u>Heliconius</u> <u>melpomene</u> and <u>Heliconius</u> <u>erato</u> have an exactly matching set of geographical races with rather less exactly matched geographical distributions. Unravelling the causes and history of this phenomenon is a worthy challenge for evolutionary genetics; the answer turns out to be basically simple, but to reveal things of general importance concerning:

*Work supported by NSF Grant B039300 and by a seed grant from HEW (BioMedical Sciences Support Grant 5S05RR07067-08 awarded to the State University of New York at Stony Brook). I am grateful for excellent butterfly-breeding facilities at the University of York, England.

†Contribution No. 110 of the Program in Ecology and Evolution of the State University of New York at Stony Brook.

185

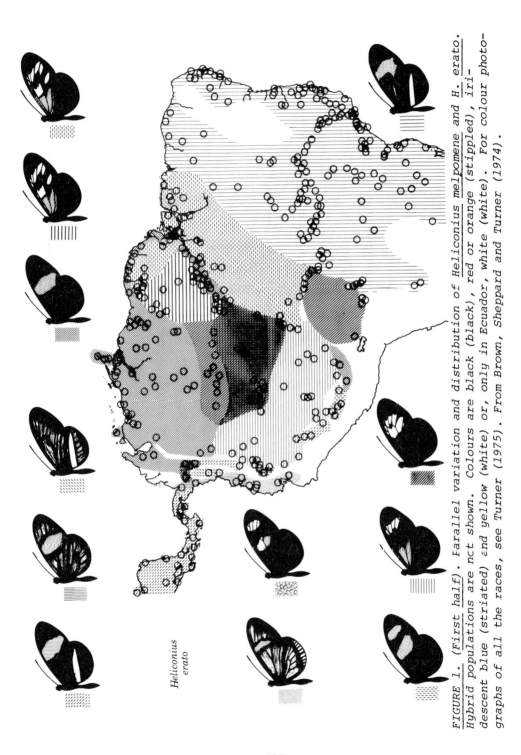

Heliconius erato

FIGURE 1. (*First half*). Parallel variation and distribution of *Heliconius melpomene* and *H. erato*. Hybrid populations are not shown. Colours are black (black), red or orange (stippled), iridescent blue (striated) and yellow (white) or, only in Ecuador, white (white). For colour photographs of all the races, see Turner (1975). From Brown, Sheppard and Turner (1974).

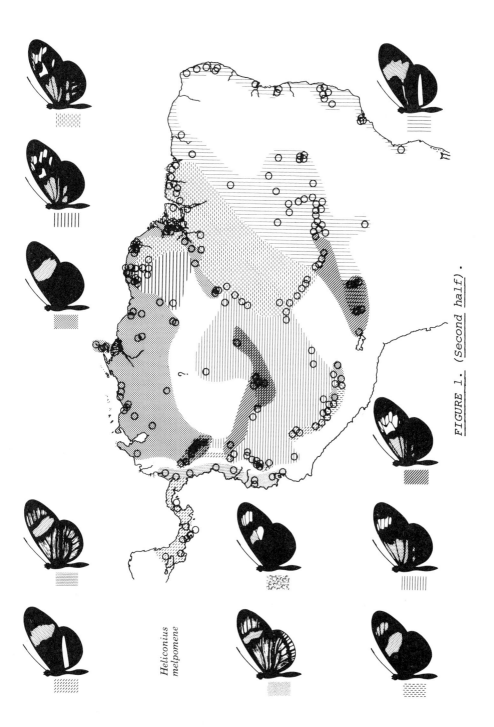

Heliconius melpomene

FIGURE 1. (*Second half*).

1 the effects of glaciations on the tropical lowlands of America

2 the importance, for genetic divergence, of the stopping of gene flow _versus_ the colonization/extinction cycle on islands

3 the timing of adaptive radiation relative to other specia-tion phenomena

4 the absolute chronology of recent evolutionary changes

5 the genetic architecture of adaptive radiation

6 independent evolution of different parts of the genome.

MIMICRY AND NATURAL SELECTION

The prime advantage in studying the patterns of these butterflies is the same as in the study of the beaks of the Galápagos finches: we know their adaptive significance. The parallel variation of _melpomene_ and _erato_ has long been recog-nized as an example of mimicry (Eltringham, 1917). Tanagers (_Ramphocelus carbo_) fed with the Trinidad race of both these species found them distasteful and learned to refuse them; they could not distinguish between them (it is not easy for a taxonomist) and would avoid one species after being conditioned to the other. This also happened with several other mimetic pairs from _Heliconius_ and the related genera (Brower, Brower and Collins, 1963). Tanagers are not known to be natural pre-dators of butterflies (although some of the experimental birds behaved as if they had already become aware of the distaste-fulness of _Heliconius_ in the wild), and insectivorous birds are so difficult to keep in captivity that no experiments have been done with them. But there is at least one field account supporting the experiments. A Jamaican naturalist's unpublished notebook records the capture of _Heliconius_ chari-tonia by a Petchary (_Myiarchus_, a tyrant flycatcher):

188

One fine summer evening, on the seacoast of St. Ann I saw
one of these birds dive from a high perch and snap up a
Zebra butterfly that had been flying close to the ground.
The bird flew back to its perch with the insect in its
beak but it had hardly settled there before it ejected the
butterfly with a violent shaking of the head. The un-
pleasantness that its capture had caused the Pitchaire was
very obviously seen in its subsequent agitation.

(quoted by Brown and Heineman, 1972).

The butterflies are obviously attacked by birds in the
wild: the Oxford University Museum has quite a few Heliconius
in the beak-mark collection, and there are many other such
specimens in European museums. Heliconius erato, although not
H. melpomene, are lethal when injected into mice (Marsh and
Rothschild, 1974); whether they are unpleasant to mice when
ingested is not known, but the readiness with which the mice
in the Stony Brook greenhouse consume my stock of both species
suggests that they are not! But erato at least clearly contains toxins, and there can be little doubt that Heliconius
are distasteful and warningly coloured; the parallel resem-
blances between the species are examples of muellerian mimicry,
in which both species are mutually protected because, roughly,
they share the load of death required in each generation to
educate and re-educate their predators.

As the keystone of muellerian mimicry is the repetition of
an identical warning to the would-be predator, we can confidently predict that deviants from the norm will tend to be
sampled by predators, and that such warning patterns are subject to normalizing selection (Fisher, 1958). In accord with
this, polymorphism is rare in Heliconius (excluding hybrid
zones, only seven species) and marked sexual dimorphism is
found only in species in which males and females fly in

different habitats (notably nattereri and probably demeter) (Turner, 1966; Brown, 1972; Brown and Meilke, 1972; Brown and and Benson, 1974). Costa Rican erato released with the red wing band blacked out were recaptured significantly less than controls carrying the same weight of dye on the black part of the wing; this strongly suggests normalizing selection by predators over the period of the experiment (Benson, 1972). Knowing the general nature of selection on these patterns is an important starting point for what follows.

PARALLEL RACES AND PARALLEL SPECIES

One of the major features of evolution has been adaptive radiation within taxonomic groups, accompanied as a matter of course by functional convergence between the members of different groups. Mimetic patterns show this effect within the confines of the genus Heliconius. In Table 1, a selection of species from the genus is subdivided on the left according to the morphology and internal anatomy of the butterflies and their pupae (the conventional taxonomic characters), and along the top according to the colour patterns. Now on first acquaintance with mimicry in Heliconius one is tempted to assume that the resemblance is due simply to common ancestry: it is not at all unusual for related species in any group to appear very similar. Common ancestry is probably the full explanation in some cases: for instance clysonymus and hortense are closely related morphologically, similar in pattern, and allopatric. They can hardly be mimics. But in general this cannot be so. To fill so many cells of the table has required divergence within phyletic groups and convergence between them. Note that melpomene and erato, paralleling one another so closely, are in separate subsections.

TABLE 1. *Convergent/divergent evolution in Heliconius. Only a few of the 50-odd species are included.*

Pattern (mimicry group)

Morphology (taxonomic subgroup)	Orange and yellow (radiate)	Red, sometimes yellow bars	Red (blotches) and yellow	Blue and yellow	Blue and white	Black and white	Tiger	Ithomiine or acraeine †	Black and yellow A	Black and yellow B (non-mimetic) †
Eueides	tales / eanes (red)	--	procula / tales	eanes (black)	--	--	isabella / lampeta	procula / pavana / vibilia	--	--
Neruda	aoede*	--	--	metharme	--	--	--	godmani	--	--
Laparus	doris (red)	--	doris (red)	doris (blue)	--	--	--	--	--	doris (green)
Heliconius Ia	erato*	erato* / hermathena*	clysonymus* / hortense*	sara* / leucadia / sapho	erato* / sapho* / eleuchia	antiochus	--	charitonia*	hewitsoni / sara	charitonia* / hermathena*
Ib	demeter*	--	ricini	--	--	--	--	--	--	--
IIa	melpomene* / timareta	melpomene*	timareta	timareta	cydno*	--	ethilla* / numata / nattereri ♀*	cydno*	pachinus	nattereri* ♂
IIb	elevatus	besckei*	--	--	--	luciana	hecale	atthis*	--	luciana*
III	xanthocles	--	--	--	--	--	--	hecuba	--	--
IV	burneyi / egeria / astraea	--	--	wallacei*	--	--	--	--	--	--

* Species discussed in text   † Heterogeneous group

The simple explanation of this pattern of evolution within the genus is that it must result from taking one step further the situation seen in melpomene and erato; if the races of both these were to become full species we would see just such a series of paired, mimetic species as we get in two rows of Table 1 (if all the other similar species, around nine in all, did the same, they would generate a whole new table). Being uniformitarian and parsimonious, we can postulate that whatever events led in the recent past to the formation of parallel races, led once or more in the more distant past to parallel speciation.

GLACIAL REFUGES IN THE TROPICS

An obvious cause for such a repeating cycle of speciation is the recurrent glaciation of the quaternary. When I suggested this in 1964 (Turner, 1965) the view was not popular, although it had been canvassed for other taxa (Fox, 1949), as it was apparently felt that the tropics were not affected. With a swing of the scientific pendulum, the use of glacial refuges is now de rigueur in South American biogeography. It seems that during glaciations on higher ground and at higher latitudes, the tropical lowlands both of Africa and South America were subject to changes both of rainfall and temperature which periodically restricted the rain forest (now almost continuous in Amazonia) to 'islands' surrounded by savannah. The most likely correlation is of cool dry spells with glacial maxima, and warm wet ones with hypsithermals, but cool wet and warm dry spells are also a possibility.

The approximate positions of the forest refuges from which the races of erato and melpomene are derived can be deduced by considering simultaneously the present distribution of the races and the current rainfall pattern (as an indicator of which areas probably remained wet during dry periods) (Figure 2) (Brown, Sheppard and Turner, 1974).

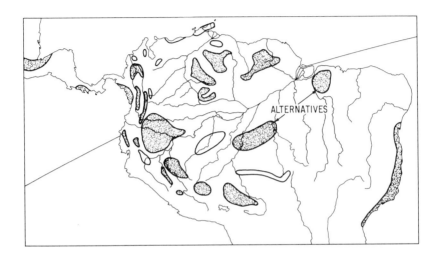

FIGURE 2. *Quaternary forest refuges deduced from the distribution of Heliconius races. Shaded refuges from melpomene and erato, remainder from other species. From Brown, Sheppard and Turner (1974).*

The hypothesis is thus that the parallel races have formed in glacial forest refuges roughly as shown, and that previous repetitions of the glacial cycle led to parallel speciation (location of refuges untraced and probably untracable, but quite possibly in roughly the same positions at each cycle). Two lines of evidence support this. First there is a close correspondence in the positions of the refuges deduced from the distributions of Amazonian birds, mammals, reptiles and plants, as well as heliconian butterflies (Haffer, 1969; Müller, 1972; Vanzolini, 1970); specifically we can derive most of the erato and melpomene races from the refuges which

Haffer (1969) proposed for Amazonian birds (the butterfly and bird refuges also match closely, but this is in itself not an independent test, as we knew the bird refuges before plotting those for the butterflies) (Brown, Sheppard and Turner, 1974).

Second, it is presumed that having differentiated in the refuges, the races have now spread with the expanding forest until they touch borders and hybridize. In that case, the suture zones should be placed where some physical or ecological barrier has slowed the spread of the species. The hybrid zones on the whole are not well known, but in three cases this prediction is confirmed. The major rivers of Amazonia limit racial distributions in several species around Manaus, and two races of _erato_ are separated by a narrow strip of savannah along the crest of the Serra do Parecis in southern Brasil (Brown and Mielke, 1972). Perhaps the most revealing is the suture zone for _melpomene_ and _erato_ in the Guianas, which is the best known because of the commercial export of 'varieties' from the hybrid populations (Joicey and Kaye, 1917); this lies at right angles to all the rivers, parallel to the coast, in a forest interspersed with small savannahs and patches of second growth, and had always seemed to have a quite arbitrary position. Not until the recent work of Janzen (1974) did it become plain that the white sand, along which this zone lies, is a major ecological barrier. White sands, according to Janzen, are exceptionally poor in nutrients even for tropical soils, and their rain forest, although similar in facies to other rain forests, is very poor in all insects, and in vines, including _Passiflora_, the food-plant of _Heliconius_. In the Guianas, the races of _melpomene_ and _erato_ from Brasil lie to the south, and those from Venezuela to the north of this essentially hostile environment.

194

GENETIC ISOLATION <u>VERSUS</u> EXTINCTION CYCLES

So far the picture is conventional: allopatric race formation leads to allopatric speciation. But there is one
obvious paradox. A <u>Heliconius</u> with a successful bad brand
image ought to keep it, and yet the races have undergone an
extraordinary diversification in the face of this stabilizing
selection (Turner, 1973). At this point conventional wisdom
about speciation fails. There is nothing that stopping gene
flow between the isolated populations of the different refuges
could do to overcome stabilizing selection on the patterns.
Genetic isolation <u>in itself</u> cannot account for the geographical
races (Brown <u>et al</u>. 1974).

Further, if restriction of gene flow were effective in this
way, races should be forming all the time throughout the
species' ranges, for <u>Heliconius</u> butterflies tend to have very
low vagility, as has been shown by mark-release-recapture
experiments on <u>erato</u> and on <u>ethilla</u> (a very close relative of
<u>melpomene</u>) (Turner, 1971c; Ehrlich and Gilbert, 1973); there
can be very little gene flow between populations separated by
only a few kilometers. It is now recognized that in general
much genetic differentiation can occur in the face of gene
flow (Bradshaw, 1971), and that conversely many species keep
a constant facies over areas so large that gene flow cannot
be the cohesive force (Ehrlich and Raven, 1969).

Now it is always possible that the patterns, or genes, have
selective effects other than mimicry (the colour of a butterfly is known to affect its rate of heating in sunlight for
instance — Watt, 1968); the divergence therefore could be due
to 'pleiotropism'. The trouble with this hypothesis is that
even if true it is unhelpful, for like special creation (a
clear alternative explanation of <u>Heliconius</u> patterns, which

suggests the fascinating possibility that the Creator is a stamp-collector) pleiotropism can explain anything.

Sheppard has proposed that we analyze the evolution of these butterflies using an extended version of the model of the evolution of muellerian mimicry developed by Fisher in The genetical theory of natural selection (Sheppard et al. 1975). In sum, this model predicts three main outcomes of selection on warningly coloured butterflies. First, a pattern differing considerably from all others is subject to nor-malizing selection and remains unaltered. Second, a pair of species which are sufficiently similar for predators sometimes to mistake extreme forms of one for the other, will converge by the selection of small genetic differences until they become muellerian mimics. Third, if a species which is not very well protected by its warning colour, either because it is not particularly distasteful or because it is scarce and hence not well known to the predators, can produce a mutation which resembles a better-protected species, then this mutation will be at an advantage to its wild-type allele and will replace it. The less protected species can then converge to the other one by means of a single mutation, even though the initial resemblance is not great. Very different patterns will tend not to converge, because this would require several simultaneous mutations.

Examples of this sytem of evolution taking place, given by Sheppard et al. (1975), include convergence by an apparent single mutation in Heliconius hermathena, and what is probably an example of polygenic convergence in H. charitonia. The Amazonian savannah species, H. hermathena, is undergoing race formation in a habitat which is divided into 'islands' by the tropical rain forest. Through most of its range it keeps its own unique pattern (Figure 3, top left), being apparently not

sufficiently similar to any other sympatric species (for
example the race of H. erato in Figure 3, top right), to converge to them. In one small area on the lower Amazonas it
flies with the races of erato and melpomene, shown at the
bottom right in Figure 3. Convergence to these is a simple
matter of removing the yellow bars, which it does by what
appears from population frequencies to be a single dominant
mutation (Figure 3, bottom left) (Brown and Benson, 1975b).

FIGURE 3. *Heliconius hermathena (left) does not mimic H. erato*
(right) when their patterns are different, even
though they fly together (above), but loses its
yellow bars to become an erato-mimic in the limited
area of the Amazon where the patterns are already
similar (below).

The Zebra (H. charitonia) flies without any other Heliconius
through much of its range in México, the U.S.A. and the
Antilles. Further south it encounters no heliconian remotely
like it in pattern. Throughout this range its pattern varies
only in minor, quantitative characters. In Ecuador and Perú,
in a moderately rich heliconian fauna, it has diverged quite
markedly in several ways from this pattern (even changing the
aspect ratio of its wings) to mimic H. atthis and a member of

another family, _Elzunia_ _pavonii_.

It is to be expected that a species which loses it habitual comimic may converge to a third species:

There is one area in which _erato_ does not mimic _melpomene_, but the related species H. _cydno_ (the race of _erato_ is third from top left in Figure 1). In this valley, _erato_ has lost _melpomene_, which is absent for undetermined reasons, and _cydno_ has also lost its usual comimic H. _sapho_ (with which it forms parallel races analogous to the _melpomene/erato_ situation); the value of their usual mimetic patterns thus weakened, _erato_ and _cydno_ turn to mimicking one another. In Panamá by contrast, all four species can be found flying together, for example along the Pipeline Road above Gamboa in the Canal Zone; here they belong to two numerically strong mimicry groups, _melpomene_ being bound in mutual mimicry to _erato_ (butterflies top left in Figure 1), and _sapho_ to _cydno_ with a quite different shared pattern of blue and white.

According to this model then, the driving force in recent race-formation in the forest refuges (and by uniformity in previous cycles of species-formation) has been differentiation of the accompanying butterfly fauna. In some instances _melpomene_ and _erato_ have apparently converged to other _Heliconius_ (many species of which share their pattern in the Amazon basin), and in others to members of other families (Acraeidae in Perú, Pieridae in Venezuela). Identifying the other convergent species is a thing still undone. In some instances, the actual pattern evolved, although not the extremely close resemblance between species, may result from selection on physiology. Thus the forms with the twin white marks on the forewing fly at over 1100 metres in the Ecuadorian Andes, and Descimon and Mast de Maeght (1971) observing that these butterflies are able to fly very suddenly in the short periods of sunlight in a predominantly cool damp climate suggest that this dark pattern has adaptive thermal properties. Selection for mimicry and selection for physiological properties of the colour pattern could clearly occur simultaneously, and are not to be treated as exclusive, rival explanations.

Now why does not the same objection hold to this model as to restricted gene flow: that races ought to form all the time in the continuous forest, for surely the fauna is always changing? Some races of course may have formed in this way (particularly in other species I am not discussing), but in general this is not happening because faunal changes in a continuous biome are 'temporary' (of the order of tens or hundreds of years) whereas the forest refuges act like true islands, subject to the classical colonization/extinction cycle (Mayr, 1965), and hence to faunal differences that last as long as the 'island' persists (of the order of 10^3 or 10^4 years). As convergent evolution between rather different patterns must depend on the occurrence and establishment of mutations (which may not be plentiful in Heliconius because of small local population size) (King, 1972), divergence/convergence is much more likely in refuges than in continuous forest.

Thus the crucial role of isolation in race and species formation in Heliconius is not the stopping of gene flow, but the mutual isolation and hence differentiation of 'island' faunas. Speciation here is an ecological rather than genetical process (cf. Sokal, 1973).

THE EFFECT OF FAUNAL POVERTY

If the evolution of muellerian mimicry depends upon a fairly close initial resemblance between the converging species, then it should be less likely to take place where there are fewer species, as the chance of initial resemblance clearly becomes greater where a large number have to partake of a large and innumerable but limited range of potential patterns.

H. erato is deprived of melpomene in two further areas, at the extreme north and south of its range; in contrast to the area already discussed where erato converges to cydno, in these places the pattern of erato remains unaltered, being here deprived of most other members of the genus with which convergence might occur (Sheppard et al. 1975).

Similarly, most of the heliconians (that is Heliconius and related genera) which inhabit open country have not formed races, or vary only in quantitative characters, despite the fact that savannah refuges should act as 'islands' during forest maxima. Benson (1971) attributes this to their greater vagility, but this can hardly be the correct explanation, just as the low vagility of Heliconius is certainly not the explanation of race formation. In fact the cause is probably in part the relative species-poverty of the savannah butterflies, and in part, and for some species only, the fact that they may not be confined to the savannahs, even though the butterfly-hunter seldom sees them in the forest. The widespread species Dryas iulia, which is comparatively uniform except for minor island races appreciable only to the specialist (Emsley, 1963), is not merely an open country butterfly, but flies also over the canopy of the rain forest (Papageorgis, 1974). It, and its comimic Heliconius aliphera, and possibly other species with this habitat, thus probably form a continuous population at all times regardless of the relative size of forests and savannahs, and are thus not subject to extinction cycles in either habitat.

Faunal poverty may account also for the rarity of butterfly

mimicry at higher latitudes.* Certainly the only temperate
species comparable to Heliconius presents a simpler picture.
The European moth Zygaena ephialtes belongs to two mimicry
groups, the northern race copying other members of its own
genus, the southern the Ctenuchid moth Amata phegea; the moths
are all distasteful and the mimicry effective to caged birds**
(Bullini, Sbordoni and Raggazzini, 1969). The hybrid zone
between the two races lies along the main fold of the mountains in Central Europe, but at least in Austria and Czecko-
slovakia is displaced to the southeast of the main watershed,
so that it may not correspond with an ecological barrier
(review of work by Reichl and by Povolný and Gregor, in
Turner, 1971a). The races have probably derived from glacial
refuges (de Lattin, 1952), in one of which ephialtes switched
from its ancestral Zygaena pattern and colour to that of
Amata, but the present distribution is difficult to interpret
because of the presence, in the southern race, of a form with
the pattern of Amata but the colour of Zygaena (red instead
of yellow). This extraordinary insect seems to be getting the
best of both worlds by mimicking both models, undergoing a
seasonal change in behaviour according to which is in flight!
(Sbordoni and Bullini, 1971). Rothschild, von Euw and
Reichstein (1973) believe that the distribution of the perfect

---

*This possibility was first suggested to me, in conversation,
by Mr. T. Mueller.

**Close reading of my review of these experiments (Turner,
1971a) may suggest to the reader that they were not adequately designed for the conclusions drawn. By a foolish
error I have misdescribed them, lumping two separate groups
of experimental birds; the design of the experiments is not
faulty.

Amata-mimic is controlled by the relative abundance of Amata and of other Zygaena species, a reasonable hypothesis but conjectural until the distribution of Amata is better known; as Bullini et al. (1969) point out, it is difficult to disentangle historical effects of the refuges from current ecology; mutually mimetic Heliconius races can certainly have rather disparate distributions today, as can be seen from certain forms of melpomene and erato.

At any rate, Zygaena ephialtes can consistently be seen as a simplified version of evolution on the same lines as Heliconius.

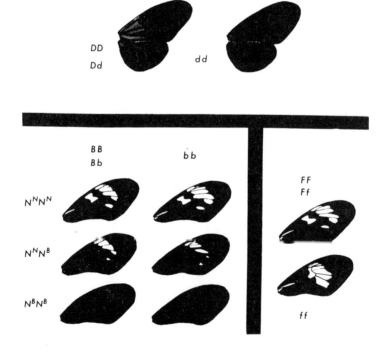

FIGURE 4. Genetics of the difference between the Suriname and Trinidad/Venezuela races of H. melpomene (after Turner, 1971b).

THE RATE OF GENE SUBSTITUTION

In accord with Sheppard's model, which suggests that the divergence of races in _melpomene_ and _erato_ results from 'switching' from one warning pattern to another, our genetical experiments show that in both species the races are differentiated by a rather small number of major loci, each altering an element of the pattern (Turner, 1972; Sheppard _et al._ 1975); the differences between two races of _melpomene_ are shown in Figure 4. The two races are both homozygous for alternative alleles at the loci illustrated. Thanks to increasing knowledge of the quaternary in South America, we can now assign tentative dates to the divergence of the races, and hence establish the rate of evolution.

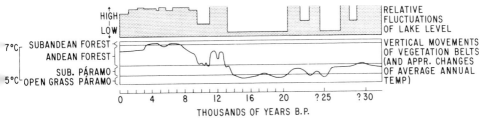

FIGURE 5. _Rainfall and approximate mean annual temperature, deduced from lake levels and pollen profiles at Lago de Fuquene in the Colombian Andes. The short cold dry period (11 - 9½ thousand years BP) is the El Abra Stadial, the longer one ending 13 thousand BP is the Fuquene Stadial. From van Geel and van der Hammen (1973)._

Figure 5 shows changes in mean annual temperature over the past 30,000 years, deduced from pollen profiles, carbon dating and sedimentology at the Lago de Fuquene, north of Bogotá (van Geel and van der Hammen, 1973). The climate has altered from warm and wet, giving maximum extent of the forest to cold

203

and dry, corresponding to forest minima. Figure 6 shows on the right, the simplest way of fitting the pattern of race formation in melpomene (and by analogy in erato) into this chronology.

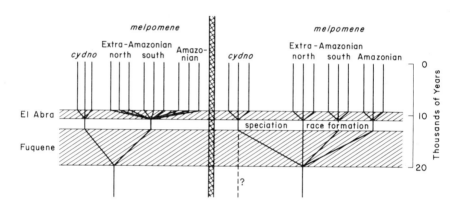

FIGURE 6. *Two of the many possible evolutionary trees connecting the races of* H. melpomene *and the related species,* H. cydno. *Cold dry periods shaded. The most rapid rate of evolution occurs in the left hand tree, where all the races of* melpomene *evolve during the short El Abra Stadial, and the speciation of* cydno *and* melpomene *occurs in the longer Fuquene Stadial. On the right, race formation in* melpomene *is a two-stage process, and* cydno *has either speciated earlier than the Fuquene Stadial (dotted line), or speciates while* melpomene *undergoes its first cycle of race formation.*

It is assumed that the races of both species which occupy the Amazon basin separated rather early from the races which now lie round the periphery of the basin, from which they differ in having an outer yellow band on the forewing and extensive orange radiate marks on both wings, as against the pattern with the red band on the forewing, with only yellow bars or

lines on the inner parts of the wings (see Figure 1). As the
differences between the Amazonian and extra-Amazonian races
are produced by an average of 4.4 complete substitutions at
the half-dozen thoroughly studied loci in each of the two
species (Table 2), whereas the differences between different
Amazonian and different extra-Amazonian races average only
2.2 substitutions, the easiest model assumes that race formation has occurred in two 'pulses', the first and longer
pulse, which took place in the Fuquene Stadial, separating
the Amazonian from the extra-Amazonian races, and the second
and shorter during the El Abra Stadial separating the individual modern races within these large groupings (as the extra-
Amazonian races are distributed in a huge 'C' around the Amazon basin, one has in addition to assume that at least the
northern and southern halves also separated early, but
diverged little). As the El Abra Stadial lasted about $1\frac{1}{2}$
thousand years, and the Fuquene about 7 thousand years, this
gives rates of allelic substitution of 4.4 in $2(7 + 1\frac{1}{2})10^3$
years or 0.26 substitutions per thousand years for the whole
period, and 2.2 in $2(1\frac{1}{2})10^3$ years or 0.73 substitutions per
thousand years for the El Abra only. The butterflies have
grossly overlapping generations, being immature larvae and
pupae for only 3 weeks and fertile butterflies for up to six
months, but something between 3 and 17 generations a year, say
about 10, seems a reasonable guess. The rate of substitution
for the whole period is thus one substitution in 386 thousand
generations. This it must be remembered is not an estimate
for the genome, but only for half a dozen loci involved in
adaptive muellerian mimicry.

By altering one's assumption, one can make the rates of
substitution slower or faster. The left hand side of Figure
6 shows the most rapid possible scheme.

TABLE 2. Number of loci differentiating colour patterns in races of Heliconius. A total of six loci is considered in each species. Data from Sheppard et al. (1975).

H. melpomene			
Amazonian		Extra - Amazonian	
Belém (5)	Suriname (4)	Trinidad (3)	Espirito Santo (11)
Belém		2	$\underset{\sim}{5}$	$\underset{\sim}{5}$
Suriname			$\underset{\sim}{4}$	$\underset{\sim}{4}$
Trinidad				2

H. erato				
Amazonian		Extra - Amazonian		
Manaus (4)	Belém (5)	Panamá (1)	Trinidad (4)	Southeast Brasil (12)
Manaus		2	$\underset{\sim}{4}$	$\underset{\sim}{3}$	$\underset{\sim}{6}$
Belém			$\underset{\sim}{4}$	$\underset{\sim}{3}$	$\underset{\sim}{6}$
Panamá				1	3
Trinidad					3

Underlined figures are comparisons between Amazonian and extra-Amazonian races. In melpomene I have ignored the C locus, about which there is only limited information. In erato, I have assumed that the D and R loci are identical (which is not proved genetically), and ignored the factors differentiating orange and red pigment. Numbers in brackets refer to the illustrations in Figure 1, numbering off from the top left.

It has the attractive feature of making race formation take place in the short stadial, and full speciation from a close relative in the longer period; it gives gene substitution rates of 1.47 per thousand years, or one allele in 68 thousand generations. Squashing the events into this time scale is awkward though. The problem of having the extra-Amazonian and Amazonian patterns diverge rapidly can be overcome by assuming that several very rapid faunal changes took place in the

appropriate refuges so that there was rapid, step-wise divergence of the patterns, or that the seemingly impossible
happened, and the patterns did diverge at a single step, or at
least in a single process. All that is required is a
'catalyst' species, with a pattern intermediate between the
two (Sheppard et al. 1975). Figure 7 represents the possible
past situation in an Amazonian refuge. The very distinct
pattern of melpomene and erato (left) can be made to converge
toward the majority pattern (right) by the presence of a
butterfly with an intermediate pattern (in the figure, a
relict, non-mimetic member of the elevatus group, IIb, which
may indeed have been such a catalyst). However, the prime
difficulty with this system is that melpomene and erato,
having speciated from their nearest relatives during the
Fuquene, are supposed to have spread throughout South America,
so as to enter all refuges in the El Abra, using the mere 2
thousand year interstadial. Faunal invasions have of course
taken place in less time!

FIGURE 7. Possible example of a 'catalyst' pattern (centre)
whose presence could cause two very different patterns (left
and right) to converge. This may have been the situation in
the Amazonian refuges, causing melpomene and erato to join
the predominant rayed group. The butterflies illustrated
(and their present locations) are from the left, melpomene
(or erato) (the relict population on the Rio Huallaga),
H. (elevatus) luciana (the relict population in southern
Venezuela), H. aoede (Amazon basin). All species now have
the right hand pattern in the Amazon basin.

STABILITY OF HYBRID ZONES

On either the slow or the fast time scale, the hybrid populations of melpomene and erato, formed by the meeting of races,
have existed at least since the start of the hypsithermal,
around 8000 BP, or for between 17 and 136 thousand generations
(we know for certain that they have existed for a mere 200
years, from early illustrations of hybrids - Turner, 1967,
1971b). This is an awkwardly long time for evolutionary
theory: either the races should have fused over a wide area,
or the Wallace effect (selection against hybrids - Murray,
1972) should have caused full speciation. There is a simple
explanation: as the hybrid zones occur at ecological barriers,
the hybrid populations form on either side of the barrier,
where the hybrid butterflies are a minority and are removed by
stabilizing selection on the patterns (all hybrid populations
of melpomene known in Suriname have a phenotypic composition
which agrees with this - Table 3). This prevents fusion, and
there is of course little that selection against hybrid phenotypes can do to prevent intermittent migration across the
barrier.

TABLE 3. Hybrid populations of H. melpomene on either side of
the ecological barrier in Suriname (Guianas). Several other
populations in the same region are known from very small
samples, to have similar gene frequencies. None is known with
a gene frequency of 50%. From Sheppard (1963) and Turner
(1971b).

Locality	Amazonian alleles	Extra-Amazonian alleles	Sample size	No. of loci detected
Moengo, north of the white sand	6 %	94 %	148	2
Brokopondo, south of the white sand	95 %	5 %	13	4

In _Zygaena ephialtes_ populations are spread throughout the hybrid zone (width about 80 km), and the puzzle remains. There is some evidence of coadaptation, in that the clines at the two loci affecting the colour pattern are displaced from one another by about 10 kilometres, suggesting natural selection against one of the recombinants (Reichl, 1958).

GENETIC ARCHITECTURE

The genetic studies of _melpomene_ and _erato_ (Sheppard _et al_. 1975) show that the mimetic patterns of the different races have evolved by what is now called H.J. Muller's 'classical' or 'beanbag' system. The races are homozygous for alternative alleles, selective substitution having taken place apparently in tandem, most of the loci being in separate linkage groups. That this simple and, as many geneticists now think, unusual form of evolution does not necessarily preclude cryptic polymorphism at these same apparently monomorphic loci is shown by _Zygaena ephialtes_, in which the 'allele' producing the yellow colour of the southern race is in fact two alleles, apparently showing a widespread polymorphism, and distinguishable by being respectively completely recessive to red, or producing an orange heterozygote (Bovey, 1966).

It is predicted by standard population genetics theory that beanbag evolution of this kind will overwhelmingly consist of the substitution of new, adaptive mutations which are dominant or partly dominant to the old wild-type; new recessive mutations have too high a probability of extinction in the generations after they appear, and initially advance too slowly under selection (Crow and Kimura, 1970). While it is clearly impossible to prove that the substituted genes have been dominant in these species, Sheppard and Turner have

recently found an aesthetically pleasing fact fully in accord
with this idea (in Sheppard et al. 1975). Both melpomene and
erato have close relatives which are non-mimetic, and which,
presumably evolving at a slower rate, are held by systematists
to represent the 'primitive' pattern. All have patterns of
yellow, longitudinal bars; the already-mentioned H. chari-
tonia and hermathena (Figure 3) are two of them. Now if one
writes out a genotype consisting of the recessive alleles at
all known loci, independently for melpomene and for erato one
comes up with a similar pattern of yellow bars! The suggestion is strong that the two species have been mimetic for a
very long time, and have evolved in parallel by the substitution of dominant genes.

As expected with beanbag evolution, the loci concerned are
mainly unlinked, or if linked have rather high crossover rates,
indicating little, if any, coadaptation. There is one very
significant exception. The pattern with a red forewing band
and a yellow hindwing bar (top left, bottom right in Figure 1,
both species) has a wide, disjunct distribution which strongly
suggests that it is close to the 'primitive' pattern at the
start of the last round of race formation (there is genetic
evidence for this as well). If so, then the yellow band of
the Amazonian races is secondary, and yet the genes for this
character are recessive to those for the red band. In both
species, the locus concerned is linked to the one whose dominant allele produces the extensive basal orange marks of the
Amazonian races. The recessive mutations have apparently
become established by 'hitch-hiking' on the chromosome that
was carrying a new, successful dominant mutation (Sheppard
et al. 1975).

The fact that both species had loci available in the same
linkage group may indicate that these are homologous descen-

dents of loci linked in the now very remote common ancestor, especially as the haploid chromosome number of the subgenus Heliconius is virtually invariant at 21, suggesting a rather small amount of chromosomal rearrangement (Suomalainen, Cook and Turner, 1972). This has been suggested as the explanation for a similar parallel linkage of genes in the Ecuadorian races of melpomene and erato (Emsley, 1965).

ADAPTIVE RADIATION

Therefore in Heliconius adaptive radiation of the mimetic colour patterns is from the first an intimate part of race and, presumably, species formation. It does not lag behind as an outcome of genetical isolation, but is one of the first processes to take place.

It is possible, and the question is much in need of investigating, that the divergence of the patterns in itself constitutes a potential isolating mechanism. One of the major signals in the courtship of erato (Trinidad population) is its own brilliant red colour (Crane, 1955); there is circumstantial evidence from the spectral response of fibres deep in the optic tract of H. sara and H. ricini that these species may respond to their own yellow colour (Swihart, 1967), and a brief observation on H. besckei also suggesting a response to yellow (Emsley, 1970). Change of mimetic pattern may thus contribute heavily to prezygotic isolation. However, the existence and persistence of the natural hybrid populations in melpomene and erato show that no strong isolating mechanisms have accompanied what is to us a spectacular phenotypic divergence between races. Isolating mechanisms seem to be lagging behind divergence a little at the species level as well: H. cydno hybridizes with H. melpomene in at least one area,

although the two look very different. Either the two separated a long time ago, and isolating mechanisms have been slow
to evolve (dotted line in Figure 2), or, less conventionally,
there has been rapid speciation between the two, leaving the
isolating mechanisms incompletely developed, possibly at the
same time as melpomene was undergoing race formation (solid
line in Figure 2).

Mimicry 'rings' (groups of mutually mimetic species) are in
some ways analogous to ecological niches. We may conclude
that change of ecological niche, mediated especially by the
faunal changes which island habitats undergo as a result of
the colonization/extinction cycle, can be an important and
early factor in causing racial divergence and after that
speciation. When the 'islands' are formed as in this case by
splitting a continuous 'land mass', founder-effects are unlikely to be very important. When the islands are newly
emerged and being colonized, then genetic differentiation by
founder effect and faunal differentiation by colonization/
extinction will both occur, and as is usual with drift and
selection processes, their effects will be closely similar
and difficult to disentangle. Colonization/extinction of
island flora and fauna is an alternative (but not exclusive)
explanation to genetic drift of the genetic differentiation of
Philaenus spumarius on islands in the Gulf of Finland described by Halkka et al. (1970). Small and outlying islands
will be the more differentiated according to both models, as
they take a smaller sample from the available faunal or genetic population.

It is also becoming clear that different parts of the
genome can evolve at different rates and in different directions. Both Pupfish and men are evolving more quickly in
loci controlling external phenotype than in allozymes or anti-

gens (B.J. Turner, 1974; Cavalli-Sforza, 1974) (the explanation that anatomists are biased by economic racism into
overstressing racial differences may apply to men, but hardly
to Pupfish — Lewontin, 1974). As we can easily anticipate
with a holometabolous insect, Heliconius larvae are evolving
in a direction different from the adults. The larvae are protected by the nauseous mulch of Passiflora in the gut, if by
nothing else; they are warningly coloured, and show muellerian
mimicry (Table 4). Like the adults, the larvae have evolved
their mimicry across the anatomical groups, but in addition,
the convergence follows a different scheme from the convergence in the butterflies; the adults of aoede and wallacei,
and of hermathena and P. pygmalion (mimetic as larvae), are
totally dissimilar!

TABLE 4. Convergent/divergent evolution among heliconiine
larvae. Considerably revised from Turner (1968), using Brown
and Holzinger (1973) and Brown and Benson (1975a,b).

Morphology (taxonomic)	Purple	White, black spots	Black, white and red transverse stripes	Yellow
Eueides	---	tales	---	---
Neruda	aoede	---	---	---
Heliconius I a	sara	erato charitonia	hermathena	---
I b	---	demeter†	---	demeter † ricini
II a	---	melpomene	---	---
IV	wallacei burneyi	---	---	---
Philaethria*	---	---	pygmalion	---

* Separate genus
† Different subspecies

It would of course be taking things too far to say that
Heliconius upholds the concept of 'genetic revolution'; half
a dozen loci affecting colour are hardly a revolution. That
one part of the genome evolves in this way gives us no ground
for believing that the whole genome does so. At first guess,
one would expect allozyme loci in these butterflies to be
behaving in the same way as the allozymes of the South American Drosophila which have surely passed through the same
forest refuges (Ayala et al. 1974). Our preliminary results
(Turner and Eanes, unpublished) show that this is so: Heliconius races are not in general homozygous for alternative
alleles at allozyme loci.

KINDS OF SPECIATION

Lewontin (1974) proposed that speciation takes place by
the rapid establishment of isolating mechanisms, involving a
few genes, followed by very slow genome divergence at the
levels of race and species; from the context he seems attracted
to this model because of the postulated desirable political
and social theories that it would entail.

It is always tempting, and usually necessary to assume that
findings with one species apply to all others. This is particularly so with speciation. Once we had platonic species.
When they collapsed, we tried for a platonic definition of the
species category (in terms, usually, of genetic isolation).
But there clearly is no single, ideal definition of 'species',
and the ways in which species are formed are probably very
diverse too. There is no single process of 'speciation'.
Similarly the term 'race formation' covers a multitude of
sins, and we can hardly expect the genetics of 'races' in
these butterflies, evolved through the effects of the diver-

gence of 'island' faunas, to be the same as the genetics of human 'races', evolved under very different circumstances and showing no evidence of 'beanbag' homozygosity for alternative alleles.

Prakash (1972) presents a convincing case for early genetic isolation by founder effects in Drosophila pseudoobscura (he does not state it as strongly as Lewontin). As far as the part of the genome concerned with pattern goes Heliconius presents us with a radically different mode of speciation. Its characteristics can be summed up as

1   a very early occurrence of adaptive radiation

2   the ecological effects of the colonization/extinction cycle, rather than stoppage of gene flow, as the driving force

3   'beanbag' evolution at a number of unlinked loci, sub-stituting new dominant alleles, probably one at a time, leading to races being homozygous for alternative alleles

4   the occasional 'hitch-hiking' of a recessive allele linked to a new dominant allele.

ACKNOWLEDGEMENTS

This paper is in large part an abstract of work performed jointly with K.S. Brown, Jr., and P.M. Sheppard. I am much in their debt for discussion and correspondence.

The drawings for this paper are by Miss Joyce Roe, except for the maps in Figure 1 which are by Mr. M. Koello and Figure 4 which is by the author.

REFERENCES

Ayala, F.J., M.L. Tracey, L.G. Barr, J.F. McDonald and S. Perez-Salas. (1974). Genetics 77: 343-384.

Benson, W.W. (1971). Am. Natur. 105: 213-226.

Benson, W.W. (1972). Science 176: 936-939.

Bovey, P. (1966). Rev. Suisse Zool. 73: 193-218.

Bradshaw, A.D. (1971). In: Ecological Genetics and Evolution (E.R. Creed, Ed.), pp. 20-50. Oxford, Blackwell Scientific.

Brower, L.P., J.Vz. Brower and C.T. Collins. (1963). Zoologica (New York), 48: 65-84.

Brown, F.M. and M. Heineman. (1972). Jamaica and its butter-flies. London, Classey.

Brown, K.S., Jr. (1972). Zoologica (New York), 57: 41-69.

Brown, K.S., Jr. and W.W. Benson. (1974). Biotropica 6: 205-228.

Brown, K.S., Jr. and W.W. Benson. (1975a). Bull. Allyn. Mus. 26: 1-19.

Brown, K.S., Jr. and W.W. Benson. (1975b). Biotropica (in press).

Brown, K.S., Jr. and H. Holzinger. (1973). Zeit. Arbeits-gemein. Österr. Entomol. 24: 44-65.

Brown, K.S., Jr. and O.H.H. Mielke. (1972). Zoologica (New York) 57: 1-40.

Brown, K.S., Jr., P.M. Sheppard and J.R.G. Turner. (1974). Proc. Roy. Soc. London B, 187: 369-378.

Bullini, L., V. Sbordoni and P. Ragazzini. (1969). Arch. Zool. Ital. 44: 181-214.

Cavalli-Sforza, L.L. (1974). Sci. Am. 231: 80-89.

Crane, J. (1955). Zoologica (New York) 40: 167-196.

Crow, J.F. and M. Kimura. (1970). An introduction to popu-lation genetics theory. New York, Harper and Row.

Descimon, H. and J. Mast de Maeght. (1971). Alexanor 7: 69-79, 121-133.

Ehrlich, P.R. and L.E. Gilbert. (1973). Biotropica 5: 69-82.

Ehrlich, P.R. and P.H. Raven. (1969). Science 165: 1228-1232.

Eltringham, H. (1917). Trans. ent. Soc. Lond. 1916, 101-148.

Emsley, M. (1963). Zoologica (New York) 48: 85-130.

Emsley, M.G. (1965). Zoologica (New York) 49: 245-286.

Emsley, M.G. (1970). J. Lepid. Soc. 24: 25.

Fisher, R.A. (1958). The genetical theory of natural selection. Second edition. Dover Books, New York.

Fox, R.M. (1949). Bull. Univ. Pittsburgh 45: 36-47.

van Geel, B. and T. van der Hammen. (1973). Palaeogeog. Palaeoclim. Palaeoecol. 14: 9-92.

Haffer, J. (1969). Science 165: 131-137.

Halkka, O., M. Raatikainen, L. Halkka and R. Lallukka. (1970). Ann. Zool. Fennici 7: 221-238.

Janzen, D.H. (1974). Biotropica 6: 69-103.

Joicey, J.J. and W.J. Kaye. (1917). Trans. ent. Soc. Lond. 1916, 412-431.

King, J.L. (1972). In: Proc. 6th Berkeley Symp. Mat. Stat. Prob. 5: 69-100. Univ. of Calif. Press, Berkeley and Los Angeles.

de Lattin, G. (1952). Ver. deutsch. Zool. Ges. 1952: 452-460.

Lewontin, R.C. (1974). The genetic basis of evolutionary change. New York and London, Columbia Univ. Press.

Marsh, H. and M. Rothschild. (1974). J. Zool. London 174: 89-122.

Mayr, E. (1965). Science 150: 1587-1588.

Müller, P. (1972). Studies in the neotropical fauna 7: 173-185.

Murray, J. (J.) (1972). Genetic diversity and natural selection. Oliver and Boyd, Edinburgh.

Papageorgis, C.A. (1974). Ph.D. Thesis, Princeton Univ.

Prakash, S. (1972). Genetics 72: 143-155.

Reichl, E.R. (1958). Zeit. Wien. ent. Ges. 43: 250-265.

Rothschild, M., J. von Euw and T. Reichstein. (1973). Proc. Roy. Soc. London B, 183: 227-247.

Sbordoni, V. and Bullini, L. (1971). Fragment. Entom. 8: 49-56.

217

Sheppard, P.M. (1963). Zoologica (New York) 48: 145-154.

Sheppard, P.M., J.R.G. Turner, K.S. Brown, Jr., and W.W. Benson. (1975). In preparation.

Sokal, R.R. (1973). Syst. Zool. 22: 360-374.

Suomalainen, E., L.M. Cook and J.R.G. Turner. (1972). Zoologica (New York) 56: 121-124.

Swihart, S.L. (1967). Zoologica (New York) 52: 1-14.

Turner, B.J. (1974). Evolution 28: 281-294.

Turner, J.R.G. (1965). Proc. XII Int. Congr. Ent. London, 1964, 267.

Turner, J.R.G. (1966). Proc. R. ent. Soc. London (B), 35: 128-132.

Turner, J.R.G. (1967). J. Linn. Soc., Lond. (Zool.), 46: 255-266.

Turner, J.R.G. (1968). J. Zool. (London), 155: 311-325.

Turner, J.R.G. (1971a). In: Ecological genetics and evolution. (E.R. Creed, Ed.), pp. 224-260. Oxford, Blackwell.

Turner, J.R.G. (1971b). Evolution 25: 471-482.

Turner, J.R.G. (1971c). Biotropica 3: 21-31.

Turner, J.R.G. (1972). Zoologica (New York) 56: 125-157.

Turner, J.R.G. (1973). Animals 15: 15-21.

Turner, J.R.G. (1975). Natural History 84: 28-37.

Vanzolini, P.E. (1970). In: Instituto Geográfico Serie Teses e Monografias, São Paulo. No.3.

Watt, W.B. (1968). Evolution 22: 437-458.

The Genetics of Haploid-Diploid Populations with Vegetative and Sexual Reproduction*

K. WÖHRMANN, R. STROBEL, and P. LANGE

The model is based on the life cycle of yeast, Saccharomyces cerevisiae (Wöhrmann, Lange and Strobel, 1974). The following characteristics of yeast are taken into account:

1. Each cell in the population is a genetically defined individual.
2. The population generally includes haploid and diploid genotypes.
3. Haploids and diploids can reproduce vegetatively by budding.
4. At most four haploid spores appear during meiosis of a diploid cell.
5. Diploids arise by fusion of two spores or two haploids.
6. Mating can occur only between genotypes of different mating types where the mating type is determined by a single locus involving two alleles (a and α) located on the IIIrd chromosome.

*This work was supported by "Deutsche Forschungsgemeinschaft".

7. Under laboratory conditions, a meiosis can only be induced by growing the populations in a special medium, poor in carbon sources (called the sporulation medium).
8. The rate of meiosis (m_i) is under genetic control.
9. Mating strictly occurs under special favorable conditions (complete medium).

Our model has been set up taking account of the foregoing biological principles (Figure 1). In each generation mating and vegetative reproduction during the vegetative phase as well as spore propagation and vegetative reproduction during the generative phase have been considered. In both phases a number of biological forces that influence the genotypic structure and the dynamics evolution of the populations may become operative.

Consider a population composed of haploid and diploid individuals with frequencies (f_i). During the first phase mating between the haploids of the a - and the α - types can occur producing diploid cells heterozygous for the mating type locus. The creation of diploids simultaneously entails a decrease of haploids.

The formation of diploids depends both on the frequency of the haploids in the population, the ratio of the mating type alleles and the mating activities of the genotypes. The level of mating activity is indicated by z_i.

Following the mating phase, haploids and diploids propagate vegetatively by budding. The relative reproduction rates are given by w_i where the heterozygote rate is normalized to unity. After the transfer of the population into sporulation medium further vegetative reproduction v_i can be observed occurring at a reduced rate. Diploids propagating sexually by meiosis do not show any budding. The frequency of the asci

220

(m_i) depends on the genotype of the diploids. The recombination rate is given by c. The relative viability of the various spores is given by u_i.

With subsequent transfer to the complete medium the spores separate from the ascus and by mating of the haploids a new generation is started. The growth of a yeast population in complete medium corresponds to a sigmoid growth curve. The maximal density will be reached in about 20 - 24 hours. By activating chemostates or by transference of the cultures daily into a fresh medium, the cultures may be maintained in the logarithmic phase. In this way the length of the phases may be varied experimentally over a certain specified range. It is possible to distinguish small differences between genotypic fitness values by altering the length of the vegetative phase and also the rate of meiosis. Figure 2 describes the production of asci from different genotypes as a function of time, cultured in sporulation medium. Again it is possible to manipulate the rate of meiosis of the genotypes by changing the length of the generative phase. In particular we find that one genotype has a much higher rate of meiosis after two days, compared to the other genotype, whereas there is no significant difference in genotypic rates of meiosis after six days. The differences described in Figure 2 are statistically significant at the 5% level. In this way it is possible to investigate the influence of distinct rates of meiosis coupled to relative fitness values to discern some genetic properties of a population, using the same genetic material. Moreover, the relative influence of the single phases expressed by the controlling parameters can be examined by varying the relative length of the phases.

In terms of this model a number of experiments have been

conducted (Strobel, 1974; Strobel and Wöhrmann, 1975a, Strobel and Wöhrmann, 1975b). The experiments done by Strobel and Wöhrmann (1972) verified the existence of two esterase loci with two alleles at each locus, differing in their electrophoretic mobilities. It was of some interest to investigate the importance of these alleles, and genes linked with them pertaining to the parameters of the distinct phases. In particular, the competitive ability of the genotypes during the vegetative phase were tested in a number of experiments. In order to reduce the influence of the genetic background, the genotypes used were backcrossed 6 or 7 times with one of the parents.

The yeast populations have been cultured in a Yeast-Extract-Peptone-Medium of 30°C. Each culture vessel was inoculated by two genotypes with the same mating type in a 1:1 relation. In every twenty-four hours 0,1 ml of the population was transferred to fresh medium. In this manner the population was kept in the log-phase for fourteen days. This time corresponds to about 150 generations, if a reproduction time of 120 minutes is assumed.

After fourteen days samples of the populations were put on petri plates and the relation of the different genotypes was determined by means of a disc electrophoresis. In another experiment the cultures were kept at the stationary phase throughout fourteen days. The observed ratios of the genotypes were χ^2 tested for significant deviations from the expected ones of 1:1 . The results are summarized in Table 1. In the first column the competing genotypes are listed, in the second column the ratios of the genotypes after cultivation under stationary conditions are given and in the third column those after daily transfer (14 days) are listed.

TABLE 1. Ratios of genotypes after cultivation for 14 days in the stationary phase resp. logarithmic phase under competition. The expected genotypic ratio is 1:1 .

genotypes	station. phase	χ^2	log. phase	χ^2
Est-1$_s$Est-2$_f$a	22		43	
+		0.40		30.08
Est-1$_s$Est-2$_s$a	18		5	
Est-1$_s$Est-2$_f$a	18		57	
+		0.40		3.80
Est-1$_f$Est-2$_f$a	22		38	
Est-1$_s$Est-2$_f$a	18		41	
+		0.40		20.48
Est-1$_f$Est-2$_s$a	22		9	
Est-1$_s$Est-2$_s$a	17		10	
+		0.90		17.16
Est-1$_f$Est-2$_f$a	23		39	
Est-1$_s$Est-2$_s$a	19		41	
+		0.10		2.04
Est-1$_f$Est-2$_s$a	21		55	
Est-1$_f$Est-2$_f$a	19		40	
+		0.10		18.00
Est-1$_f$Est-2$_s$a	21		10	

It is obvious from Table 1 that continuous cultivation in the stationary phase does not lead to a significant deviation from the starting ratios of the population. However, significant deviations do occur under competition during the log-phase. A significant deviation from the 1:1 ratio is always found in those cases in which the competing genotypes differ in the alleles of the Est-2 locus. The alleles of the Est-1 locus, on the contrary, do not have any measurable influence on the outcome of the competition experiments.

Generally the genotype with the Est-2$_f$ allele is the winner. It should be noted that only one of the loci in question has a detectable influence on the vegetative reproduction rate of the genotypes.

A further experiment has been carried out in order to test the correlation between the esterase loci and the viability of the spores. Generally the yeast forms asci with four spores. There are, however, asci with three or two spores in a defined percentage in the population. Table 2 describes the correlation between the spore viability and the esterase loci. In the first column the haploid genotypes, and the corresponding diploids are listed.

TABLE 2. *The ratio of spore genotypes in the asci with 2 and 3 spores. Expected ratio 1 : 1 : 1 : 1 .*

genotypes	spores/ ascus	$Est-1_s$ $Est-2_f$	$Est-1_s$ $Est-2_s$	$Est-1_f$ $Est-2_f$	$Est-1_f$ $Est-2_s$	Σ	χ^2
$Est-1_s Est-2_f a \times Est-1_f Est-2_s \alpha$	2	19	13	13	17	62	1.74
	3	25	20	21	24	90	0.76
$Est-1_s Est-2_s \alpha \times Est-1_f Est-2_f a$	2	13	17	13	15	58	.76
	3	22	23	19	23	87	0.49
$Est-1_s Est-2_s a \times Est-1_f Est-2_f \alpha$	3	16	17	18	15	66	0.30
$Est-1_s Est-2_f \alpha \times Est-1_f Est-2_s a$	3	10	16	16	12	54	2.00

The spores arising from these genotypes have the genotypic formulas $Est-1_s$ $Est-2_f$, $Est-1_s$ $Est-2_s$, $Est-1_f$ $Est-2_f$, $Est-1_f$ $Est-2_s$. They are expected to occur in equal proportions, if the lethality of the spores is random. Table 2 indicates that there is no significant influence of any of the esterase alleles on spore viability. The total segregation ratio is in agreement.

Regarding the genotypic construction of the populations, the ratio of meiosis is a very important parameter. Apart from the recombination processes during the meiosis, diploids

are able to double their genetic information.   There is some
information about the genetics of the rate of meiosis.

First we check the question whether one of the esterase
loci may determine the rate of meiosis.   Using methods of the
analysis of variance, the Est-1 locus indeed shows an effect.
The results are given in Table 3, indicating a significantly
higher rate of meiosis for the $Est-1_s$ $Est-1_s$ homozygote.
There are no differences between the genotypes of the Est-2
locus affecting the vegetative reproduction rate.

*TABLE 3.   Mean ascus percentages of esterase genotypes.   The
homogeneity is tested by the LSD-test.*

	$Est-1_s Est-1_s$	$Est-1_s Est-1_f$	$Est-1_f Est-1_f$
percentage	0.061	0.045	0.043
homogeneity			

	$Est-2_s Est-2_s$	$Est-2_s Est-2_f$	$Est-2_f Est-2_f$
percentage	0.056	0.059	0.060
homogeneity			

Thus the gene block, marked by the Esterase-2 locus does
appear to influence the reproduction rate and the Esterase-1
locus influences the rate of meiosis.   It is of interest to
investigate the behaviour of these loci and chromosomal segments linked to them in a population where a change between
sexual and asexual reproduction takes place.

To study these problems five populations have been cultivated for thirty generations.   Two founder genotypes with an
opposite mating type were always inoculated in the same culture vessel.   Corresponding to the generation scheme (Fig.1)

the frequencies of the genotypes and genes were estimated at the 1st, 10th, 20th and 30th generations at the termination of ascus formation. The results are presented in Figures 3-7. The x-axis corresponds to the generation number, the y-axis to the frequencies of the genotype. Altogether 17 haploid and diploid genotypes can be distinguished. In order to avoid ambiguity only those genotypes, exceeding a frequency limit of 5%, were considered.

The founder strains were of the genotypes Est-1$_s$ Est-2$_s$ α and Est-1$_f$ Est-2$_f$ a . It can be expected that some mating between these two strains occurs, resulting in the formation of heterozygotes. In agreement with these expectations we found an increasing frequency of heterozygotes during the first generations and again a decreasing frequency in later generations. With growing generation numbers an increase of diploids and a decrease of haploids can be noted. It is sur-prising that in all populations the haploids of mating type α were not detected. This is already true for the vegetative phase of the first generation. Even in later generations the haploid with the allele a is the most frequent one whenever there are available haploids in the population. With regard to the two esterase loci it is not important which kind of allele combination the haploids have.

These outcomes may be due to one or both of the following factors:

1. The mating type allele a and/or associated linked genes carry a selective advantage.

2. The haploids of mating type α are endowed with an in-creased mating activity.

The fact that the mating activity is strongly genetically controlled is shown in Table 4, where the rate of diploid formation is presented as a function of time after inoculation of both partners in the culture vessel. Depending on the genotype the diploids appear at different times. The frequency differences among the diploids after 24 hours can be accounted due to different mating abilities and different rates of reproduction.

TABLE 4. Frequencies of diploids as a function of time after incubation of the two partners. 1. and 2. indicate replications.

Genotypes		2 h	3 h	4 h	5 h	6 h	7 h	8 h	24 h
Est-1ₐ Est-2ᵣ α x Est-1ᵣ Est-2ₐ a	1.	-	-	-	0.009	0.015	0.030	0.043	0.657
	2.	-	-	-	-	0.006			
Est-1ᵣ Est-2ₐ α x Est-1ᵣ Est-2ₐ a	1.	-	-	-	-	0.001	0.001	0.001	0.020
	2.	-	-	-	-	-			
Est-1ₐ Est-2ᵣ α x Est-1ᵣ Est-2ₐ a	1.	0.002	0.005	0.008	0.017	0.024	0.056	0.052	0.387
	2.	-	0.006	0.009	0.025	0.060			

Diploids other than the double heterozygotes arise at the start only by recombination during meiosis of the double heterozygotes. Other gene combinations are easily detectable only in later generations. Haploid gene combinations differing from the founder haploids, did not become frequent. It appears that they mate soon after being transferred to complete medium.

The changes of genotypic frequencies in the first population are, in principle, the same as observed in the other populations. In populations 2 and 3 (Fig. 4 and 5) there are only slight deviations from the general behaviour while in population 4 (Fig. 6) there is a retardation. The behaviour of population 5 (Fig. 7) can be related to a strong retardation effect. Even after 30 generations this population still consists of 80% haploids, mainly of the founder type, carrying mating type allele a . We have no clear explanation for this deviation even where background effects are taken into consideration.

It is of interest to discover which of the diploids becomes most frequent in the populations. It is obvious that the populations are not at the equilibrium after 30 generations. If there are no differences between the haploid and diploid genotypes with respect to fitnesses, mating activity and rate of meiosis, then the population could consist of diploids only. Among those the double heterozygotes, the single heterozygotes and the homozygotes are expected in a ratio of 1:2:1 . The same ratio is expected for each esterase locus separately. However, there is significant deviation from these expected ratios in the observed populations (Table 5).

TABLE 5. *Frequencies of diploid genotypes after 30 generations. The populations 4a and 4b are subcultures started in the 20th generation.*

Population	Est-1 locus			Est-2 locus		
	$Est-1_s Est-1_s$	$Est-1_s Est-1_f$	$Est-1_f Est-1_f$	$Est-2_s Est-2_s$	$Est-2_s Est-2_f$	$Est-2_f Est-2_f$
1	0.08	0.34	0.41	0.16	0.34	0.31
2	0.47	0.36	0.12	0.16	0.47	0.34
3	0.29	0.19	0.21	0.08	0.24	0.37
4a	0.47	0.13	0.02	0.45	0.15	0.06
4b	0.38	0.19	0.01	0.37	0.20	0.02

With the exception of population 1 the Est-1$_s$ Est-1$_s$-genotype
is the most frequent at the Est-1 locus, and with the exception of population 4 the Est-2$_f$ Est-2$_f$-genotype is the most
frequent at the other locus. These are exactly those genotypes which have the highest rate of meiosis and reproduction
during the generative phase. Populations 4a and 4b are
subcultures started in the 20th generation. The reproducibi-
lity is quite good.

A time span of thirty generations seems to be too short
for explaining the reshuffle of the genotypic material or for
appraising the precise influence of certain genes in such a
complex model. Nevertheless, we believe that we can show
that not only fitness differences between genotypes influence
populations, but that also genes determining the mating behaviour as well as those controlling the rate of meiosis have
an important effect.

It appears that fitness differences are not the determining
factor in populations, but there are other biological factors,
e.g., sexual differences, exercising possibly greater influence on the dynamical behaviour of the populations.

An analytical treatment of the problem is difficult. We
tried to get some information by computer simulations. In
this vein we considered a two locus model involving the mating
type locus and any other locus in the genome denoted by B
(Lange, 1973; Lange and Wöhrmann, 1975). Some of the results
are recorded in Table 6. The index i runs from 1 to 8
characterizing the possible genotypes as follows:

1 : Ba , 2 : ba , 3 : Bα , 4 : bα , 5 : Ba/Bα ,

6 : Ba/bα , 7 : ba/Bα , 8 : ba/bα .

If there are no differences between the genotypes with regard
to any one of the parameters there will be no changes in the

TABLE 6. Equilibrium situations and their conditions in a two locus system in consideration of the mating type locus.

	conditions						results		
case	mating activity z_i	rate of meiosis m_i	spore viability fitness u_i	vegetative fitness w_i	recombin. frequency c	special conditions	linkage disequil. D	B-locus frequency $\hat{p}$	B-locus mating locus frequency $\hat{r}$
1.	1.	1.	1.	1.	$0 < c < .5$	$r_0 \neq .5,\ D_0 = 0$	0	p_0	r_0
2.	1.	1.	$< 1.$	1.	$0 < c < .5$	$r_0 \neq .5,\ D_0 \gtrless 0$	-	$p_0 \gtrless$	$1./0$
3.	1.	1.	$> 1.$	1.	$0 < c < .5$	$r_0 \neq .5,\ D_0 = 0$	0	p_0	$.5$
4.	1.	1.	$> 1.$	1.	$0 < c < .5$	$r_0 \neq .5,\ D_0 \gtrless 0$	0	$p_0 \gtrless$	$.5$
5.	1.	1.	$(u_a - u_b) \gtrless .5,\ u_\alpha > 1.$	1.	$.5$		0	p_0	$0 < \hat{r} < 1.$
6.	1.	1.	$u_B \neq u_b,\ u_B > 1.$	1.	$.5$		0	$1./0$	$.5$
7.	$z_1 z_3 = z_2 z_4$	1.	$> 1.$	1.	$.5$		$\neq 0$	$p_0 \gtrless$	$.5$
8.	$z_1 z_3 > z_2 z_4$	1.	$> 1.$	1.	$< .5$		$\neq 0$	$0 - .5/\ .5 - 1.$	$.5$
9.	1.	$m_5 \gtrless m_h \gtrless m_8$	$> 1.$	1.	$.5$		0	$0/1.$	$.5$
10.	1.	$m_5 \gtrless m_h \gtrless m_8$	$u = v_{dip} > v_{hap}$	1.	$.5$		0	p_0	$.5$
11.	1.		$u < v_{dip} > v_{hap}$	1.	$.5$		0	$1./0$	$.5$
12.	1.		$u > v_{dip}$	1.	$.5$		0	$0 < \hat{p} < 1.$	$.5$
13.	$0 < z > 1.$	1	1.	$w_5 > w_h > w_8,\ w_b \neq w_B$	$.5$	$\left(\frac{w_h - w_5}{w_8 - w_h}\right) \gtrless \left(\frac{1/z - 1}{w_B - w_h}\right) \times$	0	$0 < \hat{p} < 1.$	$.5$
14.	$0 < z > 1.$	1.	1.	$w_5 > w_h > w_8,\ w_a \neq w_\alpha$	$.5$	$z w_5 > w_a - (1-z) w_\alpha$	0	$1.$	$0 < \hat{r} < 1.$
15.	$0 < z > 1.$	1.	1.	$w_5 > w_h > w_8,\ w_a \neq w_\alpha$	$.5$	$z\left(w_8 + \frac{w_h - w_5}{2w_h - w_5 - w_8}\right)^2 / > w_a - (1-z) w_\alpha$	$\neq 0$	$0 < \hat{p} < 1.$	$0 < \hat{r} < 1.$

230

gene frequencies $(\hat{p}, \hat{r}, \hat{D},$ case 1).

The spore viability (u_i) : The spore viability namely the average number of spores produced by one ascus is an important parameter. In the case the number of spores per ascus is less than 2 $(u < 1$, case 2) the haploid population will fix on one mating type only, the type fixed depending on the initial frequencies. If there are more than 2 spores in the ascus (cases 3 and 4) the change in the B-gene frequency (p) depends on the initial linkage disequilibrium D_o . The frequency of the mating type at equilibrium $(\hat{r}$ or $1-\hat{r})$ is always 0.5 . It should be noted that a change in the gene frequency is observed also in the absence of fitness selection.

Genotypic differences in the viability of spores, depend on the mating type locus. The mating type frequencies may deviate from the 1:1 ratio at the equilibrium if the difference between the viability of the spores with the genotype a and α is less than 0.5 (case 5). If there are differences between the spores with respect to the B locus $(u_B \neq u_b$, case 6) the population will become fixed. There seems to be a buffer against polymorphism when viability differences are caused by different mating types.

The mating activity (z_i) : In case 7, $z_1z_3 = z_2z_4$ is assumed. Under these conditions a stable polymorphism is maintained with linkage disequilibrium and equal proportions of the mating type alleles. The frequency $\hat{p}$ may vary between 0 and 0.5 or between 0.5 and 1 depending on whether the product z_1z_3 is more or is less than z_2z_4 , (case 8). The population is then under linkage disequilibrium.

The rate of meiosis (m_i) : If the rate of meiosis of one of the homozygotes is greater than the heterozygote and the other homozygote, no polymorphism can occur at the B locus. Both mating type alleles then appear in equal proportions (case 9). Different results are obtained if the heterozygotes have the highest meiosis rate and if the spore viability is different from the vegetative reproduction rate of the diploids (v_d) in the generative phase. No change of p is observed in case 10 for $u_i = v_d > v_{hap}$ (v_{hap} = vegetative reproduction rate of the haploids during the generative phase). If $u < v_d$ (case 11) the populations will be fixed while if $u > v_d$ (case 12) they will be polymorphic.

The vegetative fitness (w_i) : Regarding fitness differences during the vegetative phase a number of situations is possible, some producing polymorphic populations while others leading to fixations. Some of these situations are important showing stable deviations from the 1:1 ratio of the mating type alleles. The conditions for such situations are indicated in Table 6 as cases 13 to 15. It is evident that the differences between the genotypes, with respect to the mating type locus, are less harmful for the maintenance of polymorphism than the differences between genotypes, with respect to any other locus.

REFERENCES

Lange, P. (1973). Dissertation, Tübingen.

Lange, P. and K. Wöhrmann. (1975). TAG 46: 7-17.

Strobel, R. and K. Wöhrmann. (1972). Genetica 43: 274-281.

Strobel, R. (1974). Dissertation, Tübingen.

Strobel, R. and K. Wöhrmann. (1975a). Genetica (in press).

Strobel, R. and K. Wöhrmann. (1975b). (In preparation).

Wöhrmann, K., P. Lange and R. Strobel. (1974). TAG 44: 1-6.

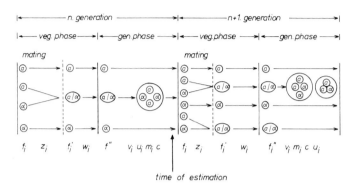

FIGURE 1.   The life cycle of the yeast.   The parameters were
explained in the text.

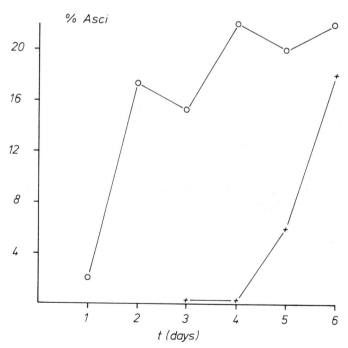

FIGURE 2.   The percentage of asci after days of incubation
in sporulation medium.

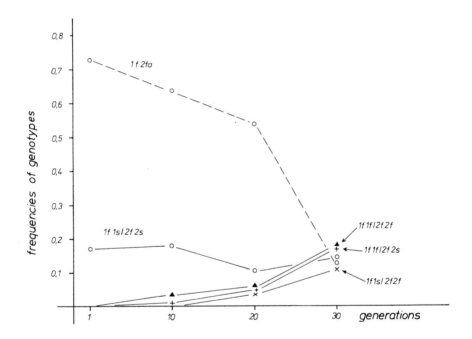

FIGURE 3. The change of frequencies of genotypes in population 1.

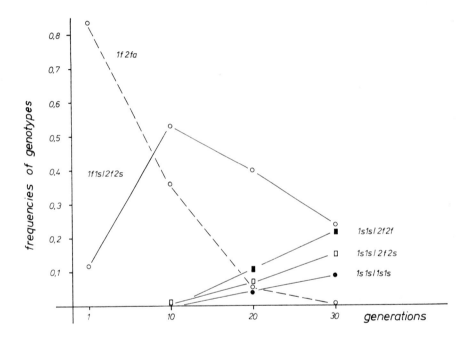

FIGURE 4.   The change of frequencies of genotypes in
population 2.

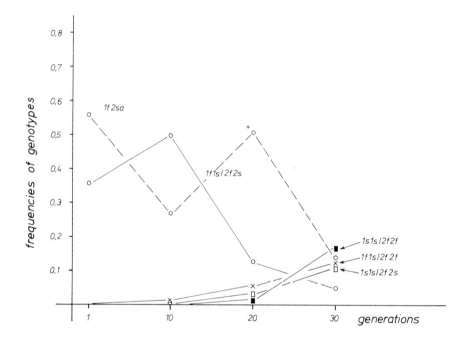

<u>FIGURE 5.</u>   The change of frequencies of genotypes in
         population 3.

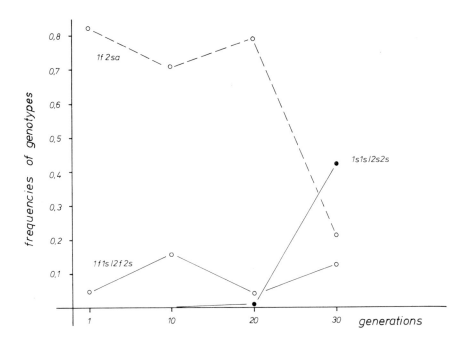

FIGURE 6.   The change of frequencies of genotypes in
population 4.

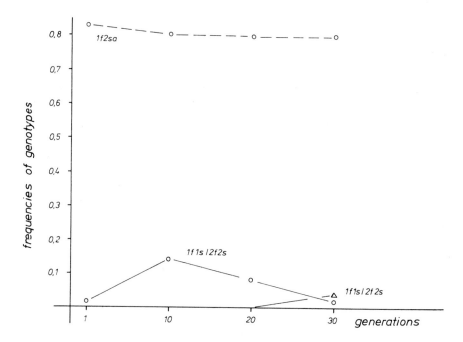

<u>FIGURE 7.</u> *The change of frequencies of genotypes in population 5.*

Morphological Variation of the Gall-Forming Aphid, *Geoica utricularia* (Homoptera) in Relation to Environmental Variation

D. WOOL and J. KOACH

INTRODUCTION

Morphological variation in natural populations has attracted the attention of biologists for centuries. Since Linné, it has been the basis for classification of organisms. Phylogenies are constructed on the basis of morphological similarity of existing species to each other and to fossils of extinct forms. Multivariate statistical techniques and electronic computers are today used in order to improve our understanding of evolutionary relationships and classification by digesting the vast amounts of information on morphological variation.

The genetics of morphological characters is often very complex. Many measurable traits are polygenic, and some gene effects are known to be pleiotropic. Morphology is affected considerably by environmental factors. In addition, only a small fraction of the genome is expressed in morphologically visible traits (Lewontin, 1974). But still, geographic variation in morphology is assumed to be brought about by natural selection. Some of the best cases supporting this view are clines in homeotherms, covered by Bergman's Rule and Allen's Rule. Convincing direct evidence on the effects of natural

selection on the morphology of poekilotherms has been given
in some cases (Kettlewell, 1961; Cain and Sheppard, 1950,
1954), but such evidence is not too abundant even today, 115
years after Darwin's Origin of Species.

The study reported here was planned to investigate the
relationship of morphological variation to environmental
variation, and, possibly, to natural selection. The research
organism, Geoica utricularia passerini, (Fordinae; Aphididae;
Homoptera) is very well suited for the purpose.

The biology of Geoica, and other Fordinae, in Israel was
investigated by Wertheim (1954, 1955), and summarized by
Bodenheimer and Swirski (1957). G. utricularia occurs in the
spring and summer on three species of Pistacia (Anacardiaceae)
in Israel : P. atlantica and P. palaestina, which are common
and may be found in the same locality, and P. khinjouk which,
in our study area, occurs only in isolated localities in the
Sinai Peninsula. The insect forms pocket-like, irregularly-
spherical galls on the leaflets of the host-tree (Figure 1).
It is known from several countries in the Middle East (Bodenheimer and Swirski, 1957), where in addition to the three
hosts mentioned above it occurs also on P. terebintus and P.
vera.

G. utricularia life history is a typical "holocycle", with
Pistacia trees as primary hosts, and the roots of many grasses
as secondary hosts (Bodenheimer and Swirski, 1957). The gall
is formed when, in early spring, a single female (fundatrix)
emerges from an overwintering fertilized egg and begins
sucking on a young leaflet as the buds come out of dormancy.
One or two generations of apterous, diploid female aphids are
produced parthenogenetically as the gall increases in volume
during the summer months. In September-November, winged forms

(alates) are produced parthenogenetically. The galls open
and the alates disperse and produce young on the ground.
These dark-cuticuled, apterous young reach the roots of
grasses and produce 1-2 generations of apterous viviparae
during the winter. The next summer, winged sexuparae are produced on the grasses, which actively migrate back to the primary host and give rise to tiny sexuals, males and females.
These mate and one egg is fertilized in each female. The egg
remains inside the dry body of the dead female, overwinters,
and the new fundatrix emerges the next spring. This is therefore a two-year cycle.

Each gall is produced by a different fundatrix, and all
individuals (up to several hundred) within a gall are identical genetically. These facts make possible the partitioning
of morphological variation into several components.
(1) Within-gall variation (which is the result of developmental "noise", microclimatic effects within the gall, and
measurement error). (2) variation among gall means on the
same tree and in the same locality, which must be mostly genetic, but environmental effects cannot be ruled out.
(3) variation among locality means, which may be the result
of natural selection, but may also be the result of random
drift and isolation. In the latter case, morphological
variation cannot be expected to show any relationship to
environmental variables.

In this paper we shall show that morphological variation of
G. utricularia is related to climatic variation, and argue
that it may be affected by natural selection.

MATERIALS AND METHODS

In September-November, 1972, 215 galls containing alates

241

were collected in 35 sites throughout the range of distribution of the host trees in Israel, from the Golan Heights to
South Western Sinai.

In the laboratory, every intact and unparasitized gall was
weighed to the nearest 0.1 mg. Then it was cut open and the
aphids were weighed to the nearest .01 mg, and preserved in
95% ethanol.

Whenever possible, 9 alates from each gall were mounted on
microscope slides. After clearing for 36 hours in 8% KOH,
they were fixed in a drop of Euparal, with wings, legs and
antennae spread out, covered and dried at 25°C.

Nineteen morphological characters were measured on each
aphid. Seventeen were linear dimensions of body parts and
two were counts of sensoria on the antennae (Table 1). In
paired organs, the right-hand side of the organism was
measured as a rule. All measurements were made on a Visopan
projection microscope (Reichert, Austria) and the linear
dimensions were measured on the screen in millimeters.

Data analysis. The approach to the analysis of geographic
variation in morphological characters was carefully worked
out by Sokal and his students in their studies of a related
aphid species, Pemphigus populitransversus, in the U.S. (e.g.
Sokal, 1952, 1962; Sokal and Rinkel, 1963; Rinkel, 1965). The
procedures suggested in these papers were adopted here, with
some modifications. The steps in the analysis were:

(1) Partitioning the variation of each character by a
"Nested" analysis of variance and calculating variance components.

(2) Correlation and factor analysis of the 19 characters,
in order to represent the variation in them by a smaller number of factors.

242

(3) Grouping the collection sites into localities and calculating averages of climatic variables for each locality.

(4) Calculating average probit scores of characters at each locality on every character factor.

(5) Multiple regression analysis of the character factors on the climatic variables.

These procedures will be described in more detail with the results. Computations were carried out on the CDC-6600 computer at the Tel-Aviv University Computation Center.

Detailed Procedures and Results

I. Partitioning of variation in each character and calculation of variance components.

A "nested" analysis of variance (Sokal and Rohlf, 1969) was performed on each character. Four levels were included: Aphids within-galls; Gall means on the same tree. Trees within collection sites; and among collection sites. The meristic characters, S1 and S2, were analyzed with a square-root transformation. The measurements of the other characters were not transformed. The variance components are listed in Table 2. Variation among gall means within trees was significant in all characters. Differences among tree means within sites were significant in all but two characters, and only four showed no significant differences among means of collection sites.

Variation within galls was generally low, although in 7 characters (WW, HW, TW, A2, A3, A4, S2) it contributed more than 20% of total variance. The lowest variance component within galls was in the long organs, WL, F1, Ti1, F2, Ti2, F3, Ti3, (less than 10%). The correlation between the mean value of a character and the variance component within galls in this character was - .706 (P < .01, 17 df). If we exclude the

243

possibility that small organs are more sensitive to developmental and microclimatic influences than long organs, this negative correlation indicates the increasing relative effect of measurement error with the decrease in organ size.

Variation among gall means on the same tree contributed 32-55% of total variation. The biology of the organism strongly suggests that this variation reflects the genetic differences among fundatrices, since environmental conditions on the same tree must be very similar for all galls (although environmental influences cannot be entirely excluded).

Variation among tree means within collection sites may be caused directly, by the influence of host condition (such as nutritional value of the sap), on the aphids in the galls, or by natural selection on the different trees. We have frequently noticed in the field that some trees are very heavily infested with galls of Geoica and other species, while adjacent trees are not infested at all. We do not have, at present, a way to measure the physiological condition of the tree, so the two alternatives are at present indistinguishable.

Since galls were collected on three host species, it was necessary to find out how different were aphids on the different hosts. In 9 of the collection sites, galls were collected on the two common hosts, P. atlantica and P. palaestina. Paired-comparison t tests (Sokal and Rohlf, 1969) indicated significant differences between hosts in only two of the 19 characters: TW and A2 (P < .05). Had more localities been sampled, perhaps more significant differences could be found, but at present we tend to regard these significant differences as representing Type 1 error, and assume that aphids on the two hosts do not differ when sampled in the same locality.

Variation among means of collection sites was significant

and considerable in all but 4 of the characters. This variation may result either from natural selection in the different
collecting sites, or from random drift and isolation. For the
following analyses we had to consider the trees-within-sites
component as part of the locality variation. The effect of
this pooling is to increase the variation within localities
where several different trees were sampled.

II.   Correlation and factor analyses of the 19 characters.

The characters were chosen by the sole criterion of being
easily measured. They certainly do not represent the insect
body adequately (for example, 9 of the 19 characters are leg
measurements). It was desirable to group correlated characters together and describe the morphological variation in
terms of a smaller number of variables, possibly representing
independent portions of the genome. The approach to this
problem is discussed in detail by Sokal (1962), Sokal and
Rinkel (1963) and Rinkel (1965).

Factor analysis with rotation to simple structure was performed on the 19 characters, using the 215 gall means of each
character as data (program BMDO3M, Dixon, 1970).

A similar analysis was run on the 19 characters using the
215 within-gall standard deviations as data - to find groups
of characters which are affected similarly by whatever affects
within-gall variation.

A third analysis of this kind was run on the 19 characters
using 34 collection-site standard deviations of gall means of
the characters as data (one collection site was omitted since
only one gall with alates was collected there).

The results of the first analysis indicated that 88% of
the variation in gall means is explained by the first 4 factors before rotation (61%, 17%, 6.4% and 3.6%, respectively).

245

When the correlations of the characters with these factors were examined, the following picture was obtained (only correlations greater than 0.6 were considered).

Factor I (M-I) was correlated with many characters - WL, WW, HW, TW, A1, F1, Ti1, F2, Ti2, F3, Ti3. This seems to be a general size factor.

Factor II (M-II) had very high correlation with only 2 characters: the sensoria, S1 and S2.

Factor III (M-III) was highly correlated with the lengths of antennal segments, A2, A3 and A4.

Factor IV (M-IV) was correlated with the lengths of the tarsi; Tar 1, Tar 2 and Tar 3.

The relationships of the 19 characters to the first 2 factors before rotation are illustrated in Figure 2.

The analysis of within-gall standard deviations yielded slightly different groups of characters. The first 4 factors before rotation explained 26%, 9.4%, 6.7% and 6.2% of the total variation, respectively. The correlations of the characters with the first 2 factors are illustrated in Figure 3.

After rotation, the character groups were as follows:

Factor I (WG-I): WL, F1, F2, Ti1, Ti2, F3, Ti3. This seems to be a measure of variation in the long organs of the insect body.

Factor II (WG-II): S1, S2. This factor is associated with variation in the number of sensoria.

Factor III (WG-III): A1, A2, A3, A4, HW, TW, WW. This factor is associated with variation in short body organs.

Factor IV (WG-IV): Tar 1, Tar 2, Tar 3. This factor is associated with variation of the tarsi, which are also short organs.

Both analyses were repeated with the addition of two gall variables as characters for each gall: Gall weight (GW) and Biomass (BM). These data were available for 171 galls.

As can be seen in Figure 2, GW and BM were not related to any of the other characters. After rotation, the two gall variables were correlated with a separate factor, with which no aphid character was correlated.

Paired-comparison t tests showed no significant differences in gall weight or biomass between galls on P. atlantica and P. palaestina, in the 9 sites in which both hosts were sampled.

The character factors obtained in the third factor analysis were different in some details from those previously described (Figure 4). Five factors were obtained, which explained together 82% of the total variation (45.9%, 13.3%, 10.6%, 6.5% and 5.9%, respectively); however, factor III represented two different groups of characters: tarsi were positively correlated with it, while antennal segments were correlated negatively with it. Therefore, for analyses intended to compare locality variation in climate with locality variation of morphological characters, six character factors were used:

Factor I (SLOC-I). This factor represents locality variation in the long segments of the legs. (F1, Ti1, F2, Ti2, F3, Ti3).

Factor II (SLOC-II). Represents locality variation in the number of sensoria. (S1, S2).

Factor III (SLOC-III). Represents locality variation in the lengths of antennal segments. (A2, A3, A4).

Factor IV (SLOC-IV). Represents locality variation in the lengths of tarsi. (Tar 1, Tar 2, Tar 3).

Factor V (SLOC-V). Represents locality variation of wings, head and thorax. (WL, WW, HW, TW).

247

Factor VI (SLOC-VI). Only the length of the third antennal segment, Al, was highly correlated with this factor.

III. Grouping the collection sites into localities for which climatic data could be obtained.

The collection localities were not selected by any criterion except that galls were found on the trees. In some, few galls with alates were collected, and some were located kilometers away from the nearest climate recording station.

Sokal and Thomas (1965), faced with the problem of too few galls per collection site, grouped them into localities with a radius of 27 miles. A circle with a radius of this magnitude almost anywhere in Israel will cover a great variety of habitats and climatic conditions. Grouping the collection sites by geographical proximity alone did not seem desirable, because within few kilometers, collection sites varied considerably in altitude. Multiple (stepwise) repression analyses (BMD 02R, Dixon, 1970) were carried out with latitude, longitude and altitude as independent variables, and collection-site means and standard deviations of six characters, in turn, as dependent variables, representing the different character factors. Latitude and longitude (taken from the map grid system of Israel) were not related to the morphological characters, but in two cases, significant regression on altitude was found: means of T13 and standard deviations of Al, both increasing with altitude. The proportion of variation explained by regression on altitude was low (20%) but it did indicate that altitude may be important.

Therefore, for the final analysis of the relationship of morphological to environmental variation, the 35 collection sites were grouped into 12 localities, taking altitude into account. The localities are listed in Table 3 and indicated on the map (Figure 5).

Climatic data

Data on six climatic variables were obtained from the records of the Israel Meteorological Service for each locality: Rainfall, Relative humidity, Daily mean, minimum and maximum temperatures, and Daily temperature range. The data were in the form of monthly averages of these variables over a number of years (ranging from a few to more than 30) from recording stations within few kilometers of each other and having similar altitude. Rain data were more abundant and measured at more locations than the other variables.

These raw data were used in the following manner. The year was divided into three seasons, which represent different stages in the life cycle of Geoica:

Season 1: April-August (for rain - only April-May, since there is no rain in the summer in Israel). In this season the aphids are inside the gall on the primary host.

Season 2: September-November. In this season, the alates form and disperse from the galls to the secondary hosts.

Season 3: December-March. In this winter season, the active aphids are in the ground on the roots of the secondary hosts.

In each locality, the mean and standard deviation of the monthly values of all climatic variables were calculated for each of the three seasons separately. For rain, the mean annual rainfall and its standard deviation were also calculated in all localities. The climatic variables were used in this form in the final analyses.

It was realized that the climatic variables may not be independent, but at the present stage of our investigation it was decided to use each variable separately and ignore interactions among them.

IV. Calculation of locality scores on each character factor

Since several characters are represented by each factor, it is desirable to find a compound measure to represent all characters in the cluster.

Sokal and Rinkel (1963) have suggested the approach which was followed here. Locality means were standardized for each character separately, and transformed to probit scores by adding 5.0 (to avoid negative numbers). These probit scores were then averaged for all characters represented by each factor. This procedure hopefully cancels out contrasting effects on individual characters and brings out the trends, if any, common to all characters in a cluster.

In the final analysis, we related these average probit scores of the 12 localities on each of the 14 character factors (described in Section II) to the means and to the standard deviations of the six climatic variables in the same localities.

V. Morphological variation and climatic variation

Stepwise multiple regression (BMD02R, Dixon, 1970) was used in this analysis. Each of the 14 character factors was run separately on the locality means and the standard deviations of each of the six climatic variables in turn. In the multiple regression, the climatic variables at seasons 1, 2 and 3 (for Rain, "4" represented annual rainfall) were used as independent variables, and each of the character factors in turn, as the dependent variable.

A concise summary of the results is given in Figure 6, indicating which environmental variable was significantly correlated (P < 0.05) with each of the character factors. Table 4 gives the details of the significant cases, such as

250

the multiple correlation coefficient R , percent variation
explained by the regression equation (R^2) , the season contributing mostly to the explained variation, and the sign of
the regression coefficient on this season.

In almost all cases, one season is responsible for
most of the morphological variation explained by the climatic
variable (Table 4). This makes the interpretation of the
climatic-morphological relationships somewhat less complex.

I. Rainfall

Locality mean annual rainfall ("Season" 4) is positively
correlated with the mean number of sensoria (M-II) and with
the within-gall S.D. of sensoria (WG-II). Mean rainfall in
season 1 is positively correlated with locality S.D. of sensoria (SLOC-II). The nature of the correlations of mean rainfall with M-III and WG-IV is not in the same direction at the
seasons involved.

The locality standard deviation of rainfall in season 1 is
positively correlated with M-I, M-III and M-IV - all linear
dimensions. (M-I and M-IV are not correlated with mean rainfall!) The mean number of sensoria, as well as the within-
gall and locality S.D. of sensoria (M-II, WG-II and SLOC-II),
are correlated positively with the standard deviation of rainfall at season 3. The locality S.D. of tarsi (SLOC-IV) is
positively correlated with standard deviation of annual rainfall ("Season" 4), and the variation in length of antennal
segments (SLOC-VI and WG-III) are negatively correlated with
this variable at season 2.

II. Relative humidity

Mean relative humidity is negatively correlated with factors M-I and M-IV, representing size and tarsi, at season 2 -
in which alates are dispersing from the opening galls. It is

also negatively correlated with SLOC-VI, representing locality standard deviation of Al, at season 1.

The standard deviation of RH is <u>positively</u> correlated with M-II and M-IV at season 2, and with SLOC-VI and WG-II at season 1.

III. <u>Mean temperature range.</u> (Trange).

This variable is negatively correlated with WG-IV at all three seasons (the factor represents variation within galls of short body parts). Its correlation with SLOC-III is not in the same direction at the seasons involved.

The standard deviation of this variable is uncorrelated with morphology.

IV. <u>Daily maximum temperature</u> (Tmax).

This variable is <u>negatively</u> correlated with factor M-III at season 3, and with SLOC-V at season 1. The correlation with SLOC-IV, is not in the same direction at seasons 1 and 2. Locality standard deviation of TMAX is <u>positively</u> correlated with M-I and M-IV (which are <u>not</u> correlated with <u>mean</u> Tmax) at season 3. These factors represent size and tarsi, respectively. It is also correlated positively with SLOC-VI at season 3 and with SLOC-IV at season 2.

V. <u>Daily minimum temperature</u> (Tmin).

This variable is <u>negatively</u> correlated with M-I, M-IV and SLOC-VI at season 3, and with M-III at season 1. The locality S.D. of Tmin is <u>positively</u> correlated with M-I, M-IV and WG-1 at season 2, and with SLOC-V and SLOC-VI at season 3.

The correlations of Tmin are quite similar to those of Tmax.

VI. <u>Daily mean temperature</u> (Tmean).

This variable, and its standard deviation, are correlated

with the same factors as Tmin (with one exception - it is not correlated with SLOC-V). But, again with one exception, the correlations are not with the same seasons. M-I, M-III, M-IV and SLOC-VI are negatively correlated with Tmean at season 1 (3 for Tmin) and the standard deviation of Tmean is correlated with these factors also in season 1 (2 or 3 for Tmin).

Similarity of localities to each other. Cluster analyses (UPGM, Sokal and Sneath, 1963) were run on the 12 localities using two sets of data as characters: (1) Locality means and standard deviations of the climatic variables at all seasons, plus latitude, longitude and altitude. (2) Locality probit scores on the 14 morphological character factors. This analysis was intended to answer the question, Are the aphids in localities with similar climate - similar also in morphology?

Since the climatic variables were measured in different units, they were all standardized before the analysis. This procedure also makes the comparison with the morphological factors more valid, since these were also standardized. Only positive correlations were used in the clustering process.

The two phenograms resulting from the clustering process are shown in Figure 7. Clearly there is no similarity between the clusters based on the different sets of data, and there is little relation to geography in either one (the localities were numbered more or less from North to South — Figure 5 — but no such order can be seen in Figure 7).

DISCUSSION

(a) The significant correlations between climatic variables and morphological factors should not be interpreted as proof of cause-and-effect relationships, which may be difficult or perhaps impossible to demonstrate in the field, but they may indicate that climatic variables have a selective effect on aphid morphology, even if the exact nature of the selective forces is unknown.

The climatic analysis was based on a rather small number of localities (12). The multiple correlation coefficient (R) need be rather large in order to be significantly different from zero at the 5% level: With one season in the equation, $r > .576$; with two, $r > .697$; and with three $r > .777$. (Rohlf and Sokal, 1969). Had more localities been available, more of the climatic variables could have been found significantly correlated with morphology. The picture presented here is, therefore, rather conservative.

Correlations among the climatic variables produce spurious correlations with morphology. Factor analysis on the climatic variables revealed that the locality standard deviations can be arranged in four groups, each related to a different factor: (a) S.D. of Tmax, Tmin and Tmean at all seasons. (b) S.D. of Rainfall, at all seasons. (c) S.D. of Trange, which was not related to morphology at all. (d) S.D. of RH, which was correlated weakly with group (a).

However, locality means of climatic variables did not show any systematic relationships to each other, and the values for the three seasons were rarely correlated. The exceptions are Rainfall, for which seasons 2, 3 and 4 (annual) were inter correlated (not with season 1!) and season 3 means of Tmax, Tmin and Tmean.

The conclusions from the climate-morphology correlations can be drawn with considerable confidence that the relationships indicated by the analyses are real. We are inclined to believe that the observed correlations are the result of selective forces exerted by the climatic factors. At the present stage of our knowledge the argument in favor of this explanation is based on indirect evidence outlined below. Much more work need be done to substantiate this claim.

(b) Climatic variables may affect aphid (alate) morphology in three ways. (1) Direct selection on alate characters. This may happen at season 2, when alates are present and disperse from the galls. (2) Indirect selection on other life-history stages at other times of the year. (3) Non-genetic, microenvironmental and physiological effects on the apterous aphids in the gall - either directly or indirectly, through the effects on the host tree. Such effects may take place only in season 1, when the aphids are in the galls.

Several climatic factors are correlated with morphology at season 1 and may be (although not necessarily have to be) included in this last category. For example, the negative correlations of Tmean with M-I, M-III and M-IV (all linear dimensions of the insect body) and the positive correlation of S.D. of Tmean with them. Even more likely to be explained in this manner is the positive correlation of S.D. of Rainfall with the same morphological factors.

The negative correlation of Relative Humidity and character factors M-I and M-IV at season 2, and the positive correlation of S.D. of RH with these factors, may indicate direct selection on alate characters (M-I and M-IV represent together all aphid characters except antennal segments and sensoria).

The fact that RH was correlated with morphology at season 2 adds weight to this argument, since inside the gall the relative humidity must be very high. Only when the galls open and the alates disperse can relative humidity affect the insects. Relative humidity was found effective when alates were held in test tubes in the laboratory: they survived longer in a higher RH (Wertheim, 1954).

But many correlations are probably best explained by the
second hypothesis. In this category we must include all
correlations of morphology with climatic variables at season
3, in which the insects live on the roots of the secondary
hosts. At season 1, the climatic variables may have a selective effect on the fundatrices at the time of founding of the
galls. This applies in particular to Rainfall, because season
1 for this variable was only the period April-May, the time of
establishment and early stages of the galls.

The correlations of within-gall variation in morphology
with some of the climatic variables, if real, must be the
result of the effects on the trees. Daily temperature Range,
S.D. of Tmean, and S.D. of rainfall are likely to affect the
aphids inside the gall in this way.

The variation among galls within localities in one character, A1, is positively correlated with the standard deviation
of all temperature measures and with the S.D. of RH. (SLOC-VI).
Other positive correlations of climatic locality S.D. with
morphological locality variation are indicated in Figure 6
and Table 4. This shows that in these characters, aphid morphology is more variable in those localities where climatic
conditions vary more. Interestingly, variation in the legs
(SLOC-I) is unrelated to climatic variation.

The dissimilarity of the two phenograms based on climatic
variables and morphological characters is the result of the
contrasting effects of different climatic variables on morphology. It may be taken to mean that the net result of many
selection pressures may create the impression that no selection is taking place.

It is interesting that morphological factors representing
linear dimensions of the aphid body (M-I, M-III and M-IV) are
generally correlated <u>negatively</u> with the locality <u>means</u> of

climatic variables (RH, Temperature) and _positively_ with the
standard deviations of these variables (this is not so for
Rainfall, with which most correlations were positive). The
optimal environment for the Aphidoidea, on a zoo-geographic
scale, is considered to be moderately warm temperatures,
moderate rainfall and moderate air humidity (Bodenheimer and
Swirski, 1957). In the present study, aphid size (represented
by the three factors listed above) seems to increase with
increasing variation in rainfall and decreasing air humidity
and temperature. If larger size is a measure of better environment, the optimal environment within the study area can
be defined as cool and not too moist. But, of course, on a
world scale, most of Israel has a rather moderate climate.

(c) The life cycle of _Geoica_ _utricularia_ is very similar
to that of _Pemphigus_ _populi-transversus_, studied by Sokal and
his students in the U.S., but differs in an important detail.
The latter has a one-year cycle: The alates disperse during
the summer, and the sexuparae return to the primary host the
same fall (Sokal, 1952). But the interrelations among morphological characters in the two organisms are quite similar. In
most cases, a general "size" factor was found in _Pemphigus_.
The antennal segments were often represented by a separate
factor. No correlation was found between gall and aphid
characters (Rinkel, 1965), as was true in _Geoica_. (This must
mean that the growth of the gall is so well synchronized with
the increase in aphid population size inside, that no deleterious effect of crowding are caused).

The within-gall variance component in _Pemphigus_ seems to be
higher than in _Geoica_: Up to 59% in continuous characters,
and even higher in meristic ones (Sokal and Rinkel, 1963). The
highest value in the present study was 30%. Some of this
difference may be due to the fact that only 4 or 2 aphids were

measured per gall in <u>Pemphigus</u>.

From a study of aphids collected in 3 different years at the same 8 localities, Sokal, Heryford and Kishpaugh (1971) concluded that considerable local differentiation of aphid populations was taking place, and that the new fundatrices each year came from secondary hosts in the same localities, to which the alates had dispersed in the summer. Heryford and Sokal (1971) found <u>seasonal</u> variation in aphid size when all alates emerging from several galls during the summer were caught and measured.

It seems to us that in <u>Geoica</u>, aphid emergence from individual galls is rather synchronized, but a long-term study is needed to verify this point.

Morphologically visible variation comprise only a small fraction of the genetic variation of organisms. It is very interesting to find out whether any patterns of relationship may be found between climatic variables and genetic variation at the molecular level in <u>Geoica</u>, as revealed by protein electrophoresis. This study prepares the necessary background for the investigation of interrelationships between morphological variation and protein variation, and between the latter and climatic variation, which we intend to undertake in the future.

SUMMARY

1. During a part of its complicated life cycle, the aphid, <u>Geoica</u> <u>utricularia</u> passerini, forms galls on Pistacia trees. Every gall is founded by one female emerging from a fertilized egg. Reproduction within the gall is parthenogenetic and all offspring are genetically identical.

The morphological variation in 19 characters was investi-

gated in 9 alate aphids from each of 215 galls collected at
35 sites throughout the range of the species in Israel.
Following factor analyses, the gall means of the 19 characters
were represented by 4 factors; the within-gall standard deviations of these characters, by 4 factors; and the locality
standard deviations of gall means, by six more factors.

The 35 sites were grouped into 12 localities by geographical
proximity and similarity in altitude. Means and standard
deviations of six climatic variables were calculated for each
locality: Rainfall; Maximum, Minimum and Mean daily temperature. Daily temperature Range; and Relative Humidity.
The calculations were made from data on monthly averages,
separately for three seasons according to the stages in aphid
life history; these variables were then used as independent
variables in multiple regression analyses, with the 14 morphological factors as dependent variables.

2. A large proportion of the variation in each character
was contributed by variation among gall means on the same
tree, which must be mostly genetic variation among fundatrices.
Significant and considerable portion of the variation was
among locality means of most characters. Within-gall variation, which must be non-genetic, comprised a relatively small
part of total variation.

A complicated network of significant correlations between
the morphological factors and the climatic variables was
found. All but one of the morphological factors were significantly correlated with at least one climatic variable. For
example, the means of temperature variables were negatively
correlated with factors representing aphid size, and the standard deviations of temperature variables were positively
correlated with them. Size was positively correlated with the
standard deviation of rainfall, and negatively with mean

259

relative humidity.

In most cases, correlation of morphology with a climate variable in one of the three seasons explained most of the co-variation. This fact contributed to the interpretation of the correlations and supported the argument that some of them, at least, may be the result of selection pressures of climate on the aphids.

Although cause-and-effect relationships between climate and aphid morphology may be hard to demonstrate in Nature, the present findings suggest that such relationships do exist.

ACKNOWLEDGEMENTS

We are most grateful to Mrs. A. Tiran who prepared the 2200 microscope slides of the aphids, and to Mrs. L. Bialer who was responsible for the enormous task of running the vast sets of data on the computer. Mr. Y. Nitzan and Mrs. R. Wool made the drawings and the galls were photographed by Mr. A. Shouv. The kind cooperation of the Staff of the Israel Meteorological Service, (Climate Department) is gratefully acknowledged. Without the help of all these people, this investigation could not have been accomplished.

REFERENCES

Bodenheimer, F.S., and E. Swirski. (1957). The Aphidoidea of the Middle East. The Weizmann Science Press of Israel.

Cain, A.J. and P.M. Sheppard. (1950). Heredity 4: 275-294.

Cain, A.J. and P.M. Sheppard. (1954). Genetics 39: 89-116.

Dixon, W.J. (1970). BMD, Biomedical Computer Programs. Univ. of California Press.

Heryford, N.N. and R.R. Sokal. (1971). J. Kansas Ent. Soc. 44: 384-390.

Lewontin, R.C. (1974). The genetic basis of evolutionary change. Columbia Univ. Press.

Kettlewell, H.B.D. (1961). Ann. Rev. Entomol. 6: 245-262.

Rinkel, R.C. (1965). Univ. Kansas Sci. Bull. 46: 167-200.

Rohlf, F.J. and R.R. Sokal. (1969). Statistical Tables. Freeman & Co.

Sokal, R.R. (1952). Evolution 6: 296-315.

Sokal, R.R. (1962). Evolution 16: 227-245.

Sokal, R.R. and R.C. Rinkel. (1963). Univ. Kansas Sci. Bull. 44: 467-507.

Sokal, R.R. and P.H.A. Sneath. (1963). Principles of Numerical Taxonomy. Freeman & Co.

Sokal, R.R. and P.A. Thomas. (1965). Univ. Kansas Sci. Bull. 46: 201-252.

Sokal, R.R. and F.J. Rohlf. (1969). Biometry. Freeman & Co.

Sokal, R.R., N.N. Heryford and J.R.L. Kishpaugh. (1971). Evolution 25: 584-590.

Wertheim, G. (1954). Trans. R. ent. Soc. Lond. 105: 79-96.

Wertheim, G. (1955). Bull. Res. Counc. Isr. 4: 392-394.

TABLE 1. *The characters measured on Geoica. Characters 9 and 10 are counts. The other characters were converted to actual length (in microns) in the table. Means and standard deviations were calculated using the 215 gall means as data.*

No.	Code	Description	Mean	S.D.
1	WL	Forewing length	1890	179
2	WW	Forewing width (at widest point)	1050	105
3	HW	Head width (at widest point)	400	28
4	TW	Thorax width (at widest point)	820	79
5	A1	Length of third antennal segment	230	23
6	A2	Length of 4th antennal segment	120	12
7	A3	Length of 5th antennal segment	100	10
8	A4	Length of terminal (6th) antennal segment	130	17
9	S1	Number of sensoria on 3rd antennal segment	17.2	7.26
10	S2	Number of sensoria on 4th antennal segment	6.3	2.45
11	F1	Femur length, foreleg	390	39
12	Fi1	Tibia length, foreleg	450	49
13	Tar 1	Tarsus length, foreleg	140	13
14	F2	Femur length, middle leg	300	30
15	Ti2	Tibia length, middle leg	450	56
16	Tar 2	Tarsus length, middle leg	140	15
17	F3	Femur length, hind leg	360	35
18	Ti3	Tibia length, hind leg	550	66
19	Tar 3	Tarsus length, hind leg	160	16

TABLE 2. Variance components at the different levels of variation of the 19 characters. Asterisks mark statistical significance at least at the 5% level.

Character	Variance Components (%)			
	within galls	galls within trees	Trees within sites	among collection sites
WL	8.6	41.5*	27.2*	22.7*
WW	30.4	23.5*	15.9*	30.2*
HW	23.8	34.6*	19.2*	22.4*
TW	21.0	37.8*	31.0*	10.1
A1	10.4	54.8*	9.8	25.0*
A2	20.1	45.8*	27.5*	6.5
A3	20.5	47.7*	24.4*	7.4
A4	27.2	48.4*	10.0	14.5
S1	13.7	42.5*	19.9*	23.9*
S2	27.3	33.5*	15.6*	23.5*
F1	8.6	35.2*	19.9*	36.3*
Ti1	7.3	32.7*	16.6*	43.3*
Tar 1	18.6	32.4*	12.7*	36.2*
F2	9.0	35.6*	24.5*	30.9*
Ti2	6.0	32.8*	13.7*	47.5*
Tar 2	14.5	33.6*	15.2*	36.7*
F3	8.2	37.5*	20.3*	34.0*
Ti3	8.0	34.2*	11.3*	46.7*
Tar 3	10.9	34.2*	14.5*	40.5*

TABLE 3. Localities used in the final analysis and the number of galls sampled in each.

no.	Name	Average altitude (m)	no. of galls sampled
1	Hula Valley	120	13
2	Gallilee (Low altitude)	430	26
3	Golan Heights	960	11
4	Zafad	810	30
5	Har Tavor	500	13
6	Carmel - (low altitude)	65	27
7	Shomron	430	14
8	Judean Hills (low altitude)	300	14
9	Hebron	950	12
10	Jerusalem	730	29
11	Mizpeh Ramon	680	13
12	Har Sarbal	1700	13
			215

TABLE 4. Details on the significant correlations between morphological variation in *Geoica utricularia* and climatic variation. (Seasons: 1 – April-August (April-May for rainfall). 2 – September-November. 3 – December-March).

Factor	Characters represented	Climatic variable	Seasons significantly correlated	Multiple R	Percent variation explained $R^2 \times 100$	Single most effective season	$R^2 \times 100$	Sign of regression on this season
A. Factors of character means								
M-I	WL, WW, HW, TW F1, Ti1, F2, Ti2 F3, Ti3	Mean RH	1, 2	0.78	61.0	2	59.1	negative
		Tmax	3	0.65	42.1	3	42.1	negative
		Tmin	1, 2, 3	0.92	85.6	3	62.0	negative
		Tmean	1, 2, 3	0.79	62.5	1	59.4	negative
		S.D. of Rainfall	3	0.84	70.0	1	60.4	positive
		S.D. of Tmax		0.62	38.2	3	38.2	positive
		S.D. of Tmin	1, 2, 3	0.79	63.0	2	41.0	positive
		S.C. of Tmean	1	0.65	43.0	1	43.0	positive
M-II	S1, S2	Mean Rainfall	1, 2, 3, 4	0.88	77.6	4	58.1	positive
		S.C. of Rainfall	1, 2, 3, 4	0.85	71.8	3	54.9	positive
		S.C. of RH	2	0.60	35.8	2	35.8	positive
M-III	A2, A3, A4	Mean Rainfall	1, 3, 4	0.79	62.6	equally effective	54.7	1, 4 positive, 3 negative
		Tmin	1, 2, 3	0.82	66.7	1	41.6	negative
		Tmean		0.64	41.6	1		negative
		S.D. of Rainfall	1, 2, 3, 4	0.93	87.2	1	54.0	positive
M-IV	Tar 1, Tar 2, Tar 3	Mean RH	1, 2, 3	0.89	79.0	2	68.7	negative
		Tmin	1, 2, 3	0.91	82.3	3	42.8	negative
		Tmean	1	0.62	38.1	1	38.1	negative
		S.D. of Rainfall	1, 2, 4	0.80	64.4	1	47.6	positive
		S.D. of RH	1	0.65	42.3	2	42.3	positive
		S.D. of Tmax	3	0.62	37.8	3	37.8	positive
		S.D. of Tmin	1, 2	0.74	54.6	2	46.5	positive
		S.D. of Tmean	1, 2	0.74	54.0	1	53.2	positive

TABLE 4 – Page two.

B. Factors of character within-gall standard deviations

Group	Characters	Factor						
WG-I	WL, Fl, Til, F2, Ti2, F3, Ti3	S.D. of Tmin	2	0.66	44.1	2	44.1	positive
		S.D. of Tmean	1	0.64	40.6	1	40.6	positive
WG-II	S1, S2	Mean of Rainfall	1, 2, 3, 4	0.90	81.0	4	62.2	positive
		S.D. of Rainfall	1, 2, 3, 4	0.91	82.0	3	66.0	positive
		S.D. of RH	1	0.58	33.9	1	33.9	positive
WG-III	A2, A3, A4	S.D. of Rainfall	1, 2, 4	0.84	71.0	2	44.0	negative
WG-IV	Tar 1, Tar 2, Tar 3 WW, HW, TW, A1	Mean of Rainfall	2, 4	0.75	56.8	equally effective	equally effective	2 positive, 4 negative
		Temp. Range	1	0.69	47.4	1	47.4	negative
		S.D. of Rainfall	1, 2, 3, 4	0.85	72.3	1, 2, 3, equally		negative
		S.D. of Tmin	2	0.66	44.1	2	44.1	positive

C. Factors of character locality standard deviations

Group	Characters	Factor						
SLOC-I	Fl, Til, F2, Ti2, F3, Ti3	none						
SLOC-II	S1, S2	Mean Rainfall	2, 3	0.58	33.1	1	33.1	positive
		S.D. of Rainfall		0.74	52.4	3	37.9	positive
SLOC-III	A2, A3, A4	Temp. Range	1, 2, 3	0.94	88.5	3	48.2	positive
						2	36.9	negative
SLOC-IV	Tar 1, Tar 2, Tar 3	Tmax	1, 2	0.71	50.6	equally effective	equally effective	1 negative, 2 positive
		S.D. of Rainfall	2, 4	0.80	63.7	4	53.1	positive
		S.D. of Tmax	2, 3	0.70	49.0	2	37.1	negative
SLOC-V	WL, WW, HW, TW	Tmax	1	0.58	33.0	1	33.0	negative
		S.D. of Tmin	3	0.61	37.8	3	37.8	positive
SLOC-VI	A1	Mean RH	1, 3	0.73	52.7	1	50.4	negative
		Tmin	2, 3	0.78	60.2	3	46.2	negative
		Tmean	1	0.65	41.8	1	41.8	negative
		S.D. of Rainfall	1, 2, 3	0.72	51.2	2	37.8	negative
		S.D. of RH	1, 2, 3	0.81	65.6	2	51.1	positive
		S.D. of Tmax	1, 2, 3	0.79	62.4	3	59.0	positive
		S.D. of Tmin	1, 2, 3	0.86	73.2	3	59.9	positive
		S.D. of Tmean	1, 2, 3	0.82	67.9	1	58.7	positive

FIGURE 1. Galls of G. utricularia. Top: on Pistacia palaestina. Bottom: on P. atlantica.

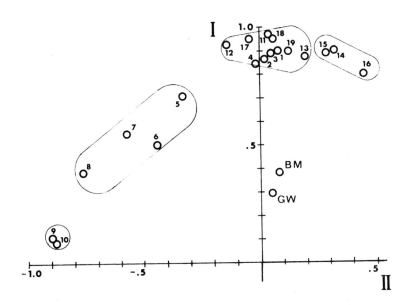

FIGURE 2.   *Factor analysis on gall means of characters.   The*
            correlations of the 19 characters with the first
            2 orthogonal factors (before rotation to simple
            structure) are illustrated.   Characters are identi-
            fied by their code numbers (Table 1).   BM = Bio-
            mass,   GW = gall weight.

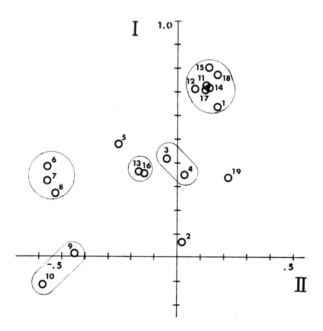

FIGURE 3. *Factor analysis on within-gall standard deviations. The correlations of the 19 characters with the first 2 orthogonal factors (before rotation to simple structure), are illustrated. Characters are identified by their code numbers (Table 1).*

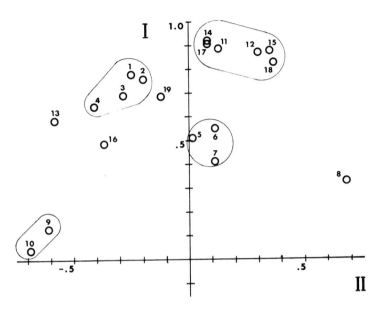

FIGURE 4. *Factor analysis on locality standard deviations of the 19 characters. The correlations of the characters with the first 2 orthogonal factors (before rotation) are illustrated. Characters are identified by their code numbers given in Table 1.*

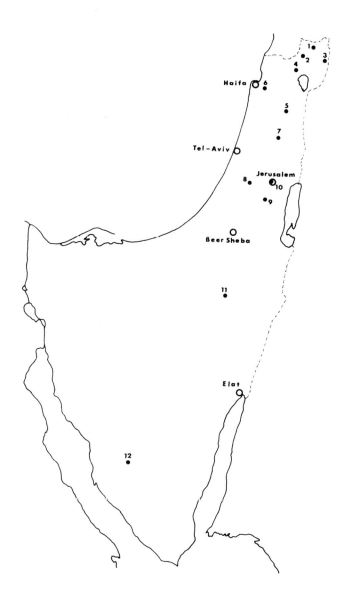

FIGURE 5. The 12 localities assembled for the final analysis
of morphological variation. Numbers correspond to
Table 3.

270

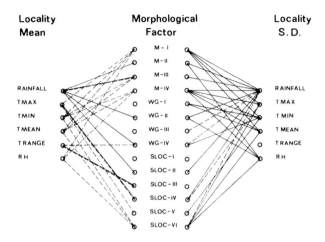

Locality Mean   **Morphological Factor**   **Locality S.D.**

FIGURE 6. Summary of the correlations of locality means and
locality standard deviations of climatic variables
with morphological character factors. Only significant correlations are plotted, (P < 0.05).
Positive correlations are indicated by solid lines,
negative ones by broken lines. In three cases,
there were significant correlations with opposite
signs at two seasons. These are indicated by a
double line - one solid, one broken.

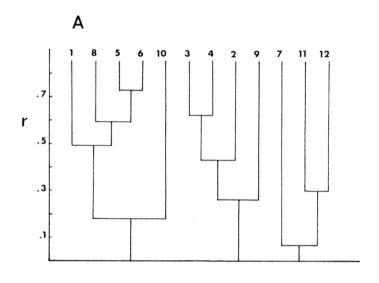

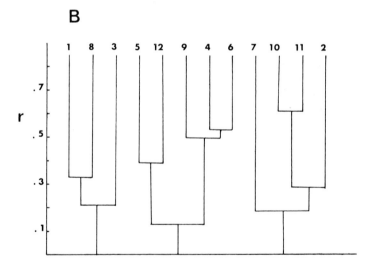

FIGURE 7. *Phenograms resulting from clustering the 12 local-*
ities, (A) based on 41 climatic and geographical
variables, (B) based on 14 morphological character
factors. The ordinate is the product-moment
correlation coefficient, r . The abscissa has no
meaning. Horizontal lines connecting two localities
or clusters indicate the correlation between them.

PART II MODELS AND EVIDENCE

The papers of this part offer a mixture of intuitive concepts, hypotheses and model studies buttressed by related data and simulations. The subjects include statistical procedures for estimating selection components, calculation of multi locus associations, some interpretations of heterosis, analyses of some cases of sexual selection, inference of selection for electrophoretic variants, the biological origin of heterosis and coadaptation of parts of the genome and testing of selective neutrality.

The individual papers are now briefly reviewed.

Christiansen and Frydenberg present a further analysis of data from the species Zoarces viviparus. They report here on samples taken over several years, and for which the data has been subjected to a procedure of selection component analysis.

Ewens and Feldman elaborate various attempts which have been made of testing the hypothesis of selective neutrality from observed allele frequency data. The four main tests dealt with are those of Ewens, of Johnson and Feldman, of Yamazaki and Maruyama, and of Lewontin and Krakauer. The last two are shown to have theoretical difficulties although all tests suffer from being based on models whose applicability to electrophoretic data is dubious.

Genetic drift, even with no selection, mutation or migration, has been shown to cause linkage disequilibrium, or nonrandom association amongst linked genes in finite population. Previous studies have dealt with pairs of loci. In his paper, Hill reports results mostly obtained by simulation on three or more loci for the disequilibrium expected in segregating populations. The multi locus association is analyzed as a multi-

273

dimensional contingency table.

Koehn and Eanes investigate the possibility that intragenic recombination may be influential in determining some of the data patterns observed for electrophoretically defined genetic polymorphism. They do this by correlating the frequencies of rare alleles with those of heterozygotes for common alleles. These correlations, it emerges, are usually high in the data sets studied so far.

Various classifications of electrophoretic enzyme loci have been introduced in recent years based on the origin of the substrate or structural and regulatory properties of genes, etc. In terms of these criteria Latter examines various selection hypotheses on a wealth of electrophoretic frequency data from Drosophila.

Models of sexual selection are described and analyzed in terms of female mating preferences for the dark, intermediate and pale phenotypes of the Arctic Skua by O'Donald. The preferred males gain a selective advantage because they breed at an earlier date in the breeding season and earlier breeding pairs fledge a greater average number of chicks. It is shown that polymorphic equilibria can be established where natural selection opposes sexual selection.

Seyffert and Forkmann formulate some problems of quantitative genetics for special traits related to biochemical action in plant populations. Related simulation results are reviewed.

The paper by Sved builds upon the experimental work which he and others, notably Sperlich and Temin, have done in connection with the fitness of chromosomal homozygotes. These experiments are discussed and the suggestion made that a profitable framework in which to evaluate the results is in terms of Wallace's hard and soft selection concepts.

Thomson, W. Bodmer and J. Bodmer discuss some theory pertinent to the HLA multi-locus system in man. A number of proposals to explicate the observed linkage disequilibrium patterns are evaluated, taking account of possible differential selection expression among the alternative haplotypes, the influence of admixture and migration pattern, historical factors, etc. Some HLA frequency data and disease resistance associations are reviewed.

A simple modification of the model of gene control in higher organisms originally proposed by Britten and Davidson can account for the observed beneficial effect of random mutations when these are present in single dose in otherwise homozygous individuals: The modified model postulates that loci possessing many sensors function most efficiently if the sequential arrangements of sensors in the control regions of the two alleles differ from one another. Wallace in his paper describes the advantage which might accompany the possession of dissimilar control regions, evidence which supports the model, the bearing of the model on classical genetic problems, its relation to the maintenance of gene-enzyme variation, and the release of variation through recombination.

Selection Component Analysis of Natural Polymorphisms using Mother-Offspring Samples of Successive Cohorts

F. B. CHRISTIANSEN and O. FRYDENBERG

INTRODUCTION

In a previous paper (Christiansen and Frydenberg, 1973),
we developed a procedure for selection component analysis of
observations comprising mother-offspring data for organisms
with discrete reproduction periods. The theory was constructed to enable the analysis of observations collected in a
single, short interval of time, and it was thus, in the stric-
test sense, only suited for organisms with a well-defined
breeding season and non-overlapping generations. Organisms
with a mature lifetime so long that an individual may reproduce in successive breeding periods will only be incompletely
analyzed in our previous mother-offspring measuring system.
However, if different generations or cohorts present in a sample cannot be identified, the mother-offspring measuring system described may still represent the best possible analytical
framework. This was the method we used on the samples of eelpouts (<u>Zoarces</u> <u>viviparus</u>) collected in 1969 and 1970 when we

---

Deceased April 7, 1975.

were unable to separate different age classes (Christiansen, Frydenberg and Simonsen, 1973). Later on, the development of a method of aging the fish (Gyldenholm and Jensen, 1975) has led to more detailed investigations of the eelpout populations during the years 1971 (Christiansen, Frydenberg, Gyldenholm and Simonsen, 1974), 1972, 1973 and 1974. The data for the later years will be thoroughly discussed in a forthcoming publication (Christiansen, Frydenberg and Simonsen, in preparation), but the observations from 1972 and 1973 have been analyzed preliminarily in order to exemplify the theoretical problems arising from mother-offspring data in which the different cohorts can be identified.

COHORT DATA

A mother-offspring data set consists of a sample of adults sorted into males, sterile females and pregnant females, where the genotype of every adult and of one random offspring from each pregnant female is determined. Suppose that during a period of Y consecutive years* at the breeding times, we sample an autosomal mother-offspring data set as shown in Table 1 and classify the adult individuals according to age, $a = 1,2,...,A$. (The notation that we use is basically that of Christiansen and Frydenberg (1973) with a few additions as shown in Table 2.)

---

*
We shall be using years as the time unit because our organism, the eelpout, breeds once a year. But the model is, of course, equally applicable to animals with other breeding patterns as long as the breeding seasons are clearly separate. The time unit will then be the interval between successive breeding seasons.

TABLE 1. *The Simple Mother-Offspring Data Set.*

	Mother-Offspring Combinations				Sterile Females	Adult Males	Genotype
	Offspring						
♀	A_1A_1	A_1A_2	A_2A_2	Σ			
A_1A_1	C_{11}	C_{12}		F_1	S_1	$A_{\delta 1}$	A_1A_1
A_1A_2	C_{21}	C_{22}	C_{23}	F_2	S_2	$A_{\delta 2}$	A_1A_2
A_2A_2		C_{32}	C_{33}	F_3	S_3	$A_{\delta 3}$	A_2A_2

TABLE 2.

List of Notations.

A	maximum age considered
a	index of age $(= 1,2,\ldots,A)$
Y	number of samples
y	index of sample $(= A+1, A+2, \ldots, A+Y)$
K	number of cohorts in data $(= A + Y)$
k	index of cohort $(= 1,2,\ldots,K)$
i	index of genotype $(1:A_1A_1, \quad 2:A_1A_2, \quad 3:A_2A_2)$
j	index of gene $(1:A_1, \quad 2:A_2)$
s	index of sex $(= ♀,♂)$
0	index designating sum or pooling
$\beta_s(a,y)$	age distribution in sample y
$\zeta_i(k)$	genotype frequency of zygotes
$\pi_j(k)$	gene frequency in zygotes
$A_{si}(a,k)$	sample of adults
$\alpha_{si}(a,k)$	genotype frequency in adults
$C_{ii'}(a,k)$	sample of mother-offspring combinations
$\gamma_{ii'}(a,k)$	frequency of mother-offspring combinations
$F_i(a,k)$	sample of mothers $(= C_{i0}(a,k))$
$\phi_i(a,k)$	genotype frequency of mothers
$\nu_j(a,k)$	gene frequency in mothers
$M_j(a,k)$	sample of transmitted male gametes
$\mu_j(a,k)$	gene frequency of transmitted male gametes
$S_i(a,k)$	sample of sterile females
$\sigma_i(a,k)$	genotype frequency of sterile females

Within a sample, each age class, $\underline{a}$, provides an observation of the cohort born $\underline{a}$ years before, so the $\underline{Y}$ samples provide information about $\underline{K} = \underline{A} + \underline{Y}$ different cohorts. Indexing each observation by the age $\underline{a}$, and the cohort $\underline{k}$, according to the year of birth, we can get a view over the data in Figure 1 where we have chosen $\underline{A} = 5$ and $\underline{Y} = 5$. The index of a sample of fish of age $\underline{a}$ and cohort $\underline{k}$ is

$$\underline{y} = \underline{a} + \underline{k} \tag{1}$$

and the total sample of fish collected in a given year is the sum over a diagonal in the Figure.

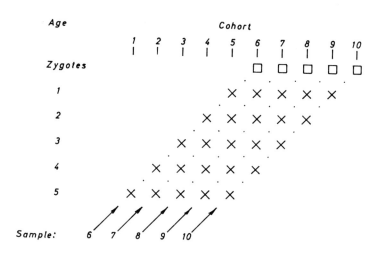

FIGURE 1.   Cohort data for $\underline{A}=5$ and $\underline{Y}=5$. Each cross
             represents a mother-offspring data, and each
             square represents an estimated zygote population.

SELECTION COMPONENT ANALYSIS

The cohort data of mother-offspring combinations of course inherit the properties of the simple mother-offspring data analyzed by Christiansen and Frydenberg (1973), namely that it is open to an analysis of the individual selection components. The components that selection may be broken into are zygotic (=viability) selection, fecundity selection, sexual selection and gametic selection. Of these components, zygotic selection is the one about which most new information is gained by using cohort data. However, a new structural feature, namely age dependence, is also added to the analysis of the "reproductive" components of selection (Christiansen et al., 1974).

The analysis of zygotic selection is made by comparing the age classes in a cohort, from zygotes up through the different ages, i.e. following a vertical in Figure 1. The analysis of the reproductive components of selection is, on the other hand, done in the contemporary populations, i.e., diagonally in Figure 1, along the direction of estimation of the zygote population. The zygote population is, with the assumption of random mating, given by

$$\zeta_1(k) = \nu_1'(k)\mu_1'(k) \quad , \tag{2a}$$

$$\zeta_2(k) = \nu_1'(k)\mu_2'(k) + \nu_2'(k)\mu_1'(k) \quad , \tag{2b}$$

$$\zeta_3(k) = \nu_2'(k)\mu_2'(k) \quad , \tag{2c}$$

where ν_j' and μ_j' are the female and male gamete frequencies in a given year:

$$\nu_j'(y) = \sum_a \beta_{\female}(a,y)\nu_j(a,y-a) \quad , \tag{3a}$$

$$\mu_j'(y) = \sum_a \beta_{\male}(a,y)\mu_j(a,y-a) \quad . \tag{3b}$$

Initially the most complicated problem is the estimation
of the genotypic proportions in the zygote population,
because this estimation requires an exhaustive knowledge of
all the reproductive components of selection and of the
mating system. In a statistical procedure the analysis of
these reproductive components must naturally precede the
estimation of the zygote population. The total analysis of
selection thus follows a scheme as shown in Figure 2.

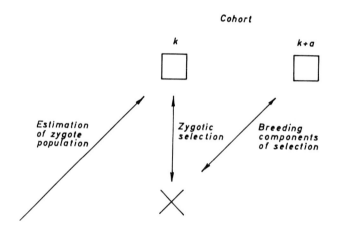

FIGURE 2. *The directions of interest for each observed*
mother-offspring data.

This figure already hints at the main statistical problems in
analyzing the cohort data: The analysis has to follow two
conflicting directions, i.e. the analysis of one aspect of
the data influences the analysis of other aspects in a far
more profound way than was the case in the simple mother-
offspring analysis. However, this conflict of analysis is

not only due to the nature of the data, but it reflects a
conceptual difficulty in the definitions of the zygotic and
the reproductive selection components in a population in
which several cohorts breed simultaneously. After the
individuals have become sexually mature, the differential
survival between breeding seasons is strictly speaking acting
as a reproductive selection component though disguised as
zygotic selection. Nevertheless, we will continue to consider this differential survival an aspect of the zygotic
selection component, but as will become apparent, the component will necessarily be included into the analysis as a
reproductive component.

The Simple Mother-Offspring Analysis

To prepare the way for the analysis of the cohort data let
us briefly review the simple mother-offspring analysis
(Table 3). This analysis consists of an examination of a
sequence of six hypotheses each of which is related to a
simple biological question. The hypotheses H_1, H_3 and H_4
are concerned with the selection components involved in breeding, while H_2 postulates random mating, and H_5 and H_6
take care of the zygotic selection. Each hypothesis is
furthermore placed in the sequence such that the assumptions
necessary to formulate the hypothesis in a simple, artless
way are tested by the previous hypotheses. From this rule
there is one exception: The procedure as formulated in Table
3 comprises no test for fecundity selection. This is because
the analysis of fecundity rests on brood size observations
which are of a form that calls for a separate conventional
analysis. Thus throughout this paper we assume that fecundity selection has been searched for and found absent. When
this is granted, the total test sequence constitutes a test

283

for the hypothesis, H_0 , that the adults occur in Hardy-Weinberg proportions, π_1^2 , $2\pi_1\pi_2$, π_2^2 , and the mother-offspring combinations arise by random union of female and male gametes both with frequencies π_1 and π_2 .

Not all the hypotheses in Table 3 are simple and nice statistically speaking. Most are canonical to the structure of the multinomial distribution, but H_3 and thereby H_5 are not, since they postulate linear rather than multiplicative structures.

TABLE 3. *Analysis of Selection Components in Mother-Offspring Data without Age Structure.*

	Hypothesis tested	Biological mechanism that may lead to rejection
H_1:	$\gamma_{22} = \frac{1}{2}\phi_2$	Gametic selection in females
H_2:	$\gamma_{11}/\phi_1 = 2\gamma_{21}/\phi_2 = \gamma_{32}/\phi_3$	Non-random mating of breeding individuals
H_3:	$\mu_1 = \alpha_{\delta1} + \frac{1}{2}\alpha_{\delta2}$	Reproductive (= sexual and/or gametic) selection in males
H_4:	$\phi_i = \sigma_i = \alpha_{\varphi i}$	Sexual selection in females
H_5:	$\alpha_{\varphi i} = \alpha_{\delta i} = \alpha_{0i}$	Different zygotic selection in sexes
H_6:	$\alpha_{0i} = \zeta_i$	Zygotic selection

Formulation of the Cohort Analysis

In the mother-offspring cohort measuring system a simple mother-offspring data set is available for each age and cohort observed, i.e. for each pair $(\underline{a},\underline{k})$. In this

material H_1, H_2 and H_4 are natural and meaningful hypotheses to investigate for each $(\underline{a},\underline{k})$. On the other hand
the hypothesis of no male reproductive selection, H_3, is
necessarily a global hypothesis on contemporary populations,
since transmitted male gametes cannot be allocated to males
of known age (Christiansen et al., 1974). The hypothesis of
no zygotic selection, $H_5 \cap H_6$, becomes a hypothesis of
equality of all age classes within a cohort. Thus the analytical conflict revealed in Figure 2 applies especially to the
analysis of male reproductive selection and zygotic selection.
In the present context the hypothesis of no male sexual and
gametic selection may be formulated

$$H_3: \mu_1'(y) = \sum_a \beta_\delta(a,y)[\alpha_{\delta 1}(a,y-a) + \tfrac{1}{2}\alpha_{\delta 2}(a,y-a)]$$

$$\text{for all } y ,$$
(4)

and the hypothesis of no zygotic selection is

$$\alpha_{\varphi i}(a,k) = \alpha_{\delta i}(a,k) = \alpha_{0i}(a,k) = \zeta_i(k)$$

$$\text{for all } (\underline{a},\underline{k}) ,$$
(5)

where $\zeta_i(\underline{k})$ is given by (2).

From (4) and (3b) we see that a meaningful investigation
of male reproductive selection requires a pooling along diagonals of male gametes, viz. we must ask the question

$$H_i: \quad \mu_j(a,y-a) = \mu_j'(y) \quad \text{for all } \underline{a} \text{ and } \underline{y} .$$
(6)

Similarly we must examine whether we can pool the males within a year, i.e. whether the genotypic proportions of any age
$\underline{a}$ equal the pooled proportions:

$$\alpha_{\delta i}(a,y-a) = \sum_{a'} \beta_\delta(a',y)\alpha_{\delta i}(a',y-a') .$$
(7)

The hypothesis H_i is in no way in conflict with the analysis of zygotic selection, but an acceptance of (7) bars the
utilization of the cohort structure in the search for zygotic
selection among adults. Hence, if we choose to regard the

analysis of the zygotic selection in the cohort structure as our primary aim, then the test of the homogeneity (7) must be done <u>after</u> the analysis of zygotic selection.

In the formulation (5), the hypothesis of no zygotic selection is a räther complex one, but it dissolves smoothly into three natural parts, namely a question of differences between adult age classes within a cohort, a question of a difference between the sexes, and, finally, a question of a difference between adults and zygotes. This may be formulated as a sequence of hypotheses:

$$H_{ii}: \quad \alpha_{\female i}(a,k) = \alpha_{\female i}(0,k) \quad \text{for all} \quad (\underline{a},\underline{k}) \quad , \tag{8}$$

$$H_{iii}: \quad \alpha_{\male i}(a,k) = \alpha_{\male i}(0,k) \quad \text{for all} \quad (\underline{a},\underline{k}) \quad , \tag{9}$$

$$H_5: \quad \alpha_{\female i}(0,k) = \alpha_{\male i}(0,k) = \alpha_{0i}(0,k) \quad \text{for all} \quad \underline{k} \quad , \tag{10}$$

$$H_6: \quad \alpha_{0i}(0,k) = \zeta_i(k) \quad \text{for all} \quad \underline{k} \quad . \tag{11}$$

In terms of the structure displayed in Figure 1, the hypotheses H_{ii} and H_{iii} correspond to a pooling of adults along the vertical direction of cohorts. In biological terms, these two hypotheses correspond to absence of zygotic selection among sexually mature individuals, i.e. no interaction between zygotic selection and reproductive selection. The residual hypotheses concerning zygotic selection, H_5 and H_6, both scrutinize the early zygotic selection, and H_6 depends upon the estimation of the genotypic proportions in the zygote population. But from the expression for $\underline{\zeta_i(k)}$, (2), and the definitions of $\underline{\nu'_i(k)}$ and $\underline{\mu'_i(k)}$, (3), it appears that this estimation is unfounded unless we have, in advance, secured the truth of H_i and of

$$\nu_j(a,y-a) = \nu'_j(y) \quad \text{for all} \quad \underline{a} \quad . \tag{12}$$

Both of these hypotheses correspond to a pooling along diagonals in Figure 1, i.e. along the direction of analysis of reproductive selection. On the female side, reproductive selection has already been tested by H_1 and H_4 and if these hypotheses have been accepted, then (12) is a hypothesis on contemporary adult female populations. If, furthermore, H_{ii} is assumed true, then the pooling of females along diagonals is equivalent to a horizontal pooling over cohorts. Thus, in a test sequence the content of (12) is expressed in

$$H_{iv}: \quad \alpha_{\varphi i}(0,k) = \alpha_{\varphi i}(0,0) \quad \text{for all } \underline{k} \ , \tag{13}$$

which is obviously a part of the biological hypothesis that the population is in equilibrium throughout the period it has been observed.

As a last prerequisite for the analysis of early zygotic selection by H_5 and H_6 , we have to analyze male reproductive selection, H_3 . We have already argued that the estimation of $\mu'_j(\underline{k})$ requires acceptance of H_i , which rests on a pooling of transmitted male gametes along diagonals in Figure 1. Thus a meaningful exploration of male reproductive selection also requires a pooling of adult males present within a year sample. This last pooling may, with the same arguments that led to H_{iv} , be formulated in the hypothesis

$$H_v: \quad \alpha_{\delta i}(0,k) = \alpha_{\delta i}(0,0) \quad \text{for all } \underline{k} \ . \tag{14}$$

Thus, in a test sequence, the test for the male reproductive selection should be executed after the pooling over cohorts has been justified by acceptance of H_v and therefore the test also calls for the additional pooling

$$H_{vi}: \quad \mu'_j(k) = \mu'_j(0) \quad \text{for all } \underline{k} \ . \tag{15}$$

After acceptance of H_v and H_{vi} , the hypothesis of no male reproductive selection takes the form

$$H_3: \quad \mu_j'(0) = \alpha_{1\delta}(0,0) + \tfrac{1}{2}\alpha_{2\delta}(0,0) \quad . \tag{16}$$

This is indeed nothing but the hypothesis known from the simple mother-offspring analysis which is here applied to the totals pooled over all ages and cohorts. The procedure proposed leads to a very late test of H_3, because this hypothesis is in conflict with the analysis of adult zygotic selection. If we try to squeeze it into the analysis before the hypothesis H_{iii} and H_v, then the parameter $\underline{\mu_j'(k)}$ contains information about the age distribution of males in the year $\underline{k}$ as seen from (4). Therefore, we have to estimate both the genotypic distributions of males and their age distributions, and we will have to carry this estimated age distribution through the analysis of the zygotic selection. These are mainly technical difficulties, but there is a more profound difficulty already discussed by Christiansen et al. (1974), namely that it is practically impossible to sample the individuals such that there will be no bias with respect to size and thereby no bias with respect to age. The information carried in $\underline{\mu_j'(k)}$ about age structure is unbiased. Thus, by the early examination of H_3 we will be carrying a number of nuisance parameters through the analysis, whereas the previously suggested analysis escapes this problem because it is performed conditional on the observed age distribution. So the early investigation of H_3 will seldom be advisable.

The Test Procedure in the Cohort Analysis

From the arguments presented above we can now construct a sequence of statistical hypotheses or biological inquiries, I_1 through I_{12} (Table 4). An analysis following this sequence will correspond statistically to a progressive pooling of the data and biologically to a step by step analysis of the selection components. The hypotheses I_1, I_2, I_4, I_{10}, I_{11} and I_{12} are identical to the hypotheses tested in the simple mother-offspring analysis, and they occur in the same sequence with the only exception that the order of H_3 and H_4 has been reversed.

TABLE 4. Analysis of Selection Components in a Mother-Offspring Data Set Classified in Cohorts.

	Hypothesis tested	Biological mechanism that may lead to rejection
I_1 (H_1):	$\gamma_{22}(a,k) = \phi_2(a,k)/2$	Gametic selection in females
I_2 (H_2):	$\gamma_{11}(a,k)/\phi_1(a,k) = 2\gamma_{21}(a,k)/\phi_2(a,k) =$	
	$\gamma_{32}(a,k)/\phi_3 = \mu_1(a,k)$	Non-random mating of breeding individuals
I_3 (H_i):	$\mu_j(a,y-a) = \mu_j(y)$	Age dependent mating behaviour
I_4 (H_u):	$\phi_i(a,k) = \sigma_i(a,k) = \alpha_{\varphi i}(a,k)$	Female sexual selection
I_5 (H_{ii}):	$\alpha_{\varphi i}(a,k) = \alpha_{\varphi i}(0,k)$	Zygotic selection among adult females
I_6 (H_{iii}):	$\alpha_{\delta i}(a,k) = \alpha_{\delta i}(0,k)$	Zygotic selection among adult males
I_7 (H_{iv}):	$\alpha_{\varphi i}(0,k) = \alpha_{\varphi i}(0,0)$	Difference in adult females among cohorts
I_8 (H_v):	$\alpha_{\delta i}(0,k) = \alpha_{\delta i}(0,0)$	Difference in adult males among cohorts
I_9 (H_{vi}):	$\mu_j(y) = \mu_j(0)$	Difference in male gametes among years
I_{10} (H_3):	$\mu_j(0) = \alpha_{\delta 1}(0,0) + \alpha_{\delta 2}(0,0)/2$	Reproductive (=sexual and/or gametic) selection in males
I_{11} (H_5):	$\alpha_{\varphi i}(0,0) = \alpha_{\delta i}(0,0) = \alpha_{0i}(0,0)$	Different zygotic selection in sexes before sexual maturity
I_{12} (H_6):	$\alpha_{0i}(0,0) = \zeta_i(0)$	Zygotic selection before sexual maturity

However, as noted by Christiansen and Frydenberg (1973), the sequential position of H_4 is arbitrary as long as it precedes H_5, so the structure of the simple mother-offspring analysis has been preserved. The additions are I_3, which is a natural part of the test for random mating (Christiansen et al., 1974), and I_5 through I_9. These latter additions constitute the two main improvements as compared to the previous analysis. I_5 and I_6 allow an analysis of the action of zygotic selection through adult life, and I_7, I_8 and I_9 challenge the genetic stability of the population through time. The addition of these hypotheses alter the interpretation of the hypotheses I_{11} (H_5) and I_{12} (H_6) somewhat. Given that the hypotheses I_1 through I_{10} are true, the intersect $I_{11} \cap I_{12}$ applies to the zygotic selection that may act among the young individuals before the first time of breeding.

The maximum likelihood estimation of the parameters in this analysis does not pose any problems other than those discussed by Christiansen and Frydenberg (1973). As in the simple mother-offspring analysis, Rao's theory (1960) will be used to test the increasingly restrictive chain of hypotheses

$$I_{on} = I_1 \cap I_2 \cap \ldots \cap I_n . \tag{17}$$

This test procedure is summarized in Table 5. Most of the testors required are identical or closely related to the $\underline{T}$-testors derived in the simple mother-offspring analysis (Christiansen and Frydenberg, 1973). Thus, $\underline{S}_1$, $\underline{S}_2$ and $\underline{S}_4$ are nothing but the respective sums of the testors $\underline{T}_1$, $\underline{T}_2$ and $\underline{T}_4$ where the summation extends over all $\underline{a}$ and $\underline{k}$. The testors $\underline{S}_{10}$, $\underline{S}_{11}$ and $\underline{S}_{12}$ are respectively $\underline{T}_3$ (with a trivial alteration), $\underline{T}_5$ and $\underline{T}_6$ applied to totals obtained by pooling over ages and cohorts.

TABLE 5.   *Summary of Test Procedure.*

Hypothesis tested	Hypothesis granted	Testor	Degrees of freedom
I_1	-	S_1	A×Y
I_{o2}	I_1	S_2	2×A×Y
I_{o3}	I_{o2}	S_3	(A-1)×Y
I_{o4}	I_{o3}	S_4	2×A×Y
I_{o5}	I_{o4}	S_5	2×(A-1)×(Y-1)
I_{o6}	I_{o5}	S_6	2×(A-1)×(Y-1)
I_{o7}	I_{o6}	S_7	2×(A+Y-2)
I_{o8}	I_{o7}	S_8	2×(A+Y-2)
I_{o9}	I_{o8}	S_9	(Y-1)
I_{o10}	I_{o9}	S_{10}	1
I_{o11}	I_{o10}	S_{11}	2
I_{o12}	I_{o11}	S_{12}	1
I_{o12}	-	Σ S	10×A×Y-1

The testor, $\underline{S}_3$, is given by Christiansen et al. (1974) and $\underline{S}_9$ has a similar structure.

$$S_9 = \sum_k F_0(0,k)\left[1 - \frac{1}{2}\hat{\alpha}_{\varphi 2}(0,k)\right] \times \frac{[\hat{\mu}_1(k) - \hat{\mu}_1(0)]^2}{\hat{\mu}_1(0)\hat{\mu}_2(0)}. \tag{18}$$

The remaining testors S_5 through S_8 are ordinary χ^2-homogeneity tests or sums thereof.

DATA FROM AN EELPOUT POPULATION

The eelpout, Zoarces viviparus L., is a marine livebearing, teleostean fish, which is common in many coastal areas of the inner Danish waters. The individuals are sexually mature at an age of one year and a half. The fish mate in late summer

291

and the females carry the young for some five months until they are delivered in mid-winter.

Mother-offspring data sets of a codominant 2-allelic esterase polymorphism, described by Simonsen and Frydenberg (1972), have been collected in the late autumn of the years 1969 through 1973. An additional sample, designated "1972", of adults was caught in the early spring, 1973. The adults from 1971 and later have been classified into age classes. The observational data appear in Table 6 and their structure is symbolized in Figure 3 for easy comparisons with the general cohort scheme in Figure 1. The adults in sample 6 and 7 of the years 1969 and 1970 are pooled and therefore do not fit easily into the analysis described in the previous section.

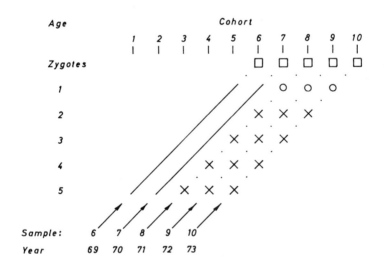

FIGURE 3. The eelpout data. In samples 6 and 7 the adults appear as a pooled class and the adolescent age class 1 has not been observed.

292

TABLE 6. Data from Eelpout Populations $(C_{ij}, S_i, A_{\delta i})$.

	1970, Σ	1971	1972	"1972"	1973	
	19 35 - 8 32	116 186 - 23 246	26 50 - 1 46	144 100	34 98 - 2 138	
	33 82 43 15 84	168 502 325 91 773	57 152 83 3 199	487 364	77 238 178 14 471	Σ
	- 62 103 11 93	- 344 569 80 783	- 87 161 5 165	451 312	- 137 303 21 439	
	52 179 146 34 209	284 1032 894 194 1802	83 289 244 9 410	1082 776	131 473 481 37 1048	

	1969, Σ	1971	1972	"1972"	1973	
	41 70 - 8 54	64 114 - 18 116	4 8 - 0 2	21 5	8 6 - 1 4	
	65 173 119 32 200	94 281 185 73 371	7 19 17 0 6	55 19	11 30 23 7 10	2
	- 127 187 29 177	- 190 307 66 350	- 12 26 2 7	42 19	- 19 42 9 8	
	106 370 306 69 431	158 585 492 157 837	11 39 43 2 15	118 43	19 55 65 17 22	

	1968, Σ	1971	1972	"1972"	1973	
		40 51 - 3 96	11 21 - 0 20	52 28	15 33 - 1 19	
1968, Σ ⟶		64 165 102 11 306	23 70 32 1 73	169 119	24 66 50 0 93	3
		- 130 201 5 319	- 34 80 1 70	168 107	- 46 91 4 85	
		104 346 303 19 721	34 125 112 2 163	389 254	39 145 141 5 197	

	1967, Σ	1971	1972	"1972"	1973	
		10 17 - 0 25	8 16 - 0 14	41 38	17 35 - 0 63	
1967, Σ ⟶		9 36 27 1 55	13 49 21 0 70	168 128	21 83 67 2 232	4
		- 15 41 3 79	- 27 32 2 58	141 106	- 45 105 3 206	
		19 68 68 4 159	21 92 53 2 142	350 272	38 163 172 5 501	

	1966, Σ	1971	1972	"1972"	1973	
		0 4 - 0 0	1 2 - 1 10	22 17	13 18 - 0 34	
1966, Σ ⟶		1 3 5 0 8	12 10 9 1 40	66 70	19 39 28 4 102	5
		0 5 9 0 14	- 10 17 0 15	77 64	- 20 41 3 106	
		1 12 14 0 22	13 21 27 2 65	165 151	32 77 69 7 242	

		1971	1972	"1972"	1973	
		1 0 - 0 5	1 3 - 0 0	6 11	1 6 - 0 16	
		0 10 5 1 18	1 4 2 1 8	24 23	2 16 7 1 28	6+
		- 3 8 3 12	- 3 5 0 11	21 13	- 4 19 2 31	
		1 13 13 4 35	2 10 7 1 19	51 47	3 26 26 3 75	

In the remaining samples, the first year class only occurs sporadically since these small, adolescent fish escape from our nets. These adolescents have therefore not been sorted out separately, but they are incorporated together with fish of undetermined age in the total catch for each year. Furthermore, each year sample contains a low proportion of fish more than 5 years old, the class 6+. This class has not been subdivided and must therefore be treated separately in the analysis.

A glance at Table 6 reveals another more technical problem, namely that of small numbers in some of the observed classes. In order to avoid lengthy and elaborate discussions on this modest fraction of the data, we have used a rough razor and omitted from each test all the samples containing expectations less than about five. However, these samples are included in later pooling, so that all the data are carried on to subsequent tests.

The analysis of the data in Table 6 according to the test procedure in Tables 4 and 5 is shown in Table 7. The special structure of the present data is reflected by the addition of six new, but rather trivial hypotheses marked by "()". The first four hypotheses, I_1 through I_4 , are tested within each observed class of mother-offspring data, viz., both in classes identified by age and cohort and in the two early classes comprising fish of all ages. After I_4 , the females in the sample "1972" caught in the early spring 1973 and the pooled pregnant and barren females in the 1972 sample become comparable and this homogeneity is accordingly tested, and after acceptance, the "1972" are then included in the 1972 females. Similarly, the males in the two comparable samples are pooled after test and acceptance of homogeneity. The hypotheses of no zygotic selection expressed in I_5 and I_6 are only formulated within the cohort data, whereas the subsequent stability hypotheses, I_7 and I_8 , may be supplemented by comparison with the additional data not sorted into cohorts. We have chosen to do this in two steps, first a pooling of the rather small data on the year class 6+, and then a comparison of the pooled cohort data, the pooled 6+ data and the data from the years 1969 and 1970.

TABLE 7. Test for Random Mating and Selection in the Eelpout Data.

	Hypothesis tested	Degrees of freedom	χ^2	P
I_1:	Gametic selection in females	15^a	14.85	.46
I_2:	Random mating per age	24^b	16.22	.88
I_3:	Age independent mating	12	20.42	.06
I_4:	Sexual selection in females	6^c	3.65	.72
( ):	72 = "72" in females	8^d	6.59	.58
( ):	72 = "72" in males	6^e	8.99	.17
I_5:	Zygotic selection among adult females	12	13.73	.32
I_6:	Zygotic selection among adult males	12	15.74	.20
I_7:	Adult females equal among cohorts	8^f	4.60	.80
( ):	Class 6+ females equal in different years	2^g	.14	.93
( ):	All adult females homogeneous	6	2.97	.81
I_8:	Adult males equal among cohorts	6^h	2.57	.86
( ):	Class 6+ males equal in different years	4	2.51	.64
( ):	All adult males homogeneous	6	6.82	.34
I_9:	Transmitted male gametes equal in all years	4	9.69	.05
I_{10}:	Male reproductive selection	1	.28	.60
I_{11}:	Sexes equal	2	.72	.70
I_{12}:	Zygotic selection before sexual maturity	1	5.46	.02
H_0:	Random mating and no selection	135^i	135.94	.46

Footnotes to TABLE 7.

Omissions due to small expectations:

a: (5,3) and 1972 age 6+ omitted.
b: (5,3), (5,4), and age 6+ in 1971, 1972, and 1973 omitted.
c: only 1969, 1970, and (2,6) included.
d: class 6+ omitted.
e: classes 2 and 6+ omitted.
f: k = 3 omitted.
g: 1971 omitted.
h: k = 3 and 8 omitted.
i: data corresponding to 54 d.f. omitted, then 135 = 190-54-1.

After acceptance of the last stability hypothesis, I_9 , all data are pooled over years and ages, and the three remaining tests are performed as if all the observations were a simple undivided mother-offspring data set.

As for test results, S_9 is weakly significant and S_{12} is genuinely significant if we choose a five per cent significance level. Before discussing these significant tests, let us observe that the total fit to the global hypothesis of random mating and no selection, H_0 , is indeed very satisfactory, yielding a χ^2 (135) of 136. Furthermore, the distribution of the testors is nice. The P 's fit the uniform (0,1) distribution very well, as shown by the testor :

$$\chi^2(36) = (-2) \times \Sigma \log P = 38.3 \qquad (19)$$

and by Figure 4. Thus, overall, there is an immediately acceptable good fit to the general hypothesis of random mating and no selection. This fact is, of course, a warning against trusting too naively an isolated significance. However, the validity of such a significance may be supported by structural consistencies in the data, which do not express themselves fully in the tests made.

FIGURE 4. *Distribution of the test probabilities of the*
tests in Table 6.

In evaluating the significance in the test for I_{12} , it
should first of all be remembered that the present analysis
is not independent of those previously made (Christiansen
et al., 1973, 1974). These tests also yielded significant
evidence for zygotic selection and it appears that the present
significance for I_{12} is due to the deviations in the older
part of the data. Hence, what the new significance tells us
is that the newer samples collected in 1972 and 1973, though
they account for about half of the observations, have been
unable to swamp the significant deviations in the older
samples. These deviations were in the direction of an excess
of homozygotes as compared to the Hardy-Weinberg expectations
and this tendency is neither supported nor denied by the newer
samples. For the time being it is thus purely a matter of
taste how much confidence one chooses to have in the evidence
for zygotic selection that our data contain.

The significance in I_9 gains no obvious support from
systematic tendencies in the data. However, the other hypothesis, I_3 , which is concerned with male reproductive
selection, has also given a somewhat extreme test value.
These two facts may suggest a structural analysis of transmitted male gamete frequencies.

DISCUSSION

The analysis expressed in the sequence of hypotheses, I_1 through I_{12} (Table 4) is an extended and refined version of the two earlier selection component analyses we have published about mother-offspring data sets.

The first analysis (Christiansen and Frydenberg, 1973) applied to a simple mother-offspring data set, i.e., a set of observations which had been collected in a single breeding season and which had not been subdivided into adult age classes. The analysis of such a data set must necessarily rest upon two major assumptions which have to be accepted a priori. The first of these assumptions is that the population studied is in gene frequency equilibrium during the period of time covering all generations represented among the caught adults. This assumption may often be bolstered up by independent observations that are more easily obtained than the mother-offspring data themselves. In our analysis of an eelpout population (Christiansen et al., 1973) we challenged, meekly but honestly, the assumption of gene frequency stability by comparing observations from two successive years. The other assumption basic to our first analysis is that the adult population is homogeneous over age classes within each sex. This second assumption remained unimpeached in our first analysis of the eelpouts, and it is indeed in general difficult to imagine that this hypothesis can gain significant support from supplementary observations as long as age classes cannot be distinguished. The importance of this assumption becomes apparent when one considers the possible properties that may prevail in populations with overlapping generations (Charlesworth, 1972). It can, for instance, be demonstrated that zygotic selection combined with age-dependent, but genotype-independent fecundities can lead to gene frequency stable populations, in which the adults, when pooled over age classes, exhibit an excess of homozygotes. The simple mother-offspring analysis will reveal only the deviation from the Hardy-Weinberg proportions and this observation is compatible with both gene frequency changing selection and, with gene frequency stabilizing selection.

Our second version of the selection component analysis (Christiansen et al., 1974) applies to a single mother-offspring data set, in which, however, the adults are classified into age classes. Also this analysis presumes gene frequency stability but the question of homogeneity among age groups may be scrutinized to a certain extent. However, the homogeneity hypotheses, which can be tested in this case, confound zygotic selection and stability not only of the gene frequency but also of the possible selection forces that may have changed the genotypic proportions in the different generations that contribute to the population at the time of sampling. Thus, despite the fact that the age classification allows the addition of a considerable amount of discriminative power to the test procedure, selection possibly indicated by the analysis can still not be allocated to the relevant biological units. The selection structure inferred from such a data set may easily emerge as a rather bewildering composite result of varying selection forces which have shifted successive generations in different ways, quantitatively or even qualitatively speaking.

The analysis presented in this paper distinguishes effectively between the questions of the zygotic selection structure and the genetic stability of the population. This desirable ability stems of course from the fact that in a seasonally breeding creature the cohorts are indeed, in space

and time, the natural population units. Different cohorts
share to a certain degree the same habitat and they are
likely to interact biologically during mating, but each
cohort may still have experienced its own selection history.
We believe that a cohort-divided mother-offspring data set is
able to yield the maximum amount of information that can be
squeezed out of mother-offspring data. It does allow estimation of those time-dependent variables of survival, reproduction, and mating that are required for a critical and more
sophisticated evaluation of the total selection that may sway
a given polymorphism (Lewontin, 1974).

Finally, let us recall a shortcoming inherent in all
mother-offspring measuring systems. They can give us only
vestigial information about the population of breeding males.
Even strong sexual or gametic selection forces among the males
may under certain circumstances completely escape our notice.
Observations linking each observed offspring to both parents
are needed in order to make the ideal analysis in which male
reproductive selection is also unable to mislead the observer.
Reliable data that couple both parents to one or more offspring
are hard to collect in nature, although it has been done in
experimental situations (Bundgaard and Christiansen, 1972).
Man himself is of course an obvious choice in which, we trust,
most scored paternities can be relied upon. But then Man's
rapidly changing environment, his complex migration patterns
and his intricately interacting intrapopulational strata may
cause new problems in evaluating the selection status of his
polymorphisms. So, maybe it may be worth while to look for
an animal where the females mate only once per breeding season and where the beasts can be caught in the act of pro-
creation.

ACKNOWLEDGEMENTS

Our eelpout researches have been generously supported by the Carlsberg Foundation of Copenhagen. Several collaborators in our department have, in one way or the other, contributed to the establishment of the eelpout data used in this paper. Finally we wish to express our gratitude to Mr. Arno Jensen and Mr. Preben Jensen for assistance in the preparation of the manuscript.

REFERENCES

Bundgaard, J. and F.B. Christiansen. (1972). Genetics 71: 439-460.

Charlesworth, B. (1972). Theor. Pop. Biol. 3: 377-395.

Christiansen, F.B. and O. Frydenberg. (1973). Theor. Pop. Biol. 4: 425-445.

Christiansen, F.B., O. Frydenberg and V. Simonsen. (1973). Hereditas 73: 291-304.

Christiansen, F.B., O. Frydenberg, A.O. Gyldenholm and V. Simonsen. (1974). Hereditas 77: 225-236.

Gyldenholm, A.O. and A. Jensen. (1975). J. Fish Biol. (In preparation).

Lewontin, R.C. (1974). The genetic basis of evolutionary change. Columbia Univ. Press, N.Y.

Rao, C.R. (1960). Sankhya, Ser. A, 23: 25-40.

Simonsen, V. and O. Frydenberg. (1972). Hereditas 70: 235-242.

The Theoretical Assessment of Selective Neutrality*

W.J. EWENS and M.W. FELDMAN

1.   INTRODUCTION

In almost every applied science which allows both a theoretical development and the accumulation of data, there will
develop controversy over how the data should best be recon-
ciled with the theory. Biology is such a science and population genetics is a part of biology in which the theory has
been very highly developed relative to our ability to observe
the phenomena that the theory purports to describe. The advance of theoretical population genetics has left us with a
subject rich in conflicting hypotheses and in the scientific
battles between their advocates. This is especially so for
the interpretation of the extensive genetic polymorphism observed in nature. Whether it is because the theory has not
advanced enough or because the experimental techniques are

*
Research supported in part by NSF Grant No. GB 37835 and NIH
Grants Nos. 10452-09-11 and GM 21135-01, and a grant from
the U.S.-Israel Binational Science Foundation.

insufficiently precise, we are apparently poor in the apparatus necessary to choose the hypothesis most compatible with
our observations of genetic polymorphism.

Before the advent of electrophoresis, speculation as to
the existence and meaning of genetic variability was based
mainly on visible mutations, lethal genes, and information
from experiments in which whole chromosomes were manipulated.
Two general schools of thought arose (Lewontin, 1974) during
this time. The "classical" view was that there existed
little genetic variation and most of what did exist was bad,
in the sense that it represented deleterious departures from
a normal state. The "balance" school on the other hand contended that there was much genetic polymorphism in nature and
that the availability of this variation mediated progressive
adaptation.

Since the initial work of Lewontin and Hubby (1966) and
Harris (1966), a vast amount of evidence has accumulated at-
testing to the ubiquity of electrophoretically detectable
genetic variation in natural populations. The average genic
heterozygosity (i.e., mean proportion of loci heterozygous
per individual) appears to be about 6% in vertebrates and
about 15% in invertebrates (Selander and Kaufman 1973, Lewontin 1974). As Lewontin (1974, pp 194-200) points out, the
finding of widespread polymorphism and heterozygosity has not
disposed of the classical theory but merely mutated it into a
neo-classical one. The neo-classical school, or "neutralists",
claim that almost all of the genetic variation is due to the
passage in and out of populations of two classes of mutations.
One class is selectively neutral, increasing or decreasing in
frequency completely at random. Another class is under
selection, but selection as viewed by the classical school;
these alleles are deleterious and are in the process of being

eliminated from the population. The neutralists allow that, in addition to these two most important types of alleles, there are very rare favorable or heterotic mutants. The balance or selectionist view remains unchanged; the genetic variation seen is preserved by natural selection and is significant for adaptive evolution. But now this position stands on the fact of polymorphism and heterozygosity rather than speculation.

What does all this polymorphism imply? Are the enzyme alleles we see under balancing selection or are they of no selective significance at all, or are they deleterious forms of some ideal gene and being expelled from the population?

In this paper we review some of the theoretical attempts to resolve this point. It is not our intention to make a comprehensive survey of the literature. Rather, we shall examine certain mathematical models of population genetics describing these processes, and review several tests of whether the data we observe are compatible with the distribution implied by these modeling assumptions.

There has been much recent work on the relationship between the biochemical properties of allozymes and isozymes and their population frequencies. In this paper functional enzyme biochemistry will not be a major concern, except insofar as it may determine the appropriate model to use and the generality of the findings.

We commence with a brief summary of the theory pertinent to each of the four most studied population genetic models. These are: (1) the classical Wright-Fisher model with its stationary distribution in the presence of mutation, subdivision and migration; (2) the infinite alleles model of Kimura and Crow (1964); (3) the infinite sites model of Karlin and McGregor (1967) and Kimura (1971); (4) the ladder-rung model of Ohta and Kimura (1973). We then evaluate the ways in which the theory of these models has been used to assess whether the allele frequency data are compatible with the neutralist or selectionist view. Here we examine the work of Ewens (1972) and the associated study of Johnson and Feldman (1973). We then discuss the method of Yamazaki and Maruyama (1973). These methods will rely on mathematical drift theory but they differ in the precision with which each adheres to its theoretical background. Section 3 is concluded by a brief discussion of several informal statistical, mathematical and biological approaches. Some of these are loosely based on drift theory, others are strongly tied to biological intuition.

Section 4 is devoted to an analysis of standardized allele frequency variances and especially their use in a test proposed by Lewontin and Krakauer (1973). Once again we feel it is necessary to clarify assumptions before applying tests based on such statistics to real data. We conclude with an evaluation of the present situation and of possible future developments.

2. ALLELE FREQUENCY MODELS

In this section (and at the beginning of Section 4) we present several mathematical models of population genetics, together with selected formulae and results derived from these models. In the following section we describe the use of these formulae to develop various tests of the neutrality theory, and also use the results to assess certain informal tests and verbal arguments relating to neutrality.

We believe that it is of the greatest importance to base

any test of selective neutrality on a genetical model which
is appropriate to the form of data used in the test. This
has not always been done in the literature, and invalid conclusions have often been drawn in this way. One of our main
aims in this paper is to discuss the relationship between
data, the assumptions involved in genetical models for this
data, and the statistical testing procedures which use the
data.

Classical models   (Wright 1931, Fisher 1930).

   The standard classical model of population genetics considers a locus A admitting two alleles, A_1 and A_2 , in a
random-mating diploid population of fixed size N . Let the
fraction of genes in generation t which are A_1 be denoted
$x(t)$. Then the model assumes that

$$Pr\{x(t+1)=x \mid x(t)=y\} = \binom{2N}{2Nx} y^{2Nx}(1-y)^{2N(1-x)} \tag{2.1}$$

This equation defines a Markov chain with absorbing states at
$x(\cdot)=0$, $x(\cdot)=1$. Many results are known for this model:
those which will be used in the sequel are

$$E\{x(t) \mid x(0)\} = x(0) \quad , \tag{2.2}$$

$$E[x(t)\{1-x(t)\} \mid x(t-1)] = \{1-(2N)^{-1}\}x(t-1)\{1-x(t-1)\} , \tag{2.3}$$

$$Var\{x(t) \mid x(0)\} = [1-\{1-(2N)^{-1}\}^{t}]x(0)\{1-x(0)\} , \tag{2.4}$$

so that

$$\frac{Var\{x(t) \mid x(0)\}}{E\{x(t) \mid x(0)\}[1-E\{x(t) \mid x(0)\}]} = [1-\{1-(2N)^{-1}\}^{t}] . \tag{2.5}$$

Suppose $x(0)=(2N)^{-1}$. Then segregation will continue until
$x(\cdot)=0$ or 1 . The mean time that the frequency x of A_1
takes values in any arbitrary range (x_1,x_2) before segregation ceases is approximately

$$\int_{x_1}^{x_2} 2\, x^{-1}\, dx \quad , \quad (2N)^{-1} \leqslant x_1 < x_2 \leqslant 1-(2N)^{-1} \quad . \qquad (2.6)$$

With mutation from A_1 to A_2 at rate u , with no reverse mutation, (2.6) should be replaced by

$$\int_{x_1}^{x_2} 2x^{-1}(1-x)^{\theta-1}\, dx \quad , \quad (2N)^{-1} \leqslant x_1 < x_2 \leqslant 1 \quad , \qquad (2.7)$$

where $\theta = 4Nu$. If back-mutation from A_2 to A_1 also exists (at rate v) , a stationary distribution for the frequency x of A_1 is defined. The form of this distribution is approximately

$$f(x) = \text{const } x^{4Nv-1} (1-x)^{4Nu-1} \quad . \qquad (2.8)$$

Further results, as well as extensions of the above to more than two alleles, are available but are not described here.

More complex models assume some form of geographical structure in the population. One such model, the "island" model, is discussed in Section 4. A second class of geographically structured models involves a collection of sub-populations with approximately isotropic migration between them. We note here two general conclusions for such models. First, if there is no selection, the total system acts effectively as one large random-mating population if one or more individuals migrate from sub-population i to sub-population j each generation. Second, for certain models of the Moran type (Moran, 1958) with no selection or mutation, the distribution of the number of heterozygotes to appear before the (certain) loss of one or the other allele is invariant to the way in which the population is subdivided (Maruyama, 1972). This motivates the neutrality test of Yamazaki and Maruyama, discussed later.

<u>Infinite alleles model</u>   (Kimura and Crow, 1964).

We now suppose that the locus A admits alleles from the infinite sequence $A_1, A_2, A_3, \ldots$ Genes mutate with probability u , and it is assumed that each mutant forms an entirely novel allelic type not currently or previously seen in the population. The most commonly used model for this process is a direct generalization of (2.1): if in any generation there exist a_i genes of allelic type A_i (i=1,2,3,...), then the probability that in the following generation there exist b_i genes of allelic type A_i (i=1,2,3,...) together with m new mutant genes, is

$$\Pr\{\underset{\sim}{b},m \mid \underset{\sim}{a}\} = \frac{(2N)!}{m! \Pi b_i!} u^m \prod_i p_i^{b_i} , \qquad (2.9)$$

where $p_i = a_i(1-u)/2N$. If we again define $\theta = 4Nu$, standard results for this model, once stationarity has been achieved, are as follows:

(i) $\Pr\{\text{homozygosity of any individual}\} \approx (1+\theta)^{-1}$ (2.10)

(ii)   mean number of alleles present with frequency in

$$(x_1, x_2) \text{ (c.f. (2.7))} \approx \theta \int_{x_1}^{x_2} x^{-1}(1-x)^{\theta-1} dx , \qquad (2.11)$$

(iii)   if the population frequencies of the alleles present in any generation are $x_1, x_2, x_3, \ldots$, then $y = \Sigma x_i^2$ is a random variable such that

$$E(y) = (1+\theta)^{-1} \qquad (2.12)$$

(see (2.10)), and further (Stewart (1975))

$$\text{Var}(y) = 2\theta/[(1+\theta)^2(2+\theta)(3+\theta)] . \qquad (2.13)$$

The distribution of y , particularly for small θ values, is skewed or even bimodal, and more rarely is unimodal and approximately symmetric (Stewart, 1975, Ewens and Gillespie,

1974).

Infinite sites model   (Karlin and McGregor, 1967, Kimura, 1971)

In this model we make the assumption, inspired by molecular genetics models of the nature of the genetic material,
that the gene consists essentially of an infinite sequence of
sites. In the population at any time, each site will be
either monomorphic or will possess two segregating "types".
Mutation occurs at the rate u per gamete per generation,
all mutations being assumed to arise at a previously monomorphic site. This model is mathematically equivalent to (2.1),
reinterpreting "locus" by "site". We do not consider the
mathematical properties of this model in any detail, since
essentially no data appropriate to it currently exists: on
the other hand, this model is the most fundamental possible
for allele frequency behavior and it is to be hoped that
ultimately such data will be available to be tested in terms
of this model.

Ladder-rung model   (Ohta and Kimura, 1973)

This model is inspired by the generation of data by electrophoresis and attempts to describe the stochastic behavior
of such data. Consider an infinitely long ladder with rungs
at the points .... -2,-1,0,1,2,.... Any allele is allocated
to a specific rung on this ladder: several alleles may belong
to the same rung. If a gene of an allelic type belonging to
ladder rung i mutates, then under the simplest possible
model it forms an entirely novel allelic type which is allocated to ladder rung i-1 (probability p), ladder rung i+1
(probability p) or ladder rung i (probability 1-2p). It
is assumed that we are capable of observing only ladder rung
frequencies, and not allele frequencies - in other words, our
apparatus is not able to distinguish alleles occupying the

310

same ladder rung. For selectively neutral alleles, essentially the only known theoretical result for this model occurs
when $p = \frac{1}{2}$ and the mutation rate is u . In this case the
probability that the two genes in any individual occupy the
same ladder rung (so that the individual appears as a homozygote), is (Ohta and Kimura, 1973),

$$Pr\{"homozygosity"\} = (1+2\theta)^{-\frac{1}{2}} \qquad (2.14)$$

An empirical result for this model (Ewens and Gillespie, 1974)
is that ladder-rung frequencies tend, under selective neutrality, to be more uniform than are allele frequencies in the
infinite alleles model. We consider later the effect of this
observation on tests of the neutrality theory.

General observations

   Before considering specific tests arising from the above
models, we make some general observations.

   First, the neutrality theory is by its nature one which
unavoidably involves stochastic processes. Any analysis of
data relating to this theory must therefore be carried out on
a statistical, rather than a deterministic, basis. In such
statistical tests the neutral theory becomes the null hypothesis and to carry out any such test the null hypothesis
(i.e., neutral theory) distribution of the test statistic
used must be derived. The alternative hypothesis, that selection exists, is normally too imprecise to yield any specific
distribution, or to allow precise power calculations for the
test. We note also that the distribution of the test statistic is determined by an appropriate sample, rather than population, theory. The former is not known for several of our
models and is approximated by the latter.

   Second, the distribution of this statistic will depend

ultimately on which population model is assumed. Here a degree of ambiguity arises, since, e.g., several 'classical' models other than (2.1) exist, such as the model of Moran (1958). Fortunately, within any class of models, the distribution of any neutrality test statistic seems invariant: if this were not so, no statistical testing of neutrality could in principle be carried out.

Third, essentially all known results assume that stationarity of the stochastic process undergone by the population has been reached. The rate at which stationarity is in fact achieved, as measured by the leading non-unit eigenvalue of some transition matrix, is usually very slow in all the models we consider. Much of the data currently collected is probably "non-stationary" and may be worthless for testing neutrality.

Fourth, the unknown parameters N and u may enter into the distribution of any test statistic, causing obvious problems of subjectivity if these parameters must be estimated a priori. Fortunately, in the cases we consider, it has proved possible to devise tests which are independent of N and u , but in further tests this will not necessarily be the case.

Fifth, it is not entirely clear which model is appropriate for certain forms of data. It is commonly accepted that the ladder-rung model is the most appropriate for electrophoretic data, yet even this assumption has been challenged (Johnson, 1974). Unfortunately, very little theory is available for this model: on the other hand, an essentially complete theory exists for the infinite alleles model. It is likely that more rapid overall progress will be made by refinement of laboratory techniques (e.g., Singh, Hubby and Lewontin, 1974, Bernstein, Throckmorton and Hubby, 1973) which will yield data in a form appropriate for the neutral alleles model, than to attempt to solve the almost impossibly difficult mathematics of the ladder-rung model.

3. ALLELE FREQUENCY-BASED TESTS OF THE NEUTRALITY HYPOTHESIS

In this section we consider several formal tests of the neutrality hypothesis, based on the above theory, as well as several less formal arguments relating to the hypothesis.

Ewens' test (1972)

This test assumes data from a population obeying the infinite alleles model, specifically of the form

$$\{2n, k; n_1, n_2, \ldots, n_k\} \quad . \tag{3.1}$$

Here $2n$ represents the number of genes observed at a certain locus, $(n \ll N)$, k is the number of different alleles noted in this sample of genes, and $n_1 \ldots n_k$ $(\Sigma\, n_i = 2n)$ are the respective numbers of genes of each allelic type observed. To test for neutrality it is necessary to find the neutral theory distribution of the vector (3.1). Equation (2.11) may be used to this end: we find, (Ewens, 1972, Karlin and McGregor, 1972)

$$\Pr\{k; n_1 \ldots n_k \,|\, 2n\} = \frac{(2n)!\ \theta^k}{k!\ n_1 \ldots n_k\ L(\theta)} \quad , \tag{3.2}$$

$$\Pr\{k \,|\, 2n\} = \ell_k\, \theta^k / L(\theta) \quad , \tag{3.3}$$

$$\Pr\{n_1 \ldots n_k \,|\, k, 2n\} = \frac{(2n)!}{k!\, \ell_k n_1 \ldots n_k} \tag{3.4}$$

where $L(\theta) = \theta(\theta+1)(\theta+2)\ldots(\theta+2n-1)$

$$= \ell_1\theta + \ell_2\theta^2 + \ell_3\theta^3 + \ldots + \ell_{2n}\theta^{2n} \quad .$$

Note from equations (3.2)-(3.4) that k is a sufficient statistic for θ : thus in discussing whether or not the vector (3.1) is consistent with a priori views on the values of N and u , attention should be focused entirely on the value of k , and $n_1 \ldots n_k$ should be ignored. This has seldom been done: this point will be taken up again later. More importantly, the sufficiency of k for θ ensures that the conditional distribution (3.4) of $n_1 \ldots n_k$, given k , is independent of θ (and hence of N and u) : thus if (3.4) is used as a basis for a test of selective neutrality, it is quite unimportant that the values of N and u are not known.

The test is based on the statistic

$$I = -\Sigma\, x_i \log x_i \quad , \tag{3.5}$$

where $x_i = n_i/2n$, whose null hypothesis distribution can be found in principle from (3.4). Approximate significance points may be calculated and the test applied to any data vector of the form (3.1). Specific details and examples of this procedure are given in Ewens (1972).

This test shares certain problems with all those involving allele frequencies. First, no test is possible when $k=1$. In this case, all that can be done is an evaluation, using (3.3), of whether $k=1$ is compatible with a priori views on the values of N and u . Second, the test is not very powerful statistically, so that neutrality may often be accepted even though selection does occur. Values of n of order at least two or three hundred are required before some measure of power is achieved: this remark is especially true for $k=2$.

Because it is based on an infinite alleles model, this test is probably not strictly appropriate for electrophoretic data (although see Johnson, 1974). If it should be appropriate, it is of interest to note that formal testing of practically all published data sets using this test yields evidence of selection at approximately 30% of loci considered. Note in this regard the discussion following equation (2.14): if the data vector yields allele frequencies significantly less uniform than that predicted by the neutrality theory, and if in fact the data are more appropriately described by the ladder-rung model, then a fortiori the data yield significant evidence of selection. In this sense, at least, Ewens' test is useful for electrophoretic data.

2.2. <u>Johnson and Feldman's test</u> (1973)

A further test based on the same theoretical background as that of Ewens has been put forward by Johnson and Feldman (1973). This test assumes data appropriate to the infinite alleles model and also uses equations (3.2)-(3.4). Johnson and Feldman form the statistic

$$E = k \Sigma x_i^2 , \qquad (3.6)$$

where $0 \leqslant E \leqslant 1$, and note that E is a measure of the evenness of allele frequencies ($E=1$ if all alleles present are equally frequent). Using the distribution (3.4), Johnson and Feldman compute the expected value of E and note that for $n \geqslant 50$, $2 \leqslant k \leqslant 10$, this expected value is an increasing function of k . These n and k values cover essentially all of the existing data. They note that in the published data, the sample values of E tend to decrease with k , so that the data do not support the neutral theory.

This observation alone does not provide a formal test of the neutral theory: however, the standard deviation of E , computed by Feldman and Johnson (unpublished), may also be used to form a reasonable testing procedure. It is found that for $k \geqslant 7$ most of the observed values of E lie more than two standard deviations from the mean. Note again that this result must be viewed with caution since the neutral alleles model may not be appropriate to the data used.

The two tests just described are based on the supposition of random-mating populations, assuming that to the extent that populations are in fact subdivided, sufficient migration occurs to make random-mating theory applicable. Unless migration is extremely rare, this assumption is justified by the empirical results of Ewens and Gillespie (1974).

Yamazaki and Maruyama's test (1972)

This test is inspired by the invariance properties of heterozygote numbers in subdivided populations obeying the classical model (2.1).

Equation (2.6) shows that the expected number of heterozygotes to appear when the frequency of A_1 (initially $(2N)^{-1}$) is x, is

$$\text{const } x^{-1}\{x(1-x)\} = \text{const } (1-x) \quad . \tag{3.7}$$

It is supposed that which of A_1 or A_2 had initial frequency $(2N)^{-1}$ is unknown: this problem is overcome by folding the frequency scale around .5 and adding. With this procedure the expected total number of heterozygotes for any x value, $0 < x < .5$, is

$$\text{const } \{1-x + x\} = \text{const } . \tag{3.8}$$

The test proceeds by visually comparing the observed total heterozygote frequencies with this predicted horizontal line.

A number of problems arise in the application of this test to present-day data which invalidate the conclusions drawn by Yamazaki and Maruyama that the data they have tested support the neutral theory.

First, if the population is subdivided into n subpopula-

tions of relative sizes $w_1, w_2, \ldots, w_n$ ($\Sigma\ w_i = 1$) , the true
frequency of heterozygotes, assuming random mating within
subpopulations, is $2\ \Sigma\ w_i x_i (1-x_i)$, where x_i is the frequency of A_1 in subpopulation i . However, the individual
w_i values are normally unknown and are assumed by Yamazaki
and Maruyama to be equal, so that the frequency of heterozygotes is assumed to be $2n^{-1}\ \Sigma\ x_i(1-x_i)$. The assumption of
equal subpopulation sizes is in fact an extreme one, as measured for example by evenness, and can result (Ewens and
Feldman, 1974) in a systematic bias in the applicability of
the test of up to 50%.

Second, and more seriously, the classical model which in-
spires the test is quite inappropriate to electrophoretic
data. There are several reasons for this. The first is that
the heterozygote invariance only relates to a mutation-free
model with two alleles, while electrophoretic models funda-
mentally involve mutation and several alleles. The second is
that the existence of mutation makes (2.6) (and hence (3.8))
inappropriate; if the classical model were appropriate, then
(2.7) should be used. The third reason is that in fact neither (2.6) nor (2.7) is appropriate for electrophoretic data.
What we require are formulae parallel to these deriving from
the ladder-rung model: such formulae are quite unknown, and
would relate to ladder-rung frequencies, rather than real
allele frequencies. Considerable amounts of real heterozygosity would then not be observed due to non-identification.

Third, the test relies on a purely visual analysis of many
data points and is subject to considerable investigator bias.
In the original data used by Yamazaki and Maruyama, a regression line fitted formally to the data points yielded a
significantly non-zero slope, even though these authors
claimed that visually no significance was discernible.

317

Finally, the theory used is population theory rather than sample theory, and the two may have quite different properties. In all, we place no value on the analysis of Yamazaki and Maruyama, at least as it is applied to electrophoretic data.

The first problem referred to above is probably unavoidable in practice. On the other hand, the remaining problems relate to inappropriate modelling, and Yamazaki and Maruyama's test will become useful if data appropriate to their model become available. Data corresponding to the infinite sites model, whose mathematical theory is very close to that of the classical model, would approximately meet this requirement. Such data is several years in the future at best. It will involve the problem of extremely tight linkage between sites, leading to extremely difficult mathematical theory of highly correlated data.

We now describe briefly several more informal and verbal assessments of neutrality, partly in the light of the above theory.

Aspinwall (1974)

Aspinwall describes a possibly unique situation regarding the behavior of pink salmon on the Pacific coast of Canada. At age exactly two years the salmon mature and spawn in the streams in which they are born, dying immediately thereafter. Their progeny return to spawn exactly two years later. This behavior produces odd year and even year populations which are genetically isolated but which nevertheless occupy exactly the same habitat. Aspinwall observes different allele frequencies at three enzyme loci between odd and even year populations, and leans towards the neutral theory to explain this phenomenon. There is no formal theory appropriate for an objective test in this unusual case.

Ayala et al. (1974)

Based on extensive sampling of a number of Central and South American Drosophila species, Ayala et al. doubt the acceptability of the neutral theory on several different grounds.

First, they compare an a priori estimate of $\theta = 4Nu$, viz. $4 \times 10^9 \times 10^{-7} = 400$, with the estimate inspired by equation (2.12), namely

$$\hat{\theta} = \{\Sigma \; x_i^2\}^{-1} - 1 \quad , \tag{3.9}$$

where $x_1 \ldots x_k$ are the sample frequencies of alleles defined after (3.5). Note that, even if the infinite alleles model is appropriate for this data, estimation of θ using (3.9) is not. The sufficiency of k for θ implies that θ should be estimated by some function of k only: once k is known, $x_1 \ldots x_k$ contain no further information about θ . Estimation of θ using (3.9) involves a variance of an order of magnitude greater than that with estimation using k . Recognizing the ladder-rung model as being probably more appropriate for the data, Ayala et al. then compare $\Sigma \; x_i^2$ to the value deriving from (2.14), viz., $(1+800)^{-1/2}$. While no theory is yet available, it again seems likely that a priori estimates of θ should be compared to some function of the number of ladder-rungs occupied, rather than their frequencies.

Second, Ayala et al. compute the empirical distribution of heterozygote frequencies. They claim that the theoretical distribution should be approximately normal, with mean $\hat{\theta}/(1+\hat{\theta})$. They again reject the neutral theory because the empirical distribution is not of this form. This argument is incorrect since, as noted above, the neutral theory distribution of heterozygosity is usually far from normal and is often skewed in precisely the same form as their observed distributions. Similarly, the empirical bimodal distribution of heterozygosity observed by Harris et al. (1974) does not immediately imply loci with two classes of mutation rate, or a selective basis, since again this form of distribution can arise under the neutral theory, with fixed mutation rate.

We do not comment a length on the third argument against neutrality used by Ayala et al. relating to the strong similarity in allele frequency configurations in different populations, and even different species, despite extensive differences in local inversion patterns. Theoretical analysis of this is very difficult and problems also arise regarding a selective explanation for this phenomenon.

Mitton and Koehn (1973)

These authors consider the linkage disequilibrium between various enzyme loci and note a clinal behavior in this disequilibrium. They infer a selective basis for this phenomenon. While this conclusion is reasonable, especially if the cline can be correlated with environmental phenomena, the appropriate neutral theory analysis of linkage disequilibrium clines is just now becoming available (Feldman and Christiansen, 1974).

Several other authors (e.g., McNaughton, 1974, Schaffer and Johnson, 1974), have correlated environmental factors with heterozygosity and other gene frequency-based measures. A complex case is reported by Christiansen and Frydenberg (1974) who note the existence of different clines at different loci in eelpout populations in the Baltic. This is forceful evidence for the selectionist viewpoint -- but again no neutral theory is available. Clearly in this example a common-sense analysis may be all that is possible.

4. TESTS BASED ON STANDARDIZED ALLELE FREQUENCY VARIANCES

Classical facts for standardized variances

If a large population is subdivided into a number n of subpopulations $P_1, P_2 \ldots P_n$ of relative sizes w_i, in each of which the Hardy-Weinberg law holds at a locus with alleles A_1 and A_2 then (Wahlund, 1928) the overall frequency of heterozygotes is reduced below Hardy-Weinberg expectation. The fractional reduction is $s^2/\bar{x}(1-\bar{x})$, where x_i is the frequency of A_1 in P_i, $\bar{x} = \Sigma\, w_i x_i$ and $s^2 = \Sigma\, w_i (x_i - \bar{x})^2$ is the weighted variance of A_1 frequencies. The fraction $s^2/\bar{x}(1-\bar{x})$, called the Wahlund's or standardized variance, is then analogous to Wright's inbreeding coefficient F.

Now consider a single finite population of size N with no linear or other pressures altering allele frequencies, non-overlapping generations and Wright-Fisher binomial sampling (2.1). From (2.3) and (2.4) we see that the variance of the A_1 allele frequency $x(t)$, at time t, conditional on the initial generation, satisfies the relation

$$\mathrm{Var}\{x(t)\,|\,x(0)\} = \{1 - \frac{1}{2N}\}\mathrm{Var}\{x(t-1)\,|\,x(0)\} + \frac{x(0)\{1-x(0)\}}{2N} \quad (4.1)$$

Thus the standardized variance, expressed as $\mathrm{Var}\,(x(t)\,|\,x(0))/x(0)[1-x(0)]$ does indeed behave like the usual probability that an individual has both his genes of identical allelic type.

A direct extension of this situation is to the island model of Wright (1943). A finite population of size N exchanges individuals at the rate m per generation with a deterministic "mainland" population whose A_1 allele frequency is fixed at x^*. The same binomial sampling formulation as in Section 2 then leads to the classical result

$$\text{Var}\{x(t) \,|\, x(0)\}/x^*(1-x^*) \;\to\; [2N\{1-(1-m)^2(1-\frac{1}{2N})\}]^{-1}$$

$$\simeq\; 1/\{1+4Nm\} \quad,$$

(4.2)

neglecting terms of order m in the denominator. [It is not widely recognized that the rate at which the limit in (4.2) is approached depends on whether x(0) , the initial fre-quency in the island, is the same as x* .]

An alternative way of looking at (4.2) involves the sta-tionary distribution from the diffusion approximation to the binomial sampling model (see e.g., Crow and Kimura, 1970, p.392). The stationary probability density under the same conditions as above is the beta distribution

$$f(x) \;=\; \text{const } x^{4Nmx^*-1}(1-x)^{4Nm(1-x^*)-1}$$

(4.3)

which has variance

$$V^* \;=\; x^*(1-x)^*/\{1+4Nm\} \quad.$$

At the stationary state,therefore, (4.2) and (4.3) produce the same value 1/(1+4Nm) for the standardized variance, which in turn is equivalent to the equilibrium probability that an individual has both genes of identical allelic type.

As has been pointed out by Jacquard (1974, p.214), it is important to note that in the above calculations Var(x(t)|x(0)) is an unknown parameter of the process. Any estimate of this parameter will presumably involve several loci and/or several populations. The observed sample stan-dardized variances used for this estimate, derived from the different loci, will differ from each other. It was sugges-ted by Cavalli-Sforza (1966) that this variability in sample standardized variances, or sample inbreeding coefficients, might form the basis for a method of assessing neutrality. The basic argument used here is that, except for differences in founder populations, drift alone should not cause substan-tial differences in the sample inbreeding coefficient between different loci.

The first formal application of this idea to a practical testing procedure was that of Lewontin and Krakauer (1973). They amplified Cavalli-Sforza's proposal, suggesting that since natural selection should operate differently for different loci and different alleles, and since the effect of breeding structure (i.e., population subdivision and sampling effects) is uniform over all loci and all alleles, the variability over loci and alleles of the estimated inbreeding coefficient should provide a test of neutrality. In this section we present our ideas on the use of this procedure, which, for brevity, we call the L-K test. We shall show that the method is valid only under stringent assumptions which seem biologically unreasonable. Our observations extend those already made by Nei and Maruyama (1975) and Robertson (1975), who first noted that a number of difficulties arise with the L-K procedure.

Island models and standardized variances

We first ask whether, under an island model, the standardized sample variance has expectation independent of the locus for which it is computed. Consider n subpopulations, $P_1 \ldots P_n$, of sizes $N_1 \ldots N_n$, exchanging individuals with a mainland population at rates $m_1 \ldots m_n$ respectively. The mainland population is deterministic, with fixed frequency x^* of A_1. Then equation (4.2) shows that, at equilibrium, the frequency x_i of A_1 in P_i satisfies

$$E(x_i) = x^*$$

$$Var(x_i) = x^*(1-x^*)/[1+4N_i m_i]$$

Now write $\bar{x} = \Sigma N_i x_i / \Sigma N_i = \Sigma w_i x_i$, and suppose the observed allele frequency variance $\hat{V}$ is computed by

$$\hat{V} = \Sigma \ (x_i - \bar{x})^2/(n-1) \quad .$$

Then

$$E\{\hat{V}\} = x^*(1-x^*) (n-1)^{-1} \sum_i \{1+4N_i m_i\}^{-1}\{1-nw_i^2\} \quad .$$

Clearly $E(\hat{V})/x^*(1-x^*)$ is independent of x^*, i.e., does not depend on the locus from which $\hat{V}$ is estimated. This conclusion is still true if an amended definition of $\hat{V}$, taking the w_i values into account, is used.

In theoretical calculations relating to the above procedure, the quantity $\hat{V}'$, defined by

$$\hat{V}' = \Sigma(x_i-x^*)^2/n \quad , \tag{4.4}$$

is often used. Clearly

$$E\{\hat{V}'\} = n^{-1} \ x^*(1-x^*) \ \Sigma \ \{1+4N_i m_i\}^{-1} \quad , \tag{4.5}$$

and the invariance of $E(\hat{V}')/x^*(1-x^*)$ to the locus considered is again evident. We remark that x^* is usually unknown to us, so that $\hat{V}'$ is not a computable quantity. We should more properly compute $E(F)$, where

$$F = \Sigma(x_i-\bar{x})^2/[(n-1) \ \bar{x} \ (1-\bar{x})] \quad , \tag{4.6}$$

and check for independence from x^*. This expectation is extremely difficult to calculate and we have followed the widespread practice of making the approximation that to a suitable degree of accuracy

$$E\{\hat{V}/\bar{x}(1-\bar{x})\} = E\{\hat{V}\}/x^*(1-x^*) \quad .$$

We continue to make this approximation in our subsequent analysis.

Historical processes and standardized variances

Consider n populations $P_1 \ldots P_n$ as above, evolving now independently with no migration to a mainland population.

Suppose each population evolves according to (2.1). If the frequency of A_1 in population i at time t is denoted $x_i(t)$, then from (2.5)

$$\text{Var}\{x_i(t)|x_i(0)\} \approx x_i(0)\{1-x_i(0)\}\{1-\exp(-t/2N_i)\} \quad . \quad (4.7)$$

If we define

$$\bar{x}(t) = \Sigma \, w_i x_i(t)$$

and

$$V_t = \sum_i \{x_i(t) - \bar{x}(t)\}^2/(n-1) \quad ,$$

we have

$$E\{V_t\} = \sum_i \{x_i(0) - \bar{x}(0)\}^2/(n-1) \, +$$

$$+ \sum_i x_i(0)\{1-x_i(0)\}\{1-\exp(-t/2N_i)\}\{1-nw_i^2\}/(n-1) . \quad (4.8)$$

Clearly, if $x_i(0) = x(0)$ for all i, so initial allele frequencies in the population are identical, use of (2.2) and (4.8) shows that

$$E(V_t)/[E\bar{x}(t)\{1-E\bar{x}(t)\}]$$

is independent of $x(0)$, i.e., is independent of the locus. But if the initial frequencies are no longer identical, such is clearly no longer the case. The same conclusion is reached if instead of (4.8) we use (c.f.(4.4)) the quantity V_t', defined as

$$V_t' = \sum_i \{x_i(t) - x_i(0)\}^2/n \quad .$$

Obviously

$$E\{V_t'\} = \sum_i x_i(0)\{1-x_i(0)\}\{1-\exp(-t/2N_i)\}n^{-1}$$

so that

$$E\{V_t'\}/[E\bar{x}(t)\{1-\bar{x}(t)\}]$$

is independent of the $x_i(0)$ if and only if the latter are all equal. Of course if, as is frequently assumed, the various subpopulations are formed by the splitting-up of a larger population, it is reasonable to assume equal initial allele frequencies between subpopulations.

We have just shown that, to a suitable degree of approximation, the expectation of a standardized variance is in some, but not all, models commonly used to describe allele frequency behavior, independent of the locus from which the alleles are drawn. Even if such independence holds, considerable problems arise in the use of standardized sample variances to test for neutrality. This arises because such tests necessarily involve further distributional properties of standardized sample variances, in particular their variance properties. We now turn to consideration of these problems, in particular in the context of the L-K procedure.

The goodness-of-fit test

Lewontin and Krakauer consider two tests of selective neutrality, which we call the goodness-of-fit test (pp 179-181 and 183-186) and the k test (pp 186-187). In this subsection we review problems associated with the former test, while in the following two subsections we review problems associated with the latter.

The goodness-of-fit test proceeds as follows. In n populations, the frequencies $x_1 \ldots x_n$ of an allele A_1 at a certain locus are computed. From these a value of F is computed according to (4.6). The same procedure is then carried out for K loci, producing K values $F_1 \ldots F_K$ of F. Defining

$$\overline{F} = K^{-1}[F_1 + \ldots + F_K] , \qquad (4.9)$$

we then calculate a collection of K standardized values

$$G_i = (n-1)F_i/\overline{F} , \quad i = 1 \ldots K .$$

The test proceeds by comparing the empirical distribution of the G_i to a chi-square distribution with n-1 degrees of freedom by a standard goodness-of-fit procedure. The neutral theory is accepted if a satisfactory fit is achieved, and rejected otherwise.

There are several real points of difficulty concerning this procedure. First, the G_i values will have an approximate chi-square distribution only if the x_i readings making up each F value are independently and identically distributed normal random variables. We believe that, in the real world, it is extremely unlikely that any one of these three requirements will be met. First, migration and/or hierarchy between subpopulations will ensure dependence between the x_i values. Second, different population sizes will ensure different distributions for the x_i values. Third, and most important, an implicit assumption is made that normality and neutrality mutually imply each other. There is no reason, either from the general theory of population genetics or the work of L-K, that this should be the case. On the other hand, the standard selective distribution for two alleles at a locus with mutation can be normal. Thus, for example, the classical stationary selective distribution formula (9.34) in Crow and Kimura (1970) is essentially normal if we put (in their notation)

$$4N_e v = 4N_e u = 1 , h = .75 , 2N_e s \text{ large.}$$

The same result is given by Ewens (1969, equation 5.16), by putting, in his notation,

$$2\beta_1 = 2\beta_2 = 1 , \alpha_2 = .75\alpha_1 , \alpha_1 \text{ large.}$$

In other words, the goodness-of-fit test may really be a test for normality of the x_i, not of neutrality. But neutrality does not imply normality, as is implied by L-K. On the contrary, the only models in theoretical population genetics which would lead to normality of the x_i values are selective models.

We thus conclude that the goodness-of-fit test is quite valueless as a test for neutrality. The assumptions of independence and identity of distribution of x_i values will hardly ever arise in practice. Finally, the assumption of normality is not appropriate to the test L-K wish to make. They realize (p.177) that certain types of selection are included in the scope of their "neutrality" null hypothesis, so that in certain circumstances the test may be 180° wrong in its implication.

We now show that similar difficulties arise with the other procedure put forward by L-K as a test of neutrality.

The kurtosis and standardized variances

As above, we suppose that a sample of n populations has yielded frequencies $x_1 \ldots x_n$ for A_1 . We define the sample inbreeding coefficient, or, (as we prefer to call it) standardized variance, as in (4.6). Assume now as before that K values of F are computed, one for each of K loci. We can define a sample average $\bar{F}$ as in (4.9) and a sample variance s_F^2 , defined by

$$s_F^2 = \Sigma (F_i - \bar{F})^2 / (K-1) \quad ,$$

from these F_i values. From these values, L-K compute a new statistic k , defined by

$$k = (n-1) s_F^2 / \bar{F}^2 \quad , \tag{4.10}$$

and base their second test for neutrality on observed values of k . Clearly the properties of the statistic k are needed to carry out such tests. On the basis of extensive simulation experiments, for several hypothetical distributions of the x_i (which are assumed independent), Lewontin and Krakauer conclude that an upper limit of 2 exists for $E(k)$, this value being reached when the x_i values are normally distributed. This conclusion is used crucially in their testing procedures. We now examine the validity of this procedure.

Suppose the x_i values used to compute any F value are independent random variables from a distribution with mean μ, variance σ^2 and fourth central moment μ_4. Clearly, from (4.10), we must find the mean and variance of F in terms of these parameters. We do this by using the definition (4.6) for F and assuming, to a sufficient order of accuracy (as discussed above),

$$E(F) = E[\Sigma(x_i - \overline{x})^2/\{(n-1)\mu(1-\mu)\}] \ ,$$
$$\text{Var}(F) = \text{Var}[\Sigma(x_i - \overline{x})^2]/[(n-1)^2\mu^2(1-\mu)^2] \ . \tag{4.11}$$

Under this approximation, standard statistical formulae give

$$E(F) = \sigma^2/\mu(1-\mu)$$
$$\text{Var}(F) = [E(F)]^2[\frac{\mu_4}{\sigma^4} - 1 + n^{-1}\{3 - \frac{\mu_4}{\sigma^4}\}]/(n-1) \ . \tag{4.12}$$

If we ignore terms of order n^{-1} and define the kurtosis γ_2 of the x distribution by

$$\gamma_2 = \frac{\mu_4}{\sigma^4} - 3 \ ,$$

then (4.12) yields immediately

$$\frac{(n-1)\text{Var}(F)}{\{E(F)\}^2} = \gamma_2 + 2 \ . \tag{4.13}$$

Before discussing this result, we note a slightly different approach adopted by Jacquard (1974, p.217). Jacquard amends the definition (4.6) of F to

$$F' = \Sigma(x_i - \mu)^2/[n\mu(1-\mu)] \ , \tag{4.14}$$

and correspondingly defines

$$k' = n \; s_F^2 / (\overline{F'})^2 \; .$$

Under this definition

$$E(F') = \sigma^2 / \mu(1-\mu) \; ,$$

$$\text{Var}(F') = \{E(F')\}^2 [\frac{\mu_4}{\sigma^4} - 1]n^{-1} \; .$$

Thus without ignoring small order terms we have, exactly,

$$\frac{n \; \text{Var} \; (F)}{\{E(F)\}^2} = \gamma_2 + 2 \; . \qquad (4.15)$$

Note that since μ is normally unknown, F' is not strictly a computable quantity, (see the above discussion concerning V'), and its interest to us is purely theoretical. We shall see later that some care is needed in judging the usefulness of F'. Both (4.13) and (4.15) may be used to evaluate the usefulness of k (defined in (4.10)), and in particular the claim that 2 is an upper bound for values of k. Clearly the value of k depends crucially on the kurtosis of the underlying x distribution. Thus for normal distributions $(\gamma_2=0)$ we expect k values of 2 : this is observed in the simulations of Lewontin and Krakauer (their Table 2). Similarly, for a rectangular x distribution $(\gamma_2 = -1.2)$ we expect k values of 0.8 : this again is observed in the simulations. But it is clear that values of k in excess of 2 can arise from certain x distributions. For example, if x has the beta distribution

$$f(x) = \text{const } x^{a-1} (1-x)^{b-1} \; ,$$

then to a suitable approximation

$$E(k) = \frac{3(a+b+1)[a^2(b+1)+b^2(a+1)+(a-b)^2]}{ab(a+b+2)(a+b+3)} - 1 \; .$$

If $a = .9$, $b = .1$, we find $E(k) = 7.61$, far in excess of the proposed upper bound of 2. Such asymmetrical beta distributions may not at all be unreasonable, in view of the stationary distributions (2.8) and (4.3).

We believe that this analysis shows that, even for independent observations, a test based on an assumed upper bound of 2 for k is not allowable. We now turn to a further situation in which k values can exceed 2, even for normal data.

The effect of correlation on standardized variances

Suppose that the underlying allele frequency distribution is normal*, with mean μ and various σ^2. Then if $x_1 \ldots x_n$ are independent observations from this distribution, $\Sigma (x_i - \bar{x})^2/\sigma^2$ has a chi-square distribution with n-1 degrees of freedom. If n is sufficiently large, this implies that to a sufficiently close approximation, $(n-1)F/E(F)$, with F defined in (4.6), has a chi-square distribution with n-1 degrees of freedom. This fact is used by L-K as the basis of one of their tests of homogeneity of F values and hence of neutrality.

There are two problems associated with this procedure. First, if $x_1 \ldots x_n$ are not independent, (and in our view samples from adjacent populations seldom will be independent, because of migration), the chi-square distribution does not hold and the test cannot be used. Second, it is necessary to make a more or less subjective assessment of the normality assumption. If this assumption cannot be made, L-K revert to a test based on k, defined above. But this test also will be affected by correlation between the x_i values, and we now turn to a consideration of the extent of this effect.

---

* Strictly speaking, the frequencies are defined in [0,1] ; we have ignored this difficulty and proceeded in the manner of L-K.

Assume that we observe a vector $x_1 \ldots x_n$ of allele frequencies, with x_i the frequency of A_1 in P_i . Suppose this vector has a multivariate normal distribution with $E(x_i) = \mu$, $Var(x_i = \sigma^2)$, and $corr(x_i, x_j) = \rho_{ij}$. $Var(F)$ must be recalculated to take account of these correlations, which were assumed absent in the calculation of (4.12). Tedious but straightforward algebra yields

$$E(k) \approx \frac{(n-1)Var(F)}{\{E(F)\}^2} =$$

$$= 2 + \frac{2}{(n-1)(1-\bar{\rho})^2} [\sum_{i \neq j} \sum (\rho_{ij} - \bar{\rho}_i)^2 - \frac{(n-1)(n-2)}{n} \sum (\bar{\rho}_i - \bar{\rho})^2]$$

(4.16)

where

$$\bar{\rho}_i = (n-1)^{-1} \sum_{j \neq i} \rho_{ij} \quad , \quad \bar{\rho} = 2\{n(n-1)\}^{-1} \sum_{i < j} \sum \rho_{ij} \quad .$$

Similarly, with F' as defined in (4.14),

$$E(k') \approx \frac{n \, Var(F')}{\{E(F')\}^2} = 2 + \frac{2}{n} \sum_{i \neq j} \sum \rho_{ij}^2 \quad .$$

(4.17)

The form of (4.16) and (4.17) enables us to assess the effect of correlations between allele frequencies, which we do by a series of remarks.

Remark 1.   In (4.17) it is clear that, to the order of accuracy we employ, the effect of correlation is always to increase $E(k')$ above 2 . If $\rho_{ij} = 1$ we have $E(k') = 2n$, as is otherwise obvious. On the other hand, if all the x_i are highly correlated, $E(k)$ is not necessarily large: indeed, if all the ρ_{ij} values are equal and large (but less than unity), then $E(k) = 2$. It is clear, from the definitions of F and F' , why this difference in behavior occurs. This makes us somewhat wary about the usefulness of the theoretical and non-computable quantity F' in discussing properties of standardized variance tests.

Remark 2. We have attempted to check whether E(k) , defined in (4.16), always exceeds 2 . This is certainly so for n=3. As a numerical example of this, suppose $\rho_{12} = .5$, $\rho_{13} = .1$, $\rho_{23} = .9$. Then E(k) = 3.72 , indicating the extent to which the claim E(k) = 2 has been violated. For values of n in excess of 3 , it remains an open question whether the right hand side in (4.16) always exceeds 2 .

Remark 3. It may eventually be possible from knowledge of migration frequencies (see e.g. Bodmer and Cavalli-Sforza, 1968) and/or genetic distances to estimate the ρ_{ij} in (4.16). It is then conceivable that k could be corrected to allow for the correlation effect.

Remark 4. All the analysis above has been made on the assumption that the variances of the allele frequencies in each population are the same. This seems to be unlikely in the real world since the variances would be closely tied to population size.

Remark 5. L-K argue, correctly, that when the x_i are normal, (and independent), (n-1)F/E(F) has approximately a chi-square distribution. Their test proceeds by calculating the sample variance of F values, s_F^2 , and from this $(n-1)s_F^2/2\bar{F}^2$, and testing for significance by referring to tables of significance of chi-square. But it is impossible that both individual F values, and also s_F^2 , should both be (up to a scale factor) distributed as chi-square. There is therefore an internal inconsistency in the application of the test.

Remark 6. In three recent studies, Nevo (1974), Kidd and Cavalli-Sforza (1974) and Bodmer, Cann and Piazza (1973), the standardized variances have been used to investigate allele frequency data for evidence of selection. In view of our analysis it seems unlikely that the assumptions necessary for the valid use of an L-K type procedure were met in these studies.

5. CONCLUSIONS

The attempts described above to assess the neutrality theory on the basis of gene frequencies and theoretical population genetics models have not in our view been successful. There are several reasons for this. Apart from the formation of tests not well founded on theory and involving internal contradictions, perhaps the main reason is that insufficient attention has been paid to using the correct mathematical model for the data used in any test. It is not even certain that real data correspond to any of the models discussed above. We believe that the gap between the real models probably satisfied by data, and those so far used, is too large to validate the testing procedures we exhibit. In this connection it should be noted that further technical advances may well require the creation of further models, or perhaps make more relevant some of the models above, in particular the infinite alleles model and the infinite sites model. It is more likely that rapid advances will be made in this direction than in the direction of deriving a full mathematical theory of the ladder-rung model. Even if model and data are eventually successfully aligned and valid tests developed, it is possible that those tests will not have sufficient statistical power to resolve the neutrality question without larger amounts of data than can realistically be acquired.

Several broader avenues than those developed above are possible. One possible approach is the development of a not necessarily genetic theory of phenotype (Lewontin, 1974) which will allow direct correlation with the environment. This clearly involves more ecology than population geneticists have been willing to use. Another approach (Lewontin, 1974) is to consider many loci and linkage disequilibria. The sampling theory for this seems extremely difficult and is a long way in the future. Problems of the power of such tests and the difficulty of determining appropriate models will, as with the above single locus theory, be encountered, and we may not be qualitatively much better off even if this enterprise is carried through.

It is possible that ultimately the most useful approaches involving gene frequencies will not involve theoretical population genetics but rather general arguments and statistical tests relating to gene frequencies and locus function. This is the kind of approach already pioneered by Gillespie and Kojima (1968) and Johnson (1973).

REFERENCES

Aspinwall, N. (1974). Evolution 28: 295-305.

Ayala, F.J., M.L. Tracey, L.G. Barr, J.F. McDonald and S. Perez-Sales. (1974). Genetics 77: 343-384.

Bernstein, S.C., L.H. Throckmorton and J.L. Hubby. (1973). Proc. Nat. Acad. Sci. 70: 3928-3931.

Bodmer, W.F. and L.L. Cavalli-Sforza. (1968). Genetics 59: 565-592.

Bodmer, W.F., H. Cann and A. Piazza. (1973). Histocompatibility Testing 1972, 753-767.

Cavalli-Sforza, L.L. (1966). Proc. Roy. Soc., Ser. B, 164: 362-379.

Cavalli-Sforza, L.L. and W.F. Bodmer. (1971). The Genetics of Human Populations. W.H. Freeman, San Francisco.

Christiansen, F.B. and O. Frydenberg. (1974). Genetics 77 : 765-770.

Crow, J.F. and M. Kimura. (1970). An Introduction to Population Genetics Theory. Harper and Row, New York.

Ewens, W.J. (1969). Population Genetics. Methuen, London.

Ewens, W.J. (1972). Theor. Pop. Biol. 3: 87-112.

Ewens, W.J. and M. Feldman. (1974). Science 183: 446-448.

Ewens, W.J. and J.H. Gillespie. (1974). Theor. Pop. Biol. 6: 35-57.

Feldman, M.W. and F.B. Christiansen. (1975). Genet. Res. 24: 151-162.

Fisher, R.A. (1930). The Genetical Theory of Natural Selection. Clarendon Press, Oxford.

Gillespie, J.H. and K. Kojima. (1968). Proc. Nat. Acad. Sci. 61: 582.

Harris, H. (1966). Proc. Roy. Soc. Ser. B, 164: 298-310.

Harris, H., D.A. Hopkinson and E.B. Robson. (1974). Ann. Hum. Genet. 37: 237-253.

Jacquard, A. (1974). The Genetic Structure of Populations. Springer-Verlag, Berlin.

Johnson, G. (1973). Nature New Biology 243: 151-153.

Johnson, G. (1974). Genetics 78: 771-776.

Johnson, G. and M.W. Feldman. (1973). Theor. Pop. Biol. 4: 209-221.

Karlin, S. and J.L. McGregor. (1967). Proc. Fifth Berkeley Symp. Math. Stat. and Prob. 4: 415-438.

Karlin, S. and J.L. McGregor. (1972). Theor. Pop. Biol. 3: 113-116.

Kidd, K. and L.L. Cavalli-Sforza. (1974). Evolution 28: 381-395.

Kimura, M. and J.F. Crow. (1964). Genetics 49: 725-738.

Lewontin, R.C. (1974). The Genetic Basis of Evolutionary Change. Columbia Univ. Press, New York.

Lewontin, R.C. and J.L. Hubby. (1966). Genetics 54: 595-609.

Lewontin, R.C. and J. Krakauer. (1973). Genetics 74: 175-195.

Maruyama, T. (1972). Genet. Res. 20: 141-149.

McNaughton, S.J. (1974). Am. Natur. 108: 616-624.

Mitton, J.B. and R.K. Koehn. (1973). Genetics 73: 487-496.

Moran, P.A.P. (1958). Proc. Camb. Phil. Soc. 54: 60-71.

Nei, M. and T. Maruyama. (1975). Genetics. (To appear).

Nevo, E., Y.J. Kim, C.R. Shaw and C.S. Thaeler. (1974). Evolution 28: 1-23.

Ohta, T. and M. Kimura. (1973). Genet. Res. 22: 201-204.

Robertson, A. (1975). Genetics. (To appear).

Schaffer, H.E. and F.M. Johnson. (1974). Genetics 77: 163-168.

Selander, R.K. and D.W. Kaufman. (1973). Proc. Nat. Acad. Sci. 70: 1875-1877.

Singh, R.S., J.L. Hubby and R.C. Lewontin. (1974). Proc. Nat. Acad. Sci. 71: 1808-1810.

Stewart, F.M. (1975). Theor. Pop. Biol. (To appear).

Wahlund, S. (1928). Heredites 11: 65-106.

Wright, S. (1943). Genetics 28: 114-138.

Wright, S. (1931). Genetics 16: 97-159.

Yamazaki, T. and T. Maruyama. (1972). Science 178: 56-57.

Non-Random Association of Neutral Linked Genes in Finite Populations

W.G. HILL

INTRODUCTION

With the widespread use of gel electrophoresis methods to estimate frequencies of polymorphic loci in natural or laboratory populations, information is now also being obtained on linkage disequilibrium between such loci. Using data on disequilibria (an expression of association of gene frequencies at different loci) some additional understanding may be obtained as to the nature of the forces of selection, mutation, drift and migration which help to maintain or reduce polymorphism. Lewontin (1974) has recently reviewed the relevant experimental work and developed these arguments further.

Theoretical and, as yet, experimental studies of linkage disequilibria have mostly been restricted to only pairs of loci. Predictions for approach to equilibria for neutral genes in infinite populations were given by Geiringer (1944) and Bennett (1954). These results have been extended to give expected values of means, variances and covariances of disequilibria for neutral genes in finite populations, but are limited to sixth moments, e.g., the mean disequilibrium among six loci or the variance of disequilibrium at three loci (Hill, 1974a,b). These expectations are computed over all populations and include those in which one or more loci have reached fixation. It is probably more important to consider disequilibria solely among populations which are segregating at all the relevant loci, and predictions of variation in disequilibria for neutral genes form the subject of this paper. Whilst it was possible to develop analytical methods for disequilibria among all populations, we now have to resort largely to Monte Carlo methods. This parallels previous two locus studies, although some approximations for two-locus disequilibria in segregating populations have been obtained analytically (Sved, 1971; Sved and Feldman, 1973; and see also the review by Kimura and Ohta, 1971).

Effects of selection are not included, the intention being to provide a basis against which selection effects can be compared and also to discuss ways in which data from populations might be analysed. Problems of selection in infinite populations have been reviewed recently for two loci by Karlin (1975) and for more loci have been discussed by Lewontin (1964a,b), Slatkin (1972); Strobeck (1973) and Feldman, Franklin and Thomson (1974). Franklin and Lewontin (1970) also considered selection effects although by simulation in a finite population.

Throughout we shall assume there are just two alleles at each locus; most of the detailed analysis is restricted to three loci, but the extension to more is illustrated and introduces no conceptual difficulties. The extension from two to three loci does introduce some problems, however.

MODEL

Consider loci A,B,C,D,... having that order on a chromosome, and let, for example, a and a' be alternative

340

alleles at the A locus. There is no mutation and all alleles are neutral with respect to fitness. Gene and chromosome frequencies are denoted p, for example p_a is the frequency of the allele a and $p_{abc'}$ the frequency of the chromosome having alleles a, b and c'. The recombination frequency between loci A and B, for example, is y_{AB}, and there is assumed to be no interference. The map length between these loci is ℓ_{AB}, and since map lengths are additive,

$$\ell_{AC} = \ell_{AB} + \ell_{BC} \ ,$$

for example. Some comparisons of map length and recombination fraction are shown in Table 1.

TABLE 1. *Comparison of* $E(r^2_{AB})$ *obtained by transition probability matrix iteration for* $N=8$ *with that predicted by Sved and Feldman. The ratio of expected values of the moments* D^2_{AB} *and* $p_a p_a, p_b p_b,$ *is also given, as is* z_{AB}, *(the likelihood ratio statistic* $x1/2N$).

ℓ_{AB}	1/4	1/8	1/16	1/32	1/64	1/128	0
L_{AB}	2	1	0.5	0.25	0.125	0.0625	0
y_{AB}	0.1967	0.1106	0.05875	0.03029	0.01538	0.007752	0
Ny_{AB}	1.5736	0.8848	0.47000	0.24232	0.12304	0.062016	0
$1/(4Ny_{AB}+1)$	0.1371	0.2203	0.3472	0.5078	0.6702	0.8012	1
$1/[1+(4N-2)y_{AB} -(2N-1)y^2_{AB}]$	0.1582	0.2419	0.3689	0.5277	0.6859	0.8119	1
$E(r^2_{AB})$	0.1614	0.2620	0.4328	0.6322	0.7898	0.8878	1
$\dfrac{E(D^2_{AB})}{E[p_a p_a, p_b p_b,]}$	0.1776	0.2880	0.4638	0.6578	0.8062	0.8970	1
z_{AB}	0.1810	0.2882	0.4688	0.6805	0.8493	0.9549	1.0766

The population comprises N diploid individuals, and products of population size and map length are denoted L, e.g., $L_{AB} = N\ell_{AB}$. Generations, t, are non-overlapping.

A haploid model is used to reproduce the populations. From the 2N chromosomes of one generation the frequency distribution is squared to give the expected genotypic frequencies after random mating including random selfing. The expected gametic output after recombination is computed, and the next generation obtained by sampling 2N chromosomes from the multinomial distribution. This procedure is used exactly in the algebraic analysis (Hill, 1974a), and in the Monte Carlo method it is simulated by sampling 2N pairs of parental chromosomes with replacement, and sampling from each pair a recombinant progeny chromosome.

REVIEW OF TWO-LOCUS THEORY

Some difficulties are encountered in defining and testing for association of gene frequency at three or more loci. Thus we consider some of the alternatives and initially review the two locus theory.

If there is dependence of frequencies at the two loci, $P_{ab} \neq P_a P_b$, and the measure of disequilibrium commonly used is

$$D_{AB} = P_{ab} - P_a P_b = P_{ab} P_{a'b'} - P_{ab'} P_{a'b} \quad . \tag{1}$$

For neutral genes in infinite populations at generation t

$$D_{AB(t)} = (1-y_{AB}) D_{AB(t-1)}$$

and for finite populations, taking expectations over a conceptual set of identical populations,

$$E(D_{AB(t)}) = (1-1/2N)(1-y_{AB}) D_{AB(t-1)} \tag{2}$$

for the haploid model (Wright, 1933). Thus the mean disequilibrium always approaches zero.

The disequilibrium can also be viewed as the covariance of gene frequencies. For example, Slatkin (1972) expressed it as

$$D_{AB} = E[(x_a - p_a)(x_b - p_b)] \qquad (3)$$

where x_a and x_b are the number (i.e., 0 or 1) of a and b genes, respectively, on the chromosome, and expression (3) is equivalent to (1). Based on the covariance concept, a useful measure of two locus disequilibrium is the correlation, r_{AB}, or squared correlation of gene frequencies

$$r_{AB}^2 = D_{AB}^2 / (p_a p_{a'}, p_b p_{b'}) \qquad , \qquad (4)$$

which is defined only for segregating populations (Hill and Robertson, 1968). The range of possible values of r_{AB}^2 is much less dependent on gene frequencies than is D_{AB}, although, as Sved (1971) has pointed out, r_{AB}^2 cannot reach values of unity for many combinations of gene frequencies. The property of r_{AB}^2 of which we shall make most use in this paper is its relation to the chi-square statistic in the contingency table test for association between alleles at the A and B locus when 2N chromosome types are identified. This statistic is $2Nr_{AB}^2$, so in the first generation of finite population started from a population in equilibrium, $2Nr_{AB}^2$, measured among the sampled parental chromosomes, has an approximately χ^2 distribution with 1 d.f., and thus r_{AB}^2 has a mean of 1/2N (Hill and Robertson, 1968). In a biological population the test for association has usually to be carried out from a sample of progeny whose numbers differ from that of the parents. Also, as a rule, the data available are on diploids so that chromosome frequencies have to be estimated by maximum likelihood. However, for codominant loci, the statistic for testing for association is nr_{AB}^2, where n are the number of chromosomes (in the haploid case) or diploid individuals sampled (Hill, 1974c).

As shown by (2), the mean disequilibrium over populations approaches zero for neutral genes, but as a result of sampling there is a variance in D_{AB} between populations. The asymptotic value of $E(r^2_{AB})$ taken over segregating populations was shown by Hill and Robertson (1968) to equal $1/4Ny_{AB}$ approximately, for large y_{AB} , and to approach unity for $y_{AB} = 0$. Subsequently, Sved (1971) found values for $E(r^2_{AB})$ using an argument which is rather hard to follow, and his result was modified by Sved and Feldman (1973) to give

$$E(r^2_{AB}) = 1/[1+(4N-2)y_{AB} - (2N-1)y^2_{AB}] \quad , \quad (5)$$

which simplifies to

$$E(r^2_{AB}) \sim 1/(1+4Ny_{AB}) \quad (6)$$

for large N and small y_{AB} . It is clear that (6) is correct either when Ny_{AB} is very large or approaches zero, but we have undertaken some numerical checks for other values. For populations with $N \leq 8$ this was done by transition probability matrix iteration, with the matrix kept to small size by utilising the symmetry of the model in a program described elsewhere (Hill, 1969); for larger values of N Monte Carlo simulation was used. Some typical exact results are given in Table 1 for $N=8$. It is clear that while the Sved-Feldman formula (5) gives a good general impression of $E(r^2_{AB})$, it is not formally correct. The largest differences occur around $Ny_{AB} = 0.25$, the Sved-Feldman values being underestimated by some 20%. The table also shows that even for this small N value, the more involved formula (5) is little better than the approximation (6). Similar results were obtained using $N=4$ and $N=6$, by matrix iteration. Simulation for $N\ell_{AB} = 0.25$ and 0.5 , corresponding to $Ny_{AB} = 0.246$ and 0.488 , respectively, was carried out with $N=20$ and 6400 replicates, giving values of $E(r_{AB}^2)$ of about 0.62 and 0.39, compared with the Sved-Feldman predictions from (5) of 0.51 and 0.35, respectively. It seems clear that, with increasing population size, the value of $E(r_{AB}^2)$ will not asymptote at $1/(1+4Ny_{AB})$. The prediction of changes with generation in $E(r_{AB}^2)$ in populations starting at equilibrium, which can be obtained from Sved and Feldman (1973), are also somewhat in error. (We are not sure where the logic of the Sved-Feldman approach breaks down, one possibility is in the assumption that the same probabilities of identity at the two loci apply in segregating and non-segregating populations.)

The moments $E(D_{AB}^2)$ and $E(p_a p_a, p_b p_b,)$ over all populations, whether segregating or not, are more readily computed than $E(r_{AB}^2)$, either by iteration of a moment generating matrix (Hill and Robertson, 1968) or by a diffusion approximation (Ohta and Kimura, 1969). As Ohta and Kimura (1969) showed by Monte Carlo simulation, however, the steady state value of the ratio $\sigma_d^2 = E(D_{AB}^2)/E(p_a p_a, p_b p_b,)$ over all populations is a very good approximation to the steady state value of $E(r_{AB}^2)$, computed only in segregating populations. A further illustration, using exact values from the transition probability matrix for $N=8$, of the similarity of these two quantities is given in Table 1. Also the approaches to the steady state values of $E(r_{AB}^2)$ and σ_d^2 with increasing generations are very similar.

Whilst not of particular relevance to the two locus review, we introduce a further measure of disequilibrium which will be useful for more loci. An alternative to the chi-square method of testing for association in a contingency table and thus for gene linkage disequilibrium is the likelihood ratio test (e.g., Sokal and Rohlf, 1969). The test statistic for chi-square may be written

$$\Sigma\,(\text{observed} - \text{expected})^2/\text{expected},$$

whereas that for the likelihood ratio is

$$2\,\Sigma\,(\text{observed})\,\log_e(\text{observed}/\text{expected}). \qquad (7)$$

In this context, the likelihood ratio statistic (7) can be written in terms of population size and frequencies as

$$2Nz_{AB} = 4N[p_{ab}\log(p_{ab}/p_a p_b) + \ldots + p_{a'b'}\log(p_{a'b'}/p_{a'}p_{b'})]$$

giving

$$z_{AB} = 2[p_{ab}\log p_{ab} + \ldots + p_{a'b'}\log p_{a'b'} - p_a\log p_a - \ldots - p_b \log p_{b'}] \quad (8)$$

(where we are dealing with $2N$ identified chromosomes). In a sample of parents from a population in linkage equilibrium, the likelihood ratio $2Nz_{AB}$ is asymptotically (for large N) distributed as χ^2 with 1 d.f. and it is easy to show that

$$z_{AB} = r_{AB}^2 + \text{terms in } D^3, D^4, \ldots$$

It turns out, however, that z_{AB} and r_{AB}^2 are numerically very similar over a wide range of parameters, and $E(z_{AB}^2)$ taken over segregating populations is closely approximated by $E(r_{AB}^2)$, as demonstrated in Table 1.

MEASURES OF DISEQUILIBRIUM FOR THREE OR MORE LOCI

A three locus disequilibrium, D_{ABC}, was defined by Bennett (1954) and equals that of Slatkin (1972). Extending (3),

$$D_{ABC} = E[(x_a - p_a)(x_b - p_b)(x_c - p_c)]$$

$$= p_{abc} - p_a p_{bc} - p_b p_{ac} - p_c p_{ab} + 2p_a p_b p_c \qquad (9)$$

It has been shown that

$$E(D_{ABC(t)}) = (1 - 1/2N)(1 - 1/N)(1 - y_{AB})(1 - y_{BC})D_{ABC(t-1)} \qquad (10)$$

(Hill, 1974a), and, of course, (10) gives the infinite population result of Bennett as $N \rightarrow \infty$.

A test for association of gene frequencies now involves a

2^3 contingency table and assuming that the gene frequencies, which are the marginal frequencies of the table, are estimated in the same analysis there are a total of 4 degrees of freedom available for testing the hypothesis of independence, $P_{abc} = P_a P_b P_c$. A partition of these 4 d.f. was suggested (in a general rather than genetic context) by Lancaster (1951) and follows the usual analysis of variance of a 2^3 factorial: use 1 d.f. each for testing pairs of loci together, i.e., the three hypotheses $P_{ab} = P_a P_b$, $P_{ac} = P_a P_c$ and $P_{bc} = P_b P_c$ in each case summing over frequencies at the third locus, and attribute the residual chi-square to the 1 d.f. for three-locus association or disequilibrium. If we denote by x^2 the chi-square statistic with 4 d.f. for testing $P_{abc} = P_a P_b P_c$, which is

$$x^2 = 2N[(P_{abc} - P_a P_b P_c)^2 / (P_a P_b P_c) + \ldots +$$
$$(P_{a'b'c'} - P_{a'} P_{b'} P_{c'})^2 / (P_{a'} P_{b'} P_{c'})] \quad , \tag{11}$$

it can be shown that

$$x^2 = 2N[r_{AB}^2 + r_{AC}^2 + r_{BC}^2 + r_{ABC}^2] \quad , \tag{12}$$

where, by analogy with (4),

$$r_{ABC}^2 = D_{ABC}^2 / (P_a P_{a'} P_b P_{b'} P_c P_{c'}) \tag{13}$$

and D_{ABC} is given by (9). The quantity r_{ABC}^2 is not a correlation as such, but a natural extension of the correlation concept to three variables. Thus it appears at first sight that Lancaster's partition can be interpreted immediately in terms of two-locus disequilibria; and in a sample of 2N chromosomes from a population in equilibrium $2N r_{ABC}^2$ is approximately distributed as χ^2 with 1 d.f.

The partition due to Lancaster has been criticised on several grounds, however, initially by Plackett (1962) who argued that Lancaster's criterion was not strictly a test of three-way (three-locus) association. He considered the necessary criterion to be that the association between, say, A and B should be the same in the group (chromosomes) having c as c' , and this criterion should be the same if we consider A with C and B with C in addition. The criterion proposed by Bartlett (1935) to define no three-way association,

$$P_{abc}P_{ab'c'}P_{a'bc'}P_{a'b'c} = P_{abc'}P_{ab'c}P_{a'bc}P_{a'b'c'} \qquad (14)$$

does satisfy Plackett's conditions: it is symmetric among the loci, and expressing the association between A and B at each level of C , (14) gives

$$\frac{P_{abc}P_{a'b'c}}{P_{ab'c}P_{a'bc}} = \frac{P_{abc'}P_{a'b'c'}}{P_{ab'c'}P_{a'bc'}} \quad .$$

An example given by Plackett (1962) illustrates that Bartlett's criterion (14) for no three-way association is not the same as that due to Lancaster, which in genetical terms is equivalent to $D_{ABC} = 0$ (from (9)). If the frequencies, each multiplied by 24, are

$$P_{abc} = 1 \quad , \quad P_{abc'} = 2 \quad , \quad P_{ab'c} = 2 \quad , \quad P_{ab'c'} = 6 \quad ,$$

$$P_{a'bc} = 3 \quad , \quad P_{a'bc'} = 2 \quad , \quad P_{a'b'c} = 4 \quad , \quad P_{a'b'c'} = 4 \quad ,$$

equation (14) is satisfied, but $D_{ABC} \neq 0$. A more extreme example, but with the same outcome, which we shall find relevant to our subsequent discussion, is $P_{a'b'c'} = 1 - P_{abc}$ with all other frequencies equal to zero.

While the definition of disequilibrium among three loci given by (9) and the partition of chi-square in (12) have some appeal they do not conform with current methods of analysis of three-way tables, which more or less uniformly are based on the Bartlett-Plackett model (see e.g., Goodman (1969) and

348

Fienberg (1970) for exposition). In the three-way model a hierarchy of levels of independence among the frequencies can be constructed which can help interpretation (Goodman, (1969); Fienberg, (1970)); and in genetical language, a typical hierarchy is shown in Table 2, which follows Hill (1975).

TABLE 2.

a. Succession of models of association of gene frequencies at three loci.

Model	p_{abc}	Fitted Association	Log likelihood[§]
0 : Complete independence of frequencies	$p_a p_b p_c$	–	$K(A)+K(B)+K(C)$
1 : Frequencies at C independent of A & B[†]	$p_{ab} p_c$	AB	$K(AB)+K(C)$
2 : Independence at B & C conditional on A[†]	$p_{ab} p_{ac}/p_a$	AB,AC	$K(AB)+K(AC)-K(A)$
3 : No three-way association	p^*_{abc}[‡]	AB,AC,BC	$K*(ABC)$
4 : All associations	p_{abc}	AB,AC,BC,ABC	$K(ABC)$

b. Succession of likelihood ratio statistics, each with 1 d.f.[††]

Source (difference in models)	Fitted Association	Log Likelihood ratio
Marginal assoc. A & B (1-0)	AB	$2Nz_{AB} = 2[K(AB)-K(A)-K(B)]$
Assoc. A & C given assoc. A & B (2-1)	AC	$2Nz_{AC} = 2[K(AC)-K(A)-K(C)]$
Assoc. B & C given assoc. A & B and A & C (3-2)	BC	$2Nz'_{BC} = 2[K*(ABC)-K(AB)-K(AC)+K(A)]$
Assoc. A, B & C given pair-wise assoc. (4-3)	ABC	$2Nz'_{ABC} = 2[K(ABC)-K*(ABC)]$
Total		$2Nz_{ABC} = 2[K(ABC)-K(A)-K(B)-K(C)]$

[†] One of three, [††] One of six, alternative hierarchies

[‡] No explicit formula for p^*_{ABC}, but satisfies (14)

[§] $K(A) = 2N(p_a \log p_a + p_{a'} \log p_{a'})$
$K(AB) = 2N(p_{ab} \log p_{ab} + \cdots + p_{a'b'} \log p_{a'b'})$
$K(ABC) = 2N\Sigma p_{abc} \log p_{abc}$, $K*(ABC) = 2N\Sigma p^*_{abc} \log p^*_{abc}$ (sum over 8 types)

349

This shows in Table 2a the values which would be taken by $p_{abc}, \ldots, p_{a'b'c'}$ when there is no association of frequencies, one pair is associated, and so on. There is no explicit formula for chromosome frequencies (p^*_{abc}) when there are all pair-wise but no three-way associations, with (14) satisfied. An iterative routine, however, given by Fienberg (1970) for example, can be used to obtain these values of p^*_{abc}. Also given in Table 2a are the log likelihoods, K , apart from constant terms, obtained by fitting the different models. For model 4 in which all two-way and three-way associations are fitted,

$$2N\Sigma \ p_{abc} \ \log \ p_{abc} = K(ABC) \qquad (15)$$

where summation is over all chromosome types. For model 2 in which, for example, B and C are independent, conditional on the gene at A , the expected frequencies satisfy $p_{abc} = p_{ab}p_{ac}/p_a$. Estimates of these pair-wise frequencies are given by marginal totals, e.g., $p_{ab} = p_{abc} + p_{abc'}$, so the log likelihood becomes, from (15),

$$2N\Sigma p_{abc} \ \log(p_{ab}p_{ac}/p_a) =$$
$$= 2N(\Sigma p_{ab} \ \log \ p_{ab} + \Sigma p_{ac} \ \log \ p_{ac} - \Sigma p_a \ \log \ p_a)$$
$$= K(AB) + K(AC) - K(A) \ , \qquad (16)$$

say, where K(AB) and K(A) denote log likelihoods computed from the specified marginal totals. For example K(A) = $p_a \ \log \ p_a + p_{a'} \ \log \ p_{a'}$. Using these likelihoods a succession of models can be fitted as shown in Table 2b, that given being one of 6 alternative sequences. For example, the test statistic $(2Nz'_{BC})$ for association between B and C, assuming association between A and B and between A and C is given by the likelihood ratio, or difference in log likelihoods, from fitting model 3 (all pair-wise but no three-way

associations) versus model 2 (B and C independent, but
conditional on the level of A , which implies possible associations between A and B and between A and C). It
turns out (see Table 2) that the likelihood ratio test for
association between a pair, A and B say, is the same if
no other pairs are fitted previously or one pair is fitted
previously. This ratio is denoted $2Nz_{AB}$, and is, of course,
equal to $2N \times$ that given by equation (8). The likelihood
ratio statistic for testing for three-way association is denoted $2Nz'_{ABC}$, and the total with 4 d.f. is denoted $2Nz_{ABC}$.
If there is no association, each statistic is asymptotically
χ^2 distributed. For further details of interpretation see
Smouse (1974) and Hill (1975).

It would be possible to undertake the same partition as
shown in Table 2a and analyse by the traditional
$[\Sigma$ (observed - expected)^2/expected] chi-square analysis. However we shall use likelihood ratios to quantify this partition
since they show more desirable properties in finite populations in which there are considerable departures from equilibrium after a few generations.

Four or more loci

The four locus disequilibrium defined by Bennett (1954) and
Hill (1974a) differs from that of Slatkin (1972), whose is
more closely related to chi-square. Extending (3) to four
loci we obtain

$$D_{ABCD(t)} = P_{abcd} - P_aP_{bcd} - \cdots - P_dP_{abc} +$$
$$+ P_aP_bP_{cd} + \cdots + P_cP_dP_{ab} - 3P_aP_bP_cP_d \quad . \quad (17)$$

Changes in $D_{ABCD(t)}$ cannot be expressed in the simple form
of (2), but expressions for change in the vector
$(D_{ABCD}, D_{AB}D_{CD}, D_{AC}D_{BD}, D_{AD}D_{BC})$ can be given (Hill, 1974a).

351

Defining $r_{ABCD}^2 = D_{ABCD}^2/(p_a \cdots p_{d'})$, the chi-square statistic extending (12) to four loci can be shown to be

$$2N[r_{AB}^2 + \ldots + r_{CD}^2 + r_{ABC}^2 + \ldots + r_{BCD}^2 + r_{ABCD}^2]$$

having 11 terms, each corresponding to 1 d.f.

The likelihood ratio partition, extending that in Table 2, is given by Goodman (1970), but in view of the computational requirement in many replicates of Monte Carlo simulation we shall restrict discussion to the total departure from equilibrium in which all possible associations are fitted. By analogy with Table 2b this quantity is

$$2[K(ABCD)-K(A)-K(B)-K(C)-K(D)] = 2Nz_{ABCD} \quad . \tag{18}$$

LIMITING PREDICTIONS

Before embarking on detailed simulation results it is useful to consider a few special cases analytically. For two loci, the ratio of expectations of moments $\sigma_d^2 = E(D_{AB}^2)/E(p_a p_{a'} p_b p_{b'})$ turned out to be a good predictor of r_{AB}^2 , the expectation of the ratio of these quantities (Ohta and Kimura, 1969; Table 1 of this paper). Unfortunately the equivalent result does not always apply for three loci, as comparison of results of Hill (1974b) and the simulation to be described will show. Thus our analytical results are very limited, but give some useful insight into the multi-locus problems.

When population size and recombination fractions are sufficiently large that $Ny_{AB} > 1$, approximately, a simple argument was used by Hill and Robertson (1968) to show that $E(r_{AB}^2) = 1/4Ny_{AB}$, approximately, at the steady state. They argued that providing $r_{AB}^2 = D_{AB}^2/p_a p_{a'} p_b p_{b'}$ was small, it was reduced to $(1-y_{AB})^2 r_{AB}^2 \sim (1-2y_{AB}) r_{AB}^2$ by recombination and increased by $1/2N$ by drift each generation, since $2Nr_{AB}^2$ is asymptotically χ^2 with 1 d.f. and has an expectation of unity in samples from populations in equilibrium. Equating the increase and loss gives $E(r_{AB}^2) = 1/4Ny_{AB} \sim 1/4L_{AB}$, the substitution of map length for recombination fraction being adequate for $y_{AB} < 0.1$ or so (Table 1). This argument can be extended to three loci. Consider $r_{ABC}^2 = D_{ABC}^2 / p_a p_{a'} p_b p_{b'} p_c p_{c'}$ with Ny_{AB} and Ny_{BC} large. The loss due to recombination is by the factor $(1-y_{AB})^2 (1-y_{BC})^2 \sim 1 - 2(y_{AB}+y_{BC})$ and, using χ^2 , the increment due to drift is $1/2N$, giving $E(r_{ABC}^2) = 1/4N(y_{AB}+y_{BC}) \sim 1/4L_{AC}$. Thus the steady state value of $E(r_{ABC}^2)$ is approximately equal to that of $E(r_{AC}^2)$, i.e., the two-locus measure of disequilibrium between the outer pair of loci on the chromosome. The value of $1/2N \times$ the total chi-square for disequilibrium at three loci is, using (11) and (12),

$$E(r_{AB}^2 + r_{BC}^2 + r_{AC}^2 + r_{ABC}^2) = (L_{AB}^{-1} + L_{BC}^{-1} + 2L_{AC}^{-1})/4 , \qquad (19)$$

which reduces to $3/2L_{AC}$ if $L_{AB} = L_{BC} = \frac{1}{2}L_{AC}$. Similar results can be obtained for four or more loci.

We now turn to the special case of no recombination amongst the loci. The population then comprises a set of different chromosomes which behave like neutral alleles, and, regardless of the number of loci, eventually only two types will remain segregating (Kimura, 1955). In some replicate populations these will comprise segregants at all the loci in question, which for three loci would be the pairs abc/a'b'c' , abc'/a'b'c' etc. Quantities such as r_{ABC}^2 and likelihood ratio statistics such as $2Nz_{ABC}$ then depend only on the frequency of the alternative chromosomes, and not on their specific configuration. Thus assume abc/a'b'c' are segregating, and let $P_{ABC} = p$. Then, from (9),

$$D_{ABC} = p(1-p)(1-2p) \quad , \tag{20}$$

and from (13),

$$r^2_{ABC} = (1-2p)^2/[p(1-p)]$$
$$= 1/p + 1/(1-p) - 4 \tag{21}$$

which is symmetric about $p = 0.5$. Single locus theory can be used to find $E(r^2_{ABC})$ in (21). Using a transition proba-bility matrix with elements q_{ij} specifying the probability the population has j chromosomes of type abc in genera-tion $t+1$ given that it had i at time t , with

$$q_{ij} = \binom{2N}{j} (i/2N)^j (1-i/2N)^{2N-j} \quad ,$$

asymptotic values of $E(r^2_{ABC})$ were obtained and are given in Table 3.

TABLE 3. *Asymptotic values of* $E(r^2_{ABC})$, $E(r^2_{ABCD})$ *and* $E(z_{AB})$ *with no recombination computed from the exact distribution or by the approximation using the uniform distribution.*

$\frac{N}{2}$		10	20	40	60
$E(r^2_{ABC})$	exact	3.17	4.34	5.58	6.33
	approx.	2.54	3.71	4.96	5.72
$E(r^2_{ABCD})$	exact	40.8	93.4	202.8	314.3
$E(z_{AB})$	exact	1.064	1.037	1.021	1.015
	approx.	1.086	1.046	1.024	1.019

The steady state distribution of unfixed classes is approxi-mately uniform (Fisher, 1930), so a simple solution is

$$E(r^2_{ABC}) = \int_{1/2N}^{1-1/2N} (\frac{1}{p} + \frac{1}{1-p} - 4) \, dp \Big/ \int_{1/2N}^{1-1/2N} dp$$

$$= 2[\log 2N + \log(1-1/2N)]/(1-1/N) - 4 \, , \tag{22}$$

which approaches $2 \log 2N-4$ as N increases. These results are compared with the exact values in Table 3. The choice of $1/2N$ and $1-1/2N$ for the bounds is somewhat arbitrary, and the approximation (22) can be improved by modifying the bounds and allowing for the slight departure from uniformity found at the ends of the distribution (Fisher, 1930). From (12), and noting that r^2_{AB}, r^2_{AC} and r^2_{BC} equal unity when there are only two chromosome types, the total chi-square expected is given by $2N[E(r^2_{ABC}) + 3]$.

The approximate and exact values of $E(r^2_{ABC})$ shown in Table 3 increase in parallel, as $\log N$. This contrasts with $E(r^2_{AB})$ for two loci which asymptotes at a value of unity for any population size. The same calculations done for four loci with no recombination give from (17) when only two chromosome types are segregating,

$$r^2_{ABCD} = \frac{1}{p^2} + \frac{1}{(1-p)^2} - \frac{4}{p(1-p)} + 9 \, .$$

Exact values of $E(r^2_{ABCD})$ obtained using the transition matrix are given in Table 3. The approximation, obtained by integrating over the uniform distribution, suggests that $E(r^2_{ABCD})$ increases in proportion to N , a result largely borne out by the exact values. The asymptotic values of the chi-square statistics for two, three and four loci thus increase as N , $N \log N$ and N^2 , respectively, when there is no recombination.

An extension of diffusion equation arguments, such as those of Hill and Robertson (1966) or Ohta and Kimura (1969), for

355

two loci to three or more, suggest that, on a time scale
inversely proportional to N, the distribution of chromosome
types is independent of N and a function of the vector of
N x recombination fractions or map lengths, providing the recombination fractions are of order $1/N$. Thus the expected
value of any quantity, such as r^2_{ABC}, which is a function of
frequencies (equation 13) and not of population size, might
be expected to be independent of N. However, as (21) shows,
in the limiting case of no recombination, r^2_{ABC} has terms in
$1/p$ and $1/(1-p)$ and takes its highest values at the ends
of the distribution. The value of $1/p$ at the end point in
a finite population with discrete classes is proportional to
N, and thus it is clear that the continuous diffusion approximation does not hold there.

The likelihood ratio statistic for testing for all departures from random association with two loci is equal to $2Nz_{AB}$
(equation 8) and when only two chromosome types are segregating with frequencies p and $1-p$, z_{AB} reduces to

$$z_{AB} = 2p \log p - 2(1-p) \log (1-p) \tag{23}$$

from (8), taking $p \log p = 0$ as $p \to 0$. Integrating z_{AB}
over the uniform distribution as in (22) and using its symmetry about $p = 0.5$, we obtain

$$E(z_{AB}) = -4 \left(\int_{1/2N}^{1-1/2N} p \log p \, dp \right) / (1-1/N)$$

$$= \left\{ [p^2(1-2 \log p)] \right\}_{1/2N}^{1-1/2N} / (1-1/N) . \tag{24}$$

The approximation in (24) can be further approximated to
$E(z_{AB}) = 1+1/N$, and tends to unity for large N, which is
also the asymptotic value of $E(r^2_{AB})$ for two loci.

Now let us consider three loci. The likelihood ratio statistic for testing for all departures from random association is $2Nz_{ABC}$ with 4 d.f. (Table 2) and with only two chromosome types segregating is proportional to

$$z_{ABC} = 2[p \log p^3 + (1-p) \log(1-p)^3 - 3p \log p - 3(1-p) \log(1-p)]$$

$$= -4[p \log p + (1-p) \log(1-p)]$$

$$= 2z_{AB} \tag{25}$$

from (23), since the expected frequency of a chromosome such as abc is p^3 if a , b and c each have frequency p . Thus, using (24), for large N

$$E(z_{ABC}) = 2 \tag{26}$$

asymptotically, when there is no recombination, and is not a function of N as are the equivalent quantities of the standard chi-square statistics: whereas z_{ABC} includes terms in p log p which tend to zero as p becomes very small, r^2_{ABC} includes terms in $1/p$. The approximation (24) using the uniform distribution is compared with the exact value using the transition matrix in Table 3. The agreement is very good, as is the further approximation $1+1/N$.

Using Table 2 we find that the likelihood ratio for the asymptotic case of three loci with no recombination and large population size would be partitioned as follows:

Source	Log likelihood ratio x (1/2N)
Marginal assoc. of A & B	$z_{AB} = 1$
Assoc. of A & C after A & B	$z_{AC} = 1$
Assoc. of B & C after A & B and A & C	$z'_{BC} = 0$
Assoc. A, B and C after all pairs	$z'_{ABC} = 0$
Total	$z_{ABC} = 2$

Of course, any other sequence of fitting the pairs would give the same partition in the sense that the first two pairs fitted would account for the total likelihood ratio.

Continuing to four loci and extending (25), we find that the total likelihood ratio statistic is three times that for two loci, and in general with no recombination the total likelihood ratio statistic is proportional to the number of pairs of "adjacent" loci, i.e., one less than the number of loci.

In view of the more desirable behaviour of the likelihood ratio statistics over chi-square statistics with change in population size at very low recombination values, as illustrated by these results for small population size, most of the remainder of the results will be restricted to them.

SIMULATION RESULTS

All simulations were started with linkage equilibrium, gene frequencies of 0.5 and, unless noted to the contrary, with 1600 replicates, and all results are plotted solely for replicates in which all loci are segregating at the specified generation. It is regrettable that most precise information is obtained on early generations, before many replicates have been fixed, and steady-state values of quantities such as likelihood ratios cannot be obtained accurately without excessive computing expenditure. Procedures which start with populations sampled from one segregating after many generations and thus representative of the steady state, have to be used with caution to avoid introducing new biases. The main advantage in commencing with equilibrium per se is that the results then show the rate at which disequilibrium accumulates by chance.

Three loci: effect of change in population size

Comparisons are given in Figure 1 of total likelihood ratio statistics (expressed as z_{ABC} , i.e., likelihood ratio /2N) for three equally spaced loci in populations of size 10, 20 and 40.

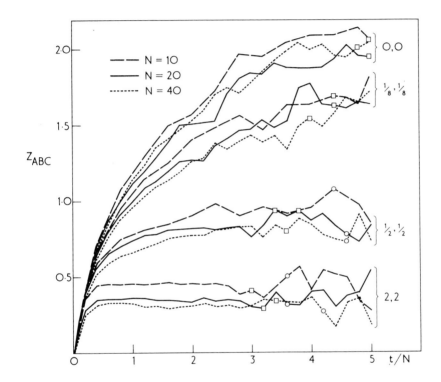

FIGURE 1. Expected values of z_{ABC} (where z_{ABC} = total likelihood ratio/2N) for three neutral loci, initially in linkage equilibrium with frequency 0.5. Generations (t) are plotted as a proportion of population size (N). Results are given for several values of N×map lengths (L_{AB}, L_{BC}) with computations made at three values of N. There are initially 1600 segregating replicates, □ , 0 denotes < 50 , < 20 segregating respectively.

Generations, on the abscissa, are plotted on a scale proportional to N , so that results for different population sizes and the same values of $L = N \times$ map length can be directly compared. It is seen that, especially at the highest distances apart $(L_{AB} = L_{BC} = 2)$ the expected value of z_{ABC} in segregating populations is rather higher at N=40 than N=10. Nevertheless the changes induced in z_{ABC} by four-fold changes in N at constant map length (e.g., N=10 or 40, $\ell_{AB} = 0.05$ giving $L_{AB} = 0.5$ or 2) are much greater than four-fold changes in N at constant N $\times$ map length (e.g., N=10 or 40, $\ell_{AB} = 0.05$ or 0.0125 giving $L_{AB} = 0.5$). Although computations at larger population sizes than 40 were undertaken for the smaller L values because of computing expense, it seems reasonably safe from the theoretical arguments and results of Figure 1 to deduce values for larger population sizes from simulations with very small ones. (This conclusion would not have held if values of r_{ABC}^2 had been plotted at low L values.) Subsequent figures in which partitions of z_{ABC} have been made, or in which more than three loci have been included have all been carried out with N=20 for low values of L or N=80 at high values of L where equilibrium is reached relatively earlier. It is noted in Figure 1 that there is closer agreement between simulation results at N=20 and N=40 than between N=20 and N=10 .

Three loci: partition of likelihood ratio

Partitions of z_{ABC} are given in Figure 2 for six different pairs of values of L_{AB} and L_{BC} using N=20 and in Figure 3 for three more pairs using N=80 and less replication (400). When two or more quantities have the same expected value, for example z_{AB} and z_{BC} when $L_{AB} = L_{BC}$, their mean is plotted.

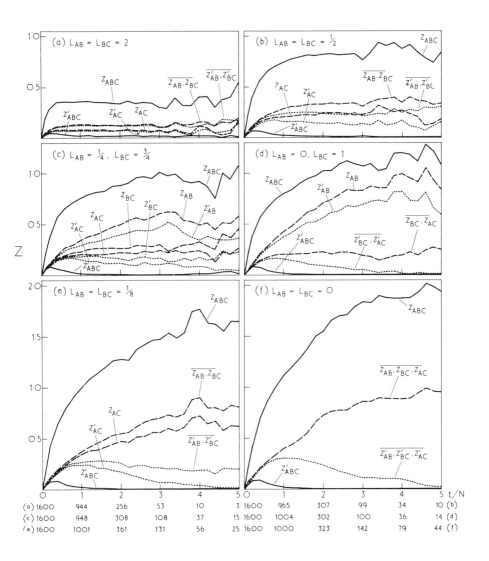

FIGURE 2. _Expected values of partitioned values of z (=like-lihood ratio/2N) plotted against t/N for three neutral loci initially in linkage equilibrium with frequency 0.5 and population size N=20. Numbers of segregating replicates are shown. Results for different values of N×map lengths (L_{AB}, L_{AC}) are a:(2,2), b:(0.5,0.5), c:(0.25,0.25), d:(0,1), e:(0.125,0.125), f:(0,0)._

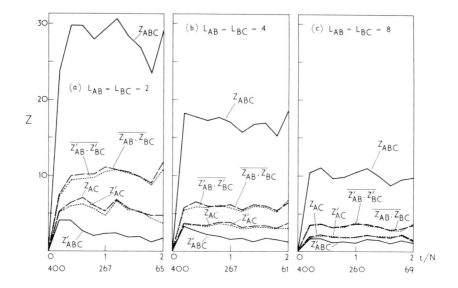

FIGURE 3. As Figure 2, but with N=80 , larger values of N × map lengths and 400 replicates. (L_{AB}, L_{BC}) are a:(2,2), b:(4,4), c:(8,8).

Consider firstly the case of no recombination $(L_{AB}=L_{BC}=0)$. In Figure 2f quantities such as z_{AB} asymptote at approximately 1.0 , z_{ABC} at 2.0 and z'_{AB} and z'_{ABC} at 0.0 as predicted by (26) and the discussion following the equation. However about 5N generations (i.e., 100 generations with N=20) are required before these asymptotic values are approached. In the first generation from equilibrium the quantities 2Nz are asymptotically χ^2 distributed, and so z_{ABC}, z_{AB}, z'_{AB} and z'_{ABC} have expected value of 4/2N, 1/2N, 1/2N and 1/2N respectively. The term for three locus association, z'_{ABC} reaches its highest value after only about 0.2N generations; whereas z'_{AC}, that for association between A and C after associations between A and B and between

362

B and C have been removed, increases together with the marginal associations, z_{AB} and z_{BC} for almost N generations. Subsequently, as only two chromosome types become of higher frequency and eventually are the only ones to remain segregating in the population, knowledge of the frequency of chromosomes carrying the specific alleles A and B and of the frequency of chromosomes carrying alleles B and C gives sufficient information to specify frequencies of chromosomes carrying alleles A and C and alleles A , B and C . Thus the residual likelihood ratio to account for the latter two types of association is zero when any two marginal associations are specified. Of course, when there is no recombination the nominal ordering of loci on the chromosome is irrelevant, so the pairs AB, AC and BC are interchangeable.

Even if there is some recombination among the loci, the quantity z'_{ABC} for three-way association never contributes an appreciable part of the total likelihood ratio after the first few generations. If a pair of the loci are completely linked, e.g., A and B in Figure 2d, z'_{ABC} contributes nothing asymptotically, and very little, or nothing, if the loci are very closely linked (e.g. Figure 2e). The marginal pair-wise associations such as z_{AB} can be obtained from two-locus theory (e.g., Table 1, but for N=8). These always exceed the corresponding conditional associations, e.g., z'_{AB} , by a large amount with tight linkage (Figure 2e) but by very little with looser linkage (Figure 2a). The results of Figure 3 show that with higher L_{AB}, L_{BC} values the differences between z_{AB} and z'_{AB} essentially vanish, and z'_{ABC} becomes of similar magnitude to z_{AC} .

Using the results given in Figures 2 and 3, estimates of steady state values of $E(z)$ have been made, and are listed in Table 4.

TABLE 4. *Estimates of partitions of E(z) at steady state for three loci (read from Figure 3 for N=80 or Figure 2 for N=20).*

L_{AB}	L_{BC}	z_{AB}	z_{BC}	z_{AC}	z'_{AB}	z'_{BC}	z'_{AC}	z'_{ABC}	z_{ABC}
N=80									
8	8	0.032	0.032	0.019	0.032	0.032	0.019	0.012	0.095
4	4	0.059	0.059	0.033	0.057	0.057	0.031	0.014	0.163
2	2	0.10	0.10	0.06	0.10	0.10	0.06	0.02	0.28
N=20									
2	2	0.13	0.13	0.08	0.12	0.12	0.07	0.02	0.35
$\frac{1}{2}$	$\frac{1}{2}$	0.34	0.34	0.22	0.26	0.26	0.14	0.01	0.83
$\frac{1}{4}$	$\frac{3}{4}$	0.55	0.30	0.23	0.41	0.16	0.09	0.00	0.94
0	1	1.00	0.23	0.23	0.77	0.00	0.00	0.00	1.23
$\frac{1}{8}$	$\frac{1}{8}$	0.81	0.81	0.62	0.20	0.20	0.01	0.00	1.63
0	0	1.00	1.00	1.00	0.00	0.00	0.00	0.00	2.00

Since the values of z in successive generations are highly autocorrelated and the number of segregating replicates falls each generation, the estimates shown in Table 4 (and subsequently in Table 5) are based on a visual assessment from the graphs and computer results of where the asymptotic values will lie, taking informally into account the conflict between the number of replicates segregating each generation and the generation number and thus proximity to the asymptote. No standard errors can therefore be attached to the estimates in the Table, but they should act as a guide to the parameter values. The estimates are likely to have lowest precision when L_{AB} and L_{BC} are small since so few replicates are segregating when the steady state values of E(z) are approached. Where, for example, $L_{AB} = L_{BC}$, the same value has been given for $E(z_{AB})$ as for $E(z_{BC})$, regardless of the

actual values in the simulation run, as in Figures 2 and 3;
and values of $E(z_{AB})$, for example, obtained in the simulation run are usually given, rather than any two-locus theoretical prediction, so that the partition of $E(z_{ABC})$ is not
too disturbed. These values in the table emphasize the small
likelihood ratio statistic due to the three-locus association,
which always is of small magnitude and only contributes a
significant proportion of the total likelihood when values of
L_{AB}, L_{BC} are large and the total amount of disequilibrium is
small.

Three loci: behaviour of r^2

For comparison, estimates of terms like $E(r^2_{AB})$ and $E(r^2_{ABC})$
at steady state are given in Table 5, based on the same computer runs as Table 4, and using the same estimation procedure.

TABLE 5. *Estimates of $E(r^2)$ at steady state for three loci
(using same simulation runs as Table 4), and*
$\alpha = E(D^2_{ABC})/E(p_a p_a, p_b p_b, p_c p_c,)$ *at steady state (computed
using a moment generating matrix).*

L_{AB}	L_{BC}	r^2_{AB}	r^2_{BC}	r^2_{AC}	r^2_{ABC}	Total	α
N=80							
8	8	0.032	0.032	0.018	0.018	0.100	0.018
4	4	0.059	0.059	0.034	0.031	0.183	0.029
2	2	0.09	0.09	0.05	0.05	0.28	0.50
N=20							
2	2	0.12	0.12	0.07	0.07	0.38	0.061
$\frac{1}{2}$	$\frac{1}{2}$	0.33	0.33	0.22	0.37	1.25	0.178
$\frac{1}{4}$	$\frac{3}{4}$	0.51	0.25	0.21	0.37	1.34	0.181
0	1	1.00	0.20	0.20	0.40	1.80	0.187
$\frac{1}{8}$	$\frac{1}{8}$	0.78	0.78	0.60	2.3	4.5	0.436
0	0	1.00	1.00	1.00	4.0	7.0	0.658

At the low values of L_{AB} and L_{BC} , $E(r_{ABC}^2)$ is more difficult to estimate than $E(z_{ABC})$ or $E(z'_{ABC})$ because there is much variation in mean level between generations. For $L_{AB} = L_{BC} = 0$, the values roughly agree with those predicted from the transition matrix (Table 3). At high values of L we see that $E(r_{AB}^2) = 1/4L_{AB}$, $E(r_{AC}^2) = 1/4L_{AC} = E(r_{ABC}^2)$ approximately, allowing for the fact that recombination fractions used approached 0.1, where map length and recombination fraction do not correspond so closely (Table 1).

While at the low values of L_{AB} and L_{BC} the behaviour of the alternative measures of three-locus association are very different, at higher values this is no longer so. Thus, as L_{AB} and L_{BC} exceed unity, there is little difference between z'_{AC} and z_{AC} , i.e. the likelihood ratio statistic for A and C after fitting AB and BC , or ignoring AB and BC ; and the total likelihood ratio statistic z_{ABC} is approximately equal to the equivalent chi-square statistic $(r_{AB}^2 + r_{AC}^2 + r_{BC}^2 + r_{ABC}^2)$. The residual d.f. for three locus association, which for the ratio test is z'_{ABC} and for chi-square is r_{ABC}^2 (each × 2N) therefore accounts for roughly the same amount of variation in each case.

The values of r^2 for two locus disequilibrium, e.g. $E(r_{AB}^2)$, given in Table 5 are conditional on segregation at all three loci. These differ from those in Table 1 which are for the two-locus model and therefore unconditional on segregation at a third locus. Comparisons between Tables 1 and 5 are also confounded with population size, however. Some three-locus simulation undertaken with N=8 corresponding to the exact transition matrix results given in Table 1 show that there are no quantitatively important differences between the values of $E(r_{AB}^2)$ unconditional and conditional on segregation at a third locus at all generations starting from

a population in equilibrium. There may be small differences
between the two models, but they could not be detected consistently with the amount of simulation which could be undertaken.

Prediction of $E(r^2_{ABC})$ from ratios of moments

The ratio of moments $E(D^2_{AB})/E(p_a p_{a'} p_b p_{b'})$ is a good predictor of $E(r^2_{AB})$ for two loci, as has been mentioned previously. Equivalent results for the asymptotic value of the
ratio $\alpha = E(D^2_{ABC})/E(p_a p_{a'} p_b p_{b'} p_c p_{c'})$ obtained from the
first eigenvector of the appropriate moment generating matrix
(Hill, 1974b) are given in Table 5. The correspondence between α and $E(r^2_{ABC})$ is seen to be very poor at low values
of L_{AB} and L_{BC}, and since α is essentially independent
of population size for given L_{AB}, L_{BC} the departure would
become greater as N is increased. However for L_{AB}, $L_{BC} > 1$
or so, there is seen to be reasonable agreement. This is not
very useful to us however, for the simple approximation
$E(r^2_{ABC}) = 1/4L_{AC}$ is also satisfactory.

Four or more loci

In view of the computer time which would be necessary in
each replicate run to enable a complete partition of the total likelihood ratio, only the total value has been computed
for some parameter sets with four and five loci. Results, as
$E(z_{ABCD})$ and $E(z_{ABCDE})$ which are $1/2N \times$ the total likelihood ratio statistics, are given for four and five loci in
Figures 4 and 5, respectively. The computational problems of
fixation of most replicates long before the steady state is
reached is even more acute with these higher numbers of loci,
so results are only approximate at high generation number.
Indeed, when the total map length is zero, the steady state

has clearly not been reached in the simulation, as shown by comparison of Figure 4 with Table 3. The total likelihood ratio ($\times$ 1/2N) can be compared using Figures 2, 4 and 5 for the outermost loci a specific distance apart, but with different numbers of intervening loci included. With L=4 between the outside pair, the asymptotic values of z are 0.08, 0.35, 0.8 and 1.6 approximately, for 2, 3, 4 and 5 loci with N=20 . In Figure 4 alternative configurations of distances among the loci, for given distance between the outside pair, are given for $L_{AD} = 1$.

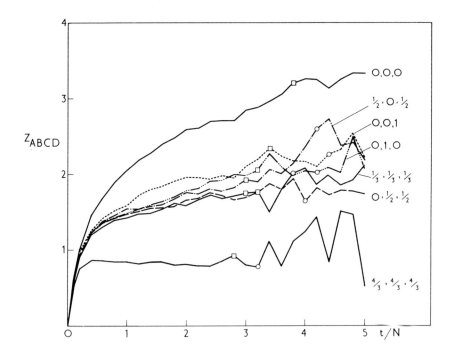

FIGURE 4. As Figure 1, but z_{ABCD} (total likelihood ratio/2N) for four loci, in each case with N=20 .

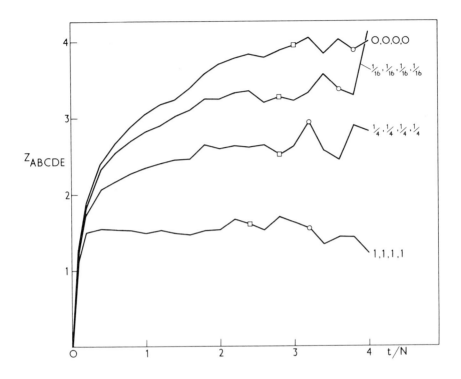

<u>FIGURE 5.</u> *As Figure 1, but z_{ABCDE}(total likelihood ratio/2N) for five loci, in each case with N=20 .*

The effect on the total value of z is not large relative to changes in the value of L_{AD} , as also found for three loci (Figure 2, Table 4). A bigger contrast would be found, however, for larger values of L_{AD} between say $L_{AB} = L_{BC} = L_{CD}$ = $L_{AD}/3$ and $L_{AB} = L_{BC} = 0$, $L_{CD} = L_{AD}$. The general pattern of the four and five locus results is similar to that for three loci.

For values of L in excess of about unity between all adjacent pairs of loci, predictions of the chi-square statistics (or quantities like r^2) can be obtained by extending the arguments used earlier (eq.19). These suggest that for four equally spaced loci the expected value of the total chi-square statistic with 11 d.f. is $2N \times 19/12L$, where L is the distance between adjacent loci.

DISCUSSION

Statistical

Our results have been presented solely in terms of chi-square or likelihood ratio statistics, usually with a scalar multiplier $(1/2N)$. These non-negative quantities may be adequate for a discussion of drift at neutral loci, where the expected values of disequilibria are zero, but are less so for discussing selection or migration where the signs of the disequilibria may be important. As Lewontin (1974) and others have pointed out, with selection as the main cause of dis-equilibrium, one might expect it to be of the same sign and magnitude in different populations. With data on animal or plant populations, however, the experimentalist or field wor-ker is likely to test for association (disequilibrium) using standard chi-square or likelihood ratio methods. If he finds evidence for the presence of such disequilibria its sign and magnitude can then be calculated, and expressed in terms of, for example, D_{AB}, r_{ABC} or the quantity which has to be added to the chromosome frequencies in order to satisfy Bart-lett's criterion (14). With neutrality, all such quantities will be distributed about zero. Of course, with symmetric selection models in infinite population disequilibria of the same magnitude but opposite sign can occur, the value found in any particular population depending on the initial fre-quencies (Bodmer and Felsenstein, 1967).

More use has been made here of likelihood ratio than chi-square statistics. When there is little departure from random association, the two methods give essentially the same values, and both are distributed as χ^2 distribution in large samples from equilibrium populations. With very large departures from random association they behave rather differently, especially when one type is very rare. For example, if in a sample of 100 chromosomes the gene frequency at each of four loci is 0.1, the rarest type has an expected frequency of 0.0001 or expected number of occurrences of 0.01 . If such a chromosome is actually obtained it contributes $(1-0.01)^2/0.01 \sim 100$ to chi-square, but only $2\log(1/0.01) = 9.2$ to the likelihood ratio. With field data, such rare classes might be pooled, but when making predictions of behaviour and making analyses over many populations as done here, arbitrary decisions to pool classes would have an uncertain effect on the results. Furthermore, pooling of classes with field data would cause some loss of information, and difficulties would clearly arise with the partition of the chi-square or likelihood ratio. In practice the most satisfactory procedure appears to be to carry out a complete partititoned likelihood ratio analysis before contemplating pooling of classes. With small numbers in the sub-classes appropriate exact tests seem to be required and some work on these is in progress. An extension of the analysis for data on diploid individuals including partition of the likelihood for several loci has been given by Hill (1975) using maximum likelihood methods and assuming random mating populations.

There remain unresolved problems in this area of analyzing multi-dimensional contingency table data, and we have already considered alternative forms of the partition. Genetic interpretations of, say, three locus associations measured by Bartlett's criterion are hard to visualize, and a fuller attempt is given elsewhere (Hill, 1975). We would not have to concern ourselves with such statistical problems of specifying three locus associations if the alternative methods did not give such radically different answers. But, as we have seen, with very tight recombination, the parameters z'_{ABC} (essentially Bartlett's criterion) and r^2_{ABC} (essentially Lancaster's criterion) approach 0 and $2\log 2N-4$ respectively. For values of L (N x population size) much in excess of unity the criteria do not differ so greatly.

Genetical

The objective of this work has been to study the process of drift for several neutral genes, so that population behaviour under the neutrality hypothesis can be determined. This is only a start, in that one has also to look at behaviour with selection and the other evolutionary forces of migration, mutation and population structure or non-random mating. Whilst there is already much information on two loci with selection, that on three or more is much more limited. In the best known study, that of Franklin and Lewontin (1970), simulation was used so there was some confounding of population size and selection effects. They concluded that, with a model of linked genes with heterozygote superiority at each locus and multiplicative effects over loci, the population would tend to only two or a few complementary chromosome types. This could correspond to the state of a population with neutral, but closely linked genes, after all but these few chromosome types had been lost by chance. With selection and heterozygote superiority, segregation at all, or most of the loci would be expected in all populations, and the indi-

372

vidual gene frequencies would tend to be intermediate. Neither of these conditions would be expected with neutrality. With selection in infinite populations, however, very many equilibrium chromosome frequency sets are possible (Feldman et al., 1974) a proportion of which are stable (Karlin, 1975). Thus, as a consequence of founder effects it would be possible for different chromosome polymorphisms to be segregating among populations. It is difficult to differentiate between migration and selection as forces maintaining similarity of frequencies in different populations. There remain problems in distinguishing between the alternative models.

With the exception of the limiting case of $L \to 0$ and for $L > 1$ we have no explicit solutions. However our observation is that, among populations segregating at three loci, the contribution (z'_{ABC}) to the likelihood ratio made by three locus association is always small in magnitude, although becoming a significant part of the total when the product of map distance between the extreme loci on the chromosome and population size is large. This may well be the most useful result to come from this study. Having established what happens with neutral genes we now require detailed information on the equivalent predictions of these likelihood ratio parameters for models of selection and other forces.

Even with two loci, little is known about the joint behaviour of drift and selection; workers have tended to study either drift with neutral genes or selection in infinite populations. It is unlikely that many analytical results will be obtained to such involved equations, but simulation of at least some parameter sets should be feasible. An illustration of the relevance of such a study is that a neutral gene as defined by Kimura and Ohta (1971) is one with a selective value of less than about $1/N$, i.e. neutrality is defined in

terms of population size. What then are expected disequilibria for pairs or more of loci with selective values of order
1/N ? The little evidence available for two loci suggests
that r^2_{AB} is similar with such selective values for heterotic loci having fitnesses combined multiplicatively to that
for neutral genes (Hill and Robertson, 1968), but the results
with epistasis or with more loci may be very different. The
tendency may be towards a reduction in the incidence of very
rare chromosome types involving several loci.

For two locus disequilibria, the critical values of recombination fraction or map length are of order 1/N : i.e.,
$r^2_{AB} \sim 1/(4Ny_{AB}+1)$, using Sved and Feldman's (1973) formula,
and just $1/4Ny_{AB}$ for $Ny_{AB} > 1$ approximately. Thus, if
Ny_{AB} exceeds unity by an order of magnitude, r^2_{AB} is very
small, and for appreciable disequilibrium to be found, populations must be of small effective size and/or linkage must
be tight. Regrettably, but not surprisingly, perhaps, we
find the same dependency for three loci, except that the three
locus association (from the likelihood ratio) is always small.
For Ny (or L) values in excess of unity, the total three
locus association is also roughly proportional to 1/Ny . Our
simulations have been based on constant population size, however, and a small founder population could induce considerable disequilibrium for long periods thereafter, even in a
large population.

As yet our discussion of the relative effects of neutrality
versus selection on multi-locus disequilibrium remains somewhat inconclusive. Is there then any point in an experimentalist or field worker estimating disequilibrium at all? Undoubtedly there is, for if he has already collected information on genotype frequencies, estimates of disequilibrium can
always be made from the same data using relevant statistical

techniques, and it will provide results on which this and any subsequent theory can be tested.

ACKNOWLEDGEMENTS

I am indebted to Mrs. Marjorie McEwan for much assistance, with both computer programming and presentation of results.

REFERENCES

Bartlett, M.S. (1935). J.R. Statist. Soc. Suppl. 2: 248-252.

Bennett, J.H. (1954). Ann. Eugen. 18: 311-317.

Bodmer, W.F. and J. Felsenstein. (1967). Genetics 57: 237-265.

Feldman, M.W., I. Franklin and G.J. Thomson. (1974). Genetics 76: 135-162.

Fienberg, S.E. (1970). Ecology 51: 419-433.

Fisher, R.A. (1930). The Genetical Theory of Natural Selection. Clarendon Press, Oxford.

Franklin, I. and R.C. Lewonton. (1970). Genetics 65: 707-734.

Geiringer, H. (1944). Ann. Math. Stat. 15: 25-57.

Goodman, L.A. (1969). J.R. Statist. Soc. B 31: 486-498.

Goodman, L.A. (1970). J. Amer. Statist. Assoc. 65: 226-256.

Hill, W.G. (1969). Jap. J. Genet. 44 (Suppl.1): 144-151.

Hill, W.G. (1974a). Theor. Pop. Biol. 5: 366-392.

Hill, W.G. (1974b). Theor. Pop. Biol. 6: 184-198.

Hill, W.G. (1974c). Heredity 33: 229-239.

Hill, W.G. (1975). Biometrics (in press).

Hill, W.G. and A. Robertson. (1966). Genet. Res. 8: 269-294.

Hill, W.G. and A. Robertson. (1968). Theor. Appl. Genet. 38: 226-231.

Karlin, S. (1975). Theor. Pop. Biol. (in press).

Kimura, M. (1955). Evolution 9: 419-435.

Kimura, M. and T. Ohta. (1971). Theoretical Aspects of Population Genetics. Princeton Univ. Press, Princeton, N.J.

Lancaster, H.O. (1951). J.R. Statist. Soc. B 13: 242-249.

Lewontin, R.C. (1964a). Genetics 49: 49-67.

Lewontin, R.C. (1964b). Genetics 50: 757-782.

Lewontin, R.C. (1974). The Genetic Basis of Evolutionary Change. Columbia Univ. Press, New York.

Ohta, T. and M. Kimura. (1969). Genet. Res. 13: 47-55.

Plackett, R.L. (1962). J.R. Statist. Soc. B 24: 162-166.

Slatkin, M. (1972). Genetics 72: 157-168.

Smouse, P.E. (1974). Genetics 76: 557-565.

Sokal, R.F. and F.J. Rohlf. (1969). Biometry. Freeman, San Francisco.

Strobeck, C. (1973). Genet. Res. 22: 195-200.

Sved, J.A. (1971). Theor. Pop. Biol. 2: 125-141.

Sved, J.A. and M.W. Feldman. (1973). Theor. Pop. Biol. 4: 129-132.

Wright, S. (1933). Proc. Nat. Acad. Sci. U.S. 19: 420-433.

An Analysis of Allelic Diversity in
Natural Populations of *Drosophila*:
The Correlation of Rare Alleles with Heterozygosity*

R.K. KOEHN and W.F. EANES

INTRODUCTION

We have examined heterozygosity of so-called common alleles and the observed number of rare alleles among enzyme loci
in twenty eight species of Drosophila. There was a strong
correlation between these two variables in a majority of the
species. This paper describes the correlated variation, proposes certain conditions that might be expected to produce
the correlation, and explores the implications our results
have on interpretations of enzyme polymorphisms in natural
populations.

Several reports have postulated and/or demonstrated recombination within cistrons of higher organisms (Ohno et.al.,
1969; McCarron et al., 1974) at commonly investigated enzyme
loci. Many investigators (Emerson, 1969; Stadler, 1973;
Holliday, 1974) are currently concerned with the mechanism(s)

---

*Contribution No.134 from the Program in Ecology and Evolu-
tion, S.U.N.Y., Stony Brook, N.Y.

of intragenic recombination, both reciprocal and non-reciprocal (gene conversion). Although the mechanisms are not presently known, intragenic recombination differs from classical intergenic recombination in that it occurs within a specific identifiable cistron, rather than between two different cistrons. In certain situations, intracistronic recombination can produce new alleles. When recombination occurs within a heterozygous locus and the alleles of the heterozygote differ at more than a single site, and if recombination occurs between two differing sites, the recombinational products will have a nucleotide sequence different from the parental alleles. The possibility of new alleles arising in populations by this mechanism has not gone unrecognized (McCarron et al., 1974; Watt, 1972), but these discussions have not typically considered the potential effect on explanations of allozyme diversity in natural populations. There are some important reasons why this consideration should be made. A major concern of contemporary population genetics is the explanation of genic heterozygosity in natural populations in terms of either neutrality or selection, the latter usually expressed in terms of qualitative and/or quantitative variations in environmental heterogeneity. All such explanations are ultimately concerned with numbers and relative frequencies of alleles (Powell, 1971, 1973; Somero and Soulé, 1974; Ewens and Feldman, 1975). Some new alleles, produced by whatever mechanism, are potentially detectable by electrophoresis and therefore have bearing on explanations of polymorphism. If, as has been suggested (see Watt, 1972), average per locus rates of intragenic recombination can be orders of magnitude greater than direct mutations, it would significantly influence considerations of the neutrality hypothesis. With the accumulation of mutational differences, recombination would dramatically increase the level of

heterozygosity at a locus (Ohno, et al., 1969). Whatever the
actual rate of intragenic recombination in nature, clearly
its combined effect with mutation must introduce new variants
at a greater rate than mutation alone. Lastly, alleles lost
from the population can be re-introduced as a rearrangement
of existing alleles.

New alleles will arise from intragenic recombination only
within a heterozygous cistron. Under the simplest consideration, the more heterozygous a locus, the more likely that new
alleles will arise as a result of intracistronic recombination. We have sought to examine this simple expectation by
estimating for data from natural populations the correlation
between heterozygosity of "common alleles" and the number of
"rare alleles" detected at a locus.

There are numerous factors that could confound a correlation between heterozygosity and the number of rare alleles,
even if intracistronic recombination were a ubiquitous event.
Briefly, these factors are as follows. For any given rate of
recombination, the rate of formation of new recombinants will
be related to both the number of nucleotide sites that differ
among alleles at a locus and the distribution of these sites
within the cistron. Among-locus heterogeneity in either recombinational rate, number of sites that differ among alleles,
and/or the distribution of these sites would tend to obscure
among-loci patterns of correlation. Secondly, new alleles
must be electrophoretically detectable in a manner that reflects the rate of their appearance. Assuming that the preceding conditions are met, detection of rare alleles is heavily
sample-size dependent, as are estimates of their relative frequencies. Perhaps most importantly, outside markers can influence the frequency of intracistronic recombination. That
is, as the genetic background is varied, the rate of intra-

cistronic recombination can likewise vary. Heterogeneity among loci in outside marker influences could thus easily obscure correlations between rate and other variables. Lastly, we cannot predict the effect of selection on a recombinational scheme. New alleles can potentially range in their relative fitnesses from zero to one. The number of generations that a recombinant allele would be expected to remain in a population will be different among various recombinant alleles and quite heterogeneous among both loci and populations. Indeed, the loss or establishment of a specific allele in a population may differ depending on when the allele was introduced. We would suspect that even if intragenic recombination were occurring, heterogeneity among loci with regard to the electrophoretic detectability of recombinant types might be sufficient to obscure the correlation we are estimating.

We must emphasize that the initial stimulus for our analyses evolved from an interest in intracistronic recombination. Some of the above factors would influence the subsequent behavior of new alleles only when we are considering their formation by intracistronic recombination, while others are pertinent to the appearance of new alleles by other mechanisms. Our results are consistent with several interpretations discussed more fully below.

DATA ANALYSIS

All data available for these analyses were in the form of allele frequencies at various loci collected from natural populations of twenty-eight species of Drosophila. We have arbitrarily defined "common alleles" as those of sample frequency .05 or greater and "rare alleles" less than .05. While many authors have made a similar distinction between "rare"

and "common" alleles, the distinction is nevertheless an arbitrary one and we do not yet know what influence this cut-off
frequency has on our results.

The detection of rare alleles is heavily sample-size dependent. Heterogeneity in sample size was minimized in our
analyses by restricting them to (1) data for which the average
number of individuals per locus was large, generally much
greater than 100, and (2) data where the number of loci studied was on the order of ten or more. For comparative purposes, we have included a few species in which the data were
not of this minimal type. Most importantly, all correlation
analyses were done with weighting of either samples or loci,
depending on the particular calculation. While individual
heterozygosity values were arc-sin transformed, this was probably not essential. The product-moment correlation computation is quite insensitive to non-normality and, in view of
the observed magnitudes of correlations, "normalization" was
probably not necessary.

For each locus of each data set, the numbers of individuals
sampled, the arc-sin transformation of the heterozygosity of
the alleles whose frequency was greater than .05, and the
number of alleles less than .05 were recorded. An example,
Drosophila willistoni from Trinidad (Ayala et al., 1972), is
given in Table 1. For loci with two alleles, where one allele
was less than .05 in frequency, heterozygosity of the locus
was nevertheless calculated, but no rare alleles were counted.
For loci with multiple alleles, but only one greater than .05,
heterozygosity was calculated between the common and second-
most common allele. Thus, a locus was considered monomorphic
only when a single electrophoretic allele had been observed.
This treatment excludes the possibility of a locus having zero
heterozygosity when alternate alleles of any frequency were

observed.

TABLE 1. A summary of data for D. willistoni from Trinidad (Ayala, et al., 1972) used to calculate the correlation between heterozygosity of common alleles and the number of observed rare alleles (< .05).

Locus	N	H*	No. of Alleles < .05
Lap-5	1072	48.24	4
Est-2	74	13.89	1
Est-4	1062	35.97	2
Est-5	984	13.89	1
Est-6	606	28.40	1
Est-7	910	45.23	4
Aph-1	962	32.11	3
Ald	42	11.42	0
Adh	132	23.59	1
Mdh-2	132	8.09	0
αGpd	132	11.42	0
G6pd	24	22.04	1
Idh	118	13.74	1
Me-1	96	13.96	0
Tpi-2	94	11.42	0
Adk-1	98	45.18	2
Adk-2	66	22.43	1
Pgm-1	90	15.74	2
Hk-1	86	11.36	1
Hk-2	102	11.36	1
Hk-3	106	8.09	0

$H = \sin^{-1} \sqrt{\Sigma 2 p_i q_i}$

A weighted, product-moment correlation coefficient was computed among loci in each data set between transformed heterozygosity and the number of rare alleles, where the weight was the number of individuals per locus. When a single locality was sampled by the original investigator, the number of rare alleles at a locus is always an integer value. However, when multiple localities were sampled, weighted average heterozygosities and weighted average numbers of rare alleles for each locus were computed, where the weight was the number per locality per locus.

RESULTS

Our results for each species are summarized in Table 2 and include the number of loci studied in each species, the average number of individuals sampled per locus, the weighted product-moment correlation coefficient determined among loci for heterozygosity of common alleles and the number of rare alleles (r), and the coefficients of determination (r^2) .

TABLE 2. *Rare alleles and heterozygosity in Drosophila.*

Species	Ref.	# of Loci	N/Locus	r	P	r^2
D. equinoxialis (Ven.)	1	30	343.2	.864	<.001	.747
D. willistoni (Ven.)	1	29	494.9	.762	<.001	.581
D. tropicalis (Ven.)	1	28	145.4	.535	<.005	.286
D. willistoni (C.)	2	25	2034.0	.908	<.001	.825
D. robusta	3	22	394.3	.885	<.001	.783
D. willistoni (Trin.)	2	21	332.8	.974	<.001	.949
D. tropicalis (S.D.)	4	21	124.5	.773	<.001	.598
D. nebulosa (Sant.)	4	20	119.0	.816	<.001	.666
D. tropicalis (Sant.)	4	20	105.6	.645	<.005	.416
D. bifaciata	5	19	368.7	.860	<.001	.740
D. nebulosa (S.D.)	4	19	74.4	.456	<.05	.208
D. tropicalis (May.)	4	19	72.1	.621	<.01	.386
D. tropicalis (Barr.)	4	17	51.3	.771	<.001	.594
D. subobscura	6	15	123.3	.273	ns	
D. obscura	7	14	1008.7	.652	<.02	.425
D. simulans	8	13	170.5	.933	<.001	.871
D. tropicalis (Yun.)	4	12	34.3	.049	ns	
D. paulistorum (A)	9	11	690.1	.566	ns	
D. pseudoobscura	10	10	206.0	.530	ns	
D. melanogaster (Raleigh)	8	10	172.7	-.333	ns	
D. athabasca	8	10	119.4	.763	<.01	.582
D. affinis	8	10	81.2	.759	<.02	.576
D. paulistorum (I)	9	9	373.4	.824	<.01	.679
D. paulistorum (T)	9	9	122.9	.390	ns	
D. pavani	11	8	846.1	.300	ns	
D. paulistorum (AB)	9	7	266.0	.203	ns	
D. mojavensis	10	6	454.3	.854	<.03	.729
D. arizonensis	10	5	205.2	.126	ns	

(1) Ayala et al, 1974
(2) Ayala et al, 1972
(3) Prakash, 1973
(4) Ayala, unpubl.
(5) Saura, 1974
(6) Saura et al, 1973

(7) Lakovaara and Saura, 1971
(8) Kojima et al, 1970
(9) Richmond, 1972
(10) Prakash et al, 1969
(11) Kojima et al, 1972

Significant correlations were observed in twenty of the twenty-eight data sets, and coefficients were high in the majority of species. A significant correlation was observed in all species where the number of loci per species exceeded seventeen.

DISCUSSION

There is no a priori reason for a relationship between the number of rare alleles and heterozygosity of common alleles in the absence of some causitive mechanism. Obviously, the total heterozygosity of a locus will increase upon the acquisition of a new allele by any mechanism. The introduction of a new allele, which is initially rare, will generally lead to a slight reduction of the heterozygosity among the alternate common alleles. However, there are characteristics of the polynomial that in certain situations could contribute to the correlation we have demonstrated. At a locus with two alleles A_1 and A_2, where $f(A_1) > f(A_2)$, the addition to the binomial distribution of a new allele (A_3) as genotypes A_1A_3 or A_3A_3 will not increase the heterozygosity contributed by A_1A_2 after the addition. The addition of the A_3 allele as A_2A_3, in contrast, will increase the heterozygosity contributed by A_1A_2, since A_1 and A_2 will be more intermediate in frequency after the addition of A_2A_3. The conditions for this concomitant increase in allelic diversity and heterozygosity between common alleles are very restricted. For example, a known condition, assuming $n_{A_1A_3}$ and $n_{A_3A_3}$ to be zero or negligible is given as:

$$p_{A_1} > \frac{3}{4} + \frac{\beta n_{A_2A_3}}{(\alpha+\beta)^2}$$

384

where p_{A_1} is the initial frequency (before addition of A_3) of the most common allele, α is the number of A_1 alleles, β is the number of A_2 alleles and $n_{A_2A_3}$ is the number of individuals of genotype A_2A_3 added to the distribution. A_3 is the rare allele and all additions to the zygotic frequency distribution are in the form of the A_2A_3 genotype where A_1 is the commonest and A_2 is the second most common allele.

The addition of individuals of A_2A_3 genotype will thus increase the heterozygosity between the common alleles (A_1 and A_2) only when the frequency of A_1 is greater than approximately .75 and the A_3 allele is added exclusively in the A_2A_3 combination. This effect is therefore dependent on both the size of $n_{A_2A_3}$ and the initial value of p_{A_1}. As $n_{A_2A_3}$ increases in value relative to a given sample size $(\alpha+\beta)$, this condition becomes more restrictive. In samples of a few hundred, the initial value of p_{A_1} must be near .95, while in samples of a few thousand, the initial value of p_{A_1} need be only approximately .76, for the addition of A_2A_3 to produce the effect.

We have no knowledge of the zygotic distributions in the original data. Since the most common allele at the majority of loci in the data we have analyzed falls within the range .75 to .95, it is formally possible that the correlation we have observed is in part due to the effect described above. That is, a locus with a common allele in the range of .9 could have associated with it a disproportionately large number of individuals heterozygous for the second most common and the rare allele. We can determine the maximum effect such distributions would have on our estimate of the correlation. For example, the allele frequencies and sample sizes from the original data were used to regenerate a Hardy-Weinberg zygotic frequency distribution. All individuals with the rare allele were then assumed to have been heterozygotes with the second most common allele in the original data. These individuals were then removed from the zygotic frequency distribution and the common allele frequencies recomputed. This generates a minimum estimate of heterozygosity among the common alleles. After this had been done for all loci of a given data set, the weighted product-moment correlation coefficient was recalculated. Where we have performed this adjustment, the correlation coefficients were only slightly reduced. We can therefore conclude that the effect is extremely small since maximizing the potential conditions does not alter the original observation.

Since total heterozygosity increases with greater numbers of alleles, the correlation we have estimated could be partially due to a concomitant increase in the number of common and rare alleles over all loci. We have reanalyzed data sets where initial correlation coefficients were significant, but included in the analysis only those loci with two common alleles. A consistent pattern of overall correlation was still evident (Table 3).

In a comparison of five species of the Drosophila willistoni group (Ayala, et al. 1974), Ayala and co-workers noted that the amount of genetic variation at a specific locus was correlated among species. The degree of genetic polymorphism at any one locus was very similar among the taxa they studied. These observations suggested that genic heterozygosity may have a greater locus-dependent component than species-dependent component. We have sought to analyze this aspect of our data by examining specific loci irrespective of the species in which they were studied. Since the same loci were

not studied by all authors, only those loci which commonly
appear in the species data could be analyzed this way.

TABLE 3. *A comparison of coefficients of determination between heterozygosity of common alleles and observed number of
rare alleles for all loci* (r_a^2) *and loci with two common
alleles* (r_x^2) *only. Data sets where* r_a *was non-significant are omitted.*

Species	r_a^2	r_x^2
D. simulans	.871	.850
D. obscura	.425	.527
D. bifaciata	.740	.795
D. robusta	.783	.524
D. affinis	.576	ns
D. athabasca	.582	ns
D. equinoxialis (Ven.)	.747	.840
D. willistoni (Trin.)	.949	.978
D. willistoni (C.)	.825	.882
D. willistoni (Ven.)	.581	.832
D. paulistorum (I)	.679	.890
D. tropicalis (Sant.)	.416	ns
D. tropicalis (S.D.)	.598	.795
D. tropicalis (May.)	.386	.752
D. tropicalis (Barr.)	.594	.837
D. tropicalis (Ven.)	.286	.592
D. nebulosa (S.D.)	.208	ns
D. nebulosa (Sant.)	.666	.870

When correlations were computed as above, but for a single
locus among species, they were very large (Table 4). These
very large correlations are perhaps due to the absence in
this analysis of many of the factors mentioned earlier that
would tend to obscure an among-locus estimate of the correlation.

TABLE 4. *Correlation between heterozygosity of common alleles and the observed number of rare alleles at seven loci irrespective of species.*

Locus	d.f.	r	r^2
Acph	1,23	.948*	.899
Adh	1,17	.926*	.857
αGpdh	1,14	.980*	.960
Idh	1,11	.770**	.593
Lap	1,19	.981*	.962
Pgm	1,17	.696*	.484
Xdh	1,15	.840*	.706

* P << .001
** P < .005

In summary, we have demonstrated widespread large correlations between the heterozygosity of common alleles and the number of observed rare alleles in various Drosophila species. This correlation occurs irrespective of the number of common alleles at a locus: as the frequencies of alleles in the common allele class become more equal, the number of rare alleles observed also increases. This is true for both among-loci comparisons within species and within-locus comparisons among species. Our results are consistent with the appearance of new alleles at a rate dependent upon the relative frequencies of the putative parental alleles. The mechanism(s) might be either intracistronic recombination and/or mutation.

The data used in our analyses were generally cited by the original investigators as evidence supporting the role of natural selection in the maintenance of polymorphism. These interpretations have been exclusively based on the frequencies of the common alleles. Since our results implicate mutation and/or intragenic recombination, rare alleles may represent "genetic noise" in allelic diversity in natural populations. If so, this diversity cannot be adequately explained by arguments that do not consider the combined roles of selective and non-selective forces.

ACKNOWLEDGEMENTS

Marc Feldman developed the mathematical argument specifying the conditional effect of genotypic additions on heterozygosity. We are both indebted and appreciative of his interest and assistance. We are very thankful to Francisco J. Ayala for furnishing much unpublished data. R. Koehn was supported by USPHS Career Award GM 28963-04 during this work.

REFERENCES

Ayala, F.J., J.R. Powell, M.L. Tracey, C.A. Mourão and S. Pérez-Salas. (1972). Genetics 70: 113-139.

Ayala, F.J., M.L. Tracey, L.G. Barr, J.F. McDonald and S. Pérez-Salas. (1974). Genetics 77: 343-384.

Ayala, F.J. (Unpublished).

Emerson, S. (1969). In Genetic Organization Vol.1; E.W. Caspari and A.W. Ravin, Eds. Academic Press, New York.

Ewens, W.J. and M.W. Feldman. (1975). This volume.

Holliday, R. (1974). Genetics 78: 273-287.

Johnson, G.B. and M.W. Feldman. (1973). Theor. Pop. Biol. 4: 209-221.

Kojima, K., J. Gillespie and Y.N. Tobari. (1970). Biochem. Genet. 4: 627-638.

Kojima, K., P. Smouse, S. Yang, P.S. Nair and D. Brnzic. (1972). Genetics 72: 721-731.

Lakovaara, S. and A. Saura. (1971). Genetics 69: 377-384.

McCarron, M., W. Gelbart and A. Chobnick. (1974). Genetics 76: 289-299.

Ohno, S., C. Stenius, L. Christian and G. Schipmann. (1969). Biochem. Genet. 3: 417-428.

Powell, J.R. (1971). Science 174: 1035-1036.

Powell, J.R. (1973). Genetics 75: 557-570.

Prakash, S., R.C. Lewontin and J.L. Hubby. (1969). Genetics 61: 841-858.

Prakash, S. (1973). Genetics 75: 347-369.

Richmond, R.C. (1972). Genetics 70: 87-112.

Saura, A., S. Lakovaara, J. Lokki and P. Lankinen. (1973). Hereditas 75: 33-46.

Saura, A. (1974). Hereditas 76: 161-172.

Somero, G.N. and M. Soulé. (1974). Nature 249: 670-672.

Stadler, D.R. (1973). Genetics 74: 113-128.

Watt, W.B. (1972). Am. Natur. 106: 737-753.

Zouros, E. (1973). Evolution 27: 601-621.

The Intensity of Selection for Electrophoretic Variants in Natural Populations of *Drosophila*

B.D.H. LATTER

INTRODUCTION

Electrophoretic surveys of enzyme variants in a number of species of Drosophila have now been reported, and the combined data allow useful comparisons to be made of the observed gene frequency distributions with those predicted by simple genetic models. An earlier examination of the data from D. pseudoobscura, D. willistoni, and D. equinoxialis has shown marked departures from the gene frequency distribution predicted by the neutral allele model (Kimura, 1968), with a significant excess of alleles at extreme gene frequencies and a corresponding deficiency of those with intermediate frequencies (Latter, 1973, 1975). The observed distribution of gene frequencies suggests a model in which mildly deleterious alleles are maintained by mutation pressure alone. Over an evolutionary time scale occasional chance fixation of slightly deleterious alleles can therefore be predicted, with periodic selective replacement by mutant alleles of fully restored activity (Ohta, 1974; Latter, 1975).

Gillespie and Kojima (1968) have drawn attention to the fact that the most commonly assayed enzymes can be divided into two groups with quite different levels of heterozygosity

391

in <u>Drosophila</u> populations. Their distinction was between
glucose metabolizing enzymes (group I), and other enzymes
(group II). Glucose metabolizing enzymes are those which
take part in glycolysis, the citric acid cycle, or the hexose
monophosphate shunt, or whose substrates are in one of these
pathways (Kojima <u>et al.</u>, 1970). Group I enzymes consistently
show far less electrophoretic variation than the enzymes of
group II (Ayala and Powell, 1972), and a similarly restricted
level of polymorphism has been reported for the glycolytic
enzymes of human brain and erythrocytes (Cohen <u>et al.</u>, 1973).

Since most group II enzymes act on substrates which are
furnished by the external environment, Kojima <u>et al.</u> (1970)
suggested that the genetic variability in this group is adaptive, and is related to the <u>substrate variability</u> of the environment. Johnson (1974) has also pointed out that <u>regula-
tory</u> enzymes, i.e., those judged to exert primary control over
the flux through a biochemical pathway, are considerably more
polymorphic than non-regulatory enzymes in <u>Drosophila</u> species,
the former having a mean level of heterozygosity of about 19%
and the latter 6%.

In this paper, the gene frequency distributions for specific substrate, variable substrate, regulatory, non-regulatory,
and glycolytic enzymes are discussed, based on the combined
data from 12 large-scale surveys of 9 species of <u>Drosophila</u>.
There are two specific objectives:-
(i)   to determine which distributions differ from expecta-
      tions based on models involving only mutation to mildly
      deleterious alleles: and
(ii) to estimate the mean selection intensity operating
      against mutant alleles in the heterozygous state, for
      those classes of enzymes whose gene frequency distribu-
      tions are satisfactorily explained by mutation - selec-

tion balance.

GENE FREQUENCY DATA

The data from the following nine species have been combined: - D. bifasciata (Saura, 1974), D. equinoxialis (Ayala et al., 1972a; 1974), D. paulistorum (Richmond, 1972), D. pavani (Kojima et al., 1972), D. pseudoobscura (Prakash et al., 1969), D. robusta (Prakash, 1973), D. subobscura (Saura et al., 1973), D. tropicalis (Ayala et al., 1974) and D. willistoni (Ayala et al., 1971; 1972b; 1974). This literature survey has been restricted to those studies in which at least three separate localities were sampled, and to those alleles reaching a frequency of 0.01 or more in at least one locality. The data for D. pseudoobscura from Bogota were not included in the analysis.

For a chosen group of enzymes the following calculations have been performed separately for each of the 12 studies. Each allele was allocated to one of five gene frequency classes, on the basis of its weighted mean frequency over all localities (with weights proportional to sample size). Table 1 sets out the details of the gene frequency classes used: class 1 includes those alleles with extreme gene frequencies, i.e., close to zero or unity, and class 5 those alleles with intermediate frequencies. Following the suggestion of Yamazaki and Maruyama (1972), the mean contribution to heterozygosity of each gene frequency class has been calculated and expressed as a proportion of the total frequency of heterozygotes, assuming Hardy-Weinberg genotypic frequencies in each locality sampled. In calculating the average contribution to heterozygosity of each gene frequency class, the weight given to each allele was $(n-1)/n$, where n is the number of alleles segregating at the locus concerned.

TABLE 1

Gene frequency classes used in summarizing Drosophila data
and the results of computer simulation.

Class	Range of gene frequencies
1	0.001 – 0.050 ; 0.951 – 0.999
2	0.051 – 0.100 ; 0.901 – 0.950
3	0.101 – 0.200 ; 0.801 – 0.900
4	0.201 – 0.300 ; 0.701 – 0.800
5	0.301 – 0.700

The proportionate contributions to heterozygosity have been
combined over the 12 studies by taking a weighted mean, with
weights proportional to the number of independent alleles involved (i.e., the total number of alleles minus the number of
loci). In the case of the superspecies D. paulistorum, the
proportions were derived separately for each semi-species and
combined by the same weighting procedure. The weighted means
for classes 1 to 5 were then finally multiplied by 10, 10, 5,
5, and 2.5 respectively, to allow for differences in the
widths of the class intervals.

The mean level of heterozygosity for a group of enzymes
has been calculated as a weighted mean over the 12 studies,
with weights proportional to the number of loci in the study
(both monomorphic and polymorphic) belonging to that particular group of enzymes.

The resulting statistics for all 25 soluble enzymes collec-
tively are presented in Table 2, by comparison with the
corresponding predictions for the neutral allele model

(Kimura, 1968). The data show a significant excess of alleles in classes 1 and 2, and a corresponding deficiency in classes 4 and 5, substantiating earlier conclusions based on a much smaller sample of data (Latter, 1975).

TABLE 2. *Relative contributions to heterozygosity of five gene frequency classes for 25 soluble enzymes in* Drosophila, *and predicted contributions based on the neutral allele model of Kimura (1968).*

	Contribution of frequency class					Mean
	1	2	3	4	5	heterozygosity
Drosophila Species	1.78±.17	1.66±.12	0.98±.12	0.64±.10	0.83±.08	0.176 ±.010
Neutral model ($N\mu$ =0.05)	0.858	0.948	0.991	1.022	1.042	0.167

Enzymes involved in glucose metabolism.

In view of the consistently different levels of heterozygosity in Drosophila populations shown by group I and group II enzymes as defined by Kojima et al. (1970), these two sets of loci have been examined separately to test for heterogeneity of the combined gene frequency data. The following enzymes are involved:

Group I. Aldolase, fumarase, glucose-6-phosphate dehydrogenase, glyceraldehyde-3-phosphate dehydrogenase, α-glycerophosphate dehydrogenase, hexokinase, isocitrate dehydrogenase, malate dehydrogenase, malic enzyme, phosphoglucoisomerase, phosphoglucomutase and triosephosphate isomerase.

Group II. Acid phosphatase, adenylate kinase, alcohol dehydrogenase, aldehyde oxidase, alkaline phosphatase, amylase, esterase, glutamate oxaloacetate aminotransferase, leucine aminopeptidase, octanol dehydrogenase, peptidase, tetrazolium oxidase and xanthine dehydrogenase.

TABLE 3. *Relative contributions to heterozygosity of five gene frequency classes for group I (glucose metabolizing) and group II (other) enzymes, compared with predictions based on the neutral allele model (Kimura, 1968).*

Enzymes	Contribution of frequency class					Mean heterozygosity
	1	2	3	4	5	
Group I	3.10 ±.45	2.40 ±.38	1.07 ±.22	0.45 ±.12	0.37 ±.14	0.102 ±.012
Neutral (N_μ=0.03)	0.893	0.966	0.996	1.016	1.029	0.107
Group II	1.47 ±.17	1.44 ±.10	0.94 ±.14	0.68 ±.12	0.96 ±.08	0.226 ±.015
Neutral (N_μ= 0.07)	0.839	0.936	0.988	1.025	1.050	0.219

The relative contribution of each of the five gene frequency classes to heterozygosity, and the mean level of heterozygosity, are given separately in Table 3 for each group of enzymes, with the corresponding expectations for the neutral allele model. Both gene frequency distributions differ significantly from the corresponding neutral allele distribution ($P < .01$), and group I differs significantly from group II ($P < .01$) having a far greater excess of alleles with extreme frequencies (i.e., frequencies falling in class 1).

It may be noted here that the omission of esterases from the group II category leads to a slight reduction in the mean level of heterozygosity from 0.226 ± .015 to 0.207 ± .021, but the relative contributions of the five gene frequency classes are not appreciably altered.

Enzymes with specific and non-specific substrates

The same set of 25 enzymes may of course be classified in other ways, and the following three-way subdivision is informative. Nine of the enzymes have variable substrates, and the remainder can be grouped as glycolytic enzymes (i.e., those catalyzing reactions involving intermediates in the glycolytic pathway) or other specific enzymes. The three groups are as follows:

Glycolytic enzymes. Aldolase, glucose-6-phosphate dehydrogenase, glyceraldehyde-3-phosphate dehydrogenase, α-glycerophosphate dehydrogenase, hexokinase, phosphoglucoisomerase, phosphoglucomutase, and triosephosphate isomerase.

Other specific enzymes. Adenylate kinase, aldehyde oxidase, fumarase, glutamate oxaloacetate aminotransferase, isocitrate dehydrogenase, malate dehydrogenase, malic enzyme, and xanthine dehydrogenase.

Non-specific enzymes. Acid phosphatase, alcohol dehydrogenase, alkaline phosphatase, amylase, esterase, leucine aminopeptidase, octanol dehydrogenase, peptidase and tetrazolium oxidase.

TABLE 4. Relative contributions to heterozygosity of glycolytic enzymes, other specific enzymes and non-specific enzymes.

Enzymes	Contribution of frequency class					Mean heterozygosity
	1	2	3	4	5	
Glycolytic	3.29 ± .47	2.56 ± .49	0.72 ± .27	0.55 ± .20	0.40 ± .15	0.100 ± .013
Other specific	1.49 ± .27	1.46 ± .17	1.04 ± .14	0.36 ± .18	1.07 ± .10	0.206 ± .016
Non-specific	1.65 ± .31	1.63 ± .17	1.06 ± .21	0.84 ± .17	0.73 ± .14	0.204 ± .016

The relative contributions to heterozygosity for these categories are presented in Table 4. It is immediately clear from this subdivision of the data that the contrast between group I and group II enzymes is not due simply to a difference between specific and variable substrate enzymes, as suggested by Kojima et al. (1970). The glycolytic enzymes and other specific enzymes have statistically different gene frequency distributions, and differ markedly in their mean levels of heterozygosity (10% and 21% respectively).

It may also be seen from a comparison of Tables 3 and 4 that the omission of non-glycolytic enzymes from group I leaves the gene frequency distribution and mean level of heterozygosity virtually unaltered. There is therefore no suggestion from these data of any heterogeneity in the glucose metabolizing category (group I).

Regulatory and non-regulatory enzymes

The classification proposed by Johnson (1974) separates out those enzymes which have variable substrates, and subdivides the remainder according to their potential for metabolic regulation. He has classified 18 of the 25 enzymes involved in this survey as follows:

Non-regulatory enzymes. Aldolase, amylase, α-glycerophosphate dehydrogenase, isocitrate dehydrogenase, fumarase, glutamate oxaloacetate aminotranferase, malate dehydrogenase and triosephosphate isomerase.

Regulatory enzymes. Adenylate kinase, alcohol dehydrogenase, aldehyde oxidase, glucose-6-phosphate dehydrogenase, glyceraldehyde-3-phosphate dehydrogenase, hexokinase, malic enzyme, phosphoglucoisomerase, phosphoglucomutase and xanthine dehydrogenase.

The relative contributions to heterozygosity of the five

gene frequency classes are given in Table 5 for these two categories, with the corresponding mean frequencies of heterozygotes. The two groups clearly have significantly different gene frequency distributions for the enzymes assayed, and need to be considered separately in any discussion of the goodness of fit of theoretical models to the gene frequency data.

TABLE 5. *Relative contributions to heterozygosity of non-regulatory and regulatory enzymes as classified by Johnson (1974).*

Enzymes	Contribution of frequency class					Mean heterozygosity
	1	2	3	4	5	
Non-regulatory	5.18 ±.99	1.89±.61	1.03±.27	0.41 ±.27	0.02±.04	0.071 ±.013
Regulatory	1.69 ±.29	1.67±.19	0.89 ±.21	0.41 ±.16	1.01±.11	0.196 ±.021

THEORETICAL MODELS

It is appropriate in this communication to give only a brief resumé of the comparisons which have been made between theoretical models and the Drosophila field data. In particular, the discussion will be restricted to panmictic, single-locus models. These models must of course be extended to include geographic structure and linkage, and the effects of these complications are currently under examination.

Infinite allele models.   It is assumed that mutation occurs with frequency μ to novel and individually distinguishable alleles, in a diploid population of constant effective breeding size N .

A. Additive fitness values

Model 1. Each mutant has a selective disadvantage s relative to the parent allele from which it was derived: i.e. if $A_1 \rightarrow A_2 \rightarrow A_3$, the fitness values of the heterozygotes are $A_1A_2 : A_1A_3 : A_2A_3 = 1\text{-}s : 1\text{-}2s : 1\text{-}3s$.

Model 2. A proportion p of newly arising mutants are neutral as regards their effects on fitness (i.e., they do not differ from their parent allele), and a proportion 1-p have a selective disadvantage s relative to their parent allele.

B. Optimum fitness models

Model 3. Natural selection is assumed to favour a fixed optimal level of enzyme activity, with fitness linearly re-lated to the deviation of activity from the optimum. The spectrum of mutant allelic effects is supposed to be a unit normal distribution, with mean displaced by m units rela-tive to the activity of the parent allele concerned. Allelic effects on activity are assumed to be additive, and the fit-ness of a genotype with activity differing by d units from the optimum is equal to $1\text{-}s|d|$.

Model 4. Natural selection is assumed to favour an opti-mal level of enzyme activity, with fitness declining as the square of the deviation of activity from optimal: i.e. the fitness of a genotype with activity differing by d units from the optimum is $1\text{-}sd^2$. Allelic effects on enzyme acti-vity are assumed to be additive.

Each mutant has an allelic effect on enzyme activity one unit less than that of its parent allele. If an allele A_0 with optimal activity gives rise to a mutant A_1 , the fit-ness values of the three genotypes are then $A_0A_0 : A_0A_1 : A_1A_1 = 1 : 1\text{-}s : 1\text{-}4s$.

Model 5. As for model 4, except that a proportion p
of newly arising mutants show no reduction in enzyme activity
and a proportion 1-p have an allelic effect one unit less
than the parent allele.

Model 6. As for model 4, except that the distribution of
mutant effects is supposed to be a unit normal distribution,
with mean displaced by m units relative to the activity of
the parental allele.

Charge class models. It is assumed that mutation occurs
with frequency μ to novel alleles, of which a proportion α
are not electrophoretically distinguishable from the parent
allele, a proportion β give rise to a polypeptide differing
by one unit of charge from the parental polypeptide, and a
proportion γ differ by two units of charge. Charge differences alone are supposed to determine relative electrophoretic mobility, but are assumed to have no effect on reproductive fitness. Positive and negative charge changes are assumed equally probable.

Values of α = 0.66, β = 0.32, and γ = 0.02 have been
used for all charge class models. These figures have been
calculated by Marshall and Brown (1975) using the "average
protein" of King and Jukes (1969), and assuming that all mutations involve single random DNA base substitutions.

Model 7. Each mutant has a selective disadvantage s
relative to the parent allele from which it was derived, with
additive fitness as in model 1. The mutant differs in charge
from its parent allele by 0, 1 or 2 units with probabilities α, β and γ .

Model 8. Each mutant has an allelic effect on enzyme activity one unit less than that of its parental allele, with
fitness equal to $1-sd^2$ as in model 4. The mutant differs

401

in charge from its parent allele by 0, 1 or 2 units with probability α, β and γ .

Model 9. The counterpart of model 6, with charge changes uncorrelated with mutational alteration in the level of enzyme activity.

COMPARISONS OF MODELS AND DATA

Group I enzymes

It has so far been possible to find parameter values for models 1, 2, 3, 4, 7 and 8, which give a statistically satisfactory fit to the group I enzyme data. These values are given in Table 6, from which it can be seen that the appropriate mean selective values for heterozygous mutants are of the order of Ns = 2 - 4. The fit using Model 2 indicates that the group I data are compatible with a frequency of neutral mutations amounting to approximately 10% of the total.

TABLE 6. *Parameter values giving a satisfactory fit to the* Drosophila *data for group I enzymes.*

Model	p	m	N_μ	Ns
1			0.14	3
2	0.10		0.14	4
3		-2.0	0.14	2
4			0.14	2
7			0.40	3
8			0.40	2

The value of Nμ = 0.14 for model 1 was chosen to give the observed level of heterozygosity, based on the theory of recurrent irreversible mutation to a single deleterious allele

(Kimura, 1964; p.45). The theory gives useful predictions of the mean frequency of heterozygotes for model 1, but not of the relative contributions of each of the five gene frequency classes. These were obtained for all models by computer simulation.

Group II enzymes

With none of the nine models has it been possible to provide a satisfactory fit to the data for group II enzymes. With model 1, for example, parameter values of $N\mu = 0.14$, $Ns = 1$ come very close to fitting the gene frequency distribution with the appropriate mean level of heterozygosity, but the data depart significantly from the model in having an excess of allele frequencies in class 5, i.e. in the range 0.3 - 0.7 . Similar difficulties arise with each of the other models studied.

It is highly likely that the group II enzymes form a heterogeneous category. It will be seen below that the data for non-specific enzymes, representing 9 of the 13 enzymes in group II, are readily explained by simple models (Table 8). A satisfactory fit to the group II data can in addition be obtained by a mixture either of neutral multiallelic loci and model 1 loci, or of overdominant loci and model 1 loci (Table 7).

TABLE 7. Mixtures of loci giving a satisfactory fit to the data for group II enzymes. (Footnote appears on bottom of following page.)

$N\mu$	Neutral loci	Overdominant loci[*]	Model 1 loci		
			Ns=1	Ns=2	Ns=3
0.14	40 %	–	–	–	60%
0.14	–	7 %	51%	42%	–

Specific and non-specific enzymes

We have previously seen that the glycolytic enzymes of Table 4 do not differ significantly from group I enzymes, of which they are a subset, in either gene frequency distribution or mean level of heterozygosity. Model 2 and model 3 have been shown to fit the data for glycolytic enzymes, with values of Ns = 4 and 2 respectively.

The data for the non-specific enzymes can also be accounted for by a number of the models under test, each with a mean selection intensity in the heterozygote of Ns = 1 (Table 8).

TABLE 8.   *Parameter values giving a satisfactory fit to the data for non-specific enzymes.*

Model	m	Nμ	Ns
1		0.14	1
3	-1.35	0.14	1
4		0.14	1
7		0.40	1

The non-glycolytic specific enzymes, by contrast, form a grouping which is extremely difficult to account for with any of the nine basic models. They show a marked excess of allele frequencies in class 5 (Table 4) which has to date been

---

*
Two-allele polymorphisms maintained by selection for the heterozygote, of sufficient intensity to keep gene frequencies almost entirely within the range 0.3 - 0.7 (class 5).

matched only by models including a small fraction of over-dominant loci. The second combination of loci in Table 7 gives a statistically acceptable fit to the data.

Regulatory and non-regulatory loci

The data for the non-regulatory loci, as classified by Johnson (1974), appear to be homogeneous and can be accounted for by the models and parameter values listed in Table 9.

The regulatory enzymes constitute a heterogeneous grouping comparable with the group II and non-glycolytic specific enzyme categories. The inclusion of over-dominant loci in the model is necessary to give a satisfactory fit to the data.

TABLE 9. Parameter values giving a satisfactory fit to the data for non-regulatory enzymes.

Model	Nμ	Ns
1	0.14	5
4	0.14	4
7	0.40	5
8	0.40	4

Loci contributing to the heterogeneity

We have seen that three overlapping subsets of the 25 loci included in this survey are heterogeneous, and not readily accounted for by simple models involving genetic variability maintained by mutation-selection balance alone. These are group II enzymes, non-glycolytic specific enzymes, and regulatory enzymes as classified by Johnson (1974). It turns out that there are three enzymes common to these groupings which do not occur in any of the other categories examined in this paper, viz., adenylate kinase, aldehyde oxidase and xanthine dehydrogenase. If any of the sample of 25 enzymes does in fact show consistent heterozygote superiority in these species of Drosophila, as suggested by the gene frequency distributions for non-glycolytic specific enzymes and the regulatory enzymes, these three enzymes are the most likely candidates.

Of the three, adenylate kinase is notable in having a deficiency of gene frequencies in classes 3 and 4 and a considerable excess in class 5, by comparison with expectations based on the neutral allele model. However, due to the manner in which this enzyme has been chosen from the total sample, it is not possible to apply statistical tests to the observed departure from neutral theory.

CONCLUSIONS

1. The gene frequency distributions for glucose metabolizing
    enzymes (group I) and other enzymes (group II) both
differ significantly from predictions based on the neutral
allele model. Group I differs significantly from group II,
having a higher proportion of alleles at extreme frequencies
and considerably fewer alleles at intermediate frequencies
(Table 3).

2. The contrast between group I and group II enzymes is not
    due simply to a difference between specific substrate en-
zymes and variable substrate enzymes, as suggested by Kojima
et al. (1970). Within the specific substrate enzyme classification, the glycolytic enzymes have a strikingly different
gene frequency distribution from the remainder, and only half
the frequency of heterozygotes (Table 4).

3. Simple genetic models involving mutation to deleterious

alleles are adequate to account for the group I enzyme data (Table 6). The mean intensity of selection against the mutant heterozygotes in these models corresponds to values of Ns in the range 2-4.

4. The data for the non-specific enzymes of group II can
    also be explained by the accumulation of slightly dele-
terious mutations, with a mean selective disadvantage in the
heterozygote corresponding to Ns = 1 (Table 8).

5. The analysis suggests that some at least of the remainder
    of the group II category are either neutral as far as fit-
ness is concerned, or subject to strong overdominant selection. The enzymes implicated are adenylate kinase, aldehyde
oxidase and xanthine dehydrogenase.

6. It may be concluded that the mean selection intensity
    operating against deleterious mutant alleles in the glu-
cose metabolizing category (group I), and the non-specific
enzyme category, are far too small to be detected experimentally. Even if effective breeding population size were as
small as N = 1000 in these species of Drosophila, the foregoing analysis would suggest selection coefficients of the
order of 0.001 - 0.004 in heterozygotes, and of 2-4 times that
magnitude in homozygotes. Selective effects of this order
can only be revealed by very extensive surveys of large numbers of enzymes.

REFERENCES

Ayala, F.J. and J.R. Powell. (1972). Biochem. Genet. 7:
    331-345.

Ayala, F.J., J.R. Powell and Th. Dobzhansky. (1971). Proc.
    Natl. Acad. Sci. U.S. 68: 2480-2483.

Ayala, F.J., J.R. Powell and M.L. Tracey. (1972a). Genet. Res. 20: 19-42.

Ayala, F.J., J.R. Powell, M.L. Tracey, C.A. Mourao and S. Perez-Salas. (1972b). Genetics 70: 113-139.

Ayala, F.J., M.L. Tracey, L.G. Barr, J.F. McDonald and S. Perez-Salas. (1974). Genetics 77: 343-384.

Cohen, P.T.W., G.S. Omenn, A.G. Motulsky, S.H. Chen and E.R. Giblett. (1973). Nature New Biology 241: 229-233.

Gillespie, J.H. and K. Kojima. (1968). Proc. Natl. Acad. Sci. U.S. 61: 582-585.

Johnson, G.B. (1974). Science 184: 28-37.

Kimura, M. (1964). J. Appl. Prob. 1: 177-232.

Kimura, M. (1968). Genet. Res. 11: 247-269.

King, J.L. and T. Jukes. (1969). Science 164: 788-798.

Kojima, K., J.H. Gillespie and Y.N. Tobari. (1970). Biochem. Genet. 4: 627-637.

Kojima, K., P. Smouse, S. Yang, P.S. Nair and D. Brncic. (1972). Genetics 72: 721-731.

Latter, B.D.H. (1973). Genetics 74:s 150-151.

Latter, B.D.H. (1975). Genetics 79: 325-331.

Marshall, D.R. and A.H.D. Brown. (1975). J. Molec. Evol. (In press).

Ohta, T. (1974). Nature 252: 351-354.

Prakash, S. (1973). Genetics 75: 347-369.

Prakash, S., R.C. Lewontin and J.L. Hubby. (1969). Genetics 61: 841-858.

Richmond, R.C. (1972). Genetics 70: 87-112.

Saura, A. (1974). Hereditas 76: 161-172.

Saura, A., S. Lakovaara, J. Lokki and P. Lankinen. (1973) Hereditas 75: 33-46.

Yamazaki, T. and T. Maruyama. (1972). Science 178: 56-57.

COMMENTS BY M. NEI

The data Dr. Latter presented are certainly in agreement with the pattern expected from his hypothesis of optimum model selection for protein loci and also Ohta's hypothesis of slightly deleterious mutation. I would like to indicate, however, that the data can also be explained by the hypothesis that a majority of polymorphic alleles are neutral but the Drosophila species from which the protein data were taken have recently gone through bottlenecks. It is known that once a population goes through a bottleneck, it takes a long time -- reciprocal of mutation rate -- for the genetic variability of the population to be recovered to the original level. Namely, if the mutation rate for electrophoretically detectable alleles is 10^{-7} per year, as is often assumed, it takes about 10 million years. Before a new balance between mutation and random genetic drift is attained, it is expected that there are many low-frequency alleles in the population, and thus the low gene frequency classes show a higher amount of heterozygosity than the classes close to 0.5. Actually, there is some reason to believe that Latter's Drosophila species have had the bottleneck effect recently. Namely, as indicated by Ayala (Proc. 6th Berkeley Symp. Math. Stat. Prob. Vol.5: 211-236, 1972), the level of average heterozygosity in these species is much lower than that expected from the mutation rate and population size.

REPLY BY B.D.H. LATTER

Dr. Nei's comment has drawn attention to an important feature of the results discussed in my paper, namely that comparisons are made with the Drosophila gene frequency data only after the attainment of equilibrium levels of hetero-zygosity in the simulated populations. Non-equilibrium populations certainly merit detailed examination, though it appears unlikely that populations having recently passed through a bottleneck will have a gene frequency distribution matching that of the Drosophila species surveyed.

If the bottleneck is so severe as to render nearly all loci homozygous, it is certainly true that gene frequency distributions with an excess of alleles at extreme frequencies will persist for an extended period of time. Less severe bottlenecks have two types of effect. Some loci are rendered homozygous and will subsequently contribute to the extreme gene frequency classes by mutation and drift as Dr. Nei suggests. The remainder are expected to have a gene frequency distribution for which alleles at extreme frequencies are _less_ common than in an equilibrium population, and alleles at _intermediate_ frequencies will therefore be expected to make a disproportionate contribution to heterozygosity at these unfixed loci.

The _Drosophila_ species included in this survey have an overall mean level of heterozygosity of 0.18, with the individual species ranging from 0.11 to 0.24; by comparison with the levels of electrophoretic variation known in other organisms, the _Drosophila_ species cannot by any means be considered to be low in heterozygosity. An hypothesis of a recent _severe_ bottleneck in population size does not therefore seem appropriate.

Mating Preferences and Their Genetic Effects in Models of Sexual Selection for Colour Phases of the Arctic Skua

P. O'DONALD

1. INTRODUCTION

O'Donald, Wedd and Davis (1974) described computer models of sexual selection which they used to analyse data on the breeding of the Arctic Skua, a predatory and piratical seabird of the Arctic and Subarctic regions. As originally suggested by Darwin (1871), selection takes place because pairs of birds that breed earlier in the breeding season fledge a higher average number of chicks than later breeding pairs. A particular male will therefore gain a selective advantage if he has characteristics that improve his chances of finding a mate and thus of breeding earlier in the season (O'Donald, 1972). The males' chances of mating may vary because some females may have preferences for males with particular characteristics of plumage or display: the males themselves may vary in their abilities to defend their territories or in their responses to the females. Such variation in the chances of finding mates is usually described in terms of mating preferences. Used in this way, mating preference is simply a descriptive term: no behavioural mechanism is implied.

Models of sexual selection can be set up to simulate the general effects of mating preferences. The genetic conse-

411

quences of these models have been analysed by computer simulation (O'Donald, 1973a, 1974). The models have also been
used to estimate the female mating preferences for the three
main phenotypes - pale, intermediate and dark - of the Arctic
Skua. Genetically, the pale phased birds are homozygous; the
intermediates are mainly the heterozygotes; but dark and intermediate birds can be misclassified, for there is a continuous range of phenotypes from intermediate to dark (O'Donald
and Davis, 1959). On average the dark males breed before the
intermediates who breed before the pales. These differences
in breeding times disappear when the males breed in subsequent years with their previous mates. No significant differences in breeding times have been observed between phases of
females. O'Donald et al. (1974) estimated the mating preferences required to produce the actual differences in the males'
breeding times. They also analysed fledging success and
showed that the sexual selection contributed to the variation
between the phenotypes.

In the models of sexual selection, it is assumed that the
birds are monogamous and the females become willing to mate
during successive periods in the breeding season. The actual
proportion of new pairs of birds breeding in a particular period is the proportion of females who came into breeding condition in that period. Among the females breeding in each
period, there are some who have preferences for particular
male phenotypes and others who simply mate at random. In the
models described by O'Donald et al., the females with the
preferences mate at the beginning of each period; the others
then mate at random with the males who are still left without
mates. The mating preferences were estimated by calculating
the theoretical distributions of the breeding times of the
males and finding the values of the mating preferences that

gave a minimum value of χ^2 .

In this paper, the earlier models are generalised so that
females with preferences can mate either before or after
those who mate at random, or simultaneously with them. The
breeding season is divided into periods of either weeks (as
in the previous calculations) or days, for the proportion of
matings taking place in a given period does affect the estimates of the preferences. Partial female preferences have
also been allowed for in the models. The maximum likelihood
estimates of the preferences have now been calculated for all
the models and log likelihood surfaces have been plotted for
some of them. Finally, the effects of the mating preferences
have been computed for successive generations of sexual selection.

2. DESCRIPTIONS OF THE MODELS

There are four basic models describing how the females'
mating preferences determine the probabilities with which the
dark, intermediate and pale males obtain mates during the
breeding season. In the simplest model, Model 1, some of the
females prefer to mate either with dark or intermediate males.
The remaining females mate at random with dark, intermediate
and pale males. The females with the preference are also
prepared to mate with pale males if no darks and intermediates
remain unmated. O'Donald (1973a) used this model to analyse
the theoretical effects of sexual selection. The selective
coefficients are frequency-dependent and may either increase
or decrease with frequency: they increase with frequency
when most of the females have the preference; when less than
about 30% of the females have the preference, they decrease.

In Model 2, the females with the preference mate with dark

males if they can; when all the dark males have found mates, these females then mate with the intermediate males; they only mate with pale males if no darks and intermediates are available. The females with no preference mate at random with available males. In Model 3, there are two groups of females with different preferences: some of them, a proportion α , prefer to mate only with dark males; others, a proportion β , prefer either dark or intermediate males indiscriminately; the remaining females, the proportion 1-α-β , mate at random. The females with preferences also mate at random if no males with the preferred phenotype remain without mates. Thus when α=0 , Model 3 is identical to Model 1: the dark allele is then dominant in its effect on mating preference. Finally, in Model 4, α of the females prefer dark males and β prefer intermediate males. This is a particular case of a more general model in which there is a separate preference for each male phenotype (O'Donald, 1974).

In fitting these models of the mating preferences to the data of the Arctic Skua on Fair Isle, O'Donald et al. considered only those models in which preferential matings take place before random matings. Models have now been programmed for computation in which the random matings take place first. These are called Models 1R, 2R, 3R and 4R. Preferential and random matings may also occur simultaneously giving Models 1S, 2S, 3S and 4S. The female preferences may be only partial: if, after a given number of attempts to mate, a female has still not encountered a male of her choice, she then mates at random regardless of her preferences even though some of the males she prefers are still available for mating. These models are called 1T, 2T, 3T and 4T. Table 1 shows the various models that have been analysed in order to estimate the mating preferences of the Arctic Skua. The S and T models

414

are more complicated than the others. Only Models 3S and 3T, 4S and 4T have been programmed for computation, as shown in Table 1: Model 1 is just a special case of Model 3; Model 2 does not fit the data of the Arctic Skua. The models in Table 1 were programmed as subroutines written in Fortran.

TABLE 1. *Classification of the Models.*

Mating preferences of the females	Preferential mating first	Random mating first (R models)	Simultaneous mating (S models)	Partial preferences (T models)
α for darks and inters	Model 1	Model 1R	-	-
α for darks or inters if no darks left	Model 2	Model 2R	-	-
α for darks β for darks and inters	Model 3	Model 3R	Model 3S	Model 3T
α for darks β for inters γ for pales	Model 4	Model 4R	Model 4S	Model 4T

Given the overall distribution of breeding times, the frequencies of the males and the female mating preferences, the subroutines compute the distributions of the breeding times of each of the three male phenotypes. The Models 1 and 4 have already been described in detail (O'Donald, 1973a, 1974). The listings of the subroutines for Models 3, 4, 3R and 4R are given in the Appendix.

The S and T models must be described in more detail. In both these models, the rate at which males are removed

from the pool of unmated males is determined simultaneously
by females choosing males preferentially or at random. The
proportions of the dark, intermediate and pale males are
assumed to be a , b and c . The total proportion of unmated
males is n = a+b+c . Thus for Model 3S, the rate of removal
of unmated males from the pool is given by the equations

$$\frac{da}{dn} = \alpha + \beta \left(\frac{a}{a+b}\right) + \delta\left(\frac{a}{n}\right)$$

$$\frac{db}{dn} = \beta \left(\frac{b}{a+b}\right) + \delta\left(\frac{b}{n}\right)$$

$$\frac{dc}{dn} = \delta\left(\frac{c}{n}\right)$$

where α is the proportion of females preferring dark males,
β the proportion preferring dark or intermediate males and
$\delta = 1-\alpha-\beta$ is the proportion mating at random. The equations
are satisfied by the solutions

$$a = n-b-c$$

$$b = Mn^{\beta+\delta} \left(1 - \frac{c}{n}\right)^{\beta/(\alpha+\beta)}$$

$$c = Kn^{\delta}$$

where $K = c_o n_o^{-\delta}$ and $M = b_o n_o^{-(\beta+\delta)} \left(1 - \frac{c_o}{n_o}\right)^{-\beta/(\alpha+\beta)}$.

In Model 4S, α, β and γ are the proportions of females
preferring dark, intermediate and pale males separately. The
differential equations are then as follows:

$$\frac{da}{dn} = \alpha + \delta\left(\frac{a}{n}\right)$$

$$\frac{db}{dn} = \beta + \delta\left(\frac{b}{n}\right)$$

$$\frac{dc}{dn} = \gamma + \delta\left(\frac{c}{n}\right)$$

which are satisfied by the solutions

$$a = \left(\frac{\alpha}{1-\delta}\right)n + \left(\frac{n}{n_o}\right)^{\delta}\left(a_o - \frac{\alpha}{1-\delta} n_o\right)$$

and so on for b and c which have similar solutions. Thus we calculate the proportions of the three types of males who secure mates in each of the periods of the breeding season.

The T models are the most realistic biologically. In these models, it is assumed that the males and females are brought together randomly. If a female then finds a male of her choice, she mates with him. If not, she returns to the pool of unmated females and the process is repeated. If a female does not find a male of the type she prefers after r random encounters, she then mates with the next male she encounters whether he is of the type preferred or not. This process is a model of the behavioural mechanism of female mating response. The female's mating behaviour is released by the specific male characteristics. If the female encounters the wrong type of male she does not respond, but the threshold of her response is lowered until eventually it reaches the point at which she is prepared to mate with the wrong type of male. If the proportions of the dark, intermediate and pale males are a , b and c , then we have the following differential equations for Model 3T.

$$\frac{da}{dn} = \alpha \left\{ 1 - \left(\frac{b+c}{n} \right)^{r+1} \right\} + \beta \left(\frac{a}{a+b} \right) \left\{ 1 - \left(\frac{c}{n} \right)^{r+1} \right\} + \delta \left(\frac{a}{n} \right)$$

$$\frac{db}{dn} = \alpha \left(\frac{b}{n} \right) \left(\frac{b+c}{n} \right)^{r} + \beta \left(\frac{b}{a+b} \right) \left\{ 1 - \left(\frac{c}{n} \right)^{r+1} \right\} + \delta \left(\frac{b}{n} \right)$$

$$\frac{dc}{dn} = \alpha \left(\frac{c}{n} \right) \left(\frac{b+c}{n} \right)^{r} + \beta \left(\frac{c}{n} \right)^{r+1} + \delta \left(\frac{c}{n} \right)$$

There is no obvious analytic solution to these equations, but given the values of a , b and c and the female preferences they are easy to integrate numerically. The Appendix gives the listing of the subroutine which calculates the distributions of the breeding times of the males.

417

In the Model 4T, the probability that a female with a pre-ference for dark males will fail to find one is $\left(1-\dfrac{a}{n}\right)^{r+1} = \left(\dfrac{b+c}{n}\right)^{r+1}$. Therefore we have the differential equations

$$\frac{da}{dn} = \alpha \left\{ 1 - \left(\frac{b+c}{n} \right)^{r+1} \right\} + \beta \left(\frac{a}{a+c} \right) \left(\frac{a+c}{n} \right)^{r+1} + \gamma \left(\frac{a}{a+b} \right) \left(\frac{a+b}{n} \right)^{r+1} + \delta \left(\frac{a}{n} \right)$$

$$\frac{db}{dn} = \alpha \left(\frac{b}{b+c} \right) \left(\frac{b+c}{n} \right)^{r+1} + \beta \left\{ 1 - \left(\frac{a+c}{n} \right)^{r+1} \right\} + \gamma \left(\frac{b}{a+b} \right) \left(\frac{a+b}{n} \right)^{r+1} + \delta \left(\frac{b}{n} \right)$$

$$\frac{dc}{dn} = \alpha \left(\frac{c}{b+c} \right) \left(\frac{b+c}{n} \right)^{r+1} + \beta \left(\frac{c}{a+c} \right) \left(\frac{a+c}{n} \right)^{r+1} + \gamma \left\{ 1 - \left(\frac{a+b}{n} \right)^{r+1} \right\} + \delta \left(\frac{c}{n} \right)$$

The Appendix gives the listing of the subroutine for Model 4T.

3. RESULTS OF FITTING THE MODELS TO THE DATA OF THE ARCTIC SKUA

Table 2 gives the distributions of breeding times of males mating with particular females for the first time. The models give the probabilities that a dark, intermediate and pale male will breed during successive periods of the breeding season.

TABLE 2. *Distribution of breeding times of colour phases of males in new pairs.*

Breeding dates in weekly intervals	Numbers breeding in the intervals			
	Darks	Inters	Pales	Total
June 10–16	1	2	1	4
June 17–23	8	24	3	35
June 24–30	16	19	6	41
July 1–7	10	24	6	40
July 8–14	3	15	5	23
July 15–21	1	2	5	8
Totals	39	86	26	151

These periods may be the weeks shown in Table 2, or each successive day. If probabilities for each day were calculated, they were lumped to give probabilities corresponding to each of the weeks in the table. The log likelihoods were then calculated for arbitrary values of the mating preferences. Random perturbations were introduced into the values of the mating preferences and the log likelihoods calculated again, the process continuing until a higher log likelihood had been found. This was repeated with smaller and smaller random perturbations until the maximum log likelihood of the model had been reached. The maximum likelihood estimates of the parameters of the models are given in Table 3.

TABLE 3. *Maximum likelihood estimates of the parameters of the models.*

Model	Parameters fitted		Likelihood
	α	β	log (base e)
3	0.086	0.403	−359.026
3R	0.067	0.256	−358.633
3S	0.071	0.343	−358.713
3T	0.070	0.343	−358.712
4	0.188	0.294	−358.813
4R	0.133	0.204	−358.568
4S	0.160	0.252	−358.653
4T	0.161	0.252	−358.651
2	0.10	−	−361.294
2R	0.06	−	−361.395

The results shown for Models 4, 4R, 4S and 4T are only given for the case when $\gamma=0$. When this is not the case, the likelihoods are very slightly increased, but the general effect is negligible. Model 4R has the highest likelihood. In general the R models have the highest likelihoods, followed by the S

and T models. However the differences in the log likelihoods
of all the models 3 and 4 are small. Only models 2 and 2R
have log likelihoods more than two units of support below the
maximum and these models can therefore be rejected.

The likelihood surfaces have been calculated for Models 3
and 3R, 4 and 4R. The results are plotted in Figures 1 and 2.

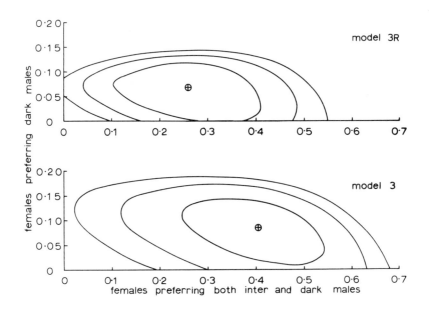

FIGURE 1. *The log likelihood surfaces of Models 3 and 3R,
showing the 1-, 2- and 3-unit support limits relative to the maximum log likelihood of Model 4R,
which has the highest likelihood. The point of
maximum likelihood is shown by the small crossed
circle.*

The contours in the figures represent the support for the
models in units of log likelihood. The log likelihoods of
the models were calculated over a wide range of values of
the parameters.

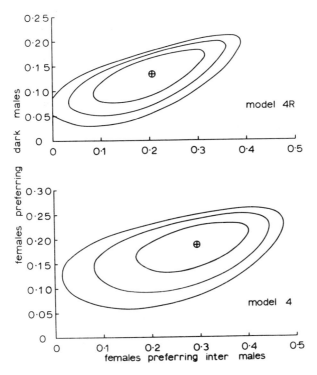

FIGURE 2. *The log likelihood surfaces of Models 4 and 4R drawn with similar values and notation to Figure 1.*

The log likelihood at the maximum (in Model 4R with $\alpha = 0.133$ and $\beta = 0.204$) was then subtracted from each of the log likelihood values to give the values of support (Edwards, 1972). The contours at the 1-unit, 2-unit and 3-unit levels of support were then interpolated to give the Figures 1 and 2. The 2-unit level of support represents approximately the 95% fiducial limit; values of the parameters are very unlikely to lie outside the 3-unit level of support. In the fitting of these models, the weekly breeding periods of Table 2 have

been used. If the breeding season is divided into daily
periods, Models 3 and 3R converge to give almost the same
estimates as 3S ; 4 and 4R converge to 4S. This is to be
expected since the preferential and random matings are then
taking place almost simultaneously.

4. SELECTION AS A RESULT OF THE MATING PREFERENCES

In models of sexual selection for polygynous species
(O'Donald, 1973b), the males who mate preferentially are also
available to mate at random. If α , β and γ are the proportions of females with preferences for genotypes AA , Aa
and aa , as for example in Model 4, then there is a stable
equilibrium point when the allele A is at a frequency

$$P_e = (\alpha + \tfrac{1}{2}\beta)/(\alpha + \beta + \gamma) \quad .$$

In the models for monogamous species like the Arctic Skua,
however, the progress of selection must be computed in
successive generations in order to determine the approach to
equilibrium. The distributions of breeding times must first
be computed for each of the male phenotypes. The sexual
selective coefficients can then be calculated from the
relationship between breeding time and fitness. The genotypic frequencies in the next generation are thus determined
and hence the new distributions of breeding times.

Unless natural selection opposes the sexual selection, the
allele A always spreads through populations in Models 3 and
3R to replace a . In Models 4 and 4R, approximately the
same equilibrium frequency, P_e , is reached as in the model
with polygyny. When the mating preferences are large, there
are slight differences at equilibrium from P_e , caused by
grouping the data on the breeding times into a few discrete

422

intervals. Table 4 shows the actual grouping used to calculate the sexual selective coefficients (O'Donald et al.
1974).

TABLE 4. Fledging success and breeding time of pairs of
Arctic Skuas.

Weeks of the breeding season x	Proportion of pairs breeding	Relative fledging success $w = 1 - \dfrac{(0.47354 - x)^2}{24.254}$
0	0.09836	0.99075
1	0.44672	0.98857
2	0.23975	0.90393
3	0.13524	0.73683
4	0.07992	0.48726

The relative fledging success is given by a quadratic
function which is a very close fit to the actual data.

The progress of sexual selection in monogamous birds is
generally frequency-dependent. This has been analyzed in
earlier papers (O'Donald, 1973a, 1974). If the dark allele,
A , is dominant in its effect on mating preference, then
selection will take place according to Model 3 or 3R with
$\alpha = 0$. Table 5 shows the final selective coefficients of
the allele a when it has nearly been eliminated from the
population by the sexual selection for A . At the smaller
values of β the selective coefficient declines from its
initial value of 0.0601: the frequency-dependence is thus
negative. At the higher values of β the frequency-
dependence is positive.

In Models 4 and 4R, a polymorphic equilibrium is always
reached. The progress of selection towards the equilibrium

423

has been calculated for the estimated values of the parameters. As shown in Table 3, these estimates are as follows:

Model 4: $\alpha = 0.188$

$\beta = 0.294$

Model 4R: $\alpha = 0.133$

$\beta = 0.204$

TABLE 5. *Final selective coefficients of the allele a at
the point of being eliminated from the population.*

Value of mating preference for intermediates β	Final value of selective coefficient	
	Model 3 s_∞	Model 3R s_∞
0.1	0.00665	0.0192
0.2	0.0147	0.0430
0.3	0.0247	0.0719
0.4	0.0374	0.107
0.5	0.0540	0.148
0.6	0.0767	0.196
0.7	0.110	0.253
0.8	0.161	0.318
0.9	0.254	0.392
1.0	0.478	0.478

Table 6 shows the rate of approach to equilibrium in the two
models when the gene frequency starts at $p_o = 0.01$. Table
7 shows the relative fitnesses during the progress to equilibrium. With these preferences ($\alpha+\beta = 0.482$ for Model 4)
there is hardly any frequency-dependent effect. But smaller
values of $\alpha+\beta$ give negative frequency-dependence and
greater values give positive frequency-dependence. The same
equilibrium is reached whatever the initial frequency.

The general reasons for these changes in fitness during
the evolution can be explained by the changes in the

frequencies of the preferred males.

TABLE 6. *Gene frequencies for the allele* $\underline{A}$ *for dark phase in models 4 and 4R.*

Generation	Gene frequency of allele $\underline{A}$	
	Model 4	Model 4R
0	0.01	0.01
10	0.0160	0.0160
20	0.0270	0.0269
40	0.0762	0.0758
60	0.195	0.195
80	0.381	0.386
100	0.557	0.570
200	0.693	0.695
300	0.695	0.696
400	0.695	0.696
	$p_e = 0.695$	$p_e = 0.697$

TABLE 7. *Relative fitnesses of the genotypes* $\underline{AA}$, $\underline{Aa}$ *and* $\underline{aa}$ *during the progress of sexual selection in Model 4.*

Generation	Relative fitnesses		
	$\underline{AA}$	$\underline{Aa}$	$\underline{aa}$
0	1.000	1.000	0.901
10	1.000	1.000	0.900
20	1.000	0.999	0.897
40	1.000	0.998	0.886
60	1.000	0.972	0.865
80	1.000	0.945	0.852
100	1.000	0.998	0.893
200	0.947	1.000	0.879
300	0.947	1.000	0.879
400	0.947	1.000	0.879

At the start of selection $p_0 = 0.01$. There are hardly any
dark males and only 2 per cent of the males are intermediate.
Thus all the preferred males, both darks and intermediates,
find mates during the first week of the breeding season. They
all have the same maximal fitness. As the gene for dark
increases in frequency, some of the intermediate males, which
at first increase in frequency much more rapidly than the
darks, are left unmated at the end of the first week. They
mate with females who have a slightly lower fledging success.
As the gene continues to increase in frequency, the intermediates become the commonest phenotype while the darks are
still relatively rare. Thus the darks remain the fittest
males, but the intermediates lose some of their advantage over
the pales. As equilibrium is approached, the frequency of the
darks exceeds the proportion of females which prefer them and
they too start to lose some of their advantage. At equilibrium, the frequency of the darks is 0.483 and that of the
intermediates is 0.424; yet only 0.188 of the females prefer
darks while 0.294 prefer intermediates. There is a greater
excess of darks over the females that prefer them than there
is of intermediates. Thus on average the intermediate males
find mates before the darks and hence have a higher fitness.
This gives rise to the heterozygous advantage that maintains
the stability of the polymorphism.

5. SEXUAL SELECTION BALANCED BY NATURAL SELECTION

In all the models a stable polymorphism can be maintained
if the sexual selection is balanced by natural selection. In
Models 3 and 3R, an equilibrium can be produced as a result
of the negative frequency-dependence of the sexual selective
coefficients when the mating preferences are small: the

sexual selective coefficients eventually decline until they exactly equal the opposing natural selective coefficients. This produces a stable equilibrium, because a further increase in the frequency of the preferred males causes their sexual selective advantage to decrease and natural selection then reduces their frequency until it has come back to equilibrium. The polymorphic equilibrium is only established if the natural selective coefficient is less than the initial value of the sexual selective coefficient. If the natural selection is initially greater than the sexual selection, then the allele that determines the preferred phenotypes must inevitably be eliminated.

In Models 4 and 4R, the equilibrium produced by sexual selection is altered by the natural selection. Table 8 shows the stable equilibrium frequencies produced when natural selection acts in opposition to the sexual selection for dark and intermediate males.

TABLE 8. *Gene frequencies at equilibrium when sexual selection is balanced by natural selection.*

Natural selective coefficient s	Equilibrium gene frequency of allele for dark			
	Model 3 $\alpha = .085$ $\beta = .405$	Model 3R $\alpha = .067$ $\beta = .258$	Model 4 $\alpha = .188$ $\beta = .294$	Model 4R $\alpha = .133$ $\beta = .204$
0	1.0	1.0	0.695	0.696
0.025	0.836	0.861	0.632	0.641
0.05	0.615	0.637	0.559	0.576
0.06	0.547	0.567	0.528	0.549
0.07	0.489	0.506	0.483	0.509
0.08	0.436	0.441	0.384	0.435
0.09	0.371	0.357	0.227	0.239
0.10	0	0	0	0

It is assumed that natural selection has an additive effect
on fitness, with the following values:

Genotypes	AA	Aa	aa
Phenotypes	dark	intermediate	pale
Relative fitness	1-2s	1-s	1

The actual gene frequency in the population of Arctic Skuas
is $p = 0.48$. Thus a natural selective coefficient of
$s = 0.07$ would be required to maintain the observed frequency
at equilibrium. However there is now good evidence that
natural selection is acting much more strongly than this in
favour of pale phases. The pale birds breed at a younger age
than the others. This gives them a greater chance of surviving to breed and by itself would produce a selective
coefficient of $s = 0.11$. Since sexual selection only
affects the proportion of birds forming new pairs, the overall
selection appears to be strongly in favour of pale. If this
is so, the population cannot be at equilibrium: it must be
evolving towards the elimination of dark in spite of the
sexual selection. (Results of calculations on data of demography and selection in the Arctic Skua will be published in
a paper with J.W.F. Davis.)

ACKNOWLEDGEMENTS

Part of the work described in this paper was made possible
by a research grant from the Natural Environment Research
Council whose support I very gratefully acknowledge. I am
also grateful to N.S. Wedd who analyzed and programmed Models
3S, 3T, 4S and 4T and fitted them to the data on the Arctic
Skua as part of work for a doctoral thesis.

SUMMARY

1.   Models of sexual selection are described and analyzed in terms of female mating preferences for the dark, intermediate and pale phenotypes of the Arctic Skua. The females with no preference are assumed to mate at random. As the females come into breeding condition during successive periods of the breeding season, they form pairs with the unmated males.

2.   The models allow for the females with preferences to mate either before or after the others who mate at random, or simultaneously with them. The preferences may be only partial. Estimates of the preferences are obtained by fitting the models to the distributions of breeding times of the Arctic Skuas; the likelihood surfaces of some of the models are illustrated graphically.

3.   The preferred males gain a selective advantage because they breed at an earlier date in the breeding season and earlier breeding pairs fledge a greater average number of chicks. The results of selection have been computed in successive generations. In the models with separate female preferences for dark and intermediate males, a polymorphic equilibrium becomes established. In other models, the gene for the sexually advantageous phenotypes becomes fixed in the population.

4.   If natural selection opposes the sexual selection, polymorphic equilibria can be established in all the models. This is a consequence of the frequency-dependence of the sexual selection. In the Arctic Skua, natural selection for pale appears to outweigh sexual selection for dark.

REFERENCES

Darwin, C. (1871). The descent of man and selection in relation to sex. John Murray, London.

Edwards, A.W.F. (1972). Likelihood. Cambridge University Press, London.

O'Donald, P. (1972). Am. Natur. 106: 368-379.

O'Donald, P. (1973a). Heredity 30: 351-368.

O'Donald, P. (1973b). Heredity 31: 145-156.

O'Donald, P. (1974). Heredity 32: 1-10.

O'Donald, P., N.S. Wedd and J.W.F. Davis. (1974). Heredity 33: 1-16.

NOTE. The appendix listing Fortran Subroutine of Models, referred to in the text, has been omitted. The author will be happy to provide a copy of this appendix on request.

Simulation of Quantitative Characters by Genes With Biochemically Definable Action
VII. Observation and Discussion of Nonlinear Relationships

W. SEYFFERT and G. FORKMANN

INTRODUCTION

A metric character, the anthocyanin content of flowers, has been investigated. We used 27 defined genotypes of Matthiola incana R. Br. resulting from the variation of 3 loci with 2 alleles each involved in the biosynthesis chain. They cause not only qualitative changes in the pigment pattern but also change the total content of the secondary metabolites (Seyffert, 1971). The 3 loci vary against an isogenic background in all possible homo- and heterozygous combinations. In a recent paper (Forkmann and Seyffert, in press), a linear model for the estimation of genetic parameters was introduced which differs from the common methods in:

(1) Instead of the population mean or the midparental value the measured value of the manifold recessive genotype is used as a reference point.

(2) Only the contributions of functional alleles which are by definition positive are estimated. The same is valid for nonallelic interactions.

(3) If functional alleles are located on different loci the interaction of the highest order is estimated. Nonallelic

interactions of lower order are possibly covered by this estimation.

The definition formulas, given in Table 1, result from a trifactorial case. The application of this model to experimental data has led to the following results (Forkmann and Seyffert, in press):

(1) Contributions of loci with 1 or 2 functional alleles are always positive. Comparable contributions of single loci are not uniform but clearly different in size.

(2) Allelic and nonallelic interactions are unidirectional negative modulo standard errors. With more functional alleles present, increased interactions occur. (Allelic interactions are defined by $uu_{ii} = b_i - 2u_i$.)

These observations permit the conclusion that the relationships between the number of alleles and the phenotypic measurements are nonlinear and that the deviation from linearity is larger when more functional alleles are present.

With increasing number of alleles the curve appears to be more flat, possibly approaching an upper limit asymptotically. This would be compatible with the case of saturation curve which characterize enzymatic reactions and some regulating events.

Using a linear model, the interaction parameters should therefore have the function of an adjustment to a saturation curve. If this were the only function, there should be expected a strong correlation between the magnitude of the interaction and the extent of contributions of the genes involved in the interaction. But as this is, by no means, always the case, it might be assumed that the estimated interaction parameters concern two different events:

(1) The adjustment to the saturation character of the curve,

and

(2) "real" genetic interactions which might be responsible for deviations from the curve.

If these two parts could be separated successfully, further information could be gained of the character's architecture.

ESTIMATION OF A SATURATION CURVE

In the simplest case we sum up the number of functional alleles irrespective of their locus and fit a saturation curve to the distribution of the measured values. An appropriate function is one of the Baule-Mitscherlich-type (Mitscherlich, 1909; 1924):

$$\frac{dy}{dx} = c(Y-y) \tag{1}$$

where x = number of alleles, y = measured value of the genotypes, Y = upper limit, and, c = constant of proportionality. Integration leads to

$$\ln(Y-y) = C-cx \quad . \tag{2}$$

If it is assumed that $y=0$ if $x=0$, it follows that

$$C = \ln Y \tag{3}$$

and therefore

$$\ln(Y-y) = \ln Y-cx \tag{4}$$

and

$$y = Y(1 - e^{-cx}) \quad . \tag{5}$$

The fitting of this half-logarithmic regression is done in the usual least squares method, assuming an approximated value for Y based on intuitive considerations. The coefficients are then estimated as follows:

$$\gamma = \frac{\Sigma x \ \Sigma \ln(Y-y) \ - \ n\Sigma x \ \ln(Y-y)}{n\Sigma x^2 \ - \ (\Sigma x)^2} \tag{6}$$

and

$$\nu = \frac{\Sigma x^2 \ \Sigma \ln(Y-y) \ - \ \Sigma x \ \Sigma x \ \ln(Y-y)}{n\Sigma x^2 \ - \ (\Sigma x)^2} \tag{7}$$

433

where γ is an estimate of c and ν an estimate of $\ln Y$.
If the estimated value γ is different from the approximated
one, the estimation has to be repeated using an improved
approximation.

RESULTS AND DISCUSSION

Our data consists of measurements on 27 defined genotypes
given in Table 2. Each value is the mean of 33 replications
of the measurement of the same genotype on different days
(Forkmann and Seyffert, in press).

Application of formulas (6) and (7) to these data results
in a function which is plotted in Figure 1 and shown in Table
3 (upper line).

In addition to the varying loci is a fourth locus, homozygous with respect to two functional alleles, which is involved
in the expression of the character also. Therefore it is
correct to estimate the saturation curve in the range $2 \leqslant x \leqslant 8$.
Now it would be possible to use the differences between measurements of defined genotypes and their expected values on
the saturation curve (see Figure 1) and to interpret them as
an expression of "real" allelic and nonallelic interactions.
But this is justified only if the increments of all the loci
are equal so that the summation over all loci is allowable.

A comparison of measurements of corresponding genotypes
(Table 2) shows, however, that this condition is apparently
not met. That is to say, there has to be assumed a specific
saturation function for each locus.

If a saturation function is estimated for each locus separately using the 3 possible genotypes, enumerated by 0 , 1
and 2 , against an otherwise recessive background, the supposition given above is verified. The results, given in Table 3

and plotted in Figure 2, show clearly the differences between
the 3 loci b , u and v .

The general function, taking account of all loci, represents
therefore merely an average trend but does not give a good
representation of the general situation.

Considering the observed data which result from 2-locus
combinations keeping the third locus homozygote recessive, one
gets the 3 possible response surfaces shown in Figure 3; i.e.,
a multidimensional response surface would represent the real
situation much better than a single saturation function. This
calls for an extension of formula (5) which considers the contributions of k different loci:

$$y = Y(1 - e^{-c_1 x})(1 - e^{-c_2 x}) \ldots (1 - e^{-c_k x}) \ . \tag{8}$$

Unfortunately, up to now, no estimation procedure suitable for
$k > 2$ could be found in literature (Patterson, 1969). It
would therefore be interesting to develop a general estimation
procedure which will enable the estimation of the parameters
$Y , c_1 , c_2 \ldots c_k$ in an appropriate manner.

Equation (8) is valid only in the case of complementary
gene action. In all other cases including the one described
above, an appropriate formula has still to be developed.

REFERENCES

Seyffert, W. (1971). Theor. Appl. Genet. 41: 285-291.

Forkmann, G. and W. Seyffert. (1975). In press.

Mitscherlich, E.A. (1909). Landwirtsch. Jahrb. 38: 537-552.

Mitscherlich, E.A. (1924). Schriften d. Konigsberger Gel. Ges.
   Nat. 3: 141.

Patterson, H.D. (1969). Biometrics 25: 159-164.

TABLE 1. *Definition of the simple model used for estimation of genetic parameters.*

genotype ijk	definition	genotype ijk	definition	genotype ijk	definition
000	= z	100	= $z+u_i$	200	= $z+b_i$
001	= $z+u_k$	101	= $z+u_i+u_k+uu_{ik}$	201	= $z+b_i+u_k+bu_{ik}$
002	= $z+b_k$	102	= $z+u_i+b_k+ub_{ik}$	202	= $z+b_i+b_k+bb_{ik}$
010	= $z+u_j$	110	= $z+u_i+u_j+uu_{ij}$	210	= $z+b_i+u_j+bu_{ij}$
011	= $z+u_j+u_k+uu_{jk}$	111	= $z+u_i+u_j+u_k+uuu_{ijk}$	211	= $z+b_i+u_j+u_k+buu_{ijk}$
012	= $z+u_j+b_k+ub_{jk}$	112	= $z+u_i+u_j+b_k+uub_{ijk}$	212	= $z+b_i+u_j+b_k+bub_{ijk}$
020	= $z+b_j$	120	= $z+u_i+b_j+ub_{ij}$	220	= $z+b_i+b_j+bb_{ij}$
021	= $z+b_j+u_k+bu_{jk}$	121	= $z+u_i+b_j+u_k+ubu_{ijk}$	221	= $z+b_i+b_j+u_k+bbu_{ijk}$
022	= $z+b_j+b_k+bb_{jk}$	122	= $z+u_i+b_j+b_k+ubb_{ijk}$	222	= $z+b_i+b_j+b_k+bbb_{ijk}$

0, 1, 2 = index of the geno- and phenotypes indicating the number of functional alleles present at locus i, j or k

$u_{i,j,k}$ = uniallel = contribution of one functional allele at locus i, j or k

$b_{i,j,k}$ = biallel = contribution of two functional alleles at locus i, j or k

$uu_{ij}, ub_{ij} \ldots bbu_{ijk}, bbb_{ijk}$ = nonallelic interactions between 2 or 3 loci

z = contribution of the manifold recessive genotype

TABLE 2. *Mean values of the genotypes and their standard errors.*

bluv	$\bar{y}$ $\pm$ $s_{\bar{y}}$		bluv	$\bar{y}$ $\pm$ $s_{\bar{y}}$		bluv	$\bar{y}$ $\pm$ $s_{\bar{y}}$	
0200	680.36	17.97	1200	1062.87	18.87	2200	1202.89	16.01
0201	743.00	16.34	1201	1006.04	10.36	2201	1112.04	13.63
0202	768.69	15.27	1202	982.63	11.79	2202	1096.24	13.27
0210	1017.25	24.63	1210	1122.30	14.90	2210	1191.49	18.76
0211	1038.23	18.11	1211	1198.51	13.46	2211	1217.79	19.62
0212	**1071.17**	14.86	1212	1154.27	8.40	2212	1198.93	15.24
0220	1059.06	26.18	1220	1121.00	15.31	2220	1057.87	11.19
0221	1124.83	14.62	1221	1209.52	6.15	2221	1250.95	15.34
0222	1108.12	14.37	1222	1242.24	13.62	2222	1349.96	16.15

0, 1, 2 = index of the geno- and phenotype with respect to the loci b, 1, u and v indicating the number of functional alleles at the corresponding locus

$\bar{y}$ = mean of the measured values of the corresponding genotypes, averaged over 33 replications

$s_{\bar{y}}$ = standard error

TABLE 3. *Estimation of the parameters of the saturation curve.*

Locus	Y	c
all loci	1372.68	.3472
b	2477.30	.1718
u	1394.01	.3782
v	782.19	1.0101

Symbols see formula (5), parameters estimated according to formula (6) and (7).

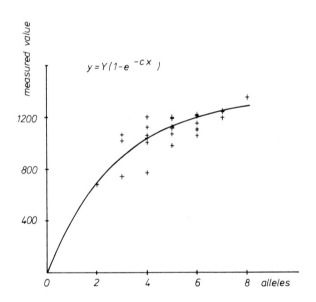

FIGURE 1. Saturation curve, estimated over 3 variated loci with 0 , 1 or 2 functional alleles each and 1 constant locus with 2 alleles.

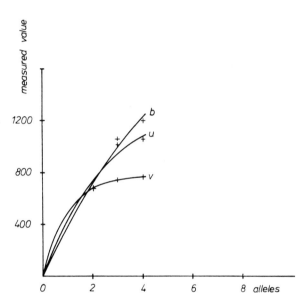

FIGURE 2. Saturation curves, estimated separately for locus
b , u and v .

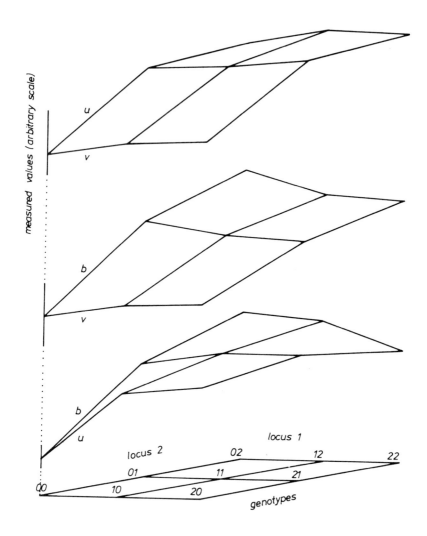

FIGURE 3. Response surfaces, observed in the three bifacto-
rial combinations bu , bv and uv . The third
locus is held in the recessive state respectively.
The measured values of the three double recessive
genotypes are therefore identical.

The Relationship between Genotype and Fitness for Heterotic Models [*]

J.A. SVED

INTRODUCTION

Over the past ten years or so there seems to have emerged a widespread acceptance that outbreeding organisms are polymorphic at an appreciable percentage of their loci. There is however far less unanimity on the question of whether this variability is selectively maintained, or whether it is a product of essentially neutral mutation and random fluctuation of gene frequencies. Furthermore amongst those who believe that selective neutrality cannot account for the nature of the variability, there are differing views on what is the dominant type of selective force.

The present discussion will be devoted to considering some of the selective consequences of the hypothesis of heterozygote advantage, which is of course potentially one of the most important forces for maintaining variability. The discussion will be centred around the concept of the relationship between genotype and fitness for large numbers of loci. A general discussion of such relationships will be given first. Following this an attempt will be made to summarize what is known about the relationship for the particular case of heterozygote advantage, using data for Drosophila.

THE RELATIONSHIP BETWEEN GENOTYPE AND FITNESS

The existence of a fitness function

The general problem of specifying a relationship between genotype and fitness is a complex one. If for example there are n loci each having two alleles, there are 3^n possible genotypes, and in general there may be a distinct fitness associated with each. A graphical representation of this situation would require $n+1$ dimensions. It is not easy to see any simplifying principles in this general case.

Obviously in order that a two-dimensional relationship can be considered, as is commonly done in discussions of this type, some simplifying assumptions must be made. In particular we require that combinations of genotypes at the various loci we are considering can be represented in some way on a linear scale. This is most easily done in the case where there are two classes of genotypes at each locus under consideration, viz. favoured and disfavoured. The genotype of an individual is then summarized by the single value $(\underline{x})$ giving the number of loci having a favoured genotype. In the discussion of this paper the favoured genotype will be assumed to be the heterozygote, but the model is equally applicable for example to the case of selective substitution of a new favoured gene at a number of loci.

If it is assumed that a fitness value can be associated with each level of heterozygosity, this implies certain things about the model. It implies that the two homozygotes (in a diallelic model) have equal selective values. Furthermore it implies that the selective effects at all loci are the same. This is perhaps not as serious a restriction as it might at first seem. Unequal effects could formally be taken into account in a slightly modified model by considering not just the number of heterozygous loci but rather some weighted measure of heterozygosity, although this will not be attempted here. More serious is the restriction on the type of epistasis which the model allows. In particular, specific interactions between particular loci are precluded. This means, for example, that we cannot consider the situation where the heterozygote at one locus under consideration is only at an advantage depending on the genotype at a second locus. Obviously restrictions such as these will sometimes be unrealistic. The applicability of models such as these must eventually be justified by obtaining empirical data to find out how important such specific interactions are in comparison to the type of general interaction implied by the model, although it is not easy to see any systematic way of obtaining such data.

The model is intended only to describe the situation in a given population within one generation. The complications caused by the reorganization of the genotype during sexual reproduction are not really relevant to the present discussion. The important frequencies in this discussion are simply the genotype frequencies before and after selection. Both viability and fertility components may be included to give a single measure of fitness from these two frequencies, provided mating is at random with respect to the genotypes concerned and the fertility of a mating is either the arithmetic or geometric mean of the fertilities of the individuals concerned (Bodmer, 1965).

The existence of a fitness function is not intended to imply that the fitness is necessarily determined by the genotype. Also the value $\underline{w}$ for any genotype is only valid for one environment, and the environment may of course be affected by the range of genotypes present in the population.

The fitness of a genotype having $\underline{x}$ heterozygous loci will be defined as $\underline{w}(\underline{x})$. The notion of favoured and disfavoured genotypes seems to imply that $\underline{w}(\underline{x})$ be an increasing function of $\underline{x}$, but models such as the optimum heterozygosity model proposed by Wallace (1958) and Mukai et al. (1965) can also formally be accommodated. Strictly speaking, $\underline{w}(\underline{x})$ is a discrete function, although it is convenient to assume a continuous model when large numbers of loci are being considered.

A second function must also be introduced. This is $\underline{f}(\underline{x})$, the frequency distribution of genotypes in a particular population. If genes at all loci are in linkage equilibrium, this will be a binomial distribution, approximating to a normal distribution. In a finite population, however, some disequilibrium is expected, leading to a skewed distribution and possibly a substantial increase in the variance (Sved, 1968).

The selective consequences of particular fitness function

To evaluate the consequences to the population of a particular fitness function it is necessary to calculate the change in frequency brought about by selection at individual loci. The obvious statistic which is relevant to such a calculation is the marginal selective value at individual loci, which has not so far been specified.

In order that the marginal selective value be the sole determinant of selective consequences at individual loci it is necessary that the effects of linkage be small, which of course will not always be true. However, the sophistication that has been possible for 2- and 3-locus models for any degree of linkage seems impossible for n-locus models. It seems necessary to make the assumption of no linkage effects as a first approximation in order to derive conclusions of any generality. Such conclusions of course will not be applicable to the type of situation envisaged by Franklin and Lewontin (1970) where the chromosome is essentially tied up in large blocks, and they may only be an approximation to what is expected in a finite population (Sved, 1968).

Calculation of the marginal selective value

The quantity $\dfrac{w(x+1)-w(x)}{w(x)}$ represents approximately the effect of substituting a heterozygote for a homozygote at one locus in an individual with $\underline{x}$ heterozygous loci. Taking into account the frequency with which such individuals are expected to occur, the average effect of substituting a heterozygote for a homozygote at one locus in the population is

$$\sum_{x=0}^{n} \underline{f}(\underline{x}) \cdot \frac{w(x+1)-w(x)}{\underline{w}(\underline{x})} \quad .$$

Passing to a continuous model, the quantity $\underline{w}(\underline{x}+1)-\underline{w}(\underline{x})$ is equivalent to the slope of the fitness function, $\underline{w}'(\underline{x})$. The marginal heterozygote advantage at a single locus is thus approximately

$$\int_{0}^{n} f(x) \cdot \frac{w'(x)}{w(x)} \, dx \quad ,$$

$$= \int_{0}^{n} f(x) \cdot \frac{d\cdot}{dx} (\log (\underline{w}(\underline{x})) \, \underline{dx} \quad .$$

If fitness is plotted on a logarithmic scale, the marginal selective value is equal to the weighted slope of the fitness function. In other words it is the slope of the fitness function in the region of the population mean which is primarily responsible for the single locus selective values.

Possible forms of the relationship

Three possible relationships between genotype and fitness whose derivation will be discussed below, are given in

Figure 1.

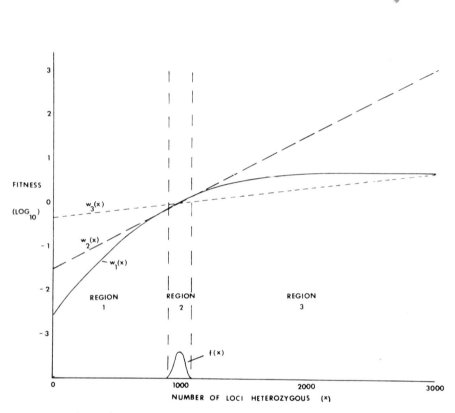

FIGURE 1. Three possible relationships connecting genotype and fitness.

The linear relationships $w_2(x)$ and $w_3(x)$ represent the case of a multiplicative relationship on a linear (non-logarithmic) scale, such as would be expected with independent selection at all loci. The function $w_1(x)$ is a cumulative normal or probit curve (King, 1967).

The x-axis of Figure 1 is labelled according to the values

suggested by Lewontin (1974, p.201) as possibly realistic for
the Drosophila genome. Three thousand polymorphic loci are
assumed, with an average level of heterozygosity per polymorphic locus of 1/3. These numbers are of course only very
roughly applicable to the present model, which assumes two
alleles at each locus and no selective differences between
homozygotes. The heterozygote advantage in a more general
model would presumably correspond to a weighted average of
heterozygote and homozygote fitnesses.

The frequency distribution $f(\underline{x})$ represents the expected
distribution of heterozygous loci per individual if all loci
are in linkage equilibrium. For the sake of convenience the
figure is divided into three regions with respect to this
distribution. Region 2 represents the region in which most
of the population lies. This is arbitrarily taken to be 99.9%
of the population, which sets the limits within 3.29 standard
deviations of the mean. The region to the left will be
referred to as Region 1, while that to the right is Region 3.

An example of some possible numerical values

Few experiments have been made which bear directly on the
form of the relationship in the region of the population,
Region 2. However, one such experiment has been carried out
by Mukai, Schaffer and Cockerham (1972), which is essentially
an elaboration of an earlier experiment by Dobzhansky and
Spassky (1953). Viability estimates were obtained from
statistical analysis of the variability from inter-crosses
amongst second chromosome lines in D. melanogaster. An estimate of the heritability of viability of approximately 0.004
was obtained, depending on assumptions of the viability of
genotypes involving the marker chromosome used in the experiment.

447

While this experiment does not directly provide evidence about the extent of heterozygote advantage, it apparently sets a limit to the selective intensity which can be achieved. The probit curve given in Figure 1 represents the curve given by Mukai et al. assuming that 20% of the overall population survives selection, which is close to the value observed in the experiment. It assumes that all selection is apportioned amongst the 3,000 heterotic loci mentioned previously.

This experiment demonstrates the fairly low level of viability selection in Drosophila, which is in agreement with the results of inbreeding experiments discussed later. However it is also claimed by the authors that this curve demonstrates that there is no difference between the linear and probit curves for such low heritabilities. This conclusion is open to some argument. The marginal heterozygous advantage per locus is this experiment, i.e. the slope in the region of the mean, comes to 0.35%. This is not a large value, but it might nevertheless have an appreciable effect for populations of effective size 1,000 to 10,000 (Robertson, 1962). On the other hand a multiplicative model with the same marginal selective value, $\underline{w}_2(\underline{x})$, is more or less indistinguishable from the probit curve in Region 2, but rises to an unrealistically high value in Region 3. Genetic load arguments based on independent locus selection would in fact show that the only feasible curve, given an upper limit of 5 as for $\underline{w}_1(\underline{x})$, is $\underline{w}_3(\underline{x})$, which of course has a considerably reduced slope.

BIOLOGICAL CONSIDERATIONS

Models of Natural Selection

In his discussion of models for the action of natural selection, Wallace (1968, 1970) introduced the terms 'hard' and 'soft' selection, which will be used in this discussion.

The terms describe models closely related to the two viability models analyzed by Sved, Reed and Bodmer (1967). Hard selection refers to death or lowering of fitness attributable to an individual's genotype, independently of the genotypes of other individuals in the population. Soft selection refers to death or lowering of fitness of an individual which would by nature of its genotype be quite capable of surviving and reproducing, but which is prevented from doing so by an overall limitation of 'resources', and the usurping of these resources by other individuals in the population. Soft selection usually directly or indirectly involves the notion of intraspecific competition.

The concepts of hard and soft selection are closely related to the concepts of density-independent and density-dependent population control. There might however be some confusion if the ecological terms were used in this discussion, since the two sets of terms are by no means always mutually consistent. Even in populations which are regulated in a density-dependent manner, hard selection is expected to occur against some genotypes, e.g. lethals, a situation envisaged by Crow (1970) in his discussion of gene substitution in a population of constant size. On the other hand, the concept of soft selection appears to depend on the existence of density-dependent population regulation.

Models of natural selection and fitness functions

It will be argued in this section that the models of hard and soft selection lead to the linear and probit fitness functions respectively. In neither case is the relationship a direct one, and it is important to decide what aspects of the fitness functions follow from what assumptions of the models of natural selection, and how robust these derivations are.

449

The multiplicative relationship for hard selection is clear for the case of lethal genes, where the overall probability of survival of an individual is the product of the probabilities of not having the lethal genotype at any of the constituent loci. It is not as easy to suggest examples of hard selection with non-lethal genes, but it has generally been assumed that as a first approximation, independence of action of the genes implies independence on the scale of fitness.

The derivation of the probit curve from the model of soft selection requires the introduction of the more explicit threshold model. It is necessary to introduce this in some detail, as threshold models have been criticized by a number of authors (e.g., Crow and Kimura, 1970, p.307; Nei, 1971; Mukai and Maruyama, 1971) as being unrealistic and unjustified. Some of these arguments are however based on a rigid interpretation of the most extreme threshold model. The multiplicative model would seem equally unrealistic based on similar criteria.

The simplest model of soft selection is the extreme case where competition for a limiting resource is completely determined by the genotype at a series of loci. Surviving individuals are thus those with the highest number of favourable loci, or in heterotic models, the most heterozygous individuals. The probit curve in this case takes its limiting form, viz., a step-function, where all individuals below the cut-off point have zero fitness and all above the point have maximum fitness. Extreme models of this type have in fact been used by Milkman (1967), Maynard-Smith (1968) and Wills, Crenshaw and Vitale (1970). This model is useful in that it gives an upper limit to the amount of selection which can be expected for a given death rate in the population (Maynard-

Smith, 1968). However without some modification it cannot be regarded as realistic.

Chance or environmental factors have been explicitly introduced into the model by King (1967). This has been done by assuming that such factors can be subsumed on an underlying scale, towards which genotype also contributes, and that a truncation point or threshold once again applies. It could equivalently be done by assuming a gradual rather than sharp threshold. However neither of these necessarily gives an accurate physical picture of the way in which soft selection acts. As emphasized by King, the threshold model is an intellectual construction which is not intended to describe any real system, but is a convenient statistical approach which can be applied pending a better understanding of the true action of natural selection.

It is in fact possible to identify one aspect of the action of soft selection which is not accurately modelled by the threshold model. In a population which is regulated owing to some limiting resource, it is likely that some individuals will by chance, quite independently of genotype, find supplies of the resource which will enable them to survive and reproduce. This possibility is really the essence of the soft selection model. This cannot be simulated by a threshold model, in which the genotype always makes some contribution to the fitness. This reservation is probably of minor importance in Regions 2 and 3 of the fitness curve, but as discussed later it may be important in the prediction of fitness of highly homozygous genotypes.

The value of the threshold model is that it makes possible a simple calculation of the marginal selective value for a given overall intensity of selection. However a strong case can be made for a probit type of curve from the assumption of

451

soft selection, without recourse to a threshold model (e.g., Sved et al., 1967).

There are two significant attributes of the probit curve. First it levels off to a reasonable upper limit in Region 3. Secondly it has a reasonably high slope in the region of the population (Region 2). The first of these attributes is trivial. For a viability model it follows simply from the fact that the survival probability cannot be greater than unity (Wallace, 1970). For a fertility model it follows from the fact that there is a physiological limit to the number of offspring which an individual can produce.

The second attribute is the more important one. Starting from the upper regions of the curve, the reduction in fitness with increased homozygosity is negligible, and it becomes appreciable just in the region of the population mean. This is in contrast to the multiplicative model where the reduction is constant in all regions of the curve, and to the inverse square model proposed by O'Donald (1969), where the increase in slope is continuous and independent of $f(x)$. The reason for this aspect of the probit curve follows directly from the assumptions of soft selection. An individual with 2,000 loci heterozygous in Figure 1 would have a reduced fitness if it had to compete with more heterozygous individuals. But since there are essentially no individuals with more than 2,000 loci heterozygous, such competition is non-existent. Therefore by the assumptions of soft selection, any (rare) individual with 2,000 loci heterozygous would suffer no impairment of fitness. Only in the region of the population does such competition come into play.

It may seem like a futile exercise to be discussing the fitness of essentially non-existent individuals. However the soft selection model is less restrictive on this point, since under this model it does not really matter what the fitness of these individuals is. Under the multiplicative or inverse square model however, a very definite relationship between the fitness of non-existent individuals must be assumed.

Mixed models with hard and soft selection

A realistic fitness function should be based on the assumption that some genotypes will be subject to hard selection, while others will only be subject to soft selection. However, the arithmetic of the probit model compared to the linear model is such that if comparable selective intensities are devoted to hard and soft selection, the overall effect of the soft selection on marginal selective values will be much greater. For example, considering the 3,000 loci of Figure 1, a 1% marginal heterozygote advantage can be attained from a threshold model with a 50% survival rate and only a 10% heritability on the underlying scale. The same marginal heterozygous advantage and overall survival rate can be attained for only 100 loci under the multiplicative model. The overall shape of the fitness curve for this combined model with equal selective intensity devoted to hard and soft selection would not be expected to be very different from the shape of the probit curve by itself.

The relative contributions of hard and soft selection

Inbreeding experiments provide probably the most illuminating information on the relationship between genotype and fitness for the particular case of heterotic loci, since large numbers of loci can be made homozygous simultaneously. Information is available for several types of such experiment in Drosophila, and the information from two experiments for Chromosome II in D. melanogaster will be considered here.

Temin (1966) has tested the viability of a large number of

chromosomes, giving a mean reduction in viability of chromosome homozygotes of approximately 34%. The majority of this
effect is attributable to lethal genes, which are not of
particular interest in this context. The inbreeding
depression due to non-lethal genes is only about 14%. The
impairment of fertility was also tested in this experiment.
Once again the majority comes from severe detrimentals causing
complete sterility, leaving only about 5% attributable to
genes of small effect. Overall therefore, non-lethal non-
sterile chromosome homozygotes have a reduction in fitness of
about 19% in this experiment, compared to wild-type heterozygotes.

By contrast, a population cage experiment carried out by
Sved (1971) yielded much lower mean fitness estimates for
twenty-four randomly chosen non-lethal Chromosome II homozygotes in D. melanogaster. Of these twenty-four chromosomes,
which included some causing complete sterility, nine were
chosen early in the experiment as having the highest combination of viability and male and female fertility. The estimated mean fitness of these nine non-lethal non-sterile
chromosome homozygotes came to 21%. Presumably the essential
difference between these two experiments, one yielding an
inbreeding depression of 19% and the other of 79%, is that
the latter experiment involves competitive conditions and
thus can be expected to take into account both soft and hard
selection, while the former takes into account only hard
selection.

Which component, or components, of fitness is responsible
for the very large differences between these two estimates
remains something of a mystery at the moment. However indirect evidence from two sources suggests that the fertility
component may be considerably more important in the population

cage experiment thatn would be suggested by the non-competitive measures of viability and fertility. The first of these is the experiment on Chromosome IV systems of Bundgaard and Christiansen (1972), which strongly suggests the importance of the sexual component, particularly the male competition component.

The second line of evidence concerns the failure of experiments to influence the viability ratio by crowding. One such experiment carried out in D. pseudoobscura (unpublished) used counted numbers of eggs and measured amounts of yeast as a limiting resource. Although large differences in overall selective intensity were attained, the viability ratios for three chromosome homozygotes remained unaffected. Temin et al.(1969) also reported similar findings in crowded and uncrowded cultures of D. melanogaster. Taken together these results suggest the rather surprising conclusion that despite the apparent importance of soft selection in determining overall fitness, the major component of pure viability selection is attributable to hard selection.

Estimates of the marginal heterozygote advantage from inbreeding experiments

The functions $w_1(x)$ and $w_2(x)$ of Figure 1 are based on viability estimates which are not necessarily attributable to heterozygote advantage. The curves may however be useful in illustrating the present discussion since the inbreeding depressions which they predict are of the right order of magnitude for the population cage experiment discussed above. Homozygosity of the second chromosome of D. melanogaster is expected to reduce the overall heterozygosity by about 40%, counting the X-chromosome as having half the effect of each of the major autosomes. The fitness corresponding to this level of heterozygosity is approximately 0.25 for the multiplicative function $\underline{w}_2(\underline{x})$, and 0.20 for the probit function $\underline{w}_1(\underline{x})$. Thus a marginal heterozygote advantage of somewhat less than half of one percent is indicated by this experiment. Of course it must be remembered that deleterious recessives also contribute to an unknown extent to the inbreeding depression.

The genotypes constructed in such inbreeding experiments are well outside the range of genotypes commonly expected in the population. Therefore the predictions of inbreeding depression are by no means necessary predictions given the marginal heterozygote advantage, since any shape of the curve in Region 1 would be consistent with this. As mentioned previously the probit model becomes unrealistic in its prediction of extreme low fitnesses, although 20% is probably not sufficiently low for this reservation to apply. Some comfort can perhaps be taken from the fact that the predictions of the two models are not very different.

In trying to determine the shape of the fitness curve in Region 2, it would clearly be very useful to have comparable inbreeding estimates for shorter chromosome lengths. The technical problems of producing partial chromosome homozygotes seem formidable however, particularly in view of the necessity to ensure linkage equilibrium between the homozygous and non-homozygous chromosome portions. The experiment of Vann (1966) on newly produced inversions provides a considerable amount of closely related information, although it is unfortunately difficult to relate to direct chromosome homozygosity experiments. Information is also available on the viability of strains homozygous for more than one chromosome, but the problems of extrapolation from this type of genotype are even greater than for single chromosome homozygotes.

Ecological considerations of the soft selection model

The discussion of soft selection has so far been in terms of an imagined limiting 'resource'. Whether this is a realistic concept for most natural populations seems a moot point at present. Intuitively it would seem that if populations need a number of different 'resources', e.g. different kinds of food, space, etc., that only one should ultimately be limiting for any population. Shortage of food may in fact limit the numbers of many or perhaps even most organisms (see e.g., Lack, 1954; Colinvaux, 1973).

Competition for a small number of independent resources in fact would not change the nature of the results greatly. For example a model with 50% survival at two thresholds, each involving half of the loci, can be compared with 25% survival at a single threshold. Numerical calculations of this kind have been made for a range of heritabilities, with the other parameters being as in Figure 1. The marginal heterozygote advantage for two independent thresholds is reduced to about 91% of the value for a single threshold, while for five and ten independent thresholds the comparable reductions are to 73% and 61% respectively.

Whatever the nature of the limiting resource (or resources), it is difficult to argue that any aspect of an individual's phenotype is irrelevant in the competition for that resource. In a real sense, an individual's competitive ability may be a function of many different characters influenced by otherwise unrelated genes. The same arguments apply also to competition for mating, which is probably of crucial importance in males of a wide variety of species.

Nei (1971) argues against the idea that competitive selection will lead to a threshold model. His arguments are based

on a single locus model, which is not really relevant to the
present discussion. However he extends this to multiple
locus models, claiming, in contrast to the above views, that
competition will generally involve different genes and
characters independently. While it may be difficult to
resolve these sorts of arguments using experimental evidence,
it seems that the model of independent selection against a
large number of characters is not really compatible with the
idea of density-dependent regulation, which implies some sort
of overall population control.

Ecological considerations are important when we try to
interpret the results of laboratory experiments with Drosophila. Although the ecology of Drosophila is not generally
well understood, it seems certain that adult Drosophila live
a much more complex life in the field than in the population
cage or bottle. If the essence of fitness in the field involves success in finding food, shelter, mates, egg-laying
sites, etc. it should not be surprising if the potential competitive advantages of some genotypes are reduced if the need
to carry out these activities is removed.

A possible demonstration of this type of effect comes from
an experiment of J. Mandryk (personal communication), who has
estimated the fitness of chromosome homozygotes using population cages and serial transfer in bottles. The latter
situation in a simplistic sense maximizes competition in that
more adults compete with each other per unit volume of space
and food. However in nearly all chromosome lines the advantage of the heterozygote was greater in the population cage,
which presumably reflects the greater complexity of environment in the cage compared to the bottle.

These sorts of considerations are also important in comparisons of the overall importance of soft and hard selection

in inbreeding experiments. It is comparatively easy to
measure the effects of hard selection against particular
genotypes. The measurements of Temin (1966) discussed above
probably reflect with a reasonable degree of accuracy the
relative advantages of chromosome heterozygotes over homozygotes which can be expected from hard selection under conditions of temperature, humidity, etc. which are not too
different from those of the laboratory. On the other hand,
it is much more difficult to achieve conditions which ensure
a realistic contribution from soft selection. There are probably few aspects of an individual's phenotype, including
behavioural aspects, which are not potentially of advantage
under some conditions, but unless the experiment is set up to
include such conditions these aspects of fitness will presumably escape detection.

DISCUSSION

A non-answer to a pseudo-problem?

Low fitnesses of chromosome homozygotes have been demonstrated in population cage experiments with Drosophila, low
enough perhaps to accord with predictions from the hypothesis
of widespread heterozygote advantage. However it would be
rash to make positive assertions about the maintenance of
variability from just this observation. It must first be
acknowledged that the simple two-allele model of heterozygote
advantage considered in this paper could not account for the
complex multi-allele polymorphisms observed. A second important reservation is the possibility that the low fitness of
chromosome homozygotes is due merely to deleterious recessives
rather than to genuine overdominance.

Conflicting evidence also comes from statistical analyses
of the patterns of electrophoretic variability (see the dis-

cussions by Latter and by Ewens and Feldman in this volume),
which seem to indicate that selection only needs to be invoked for the maintenance of a minority of polymorphisms. The
unpredictability of linkage-selection effects (see e.g., Sved,
1968) might however be a complicating factor in such analyses.

At the same time it may be pertinent to enquire into the
reasons for wishing to demonstrate whether or not most variability is maintained by heterozygote advantage. Of course
this is a question of intrinsic interest. Also the finding
of high levels of heterozygote advantage might be of some
importance per se, for example in designing breeding programs
for artificial selection. Nevertheless, there are some
grounds for believing that unequivocally answering this question would not greatly increase our understanding of the
genetical structure of populations. Two lines of argument
will be considered here, the first related to genetic load,
and the second to long-term genetic flexibility under the
'balance hypothesis' (see Lewontin, 1974, for a discussion of
opposing arguments on this hypothesis).

The genetic load argument carried with it the connotation
that selection in favour of heterozygotes has important consequences for the fitness of the population. It implied that
there is a sharp and consequential difference between selection and non-selection, particularly for the case of selective gene substitution. Under the soft selection model however, these distinctions become blurred. Even if high levels
of selection can be demonstrated, this does not imply that
the fitness of the population is in any sense affected. It
is the population structure which determines the selective
intensities at individual loci, rather than vice versa. Thus
if it is accepted that the soft selection model has removed,
at least conceptually, some of the difficulties posed by the

genetic load argument, it might also be accepted that the
importance of demonstrating whether variability is maintained
through heterozygote advantage has been correspondingly
reduced.

The second question concerns the possible long-term advantages to the population of a supply of variability. From an
evolutionary point of view the existence of a supply of variability seems crucial. But the importance here would seem to
be the existence of such variability, rather than the question
of whether it is selectively maintained. The finding of high
levels of variability at the protein level, and perhaps more
importantly the demonstration that populations are able to
respond to different types of artificial selection, have
provided important support for the hypothesis of long-term
evolutionary flexibility. What needs to be asked is whether
these findings would be affected by a demonstration that the
variability is maintained by heterozygote advantage. Such a
demonstration would make it seem little more than coincidental
that the variability exists to be utilized for evolutionary
change. It would tell us no more about a population's ability to respond to environmental changes than if the variability had been shown to be selectively neutral.

One proviso, at least, should be given to this argument.
A model may be proposed in which, for example, two alleles at
a locus code for iso-enzymes with different temperature
optima. If the heterozygote produced a mixture of these two
enzymes, it might then be better able to cope with a range of
temperature, leading to a selective advantage and to the
establishment of a polymorphism. Such a system would then be
in a position to respond to permanent change in the environment by an increase in the frequency of whichever homozygote
is favoured. In this model the ability to respond to

environmental changes is in some sense predicted by the fact that the heterozygote is at an advantage. But the model is a rather simplistic one, and in any case is unlikely to explain the occurrence of inbreeding depression in the relatively homogeneous environment in which most inbreeding experiments are carried out. An example of a more plausible model is one where heterozygote advantage is due to the formation of hybrid enzymes. Any evolutionary flexibility in such a case would be unrelated to the present heterozygote advantage. Thus it seems that the relationship between short-term and long-term aspects of the balance hypothesis needs to be clarified before the demonstration of heterozygote advantage can be said to be of evolutionary significance.

REFERENCES

Bodmer, W.F. (1965). Genetics 51: 411-424.

Bundgaard, J. and F.B. Christiansen. (1972). Genetics 71: 439-460.

Colinvaux, P. (1973). Introduction to Ecology. J. Wiley, New York.

Crow, J.F. (1970). In Mathematical topics in population genetics. K. Kojima (Ed.), Springer-Verlag, Berlin.

Crow, J.F. and M. Kimura. (1970). An introduction to population genetics theory. Harper and Row, N.Y.

Dobzhansky, Th. and B. Spassky. (1953). Genetics 38: 471-484.

Franklin, I.R. and R.C. Lewontin. (1970). Genetics 65: 707-734.

King, J.L. (1967). Genetics 55: 483-492.

Lack, D. (1954). The natural regulation of animal numbers. Oxford Univ. Press, London.

Lewontin, R.C. (1974). The genetic basis of evolutionary change. Columbia Univ. Press, N.Y.

Maynard-Smith, J. (1968). Nature 219: 1114-1116.

Milkman, R.D. (1967). Genetics 55: 493-495.

Mukai, T., S. Chigusa and I. Yoshikawa. (1965). Genetics 52: 493-501.

Mukai, T. and T. Maruyama. (1971). Genetics 68: 105-126.

Mukai, T., H.E. Schaffer and C.C. Cockerham. (1972). Genetics 72: 763-769.

Nei, M. (1971). Genetics 68: 169-184.

O'Donald, P. (1969). Nature 221: 815-816.

Robertson, A. (1962). Genetics 47: 1291-1300.

Sved, J.A. (1968). Genetics 59: 543-563.

Sved, J.A. (1971). Genet. Res. 18: 97-105.

Sved, J.A., T.E. Reed and W.F. Bodmer.(1967). Genetics 55: 469-481.

Temin, R.G. (1966). Genetics 53: 27-46.

Temin, R.G., H.U. Meyer, P.S. Dawson and J.F. Crow. (1969). Genetics 61: 497-519.

Wallace, B. (1958). Proc. Xth. Intl. Cong. Genet. 1: 408-419.

Wallace, B. (1968). Topics in population genetics. W.W. Norton, N.Y.

Wallace, B. (1970). Genetic Load. Its biological and conceptual aspects. Prentice-Hall, N.J.

Vann, E. (1966). Am. Natur. 100: 425-449.

Wills, C., J. Crenshaw and J. Vitale. (1970). Genetics 64: 107-123.

The HL-A System as a Model for Studying
the Interaction between Selection, Migration, and Linkage

G. THOMSON, W.F. BODMER, and J. BODMER

INTRODUCTION

The HL-A system is an antigenic polymorphism detectable on almost all human tissues and is the major human histocompatibility system. Most of the serologically detected antigens of the HL-A system described up to now behave as if they were controlled by multiple alleles at two loci, the LA locus, and the 4 locus. These two loci are separated by a recombination fraction of about 0.8%.

The large number of alleles at the LA and 4 loci make the HL-A system by far the most variable polymorphism known in man. At least 14 alleles have been described for the LA locus and 17 for the 4 locus. The possible number of haplotypes is thus 270, the number of genotypes 36,585 and the number of distinct phenotypes 16,324. The frequency of heterozygotes in a population provides a measure of the extent of polymorphism. In Caucasian populations, for example, about 75% of individuals are heterozygous at both the LA and 4 loci while only 2 to 3% are homozygous at both loci. For comparison, the proportion of homozygotes for the Rhesus blood group system, the most polymorphic of the red cell blood groups is about 36% while that for the ABO system is about 56%.

It is possible to calculate, approximately, how many genes correspond to a recombination fraction of 0.8% (Bodmer, 1972). Even if only a fraction of the DNA is active it is probable that there exist hundreds of cistrons between the LA and 4 loci. What are the functions of these genes? A certain amount is already known about this, particularly if comparisons are drawn with the mouse histocompatibility system H-2. Major histocompatibility systems, such as HL-A, are presumed to have their counterparts in all mammalian species. The striking homology between the HL-A system of man and the H-2 system of the mouse has been emphasized repeatedly. The homology extends to the two locus structure for most of the major serologically detected antigens and to the probable arrangement of the genes controlling immune response, mixed lymphocyte culture (MLC) response and other functions, including disease susceptibility, graft versus host response and graft rejection. Pedigree studies in mouse and man suggest that these functions are genetically separable from the genes which determine the serologically detectable antigens. (For references and general background on the HL-A and H-2 system see e.g., Bodmer 1973b, 1975.)

It is of paramount importance when discussing the evolution of the HL-A system to consider it as a linked complex of many genes. Any attempts at understanding the evolutionary significance of this complex must depend on an understanding of the population genetics of linked loci and the types of selective forces that may favour interactions between closely linked loci. In the following sections the HL-A system will be considered in this context.

With regard to population studies the two main features of

the HL-A system which are of interest are the high level of polymorphism and the existence of significant linkage disequilibrium between alleles of the two loci which determine certain pairs of antigens. These observations are generally taken as evidence of selection acting directly on the HL-A system (see Bodmer, Cann and Piazza, 1973, and Bodmer 1973c for discussion).

In the following consideration will first of all be given to the population data. The data will then be discussed in terms of mechanisms known to cause linkage disequilibrium. Finally the action of these mechanisms within the HL-A system will be discussed.

Population Data

Average gene frequencies for the alleles of the LA and 4 loci in the major racial groups of man are given in Table 1. There are, as expected, a number of substantial differences between the population groups. For example, the antigen HL-A1 occurs predominantly in Caucasoids, its presence in Negroids probably being accounted for largely by Caucasoid admixture. The antigens HL-A11 and W27 are also absent from Negroids while on the other hand W17 and W30 have a substantially higher frequency in Negroids than in other populations. The antigens HL-A3, W14 and W18 are virtually absent from Mongoloids, while HL-A9 has its highest frequency in these populations (especially in Oceana and Australasia).

As already mentioned one of the main features of the HL-A system is the existence of significant linkage disequilibrium, which we will denote by D . To define D consider a two locus, two allele (A,a and B,b) model. Ceppellini and co-workers (1967) introduced the term haplotype (from haploid

467

genotype) for the possible chromosome or gametic types.

TABLE 1. *Average LA and 4 gene frequencies (in per cent) in various human racial groups (based mainly on Histocompatibility Testing 1972).*

HL-A alleles		Europe	Middle East	Asia	Negroids	Asian Mongoloids	Oceana & Austr- alasia	American Indian
			Caucasoids					
LA	1	17	12	12	4	4	O	O.5
series	2	28	19	15	17	24	8	48
	3	15	10	7	5	2	O	1
	9	9	20	14	9	25	42	25
	10	6	6	7	8	6	15	O
	11	6	7	18	O	17	17	1
	W28	5	5	5	9	2	O	9
	W29	4	2	1	6	1	O	O
	W30	2	4	1	21	3	1	2
	W31	2	1	1	2	1	O	6
	W32	4	4	3	1	1	1	O
	Bla	2	10	16	18	14	16	7.5
4	5	5	14	19	3	11	O	11
series	7	13	4	5	7	2	O	1
	8	11	4	8	3	O	O	O
	12	16	10	10	15	4	O	1
	13	2	2	2	2	3	9	O
	W5	10	13	12	11	5	O	23
	W10	6	4	6	3	14	27	13
	W14	4	4	O	3	O	O	1
	W15	6	3	8	2	16	16	13
	W16	4	6	1	3	9	2	12
	W17	3	4	8	15	6	O	0.5
	W18	5	5	2	6	1	2	0.5
	W21	2	7	1	2	O	O	4
	W22	3	3	3	0.5	4	36	O
	W27	4	3	2	O	4	4	3
	Bla	6	14	13	24.5	21	4	17

Negroid average includes Uganda population typed in Oxford; Middle
East Caucasoids include Armenians and Libyan Jews typed in
Oxford; all other data from Bodmer, J. G. et al 1973.

In the present model there are four haplotypes $\underline{AB}$, $\underline{Ab}$, $\underline{aB}$ and $\underline{ab}$ and we denote their frequencies by X_1, X_2, X_3 and X_4 respectively. Defining the linkage disequilibrium D as $D = X_1 X_4 - X_2 X_3$ then the haplotype frequencies can be written in the well-known form

$$X_1 = P_A \, P_B + D$$

$$X_2 = P_A \, P_b - D$$

$$X_3 = P_a \, P_B - D$$

$$X_4 = P_a \, P_b + D$$

where P_A, P_B are the frequencies of the alleles A and B respectively, etc. For the HL-A system this representation is readily extended to a two locus multiallele model. The frequency of the haplotype ij , determining antigens i of the LA series and j of the 4 series can be expressed in the form

$$x_{ij} = P_{Ai} \, P_{Bj} + D_{ij}$$

where P_{Ai} is the frequency of the antigen allele i and P_{Bj} is the frequency of the antigen allele j . The D_{ij} can be estimated directly from the 2 × 2 association table for the i and j antigens (see Mattiuz et al.,1970).

Perhaps the best known haplotype with a well established significant D is HL-A1,8. In a typical Northern European population the frequency of HL-A1 is 0.16 and of HL-A8 is 0.13, while the frequency of the haplotype HL-A1,8 is 0.08 giving D = 0.06 . The magnitude of this value is emphasized when one considers that the maximum possible value of linkage disequilibrium, for these particular allele frequencies, is approximately 0.1 . Assuming a recombination fraction of 0.008 it can be shown that, in the absence of selection, D would decrease by a factor of 5 , and so to an effectively insignificant value, in about 200 generations (approximately 5,000 years). The evidence on the distribution of the HL-A1,8 haplotype throughout Europe suggests that it has been present at reasonable frequencies since pre-agricultural times. Thus the high observed level of linkage disequilibrium

cannot readily be explained by inadequate time to reach equilibrium in the absence of selection (see Bodmer, 1973a). One
of the more interesting problems is to explain this persistent occurrence of linkage disequilibrium within the HL-A
system. (For further data on the HL-A system, see Histocompatibility Testing 1972, especially the joint report by
Bodmer, J.G. et al., 1973, which contains the results of the
Fifth International Histocompatibility Testing Workshop
during which some 27 laboratories typed a total of 54 populations from all over the world.) Possible ways of maintaining disequilibrium will be discussed in the next sections.

Migration and Admixture

A comparison of the HL-A frequencies between the major
racial groups clearly hides some significant variations within
the groups. One of these is the evidence for clines in the
frequencies of antigens and haplotypes. For example, the
frequency of the haplotype HL-A1,8 is highest in Northern
Europe, decreases substantially in Southern Europe and virtually disappears in the Middle East (see Table 2). It seems
likely that such clines in gene frequency are correlated with
the advance of the Neolithic peoples from their centre of
origin in the Middle East, as suggested by Cavalli-Sforza
(1972). The haplotype HL-A3,7 shows a pattern of distribution very similar to that of HL-A1,8, while other haplotypes such as HL-A1,W17 and HL-A11,5 tend to show a complementary distribution, with high values in the Middle and Far
East and much lower values in Northern Europe (see Table 2).

It is well known that transient clines in gene frequencies
may be established due to the arrival of a foreign population
in an already inhabited area, occupied by people with
different gene frequencies. It is possible that many existing

470

clines originated in this way (see Cavalli-Sforza and Bodmer, 1971, pages 483-490). A characteristic of a cline created by migration, such as the advance of the Neolithic peoples, is that it should affect all polymorphisms in the same way.

TABLE 2. *Some Caucasoid HL-A haplotype frequencies (frequency × 10^3).*

Populations	1,8	3,7	1,W17	11,5
N.Scotland (Hebrides)	121	36	22	-
English	87	55	12	2
French	46	53	11	7
Italians (Ferrara)	42	20	7	-
Basques	31	23	51	5
Turks	10	-	16	13
Lebanese	4	6	17	24
Arabs (Israel)	28	6	9	45
Pakistan	-	8	43	80
India	7	11	34	45

Data are based on Bodmer, J. G. et al (1973). The populations are arranged in approximate geographical sequence from North to South. Underlined values are those which are significant at least at a 5 per cent level; - means negligible estimated frequency.

So, if there is no selection acting, one would expect to see a parallelism of clines for all loci. We plan to analyse human data on various blood group systems in this respect. Linkage disequilibrium may also be created by this mixing of two distinct populations. Consider a two locus, two allele model and let the frequencies of alleles A and a at the first locus be p_1 and q_1 in the first population and P_1 and Q_1 in the second population. Similarly let p_2, q_2 and P_2, Q_2 be the corresponding frequencies of alleles B and b at the second locus. If the populations are mixed in the proportions m to 1-m then in the mixed population the linkage disequilibrium D between the two loci A and B is given by

$$D = mD_1 + (1-m)D_2 + m(1-m) \ (p_1 - P_1) \ (p_2 - P_2) \qquad (1)$$

where D_1 and D_2 denote the linkage disequilibrium in the two populations considered separately (see Cavalli-Sforza and Bodmer, 1971, page 69; and Nei and Li, 1973). $D \neq 0$ even if $D_1 = D_2 = 0$, provided the allele frequencies in the two populations are different, that is, $p_1 \neq P_1$ and $p_2 \neq P_2$. It should be noted that the linkage disequilibrium in the mixed population is very strongly dependent on the actual difference in gene frequencies. The maximum effect is of course produced when the alleles under consideration are at a frequency of 1.0 in one of the populations and absent in the other. If this model is applied to migration between two populations we see that if the migration rate is not too large, as is the usual case, the amount of linkage disequilibrium created will not be large. This is illustrated in Table 3.

The above is a static model of migration which does not take into account the more realistic situation where in each generation there is a small amount of migration occurring. This latter case has been considered by Nei and Li, 1973 and Feldman and Christiansen, 1975. Feldman and Christiansen, 1975, studied the effects of population subdivision on the evolution of two linked loci without selection. They considered the equilibrium properties and rates of approach to equilibrium of two deterministic models of population subdivision. The first model, termed the simple stepping stone model, is as follows. There are L subpopulations of equal size arrayed along a line and after migration each subpopulation consists of a proportion m of immigrants from the population on the left and a proportion m of immigrants from the population on the right and the remaining individuals come from the subpopulation under consideration. Mating is at random within each of the subpopulations and it is assumed that there is no selection.

TABLE 3. *Examples of the linkage disequilibrium D created in a mixed population (see equation (1) for definition of parameters. In all the examples below* $D_1 = D_2 = 0$).

m	p_1	p_2	P_1	P_2	D
0.05	1.0	1.0	0.0	0.0	0.048
0.05	0.5	0.5	0.0	0.0	0.012
0.05	0.4	0.3	0.3	0.1	0.001
0.01	0.5	0.5	0.0	0.0	0.003

m is the migration rate, p_1, p_2 the allele frequencies in the first population and P_1, P_2 those in the second.

The second model, termed the stepping stone cline model, is essentially as above except that it is assumed that at the left and right hand ends of the array there are two large populations in which the gene frequencies and disequilibrium are constant. Feldman and Christiansen (1975) have shown that at equilibrium in this second model there will be a

linear cline in gene frequencies connecting the two large
constant populations. This cline will also be accompanied by
a "cline" of linkage disequilibrium values. In the simple
stepping stone model at equilibrium the gene frequencies in
each population are equal and there is zero linkage disequilibrium in each population. However, clines in gene frequencies and significant linkage disequilibrium can be maintained for relatively long periods before equilibrium is
reached. This is illustrated in Tables 4 and 5 which show
the results of numerical simulations taking the extreme case
where an allele is present in only one population initially,
from which it spreads to the other populations by migration.

TABLE 4. A simple two locus, two allele migration model,
without selection (the single stepping stone model of Feldman
and Christiansen, 1975). There are 6 population groups,
arranged linearly, with a migration rate of 1% between neighbouring population groups each generation. For each popula-
tion the upper value is the allele frequency of $\underline{A}$ and $\underline{B}$,
denoted P_A and P_B and assumed equal, and the lower value
is the linkage disequilibrium D . For example, in population
1 initially $P_A = P_B = 1.0$, and $D=0$ while at generation
200 $P_A = 0.385$, $P_B = 0.385$ and $D = 0.054$. The recombination fraction between the two loci is 0.01 . At equilibrium the gene frequencies will all be equal to 0.167 and D
will be zero in all populations (Feldman and Christiansen,
1975).

Gener-ation	Pop_1	Pop_2	Pop_3	Pop_4	Pop_5	Pop_6
0	1.000	0.000	0.000	0.000	0.000	0.000
	0.000	0.000	0.000	0.000	0.000	0.000
100	0.523	0.309	0.123	0.036	0.008	0.002
	0.126	0.113	0.057	0.017	0.004	0.001
200	0.385	0.296	0.179	0.088	0.036	0.016
	0.054	0.053	0.040	0.021	0.009	0.004
300	0.318	0.268	0.191	0.118	0.065	0.039
	0.023	0.024	0.021	0.015	0.008	0.004
400	0.278	0.244	0.191	0.134	0.089	0.065
	0.010	0.011	0.011	0.009	0.006	0.004

TABLE 5. Transient linkage disequilibrium due to migration.

As for Table 4 except that the initial allele frequencies in population 1 are 0.5 instead of 1 . For this example, at equilibrium the gene frequencies will be 0.083 and D will be zero in all populations (Feldman and Christiansen, 1975).

Generation	$Pop^n 1$	$Pop^n 2$	$Pop^n 3$	$Pop^n 4$	$Pop^n 5$	$Pop^n 6$
0	0.500	0.000	0.000	0.000	0.000	0.000
	0.000	0.000	0.000	0.000	0.000	0.000
100	0.261	0.155	0.061	0.018	0.004	0.001
	0.032	0.028	0.014	0.004	0.001	0.000
200	0.193	0.148	0.090	0.044	0.018	0.008
	0.014	0.013	0.010	0.005	0.002	0.001
300	0.159	0.134	0.096	0.059	0.033	0.020
	0.006	0.006	0.005	0.004	0.002	0.001
400	0.139	0.122	0.095	0.067	0.044	0.032
	0.003	0.003	0.003	0.002	0.002	0.001

In Table 4 significant disequilibrium is created by the migration for about 300 generations, which corresponds to approximately 7,500 years in human populations.

It should be emphasized that the model considered is a very simplistic one and not a realistic representation of the advance of agriculture in Europe. However, it illustrates quite adequately the points under discussion. The extent of the clines and the amount of linkage disequilibrium generated by migration depend very strongly on the differences in gene frequencies in the initial populations, the recombination fraction between the two loci and on the rate of migration. The strong dependence on the difference in initial gene frequencies is illustrated by comparing Table 5 (where the initial gene frequencies in population 1 are 0.5) with Table 4 (where the corresponding frequencies are 1.0). In Table 5 the maximum D is only about half that in Table 4, and reaches insignificant levels in about half the time.

475

It remains an open question whether or not the genetic differences between the mesolithic hunters and gatherers living in Europe and the farmers who gradually invaded coming from the Near East, were sufficient to create a cline of gene frequencies as observed. There is, however, evidence that the original genetic differences could have been quite significant (Cavalli-Sforza, 1972). On the other hand, because the HL-A system is so variable it seems very unlikely that the difference in frequencies for a particular antigen between two populations would be large enough to create the linkage disequilibrium observed for, say, the haplotype HL-A1,8. In present populations the maximum observed frequencies for the alleles HL-A1 and HL-A8 are about 0.23 and 0.16 respectively and no population group has any allele frequency greater than 0.5 (Histocompatibility Testing, 1972).

Related to the study of migration patterns is the question of admixture. Gene frequencies can be studied in populations, such as the American blacks, which are known to be mixtures of two or more reasonably well defined populations, in this case Caucasians and African blacks. The mixture is due to a persistent though small level of intermixture between the original populations over some period of time. This approach has been used in the search for selection (see e.g. Reed (1969) and Cavalli-Sforza and Bodmer (1971) pages 490-497). Previous studies have only considered one locus models of admixture. In relation to the HL-A system a two locus model should be studied. An appropriate model is discussed in the following section.

Admixture Model

Consider two populations, say Caucasian and Negro, referred to as C and N. All progeny from C × N and N × N

matings are classified as Negro. Each generation a proportion α of matings are $C \times N$ and a proportion $(1-\alpha)$ are
$N \times N$. We consider a two locus model, with two alleles at
each locus, denoted $\underline{A},\underline{a}$ and $\underline{B},\underline{b}$. The gene frequencies
and linkage disequilibrium are assumed to remain constant
each generation in the Caucasian population. This model is
not, therefore, realistic in a limiting sense as it does not
allow for gene flow from the Negro population to the Caucasian population.

The frequencies of the four gametic types $\underline{AB}$, $\underline{Ab}$, $\underline{aB}$ and
$\underline{ab}$ in the Negro population at generation n are denoted by
$X_1^{N(n)}$, $X_2^{N(n)}$, $X_3^{N(n)}$ and $X_4^{N(n)}$. The four gametic
frequencies in the Caucasian population are denoted by X_1^C ,
X_2^C , X_3^C and X_4^C and as stated above these are assumed
constant over time. The linkage disequilibrium in the Negro
population at generation n is denoted by $D^{N(n)}$
$(= X_1^{N(n)} X_4^{N(n)} - X_2^{N(n)} X_3^{N(n)})$. D^C denotes the (constant)
linkage disequilibrium in the Caucasian population.

For $N \times N$ matings, which contribute a proportion $1-\alpha$
of the offspring to the next generation, their contribution
of, for example, $\underline{AB}$ haplotypes is

$$X_1^{N(n)} - rD^{N(n)}$$

where r is the recombination fraction between the A and
B loci. Similar expressions hold for $X_2^{N(n)}$, $X_3^{N(n)}$ and
$X_4^{N(n)}$. The $C \times N$ matings provide a combined gametic pool
with the gametic types $\underline{AB}$, $\underline{Ab}$, $\underline{aB}$ and $\underline{ab}$ in proportions
$Y_1^{(n)}$, $Y_2^{(n)}$, $Y_3^{(n)}$ and $Y_4^{(n)}$ where

$$Y_i^{(n)} = \tfrac{1}{2}(X_i^{N(n)} + X_i^C) \quad , \quad i = 1,;4 \quad .$$

These provide a proportion α of the offspring for the next
generation giving

$$Y_1^{(n)} - rD^{Y(n)}$$

as the contribution to $\underline{AB}$ haplotypes, where

$$D^{Y(n)} = Y_1^{(n)} Y_4^{(n)} - Y_2^{(n)} Y_3^{(n)} .$$

Combining the outputs from the two types of matings and extending the above expressions to the other haplotypes gives the following recursion system

$$X_1^{N(n+1)} = (1-\alpha)(X_1^{N(n)} - rD^{N(n)}) + \alpha(Y_1^{(N)} - rD^{Y(n)}) \qquad 2(i)$$

$$X_2^{N(n+1)} = (1-\alpha)(X_2^{N(n)} + rD^{N(n)}) + \alpha(Y_2^{(n)} + rD^{Y(n)}) \qquad 2(ii)$$

$$X_3^{N(n+1)} = (1-\alpha)(X_3^{N(n)} + rD^{N(n)}) + \alpha(Y_3^{(n)} + rD^{Y(n)}) \qquad 2(iii)$$

$$X_4^{N(n+1)} = (1-\alpha)(X_4^{N(n)} - rD^{N(n)}) + \alpha(Y_4^{(n)} - rD^{Y(n)}) \qquad 2(iv)$$

Thus it is easy to see that the frequency of the $\underline{A}$ gene in the Negro population at generation n, denoted $P_A^{N(n)}$ $(= X_1^{N(n)} + X_2^{N(n)})$, follows the recursion system

$$P_A^{N(n+1)} = (1-\alpha)P_A^{N(n)} + \frac{\alpha}{2}(P_A^{C} + P_A^{N(n)}) \qquad (3)$$

where P_A^{C} denotes the (constant) frequency of the $\underline{A}$ gene in the Caucasian population. It follows that

$$P_A^{N(n)} = M_n P_A^{C} + (1-M_n) P_A^{N(0)} \qquad (4)$$

where $M_n = 1 - (1 - \frac{\alpha}{2})^n$ is the total Caucasian admixture in the Negro population after n generations and $P_A^{N(0)}$ is the frequency of the $\underline{A}$ gene in the original African population. Similarly, for the $\underline{B}$ gene,

$$P_B^{N(n)} = M_n P_B^{C} + (1-M_n) P_B^{N(0)} . \qquad (5)$$

The linkage disequilibrium in the Negro population is given by

$$D^{N(n+1)} = X_1^{N(n+1)} X_4^{N(n+1)} - X_2^{N(n+1)} X_3^{N(n+1)}$$

$$= ((1 - \frac{\alpha}{2})X_1^{N(n)} + \frac{\alpha}{2}X_1^{C} - Z)((1 - \frac{\alpha}{2})X_4^{N(n)} + \frac{\alpha}{2}X_4^{C} - Z)$$

$$- ((1 - \frac{\alpha}{2})X_3^{N(n)} + \frac{\alpha}{2}X_3^{C} + Z)((1 - \frac{\alpha}{2})X_2^{N(n)} + \frac{\alpha}{2}X_2^{C} + Z)$$

where $Z = (1-\alpha)rD^{N(n)} + \alpha rD^{Y(n)}$.

It is easy to show that (equation 1 with $m=\frac{1}{2}$)

$$D^{Y(n)} = \frac{1}{2}D^{N(n)} + \frac{1}{2}D^{C} + 1/4(P_A^{N(n)} - P_A^{C})(P_B^{N(n)} - P_B^{C}) \ .$$

Thus $D^{N(n+1)} = (1 - \frac{\alpha}{2})^2 D^{N(n)} + \frac{\alpha^2}{4}D^{C} - Z$

$$+ \frac{\alpha}{2}(1 - \frac{\alpha}{2})(X_1^{N(n)}X_4^{C} + X_4^{N(n)}X_1^{C}$$

$$- X_2^{N(n)}X_3^{C} - X_3^{N(n)}X_2^{C})$$

$$= (1 - \frac{\alpha}{2})^2 D^{N(n)} + \frac{\alpha^2}{4}D^{C} - (1-\alpha)rD^{N(n)}$$

$$- \frac{\alpha r}{4}(2D^{N(n)} + 2D^{C} + (P_A^{N(n)} - P_A^{C})(P_B^{N(n)} - P_B^{C})$$

$$+ \frac{\alpha}{2}(1 - \frac{\alpha}{2})(D^{N(n)} + D^{C} + (P_A^{N(n)} - P_A^{C})(P_B^{N(n)} - P_B^{C}))$$

and so

$$D^{N(n+1)} = (1 - \frac{\alpha}{2})(1-r)D^{N(n)} + \frac{\alpha}{2}(1-r)D^{C}$$

$$+ \frac{\alpha}{2}(1 - \frac{\alpha}{2} - \frac{r}{2})(P_A^{N(n)} - P_A^{C})(P_B^{N(n)} - P_B^{C}) \ .$$

This latter expression can also be written in the form

$$D^{N(n)} = (1 - \frac{\alpha}{2})(1-r)D^{N(n-1)} + \frac{\alpha}{2}(1-r)D^{C}$$

$$+ \frac{\alpha}{2}(1 - \frac{\alpha}{2} - \frac{r}{2})(1 - \frac{\alpha}{2})^{2(n-1)}(P_A^{N(0)} - P_A^{C})(P_B^{N(0)} - P_B^{C})$$

$$(8)$$

using equations (4) and (5). $D^{N(n)}$ is very dependent on the difference between the gene frequencies in the two original populations, as was the case for the migration model previously discussed. Even if $D^C = 0$ and $D^{N(0)} = 0$, $D^{N(n)}$ will, however, build up in value initially.

The data of Albert, Mickey and Terasaki (1973) on the HL-A frequencies in an American Negro population has been considered in this context. The gene frequency for HL-A1 in Negroes is 0.058 and for HL-A8 is 0.041, and D = 0.0136 . The values in the Caucasian population are HL-A1 (0.142), HL-A8 (0.102) and D = 0.053. To apply the above admixture model we need to know the gene frequencies and linkage disequilibrium in the original African population. This presents something of a difficulty, especially as there are no data on HL-A frequencies in the relevant present day populations of West Africa. Estimates of the amount of admixture (M_n) in Negro populations vary from about 0.2 to 0.27 (Reed, 1969).

What we shall do is take $M_n = 0.25$ and then calculate what the original African frequencies should have been to give rise to the frequencies observed today, assuming no selection and using equations (4) and (5). Doing this, the original African gene frequencies are estimated to be 0.03 for HL-A1 and 0.021 for HL-A8. These values would be in agreement with the idea that HL-A1 and HL-A8 are predominantly found in Caucasian populations. The fact that the estimated frequencies are non-zero could be accounted for by previous Caucasian admixture (that is, to much earlier contacts between Caucasoids and Negroids in Africa). If the linkage disequilibrium in the original African population was zero, that is $D^{N(0)} = 0$, would admixture be sufficient to explain the observed disequilibrium of 0.0136 in the Negro population? It is generally agreed that between 10 and 14 generations must be considered since the time of the original African to the present Negro populations. If we take the number of generations as n = 10 and $M_n = 0.25$ we require $\alpha = 0.0568$. So for

480

the example under consideration we have these values and

$r = 0.008$, $P_A^C = 0.142$, $P_B^C = 0.102$, $D^C = 0.053$,

$P_A^{N(0)} = 0.03$, $P_B^{N(0)} = 0.021$, $D^{N(0)} = 0$. These values
give after 10 generations, the gene frequencies observed in
the Negro population, and a predicted D value of 0.0144
(calculated by iterating equation (8)), which is very close
to the observed value of 0.0136. Different values of M_n
and the number of generations of admixture have been considered, for example (M = 0.25, n = 14), (M = 0.2, n = 10)
and (M = 0.2, n = 14). The respective estimates of $D^{N(n)}$
in these cases are 0.0141, 0.0114 and 0.0112. Thus the model
is fairly insensitive to these changes in the parameters.
The existence of linkage disequilibrium for the haplotype
HL-A1,8 in American Negroes appears thus to be quite adequately explained by admixture. This of course gives no
explanation for the existence of persistent linkage disequilibrium for the haplotype HL-A1,8 in the Caucasian population.

While the above model is of general importance in showing
how linkage disequilibrium can be created due to admixture it
will probably be of little use in attempts to measure selection in the particular case of the HL-A system. The reason
for this is the extreme difficulty of estimating the gene
frequencies in the original African populations, again due to
the extreme variability of the system.

There is so far no direct evidence for selection acting in
the HL-A system, as measured for example by an effect of HL-A
antigen on viability, fertility or foetal-maternal incompatibility (Mattiuz et al., 1970). Various aspects of the effect
of selection and the maintenance of linkage disequilibrium
are discussed in the next section.

Selection and Maintenace of Linkage Disequilibrium

It is well known that certain types of selection interaction between genes can lead to equilibria with $D \neq 0$.
This was originally suggested by Fisher in 1930 and many
theoretical studies have since confirmed and extended Fisher's
ideas (Kimura, 1956; Lewontin and Kojima, 1960; Bodmer and
Felsenstein, 1967; Karlin and Feldman, 1970). The interactions of linkage and selection are, however, not only
important for equilibrium properties of populations but must
also be considered in terms of the conditions required for
the initial increase of new gametic combinations. Bodmer and
Parsons (1962) were the first to emphasize this approach. It
should also be remembered that even if the selective regime
is such that the equilibrium value of D will be zero, it is
nevertheless possible that D will be maintained at a non-
zero value for quite a considerable length of time while the
genes are evolving towards their equilibrium values (Thomson,
1975).

As mentioned previously, we are interested in looking at
the HL-A system as a linked complex of genes. The next step,
as far as the theory is concerned, is to consider a three
locus model. The rationale is that we should then be able to
study what effect the evolution of a third locus has on the
two major serologically detected loci of the HL-A system,
namely, LA and 4 . We use a simple extension of the two
locus, two allele representation of haplotype frequencies.
Thus, for a three locus model, with two alleles at each locus,
we have eight possible haplotypes. The haplotype frequencies
can be completely specified by three gene frequencies and
four disequilibrium parameters. Three of the disequilibrium
parameters measure pairwise disequilibrium while the fourth
is a measure of triple association after taking account of

the pairwise associations. The representation of the haplotype frequencies is given in Table 6.

TABLE 6. *Haplotype frequencies in terms of gene frequencies
and disequilibrium parameters for a three locus model, with
two alleles at each locus. At the first locus the alleles
are* $\underline{A}$ *and* $\underline{a}$ *, at the second* $\underline{B}$ *and* $\underline{b}$ *and at the third
* $\underline{C}$ *and* $\underline{c}$. P_A *,* P_B *etc. denote the frequencies of the
alleles* $\underline{A}$ *and* $\underline{B}$ *respectively, etc.* ($P_A + P_a = 1$, $P_B + P_b = 1$,
$P_C + P_c = 1$). D_{BC} *denotes the pairwise linkage disequilibrium
between the loci* B *and* C *and similarly for* D_{AB} *and
* D_{AC} . D_{ABC} *is the third order linkage disequilibrium after
taking account of the pairwise associations and is defined
implicitly by the equations given below. The eight haplotype frequencies can be written as follows:*

$$f(ABC) = P_A P_B P_C + P_A D_{BC} + P_B D_{AC} + P_C D_{AB} + D_{ABC}$$

$$f(ABc) = P_A P_B P_c - P_A D_{BC} - P_B D_{AC} + P_c D_{AB} - D_{ABC}$$

$$f(AbC) = P_A P_b P_C - P_A D_{BC} + P_b D_{AC} - P_C D_{AB} - D_{ABC}$$

$$f(Abc) = P_A P_b P_c + P_A D_{BC} - P_b D_{AC} - P_c D_{AB} + D_{ABC}$$

$$f(aBC) = P_a P_B P_C + P_a D_{BC} - P_B D_{AC} - P_C D_{AB} - D_{ABC}$$

$$f(aBc) = P_a P_B P_c - P_a D_{BC} + P_B D_{AC} - P_c D_{AB} + D_{ABC}$$

$$f(abC) = P_a P_b P_C - P_a D_{BC} - P_b D_{AC} + P_C D_{AB} + D_{ABC}$$

$$f(abc) = P_a P_b P_c + P_a D_{BC} + P_b D_{AC} + P_c D_{AB} - D_{ABC}$$

The four disequilibrium parameters can be estimated from the
$2 \times 2 \times 2$ association table by a straightforward extension
of the approach used for two loci (Manuscript in preparation).

A three locus model of particular relevance to the HL-A
system is that where we consider the effect of a selected
locus on closely linked neutral loci (see Thomson, 1975). The

analogy is that selection could be acting on, say, the immune response genes within the HL-A system and not on the LA and 4 loci themselves. It turns out that as a selected locus evolves towards its equilibrium value quite large linkage disequilibrium values can be created at closely linked neutral loci, and remain for considerable lengths of time. We consider three loci A, B and C with two alleles at each locus, denoted $\underline{A}$, $\underline{a}$ etc. The B locus is selected (heterosis), the fitness scheme being

$$\begin{array}{ccc} \text{BB} & \text{Bb} & \text{bb} \\ \hline 1-S_1 & 1 & 1-S_2 \end{array} \quad . \qquad (9)$$

The A and C loci are selectively neutral. We assume that a new mutation $\underline{b}$ has just recently occurred. The frequency of the $\underline{b}$ gene will be increasing each generation, until equilibrium is reached. If we look at what is happening at the neutral A and C loci we see that in general the gene frequencies there are also changing. Denoting frequencies in the next generation by primes then it can be shown that

$$P_A' = P_A - \frac{(S_1 P_B - S_2 P_b)}{\overline{w}} \cdot D_{AB} \qquad (10)$$

where $\overline{w} = 1 - S_1 P_B{}^2 - S_2 P_b{}^2$ is the mean fitness of the population and all other parameters are as defined in Table 6. Also,

$$D_{AB}' = (\frac{P_B' P_b'}{P_B P_b} - \frac{R_1}{\overline{w}}) \cdot D_{AB} \qquad (11)$$

where R_1 is the recombination fraction between the A and B loci. Similar expressions hold for P_C and D_{BC}. The term $\frac{P_B' P_b'}{P_B P_b} > 1$ until $P_B = P_b = \frac{1}{2}$. $\frac{R_1}{\overline{w}}$ is always greater than R_1. So there exist two opposing forces, the first term is initially increasing the linkage disequilibrium value while the second term is decreasing it at a rate greater than under neutrality. Quite significant levels of linkage disequilibrium can be generated in this way. A point of more interest is that while the $\underline{b}$ gene is evolving towards its equilibrium value significant linkage disequilibrium may also be created between the two neutral loci A and C . Even when the disequilibrium between the neutral loci is initially zero it can build up in value. These points are illustrated in Table 7 by a numerical example based on iterating the appropriate recursion system (Thomson, 1975).

TABLE 7. *The effect of a selected locus on closely linked neutral loci. There are three loci A, B and C. Locus B is halfway between A and C , with recombination fractions for intervals AB and BC both 0.005. The B locus is selected; the fitness scheme is as in (9) with $S_1=S_2=0.05$. The notation for gene frequencies and linkage disequilibrium is as described in Table 6. Allele $\underline{b}$ is a new mutant occurring on the chromosome $\underline{aBc}$ to produce gamete $\underline{abc}$, and we consider the evolution of the three loci as the $\underline{b}$ gene increases in frequency from 0.01 to 0.49 (near its equilibrium value). Loci A and C are assumed to be already polymorphic as indicated by the initial (haplotype) frequencies.*

Assumed Initial Haplotype Frequencies

ABC	ABc	AbC	Abc	aBC	aBc	abC	abc
.665	.285	.000	.000	.035	.005	.000	.010

Gen.	P_b	P_a	P_c	D_{AB}	D_{AC}	D_{BC}	D_{ABC}
0	.01	.05	.30	.0095	.000	.007	-.007
80	.23	.22	.42	.115	.052	.085	-.030
160	.46	.34	.52	.106	.032	.078	-.003
213	.49	.35	.52	,081	.019	.060	-.001

Haplotype frequencies at generation 213

ABC	ABc	AbC	Abc	aBC	aBc	abC	abc
.244	.167	.084	.152	.059	.040	.090	.164

In the example in Table 7 the disequilibrium between the (polymorphic) neutral loci A and C , D_{AC} is initially zero and the selected locus B is half way between A and C , with recombination fractions for AB and BC both 0.005. As the b gene increases in frequency from 0.01 to 0.49 (near its equilibrium value) D_{AC} builds up in value and then decreases. At generation 80 for example $D_{AC} = 0.052$ ($P_a = 0.22$, $P_c = 0.42$) while at generation 160 $D_{AC} = 0.032$ ($P_a = 0.34$, $P_c = 0.52$).

The effect of the selected locus on closely linked neutral loci lasts for as long as it takes the selected locus to reach its equilibrium value. For smaller selective values the effect will last a longer time but will not be as strong as there is also more time for recombination to reduce the effect of the pulling along of the gene frequencies together, which is what creates the linkage disequilibrium. In Tables 8 and 9 the effects of different selective values are considered. In the example in Table 8 there is 10% selection acting on the B locus while in Table 9 the selection is of order 1%. All the initial conditions are the same as in Table 7 to facilitate a direct comparison. The disequilibrium D_{AC} generated between the neutral loci A and C is much greater for the case of stronger selection. Thus, for 10% selection $D_{AC} = 0.08$ at generation 40 (Table 8) whereas for 5% selection the value of D_{AC} does not get as large as this. However, the effect of the selected locus on the whole lasts for a longer time with the lower value of selection. In all cases D_{AC} eventually goes to zero but usually well after the gene frequencies have reached their equilibrium values (see Table 7). For more discussion on this model see Thomson (1975).

TABLE 8. *Effect of selected locus on linked neutral loci.*

Exactly as for Table 7 except that the selection coefficients for locus B are now $S_1 = S_2 = 0.10$.

Assumed initial haplotype frequencies

ABC	ABc	AbC	Abc	aBC	aBc	abC	abc
0.665	0.285	0.000	0.000	0.035	0.005	0.000	0.01

Gen^n	P_b	P_a	P_c	D_{AB}	D_{AC}	D_{BC}	D_{ABC}
0	0.01	0.05	0.30	0.0095	0.000	0.007	-0.007
40	0.25	0.25	0.45	0.145	0.080	0.107	-0.043
80	0.47	0.41	0.57	0.158	0.072	0.116	-0.006
103	0.49	0.42	0.58	0.140	0.057	0.103	-0.002

Haplotype frequencies at generation 103

ABC	ABc	AbC	Abc	aBC	aBc	abC	abc
0.271	0.163	0.030	0.112	0.049	0.027	0.074	0.274

TABLE 9. *Effect of selected locus on linked neutral loci.*

Exactly as for Table 7 but with $S_1 = S_2 = 0.01$.

Assumed initial haplotype frequencies

ABC	ABc	AbC	Abc	aBC	aBc	abC	abc
0.665	0.285	0.000	0.000	0.035	0.005	0.000	0.01

Gen^n	P_b	P_a	P_c	D_{AB}	D_{AC}	D_{BC}	D_{ABC}
0	0.01	0.05	0.3	0.0095	0.000	0.007	-0.007
200	0.06	0.08	0.32	0.021	0.005	0.015	-0.005
400	0.23	0.11	0.35	0.023	0.002	0.017	-0.001
600	0.38	0.13	0.36	0.011	0.000	0.008	-0.000
800	0.46	0.13	0.36	0.004	0.000	0.003	-0.000
1094	0.49	0.13	0.36	0.001	0.000	0.001	0

Haplotype frequencies at generation 1094

ABC	ABc	AbC	Abc	aBC	aBc	abC	abc
0.287	0.159	0.273	0.153	0.042	0.023	0.041	0.023

Based on the above considerations it seems possible that
the disequilibrium observed between certain alleles of the
HL-A system is not necessarily an indication of selection
acting directly on the LA and 4 loci. It could merely indicate selection acting on closely linked loci. Selection of
5% can generate linkage disequilibrium comparable with values
observed in the HL-A system, though in the numerical examples
given here this is not sustained for the minimum of 5,000 to
10,000 years (200 to 400 generations) that are needed to
account for the persistence of the HL-A1,8 haplotype. This
model has, however, considerable flexibility in terms of ways
of creating linkage disequilibrium. The selection need not
be acting solely on one locus. One must also consider the
possibility of the cumulative effect of weak selection acting
on a number of loci. In this context one should also remem-
ber that it is generally very difficult if not impossible to
distinguish the selective effects of a given detectable gene
locus from those that may be due to closely linked loci, the
products of which are not readily detectable.

Disease Association

Another direct approach to the search for selection on the
HL-A system is to look for associations between HL-A antigens
and diseases that may be of selective significance. A rationale for the existence of such possible associations comes
from the discovery, in several animal species, of specific
immune response genes closely linked to the species' major
histocompatibility system. Although early studies showed
only weak associations with Hodgkin's and some other diseases,
subsequent studies have shown very striking associations with
a number of diseases having a presumptive or suspected auto-
immune aetiology, including especially ankylosing spondylitis,

psoriasis, coeliac disease, myasthenia gravis and multiple
sclerosis.

Some data on HL-A and disease association are shown in
Table 10 (based on McDevitt and Bodmer, 1974; and Bodmer and
Bodmer, 1974).

TABLE 10. HL-A and disease associations.

Disease	No. of studies	Antigen	Frequency in patients %	Frequency in controls %	Average relative risk
Ankylosing spondylitis	5	W27	90	7	141
Reiter's disease	3	W27	76	6	46.6
Acute anterior uveitis	2	W27	55	8	16.7
	(6	HL-A13	18	4	5.0
Psoriasis	(6	W17	29	8	5.0
	(4	W16	15	5	2.9
Graves' disease	1	HL-A8	47	21	3.3
Coeliac disease	6	HL-A8	78	24	10.4
Dermatitis herpetiformis	3	HL-A8	62	27	4.5
Myasthenia gravis	5	HL-A8	52	24	4.6
S.L.E.	2	W15	33	8	5.1
Multiple sclerosis	4	HL-A3	36	25	1.7
		HL-A7	36	25	1.5
Acute lymphatic leukaemia	7	HL-A2	63	37	1.7
	(8	4c(W5)	25	16	1.6
Hodgkin's disease	(7	HL-A1	39	32	1.3
	(7	HL-A8	26	22	1.3
Chronic hepatitis	1	HL-A8	68	18	9.5
Ragweed hayfever, Ra5 sensitivity	1	HL-A7	50	19	4

The relative risk is $\dfrac{pd(1-pc)}{pc(1-pd)}$ where pd = frequency in diseased and pc = frequency in controls.

See Bodmer and Bodmer (1974) and McDevitt and Bodmer (1974) for detailed references on the
origin of this data.

The table shows, for each disease, the number of studies on
which the data are based, the HL-A antigen involved, the
antigen frequency in diseased patients and controls and a
measure of its association with the disease. By far the most
significant association shown is between the antigen W27 and
ankylosing spondylitis. A number of studies agree, with
remarkable consistency, in showing a frequency of about 90%
for W27 in patients with ankylosing spondylitis, whereas the

frequency of this antigen in the control population is only 5-10%. Two other diseases (Reiter's and Uveitis) which are often associated with ankylosing spondylitis also show striking increases in the frequency of antigen W27.

If associations between HL-A and disease are not due to the direct effects of the serologically determined HL-A antigens but, as suggested by McDevitt and Bodmer (1972) and others, to the effects of closely linked loci, then there must be significant linkage disequilibrium between the alleles at these loci and those at the LA and 4 loci. Associations due to linkage disequilibrium are likely to be much more complex than those resulting from a simple direct antigenic effect, and may vary substantially from one population to another with variation in haplotype frequencies. Even if a disease susceptibility is associated with the effects of a gene in the HL-A region this will not show up as a population association with the serologically detected antigens unless the disease susceptibility locus and the LA or 4 loci are in significant linkage disequilibrium.

It is important to point out that even though most people with ankylosing spondylitis are W27, nevertheless the comparatively low incidence of the disease (0.1 to 0.4 per cent in males and about 0.025 to 0.1 per cent in females, for Caucasians) means that at most about 1 to 3 per cent of males and 0.2 to 0.9 per cent of females who are W27 actually get the disease. Some W27 individuals, of course, may get other associated diseases, such as Reiter's and Uveitis, but this, as a proportion of all those who are W27, will also be quite small. If the disease association with W27 is due to a closely linked immune response (Ir) gene, there are three major (not mutually exclusive) explanations for the fact that only a relatively small proportion of W27 individuals suffer

490

from one of the relevant diseases:

(1) Environmental factors, such as exposure to particular viruses, bacteria or other antigens may be a part of the cause of the disease and only a small proportion of individuals with the appropriate genotype may have had the appropriate environmental exposure.

(2) Other genes, not closely linked to the HL-A system may contribute to the disease susceptibility.

(3) Only a minority of W27 carrying haplotypes may carry the relevant Ir locus allele.

It would clearly be of interest to use data on, for example, the frequency of W27 in diseased and control individuals to estimate the frequency of haplotypes carrying various combinations of the W27 and Ir locus alleles. Under certain reasonable assumptions one can use such data to estimate the proportion of Ir allele haplotypes which carry W27 (see Bodmer and Bodmer, 1974, Figure 1). More information, however, is needed to distinguish the three explanations suggested above for why only a minority of W27 individuals are diseased.

Family data should help to resolve this question. Thus, if a relatively high proportion of W27 relatives of W27 individuals with disease are themselves diseased, then the third explanation, namely that the Ir allele is present in a minority of W27 haplotypes, would be most likely. On the other hand, if the proportion of such W27 relatives with disease was not much different from that among W27 individuals in the population as a whole, explanations (1) and (2) would be more likely and the Ir allele might be present in most W27 carrying haplotypes. The available data on the heritability of ankylosing spondylitis suggest that the latter

is not the case and point to the third explanation. (A further analysis of this question will be published elsewhere -
for a recent review on W27 and its associated diseases see
Brewerton, 1975).

Most of the diseases whose HL-A association has been
established are either very rare, or occur late in life, or
have little effect on viability or fertility. It therefore
seems unlikely that any of the associations reported to date
will themselves be of much selective significance. If, however, associations were found with important infectious
diseases, such as smallpox or cholera, that have now, or had
in the past, a relatively high incidence, then these could
well be of great significance for the evolution of the HL-A
system. Ceppellini and co-workers (Piazza et al., 1973) have
for example provided some suggestive evidence for an association of HL-A with malarial incidence based on variations
in the HL-A antigen frequencies in Sardinian villages.

Discussion

The diseases associated with the HL-A system include auto-
immune diseases and cancers, as well as those possibly due to
infectious agents. As already emphasized one cause of such
associations may be histocompatibility linked immune response
genes. Additional possible mechanisms for HL-A and disease
association are: (see for example McDevitt and Bodmer, 1974)
(1) that histocompatibility antigens could serve as receptors for, or interact with receptors for specific viruses or
other pathogenic agents;
(2) that such agents might incorporate into their coat a
portion of the membrane from the cell they are attacking
which in turn carried histocompatibility antigens;
(3) that histocompatibility antigens could induce tolerance

to foreign antigens with which they cross react.

It is interesting to note that all these mechanisms for explaining HL-A -disease associations, including linked Ir genes, imply that a pathogen may have to adapt to the antigenic constitution of the host. If the host's antigenic constitution changes, then the pathogen will take time to adapt to this change and so new antigens should initially confer an advantage on the individual that carries them. In general one might expect that the greater the population heterogeneity with respect to cell surface antigens the more difficult it will be for a pathogen to adapt successfully to infect the majority of the population. These mechanisms therefore lead to a form of frequency dependent selection with new antigenic types having an advantage (Bodmer, 1972).

The most significant aspect of this is that it is a well known mechanism for giving rise to polymorphism. A number of the mechanisms discussed previously could possibly account for the observed values of linkage disequilibrium, but none of them provides a mechanism for estbalishing such large amounts of heterozygosity as observed in the HL-A system. A model of frequency dependent selection would, perhaps, be the most likely to be compatible with such levels of heterozygosity.

There are a number of possible (not mutually exclusive) mechanisms which could give rise to significant linkage disequilibrium, in addition to the effects of migration and selection which we have discussed in this paper. The following is an attempt to identify all these possible mechanisms:

(1) Selection

(2) Time - i.e. recent origin of a haplotype so that D has not had time to decrease to insignificant levels.

(3) Migration and admixture

(4) Inbreeding

(5) Random drift effects

(6) Sampling bias due to picking out the extreme D value
     from a relatively large number of combinations.

   It is clearly difficult to be sure in any particular case
which explanation, or combination of explanations is the most
appropriate. Intuitively it seems possible, as emphasized by
Bodmer (1973a), that persistent linkage disequilibrium,
beyond that accounted for by possibility (2) and such as is
observed in the case of the HL-A1,8 haplotype, depends on the
existence of some sort of selective effects. Inbreeding
levels in most human populations are sufficiently low, and
effective population sizes sufficiently large to make explanations (4) or (5) quite unlikely (see Bodmer, 1973a). Karlin and McGregor (1968) pointed out that the variance of the
linkage disequilibrium tends to zero at a slower geometric
rate than the expected value of the linkage disequilibrium.
This would suggest that although the average value of D may
be close to zero the actual observed values could be different
from zero. The finding of similar levels of linkage disequilibrium for HL-A1,8 in several related populations makes
this and also explanation (6) extremely unlikely. In this
paper we have shown that migration or admixture are most probably not sufficiently strong forces to maintain the values
of D observed for a sufficiently long period of time. In
this regard Karlin (personal communication) has also pointed
out that the discrete stepping stone model may give a slower
rate of decay than a continuous model of migration. This
leaves some form of selection as the most likely explanation.
A particular feature of our analysis is the demonstration of
substantial, relatively long term, transient effects. These

494

make it difficult to distinguish between relatively weak interactive selection involving alleles at two or more loci, and simple relatively strong selection acting on just one linked locus in the region. Although the models we have considered cannot give a definitive answer in a particular situation they are of considerable value in giving estimates of the effect that will be produced in a population under given assumptions, and so provide an appropriate basis for comparing theory and observation, as in the case of the HL-A system.

Summary

The major serologically detected antigens of the HL-A system are determined by two closely linked, highly polymorphic loci LA and 4. The system as a whole appears to involve a complex of many closely linked loci, with possibly related functions on the cell surface. A particular feature of data on HL-A antigen frequencies is the existence of substantial linkage disequilibrium between certain alleles at the LA and 4 loci. The pattern of haplotype frequencies in Caucasian populations suggests that this disequilibrium has persisted for at least 5 to 10 thousand years.

A review of migration models for two linked loci, together with some numerical simulations of the models, suggests that the observed levels of disequilibrium for the HL-A1,8 haplotype are quite unlikely to be generated by migration effects. Application of a two locus admixture model to HL-A data on Caucasoid-Negroid admixture, suggests, however, that the frequency of the HL-A1,8 haplotype, and the magnitude of its associated disequilibrium value, in American Blacks can be adequately accounted for by simple admixture in the absence of selective effects.

Analysis of a three linked locus model, in which only one locus is subject to selection, shows that selection acting on a single locus can generate relatively long term linkage disequilibrium between two closely linked neutral loci. Such a mechanism is suggested by data on HL-A and disease associations which can best be interpreted in terms of the effects of immune response genes which are known to be located in the HL-A region.

It seems possible that the level of heterozygosity for the HL-A system may be explained by a form of frequency dependent selection, again suggested by the HL-A and disease association data. An overall review of possible mechanisms which can give rise to significant linkage disequilibrium suggests that the case of the persistent HL-A1,8 haplotype is best explained by some sort of selective effect.

ACKNOWLEDGEMENTS

We should like to thank Ellen Solomon, Cathy Falk, Peter Goodfellow and Jenny Brown for many helpful discussions and comments in the course of preparing this paper. We should also like to thank Penny Pickbourne for her help with computing. This work was supported, in part, by a grant from the Medical Research Council. Glenys Thomson is an 1851 Research Fellow.

REFERENCES

Albert, E.D., M.R. Mickey and P.I. Terasaki. (1973). In Histocompatibility Testing 1972, pp. 233-240; Edited by Jean Dausset and Jacques Colombani. Munksgaard, Copenhagen.

Bodmer, J.G., P. Rocques, W.F. Bodmer, J. Colombani,
L. Degos, J. Dausset and A. Piazza. (1973). Joint report
of the Fifth Histocompatibility Workshop Histocompatibility
Testing 1972, p.619-667. Munksgaard Copenhagen.

Bodmer, W.F. (1972). Nature 237: 139-145.

Bodmer, W.F. (1973a). In Histocompatibility Testing 1972,
pp. 611-617; Edited by Jean Dausset and Jacques Colombani.
Munksgaard, Copenhagen.

Bodmer, W.F. (1973b). In Defence and Recognition. Edited by
R.R. Porter. Butterworths, London (in press).

Bodmer, W.F. (1973c). Israel J. of Med. Sci. 9: 1503-1518.

Bodmer, W.F. (1975). Transpl. Proc. 7, No.1, 1-4.

Bodmer, W.F. and J.G. Bodmer. (1974). In Tenth Symp. in Adv.
Med. pp. 157-174. Pitman Medical Press, London.

Bodmer, W.F., H. Cann and A. Piazza. (1973). In Histocompatibility Testing 1972, pp. 753-767. Edited by Jean
Dausset and Jacques Colombani, Munksgaard, Copenhagen.

Bodmer, W.F. and J. Felsenstein. (1967). Genetics 57/2,
237-265.

Bodmer, W.F. and P.A. Parsons. (1962). Adv. Genet. 11: 1-100.

Brewerton, D.A. (Ed.) (1975). Symposium on Histocompatibility and Rheumatic Disease. Ann. Rhe. Dis. 34, Suppl.No.1.

Cavalli-Sforza, L.L. (1972). In Human Genetics. Edited by
De Grouchy, J. de, F.J.G. Ebling and I.W. Henderson.
Excerpta Medica, Amsterdam, pp.79-95.

Cavalli-Sforza, L.L. and W.F. Bodmer. (1971). The Genetics
of Human Populations. W.H. Freeman and Co.

Ceppellini, I.R., E.S. Curtoni, P.L. Mattiuz, V. Miggiano,
G. Scudeller and A. Serra. (1967). In Histocompatibility
Testing 1967. Edited by E.S. Curtoni, P.L. Mattiuz and
R.M. Tosi. Munksgaard, Copenhagen.

Feldman, M.W. and F.B. Christiansen. (1975). Genet. Res.
(In press).

Fisher, R.A. (1930). The Genetical Theory of Natural Selection. (2nd Edition). New York. Dover 1958.

Histocompatibility Testing 1972. Edited by Jean Dausset and
Jacques Colombani. Munksgaard, Copenhagen.

Karlin, S. and M.W. Feldman. (1970). Theor. Pop. Biol. 1: 39-71.

Karlin, S. and J. McGregor. (1968). Genetics 58: 141-159.

Kimura, M. (1956). Evolution 3: 278-287.

Lewontin, R.C. and K. Kojima. (1960). Evolution 14: 458-472.

Mattiuz, P.L., D. Ihde, A. Piazza, R. Ceppellini and W.F. Bodmer. (1970). Histocompatibility Testing 1970, pp.193-205. Edited by P. Terasaki. Munksgaard, Copenhagen.

McDevitt, H.O. and W.F. Bodmer. (1972). Am. J. of Med. 52: 1-8.

McDevitt, H.O. and W.F. Bodmer. (1974). The Lancet, pp. 1269-1275.

Nei, M. and W. Li. (1973). Genetics 75: 213-219.

Piazza, A., M.C. Belvedere, D. Bernoco, C. Conighi, L. Contu, E.S. Curtoni, P.L. Mattiuz, W. Mayr, P. Richiardi, G. Scudeller, and R. Ceppellini. (1973). In Histocompatibility Testing 1972. Edited by Jean Dausset and Jacques Colombani. pp.73-84. Munksgaard, Copenhagen.

Reed, T.E. (1969). Science 165: 762-768.

Thomson, G. (1975). (In preparation).

The Structure of Gene Control Regions
and Its Bearing on Diverse Aspects of Population Genetics

B. WALLACE

INTRODUCTION

The population geneticist, according to Lewontin (1965),
"has been employed for the last thirty years, in polishing
with finer and finer grades of jeweller's rouge those three
colossal monuments of mathematical biology The Causes of
Evolution, The Genetical Theory of Natural Selection, and
Evolution in Mendelian Populations." "It is true, of course,"
Lewontin continues, "that these early formulations made very
many simplifying assumptions about the nature of the genetic
system, on the one hand, and the nature of the environment on
the other." He then continues by discussing the need to investigate variable environments and multi-gene complexes in
order to test the adequacy of the edifices erected by Haldane,
Fisher, and Wright.

Today I intend to discuss the nature, not of the genetic
system, but of the gene. Specifically, I want to discuss what
may lie behind the symbols - the A's and a's , the a_1's ,
a_2's, a_3's, and the a_i's and a_j's - of population genetics,
and to ask whether a clearer understanding of the physical nature of the gene locus might not aid in understanding abstract
symbols and their manipulations.

An understanding of the reality that lies beneath an abstraction, some would claim, is unnecessary. Reality, according to these persons, in no way influences the logical manipulation of symbols. I disagree with this view. The general features of the Watson-Crick model of DNA, it may be recalled, met the requirements of the gene which had been postulated by H.J. Muller (1929) and other early geneticists; had they not done so, DNA would not have been recognized as the genetic material. Nevertheless, once identified, the structure of DNA opened previously unsuspected vistas to both theoretical and experimental geneticists. I hope that a similar (but somewhat more modest) revelation would follow from an appreciation of gene control mechanisms.

The elaborated Britten and Davidson model

Britten and Davidson (1969) are largely responsible for the concept of hierarchal or cascade control of gene action in higher organisms. They suggested that the RNA transcribed at certain gene loci in higher eucaryotes is capable of activating or inhibiting the transcription of structural genes at other loci. Under this view, development is seen as the successive activation of sets of genes and the turning off of others through a system of communicating signals. Crick (1971) elaborated on this hypothetical scheme by postulating that single strands of DNA are most likely to contain the necessary receptor sites.

Wallace and Kass (1974) argued that gene loci possessing many sensors would operate more efficiently if the sequential orders of the sensors of the two alleles were not identical. As a direct consequence of their argument, they were forced to postulate a structure of the DNA molecule that would permit sensors to exchange position. The structure which might serve that function consists of tandemly duplicated reverse repeats or, in more modern terminology, sequences of overlapping palindromes. Such a structure is shown in Figure 1.

······a b c d O e f g h— h́ǵf́éOd́ćb́á́—a b c d O e f g h— h́ǵf́éOd́ćb́á́······

FIGURE 1. *The repetitive pattern of overlapping palindromes (reverse repeats, or rotational symmetries) that permits gene control sensors (0) to exchange positions within the control region of a gene locus.*

On the advantage of dissimilar sequences of sensors at homologous loci

The advantage which is conferred by alleles that possess activating sensors which are arrnaged in dissimilar orders is to be found, according to Wallace and Kass (1974) and Wallace (1974), in two characteristics of such loci: (1) A set of genes which are distributed among a number of loci and which respond to a given regulatory signal, cannot be considered as having responded until the products (mRNA transcripts or, later, the polypeptides which they specify) of all genes are in production. The elapsed time between the signal and the response is systematically longer if the two alleles at each locus have identical sequences of sensors; almost invariably, one gene among those of the set will be activated by a physically remote sensor and, consequently, will delay the coordinated response of the entire set. Simple randomization of the positions occupied by sensors within the control regions of each pair of alleles effectively minimizes this delay. (2) Should the proper functioning of a set of genes depend upon the more-or-less simultaneous production of mRNA by the structural gene of each locus, the randomization of the positions of sensors in the control regions of each pair of alleles greatly increases the probability that at least one member of each pair of alleles within the set will be functioning simultaneously with all others.

The points made above are illustrated in Figures 2 and 3.

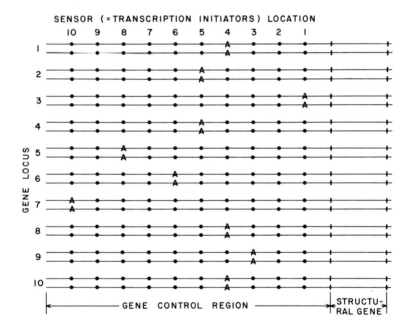

FIGURE 2. *A hypothetical example consisting of 10 gene loci each of which is under the control of 10 control systems of which only* A *is common to all 10 loci. In this example, it is assumed that the sequences of sensors controlling the two alleles at a given locus are identical.*

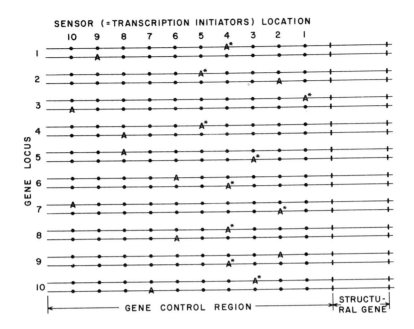

FIGURE 3. *A hypothetical example identical to that of Figure 2 except the sequences of sensors controlling the two alleles at a given locus are assumed to differ. The asterisks (*) denote A-sensors that would lead to the most nearly simultaneous production of mRNA at all 10 loci following a call from the A control system.*

Each figure illustrates ten gene loci each of which bears ten sensors. One regulatory system, A , is shown as calling for the operation of all ten genes. In Figure 2, sensor A is shown as occupying the identical sensor position in the gene control region of each allele. In contrast, it is shown as occupying randomly chosen locations in the control regions of the two alleles of each locus in Figure 3.

Suppose that system A is not functional until the gene

products of all ten loci are in production. Under the
arrangement shown in Figure 2, the system would not be functional until transcription at locus 7 had proceeded from the
furthermost (10th) sensor location to the structural gene at
that locus. In Figure 3, locus #4 would be the last to respond; even there, however, transcription must proceed only
from the fifth sensor location.

Should it be important that the gene products of all loci
be produced more-or-less simultaneously, a glance reveals
that the randomization of sensors shown in Figure 3 would be
much more efficient than the scheme illustrated in Figure 2.
The distribution of A-sensors in Figure 3 generates 1024
possible arrays many of which (such as the one designated by
asterisks) will be much less dispersed among sensor locations
than is the one illustrated in Figure 2.

Summary of evidence, existing or possible, suggesting that the proposed model has a basis in fact

a. Mutable loci such as scute.

If the sensors which are responsible for the transcription
of genes are embedded in a matrix of overlapping palindromes,
then the accidental loss of an odd number of sensors leaves a
control region that contains a tandem duplication which is
comparable in size to a reverse repeat. Intrastrand recombination during the development of the individual carrying such
a locus may cause the loss of still an additional sensor
(Figure 4). Should the time of loss be related to the position of the sensor within the gene control region, a one-to-
one correspondence may be found between various phenotypes
and chromosomal positions. Such a relationship was discussed
in detail for the various scute alleles (the so-called

"subgene" hypothesis) by Dubinin (1932).

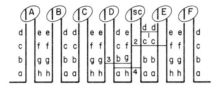

OUTCOME OF RECOMBINATION:

1)	A	B	C	D	sc	E	F
2)	A	B	C	D	E	sc	F
3)	A	B	C	*	sc	E	F
4)	A	B	C	D	*	E	F

FIGURE 4. _A diagram illustrating that an error in the pattern_
of overlapping palindromes might lead to further
errors such as the loss of a sensor (* in 3 and 4)
as the result of somatic intra-chromosomal recom-
bination.

b. Transposable gene control elements.

Focusing attention not on the unstable locus but on the
excised fragment, it is seen that once it is excised from the
chromosome it becomes an episome-like particle bearing a specific sensor (Figure 5). Because the repetitious DNA of
different loci may be largely homologous, this element may be
inserted within the control region of a second locus. Once
inserted, the new sensor may give rise to an unstable mutant
allele; stability can be restored by the elimination of either
of two sensors: one outcome results in a reversion to the
(stable) wild type allele and the other in a stable mutant
one.

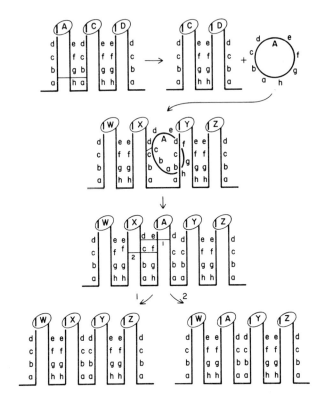

FIGURE 5. A diagram illustrating the possible fate of a sensor that is excised from a control region because
of an error in the pattern of overlapping palindromes. If the repetitive sequences of DNA at
many loci are homologous, the excised sensor may
become incorporated in a foreign locus, only to
produce an instability at that locus. The alternatives illustrated at the bottom of the figure
represent (1) the reversion of the mutable locus
to its original (wild type) state through the loss
of the foreign sensor, or (2) the stabilization of
the new mutant allele through the loss of a native
sensor (X) .

c. The palindromic configuration of operator loci.

The overlapping palindromes which have been postulated as a means by which adjacent sensors in a gene-control region may exchange places result in the rotation of the sensor with each move. In order that transcription of the gene can occur despite the rotation of the sensor region, it has been necessary to postulate further that the sensor itself is a palindrome. Maniatis and Ptashne (1973) and Gilbert and Maxam (1973) have in the meanwhile found that the lamda and lac operators do indeed have symmetrical, palindromic configurations.

d. The variable length of HnRNA associated with the mRNA of a given locus, and its variation among inbred strains.

The following account is a prediction rather than a report of past findings. The prediction is a mere restatement of previous claims: The same mRNA can be called for by different gene control systems each of which has a sensor in one of several sensor locations; therefore, the length of the transcribed RNA molecule will vary with the location of the replication initiation site (sensor). Furthermore, because the same sensor can take up various locations, different inbred lines, within which a given sensor may be found at different locations, should produce different sized HnRNA's in response to the same signal (Figure 6).

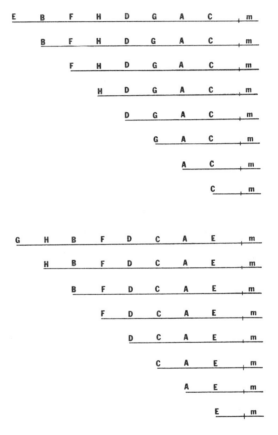

FIGURE 6. Predictions regarding the transcribed RNA (mRNA plus the still unprocessed heterogeneous nuclear RNA) made by the Wallace-Kass model. First, the amount of transcribed RNA attached to the mRNA will vary according to the signal which calls for gene action. Second, the amount of transcribed RNA attached to the mRNA of the two alleles following a given signal will frequently differ.

Coadaptation of different gene arrangements in local populations.

The dissimilar arrangement of sensors within the gene control regions of allelic genes can give rise to duplications
and deficiencies for specific sensors as a result of inter-
chromosomal recombination. The lack of a sensor at a given
locus would be a serious genetic defect. What devices are
available to organisms to extract the greatest advantage from
the dissimilar sequences with the least cost from recombination? Inverted gene sequences provide one such device.
Recombination within an inverted sequence leads to the formation of dicentric and acentric chromosomal fragments. The
dicentric chromosomes, crossover chromosomes within which the
duplication-deficiency control regions would be found, tend
to be eliminated from functional gametes (Sturtevant and
Beadle, 1936; McClintock, 1941). Thus, inversion polymorphisms may be regarded as means by which populations take
advantage of the proposed model. Furthermore, the advantage
is based on complementary sensor sequences at loci that are
carried by chromosomes with different gene arrangements and
which are built up within local populations or populations
inhabiting limited geographic regions. This account closely
resembles Dobzhansky's (1948) description of the coadaptation
of gene arrangements within local populations of Drosophila
pseudoobscura.

The neutralist-selectionist controversy

Given that a gene locus has a control region consisting of
a given number (n) of sensors, these can be ordered into a
finite number $(n!)$ of different sequences. These sequences,
in turn, have various frequencies $(p, q, r, s, \ldots)$.

Now, if at this locus there exist two alleles $(a_s$ and $a_f)$ whose gene products ($\underline{S}$ and $\underline{F}$) migrate at different rates in electrophoretic gels, individuals which are heterozygous for these alleles most likely possess a selective advantage even though the gene products themselves may be neutral or nearly so.

The basis for the above statement is the earlier claim that the possession of dissimilar sequences of sensors is selectively advantageous. Although the average frequency of the various sequential orders of sensors may be $\underline{p}, \underline{q}, \underline{r}, \ldots,$ the frequencies of sequences actually associated with the alleles $\underline{a}_s$ and $\underline{a}_f$ at a given moment may be written $\underline{p}_s$, $\underline{p}_f$, $\underline{q}_s$, $\underline{q}_f$, $\ldots$ where $\underline{p}_s \approx \underline{p} \pm \underline{d}_1$, $\underline{p}_f \approx \underline{p} \mp \underline{d}_1$, $\underline{q}_s \approx \underline{q} \pm \underline{d}_2$, $\underline{q}_f \approx \underline{q} \mp \underline{d}_2$, etc. The average proportion of individuals carrying identical sequences among $\underline{a}_s/\underline{a}_s$ and $\underline{a}_f/\underline{a}_f$ homozygotes equals $\underline{p}^2 + \underline{q}^2 + \underline{r}^2 + \ldots + (\underline{d}_1^2 + \underline{d}_2^2 + \underline{d}_3^2 \ldots)$ whereas that among $\underline{a}_s/\underline{a}_f$ heterozygotes equals $\underline{p}^2 + \underline{q}^2 + \underline{r}^2 + \ldots - (\underline{d}_1^2 + \underline{d}_2^2 + \underline{d}_3^2 \ldots)$. Thus, despite the possible (but, in my opinion, unlikely) neutrality of the gene products themselves, heterozygotes may nevertheless possess a selective advantage because of the composition of the associated gene control regions. Indeed, one could argue that loci with large numbers of sensors are more likely to possess allozyme polymorphisms than are loci at which the number of sensors (that is, the need for complex genetic control) is small.

On the release of variability through recombination

In a series of coordinated experiments (Spassky, et al., 1958; Dobzhansky, et al., 1959; Spiess, 1959; and Krimbas, 1961), Dobzhansky and his colleagues studied the mean viability and the variance in viability of flies (D. pseudo-obscura, D. prosaltaus, D. persimilis, and D. willistoni) homozygous for recombinant chromosomes obtained from females which were heterozygous for different pairs of quasi-normal chromosomes. The means of the recombinant homozygotes consistently regressed toward the mean viability of flies homozygous for quasi-normal chromosomes sampled from the original wild populations. Depending upon the species, from one third to two thirds of the total variation observed among the original sample of quasi-normal chromosomes were regenerated among the recombination products of the tested quasi-normals.

A similar experiment can be simulated using the five control systems and seven gene loci depicted in Figures 7 and 8.

REGULATORY SYSTEMS:	CALL FOR GENES:
A	1 , 3 , 4 , 5 , 7
B	2 , 3 , 4 , 5 , 7
C	1 , 2 , 3 , 6
D	1 , 2 , 4 , 6 , 7
E	2 , 3 , 5 , 6 , 7

a.

STRUCTURAL GENES:	POSSESS SENSORS FOR SYSTEMS:
1	A , C , D
2	B , C , D , E
3	A , B , C , E
4	A , B , D
5	A , B , E
6	C , D , E
7	A , B , D , E

b.

FIGURE 7. *A hypothetical example in which each of five control systems (A-E) calls for a different set of genes from among a total of seven loci (1-7). (a) The genes called for by each control system are identified. (b) The sensors that each gene must possess in order to satisfy (a) are listed.*

511

GENE	ASSOCIATED SENSORS	I	II	III	IV	V
1	A , C , D	C A D	C A D	A D C	A D C	A C D
2	B , C , D , E	B D E C	E D C B	E B D C	E C D B	C D B E
3	A , B , C , E	E A C B	B A C E	E B C A	A E C B	C B A E
4	A , B , D	D A B	A D B	A D B	D A B	D A B
5	A , B , E	E B A	B E A	B E A	B E A	B A E
6	C , D , E	D C E	D E C	D C E	C E D	E C D
7	A , B , D , E	D A B E	D A B E	E A B D	E B D A	B A D E

FIGURE 8. *The information listed in Figure 7b is repeated. In addition, five (I-V) combinations of randomly generated sequences of sensors at each of the seven loci are also listed (for explanation, see text).*

Figure 7a merely lists the five systems (A-E) and identifies the genes which each controls; in Figure 7b the same information is listed in the form of sensors which are needed in the control regions of each of the seven gene loci. In Figure 8 are shown five combinations of control regions whose sensor positions have been arrived at through the use of random digits.

In this hypothetical example, developmental time in arbitrary units is said to equal the sum of the times required for each control system to run its course. This time, in turn, is determined for each system by the remoteness from its structural gene of the most remote sensor; this is the limiting locus of each system. Thus, an individual that was homozygous for combination IV would require four time units for system A , four for B , three for C , three for D , and two for E ; total time in this case equals 16. For combination V, the elapsed times would be three for A , three for

B , two for C , three for D , and four for E ; the total equals 15. On the other hand, a heterozygote that carried both combination IV and V would require the following times for the operation of the five control systems: two for A , three for B , two for C , three for D , and two for E ; the total in this case equals 12. These numbers merely illustrate the earlier argument in which an advantage was claimed for individuals heterozygous for allelic control regions with dissimilar sequences of sensors.

Data of the sort computed in the preceding paragraph have been obtained by the use of a computer. One thousand (rather than five) combinations were generated (nearly 18 million are possible). The development time was then calculated for 1000 homozygotes (i/i) and for 999 heterozygotes (i/j where $j = i+1$) . The results are shown in Figure 9. The mean developmental time of heterozygotes is systematically lower than that of homozygotes. Similar pairs of curves are typical of viability, fertility, and developmental studies of Drosophila homozygous for chromosomes sampled from wild populations or heterozygous for random combinations of these same chromosomes.

Of the 1000 combinations of sensor sequences analyzed above, a sample of 100 homozygotes and 99 heterozygotes were printed out in detail by the computer. Twenty-one of the 100 homozygotes had developmental times of 16 arbitrary units. Three of these (the first, last, and center one from the printed list) were chosen as quasi-normal chromosomes which were to be permitted to recombine, and whose recombinant combinations were to be examined for their effects on viability. With seven loci (assumed to be linked to this study) there are six levels at which recombination can occur; consequently, three heterozygous combinations generate 36 recombinant "chromosomes".

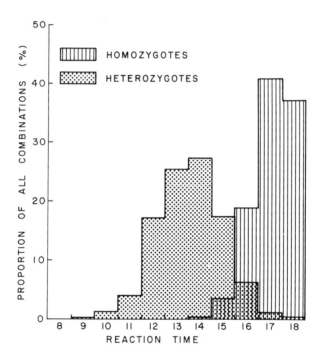

FIGURE 9. *The distributions of total reaction times for homozygous (I/I, II/II, etc.) and heterozygous (I/II, II/III, etc.) combinations of sensors such as those illustrated in Figure 8.*

The results of this study are shown in Figure 10. The mean developmental time of the 36 possible recombinants equals 16.61 units rather than the 16.00 of the three selected "chromosomes". Thus, the mean of the recombinant homozygotes had regressed toward the population mean of 17.07.

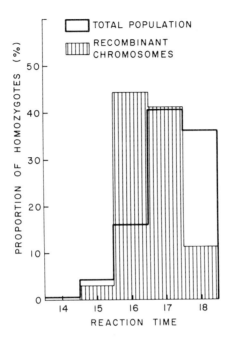

FIGURE 10. *The variation in total reaction times of homozygous recombinant chromosomes that were generated by crossing-over between three "chromosomes"
whose homozygous reaction times were 16. The
distribution labeled "total population" is that
of homozygotes shown in Figure 9.*

The variance in developmental time of the recombinant homozygotes equalled 0.516; that of the original population of
homozygotes equalled 0.750. The three tested quasi-normal
chromosomes, when permitted to produce recombinant chromosomes, regenerate two-thirds of the variance exhibited by the
entire population of homozygotes. In both the regression of
the mean and the release of variation, the simulated system
mimics closely the results observed in the case of D. pseudo-

obscura.

In two early studies on the release of viability through recombination (Dobzhansky, 1946; Wallace, et al., 1953), the viabilities of individuals homozygous for recombinant chromosomes frequently failed to form symmetrical distributions; symmetry would be expected if the viability effects of recombinant chromosomes merely reflected the gain or loss of relatively stable or predictable viability modifiers. In no instance did recombination in the simulated experiment generate symmetrical arrays of recombinant viabilities. The observed patterns were as follows:

	15	16	17	18
a/b	-	7	5	-
b/c	-	6	6	-
a/c	1	3	4	4

In this respect the simulated data also resemble earlier experimental observations.

On the loss of fitness under artificial selection

When subjected to artificial selection, populations of both plants and animals suffer from a loss of fitness. If selection is relaxed, fitness is restored; simultaneously, gains made under selection are lost to some extent, in some cases completely. Alternatively, if selection is continued for a considerable time, the fitness of the selected population improves. These phenomena were the subject of Lerner's (1954) essay entitled Genetic Homeostasis. The model for gene control that has been described above accounts for both the loss and, subsequently, the recovery of fitness under selection. The subsequent recovery of fitness is, of course, an essential element of natural selection because, otherwise,

continuous adaptation to a changing environment would occur
only at an enormous cost to the evolving population.

As a first approximation, we have assumed that sensors
take up random positions within allelic control regions.
While it is true that the heterozygosity which results from
randomization results in more efficient gene action than does
homozygosity for identical control regions, certain combinations of sensors are more efficient than others. The most
efficient combinations are those in which the arrangements of
sensors tend to be mirror images of one another (ABCD/DCBA).
On the other hand, such arrangements are those which are most
likely to yield (1) duplication-deficiency cross-over products and (2) the least efficient homozygotes if the loci
involved are not linked. Consequently, within any one
locality, selection should lead to an array of sensor
sequences which maximizes the fitness of surviving individuals.

Selection, either artificial or natural, can be viewed as
the imposition upon a population of a more-or-less urgent
need to meet demands which previously had been trivial or
even non-existent. These demands, in turn, can be visualized
as specifying that specific sensors occupy specific locations
within the control regions of various genes.

Because selection, visualized in this manner, favors a
small portion of the chromosomes originally present in the
population - a subset of chromosomes which in their entirety
previously generated the highest possible average fitness,
the average fitness of the selected population (when measured
according to conventional standards) must decrease.

No matter what demands are made regarding the position of
certain sensors with control regions of various gene loci,
there are arrangements of the sensors at other loci (and the
remaining sensors at the specified loci) which maximize the

average fitness of the selected population once more. These
events are illustrated in Figure 11.

GENE	(A) ORIGINAL CHROMOSOMES 17	18	19	(B) RECOMBINANTS GIVING MOST RAPID RESPONSES		(C) HIGH FITNESS RESTORED TO SELEC- TED POPULATION	
1	A C D	D A C	C A D	D A C	D A C	D A C	C A D
2	D B E C	E C B D	B C D E	E C B D	E C B D	E C B D	D B C E
3	E C A B	A C B E	B C E A	A C B E	A C B E	A C B E	A E B C
4	A D B	B A D	D B A	B A D	D B A	D B A	A B D
5	A B E	E B A	B A E	B A E	B A E	B A E	B E A
6	D E C	D C E	E C D	D E C	D E C	D E C	E D C
7	D E B A	A B E D	A D B E	D E B A	D E B A	D E B A	A B E D

FIGURE 11. *A hypothetical example illustrating the genetic
response of a population to a program of artificial selection and the further changes which
would lead once more to high fitness (small reaction time) for the selected population (for
explanation, see text).*

In generating the events which are illustrated in Figure
11, I first specified that the selection program would require that sensor A occupy position 1 of gene 3 , that
B would occupy position 1 of gene 5 , and that C occupy
position 3 of gene 6 .

Second, without examining the actual combinations of sensors, I determined from the computer printout that combinations 17-18 and 18-19 were among those with the shortest
total developmental times. The arrangements of sensors in
these three combinations are shown in Figure 11a.

By chance, the restriction imposed by selection can be met

by merely recombining portions of the three selected combinations (once more I treat the seven loci as if they are linked). Two obvious recombinants are shown in Figure 11b; each of these is equally likely and each satisfies the conditions that were stipulated in advance.

The developmental time of individuals homozygous for either of the selected recombinants or heterozygous for the pair is 18, a much greater time than that for 18/19 (12), 17/18 (11), or 17/19 (14). The greater developmental time corresponds to the loss of fitness commonly seen in selected populations.

On the other hand, the selected population can recover as shown in Figure 11c. Even if homozygosis for the positions of A , B , and C in genes 3 , 5 and 6 is necessary for selection to be most effective, the rearrangement of sensors at other loci can shorten developmental time of some selected individuals to 11, a considerable improvement over that of the individuals who responded immediately to the selective pressure. Consequently, although an immediate loss of fitness may accompany the response of a population to a novel regime of artificial (or natural) selection, high fitness may be restored by an appropriate re-shifting of sensors within gene control regions.

Concluding remarks

Upon hearing of the model I have described in this paper, a good friend shook his head and declared his lack of interest in what I was saying.

"Your model is much too powerful", he objected. "When gene action can be modified by the position of sensors within the control regions at individual loci and when sensors can move from locus to locus bringing any gene under the control of any regulatory system, one has a model that can explain everything."

The same objection, I have learned (Peter Thompson, personal communication), was raised by Professor Charles Zelleny against the side-chain theory of the structure of the gene proposed by David H. Thompson (1931). Modern techniques, however, are able to probe the structure of gene control regions. Much of what is being found seems, at this still early stage, to agree with predictions based on the Wallace-Kass model. Should the data continue to support the model, it would appear that the observed physical structure of DNA exists because it meets the demands for which the model was created, because it allows transcription initiation sites (sensors) to move about within and among gene control regions.

Under any model of developmental genetics, some kind of mutual adjustment or "coadaptation" of genes and gene products is required. The presently popular concern with electrophoretically detectable protein variation focuses our attention on the need for a harmonious action of differing substances. The Wallace-Kass model is considerably simpler because it deals primarily with the harmony that arises from the synchronous timing of interacting gene systems. On the face of it, harmonizing time seems a less formidable task than harmonizing the effects of complex molecules; indeed, temporal synchrony must precede material coadaptation under any postulated model.

REFERENCES

Britten, R.J. and E.H. Davidson. (1969). Science 165: 349-357.

Crick, Francis. (1971). Nature 234: 25-27.

Dobzhansky, Th. (1946). *Genetics* 31: 269-290.

Dobzhansky, Th. (1948). *Genetics* 33: 588-602.

Dobzhansky, Th., H. Levene, B. Spassky and N. Spassky. (1959). *Genetics* 44: 75-92.

Dubinin, N.P. (1932). *J. Genet.* 26: 37-58.

Gilbert, W. and A. Maxam. (1973). *Proc. Nat. Acad. Sci. U.S.* 70: 3581-3584.

Krimbas, C.B. (1961). *Genetics* 46: 1323-1334.

Lerner, I.M. (1954). *Genetic Homeostasis.* Oliver and Boyd, Edinburgh.

Lewontin, R.C. (1965). *Proc. 11th Int. Cong. Genet.* 3: 517-525.

Maniatis, T. and Mark Ptashne. (1973). *Nature* 246: 133-136.

McClintock, B. (1941). *Genetics* 26: 234-282.

Muller, H.J. (1929). *Proc. Int. Cong. Plant Sciences* 1: 897-921.

Spassky, B., N. Spassky, H. Levene and Th. Dobzhansky. (1958). *Genetics* 43: 844-867.

Spiess, E.B. (1959). *Genetics* 44: 43-58.

Sturtevant, A.H. and G.W. Beadle. (1936). *Genetics* 21: 554-604.

Thompson, D.H. (1931). *Genetics* 16: 267-290.

Wallace, B. (1975). *Evolution* 29: 193-202.

Wallace, B. and T.L. Kass. (1974). *Genetics* 77: 541-558.

Wallace, B., J.C. King, C.V. Madden, B. Kaufmann and E.C. McGunnigle. (1953). *Genetics* 38: 272-307.

PART III THEORETICAL STUDIES

The emphasis in this part is on the consequences and implications derived from theoretical models based primarily on
analytic methodology. The topics cover migration selection
interaction, selection effects varying randomly in time, the
rate of advance of a mutant, aspects of modifier genes and
multi-locus systems, density dependent selection, genetic distance and speciation, identity by descent, altruism and kin
selection.

A brief summary of the individual papers is now presented.

Cockerham exploits the methodology of identity by descent
to evaluate the number of heterozygotes left by neutral genes
for a number of mating patterns.

A model of altruistic or spiteful relations between individuals in populations is developed by Eshel and Cohen taking
into account genetical relatedness, inherent competivity, and
reproductive potential of the individuals concerned. The
model extends and refines ideas of Hamilton. Depending on
initial conditions, cases exist where both altruism and spite
can be established. It is also shown that stable altruism is
possible involving unrelated individuals.

Feldman and Krakauer review and present some new results on
the recent theory of selectively neutral genes which modify
evolutionarily important features of other gene loci. Among
the new results is one on polymorphisms for neutral genetic
modifiers. The theory of modification of segregation distortion is discussed including some findings pertaining to the
evolution of the sex-ratio in the XX-XY system of sex determination.

Hadeler summarizes recent work on the problem of the rate

of advance of an advantageous mutant. Parallels are drawn to problems pertaining to the spread of an epidemic and nerve transmission.

Hartl and Cook contribute to the study of the evolution of infinite diploid populations subject to selection intensities that can vary systematically and randomly in time. A number of calculations of cases of stochastic polymorphisms are carried out. The nature of the dependence on the parameters are highlighted and interpreted.

A multi deme population genetics model involves three principal components: The environmental selection gradient, the migration patterns, and population structure. Karlin introduced two concepts for comparing forms of environmental heterogeneity. The comparison of hard versus soft selection fits into this framework. The influence of migration structure and variable deme sizes on the maintenance of a protected polymorphism are also analyzed.

The paper of Karlin and Richter-Dyn seeks to clarify, for cline models, the interrelations between patterns of geographically differentiated gene frequency and the background selection and gene flow structure. In particular, the effects of unequal local rates of migration, different relative deme sizes, the multiplicity of demes, the length of the neutral region and other factors are investigated.

Matessi and Jayakar analyze a mathematical model in which competition is represented by the classical Lotka-Volterra system thereby entailing selective pressures both density and frequency dependent. The results are interpreted as a process of niche expansion and a theoretical description of "ecological release".

Speciation depends on the acquisition of reproductive isolation and ecological compatibility between the diverging

populations. Nei proposes an evolutionary scheme of ethological isolation by means of random fixation of mutant genes.
Some mathematical models of genetic differentiation of populations are also presented taking into account mutation,
random genetic drift, and selection. The application of the
theory to protein polymorphism suggests that the number of
electrophoretically detectable codon differences per locus
(genetic distance) is generally less than 0.05 between races,
0.02 ~ 0.05 between sub-species, 0.1 ~ 2.0 between species
and more than 1.0 between genera.

Slatkin proposes a simple model in order to calculate the
difference in the times until fixation of an advantageous
allele in the different parts of a population's range. It is
argued that occasional long distance migrants can be much
more effective than short distance migrants in dispersing an
advantageous allele.

Strobeck examines a special multiplicative viability regime
with 3 loci. Comparisons are made between the stability conditions for the internal Hardy-Weinberg equilibrium having all
three loci segregating and the boundary Hardy-Weinberg equilibrium involving only two segregating loci. Implications for
the crystallization of the genome are suggested.

Number of Heterozygotes Left by Neutral Genes*

C.C. COCKERHAM

That the number of heterozygotes left by a neutral gene in finite populations is invariant of the structure (Maruyama, 1971) is not an obvious result. Maruyama used what could be construed to be a haploid model and introduced migration in a stochastic manner such that the same individual or gene could migrate to more than one colony. In this note it is shown that his conclusion holds when individuals actually migrate in an arbitrary fashion. I shall consider also the cases of two gametes per parent, separate sexes, and the avoidance of mating relatives.

In all cases generation time is discrete and the total population size, N , remains constant over time, indexed by t . Solutions are much simplified if one can find an average identity by descent measure, G , such that

*Contribution from the North Carolina Agricultural Experiment Station. This investigation was supported by NIH research grant number GM 11546 from the National Institute of General Medical Sciences.

$$G^{(t+1)} = G^{(t)} + \frac{1-F^{(t)}}{2X} \tag{1}$$

where F is the average of the N inbreeding coefficients of individuals and X is some number. Then, the total number of heterozygotes,

$$H = \sum_{t=0}^{\infty} N\left(1-F^{(t)}\right) ,$$

left by an initial mutant gene during the course of fixation, without further mutation, is found simply as

$$H = 2XN\left(G^{(\infty)} - G^{(0)}\right) .$$

One candidate for G is the mean identity measure (F_{ℓ} in Cockerham, 1970) for all pairs of alleles,

$$G^{(t)} = \frac{F^{(t)}}{2N-1} + \frac{2N-2}{2N-1} \bar{\theta}^{(t)} , \tag{2}$$

where $\bar{\theta}$ is the average of the identities by descent for all $2N(N-1)$ pairs of alleles among individuals. Note that G is the actual probability of a random pair of alleles in the population being identical by descent. An alternate representation of G (denoted as Ψ in Cockerham, 1970) takes into account the gametic sampling in the transition from one generation to the next,

$$G^{(t+1)} = \frac{1+F^{(t)}}{2N_e} + \left(1 - \frac{1}{N_e}\right)\bar{\theta}^{(t)} , \tag{3}$$

where N_e is the reciprocal of the probability of a random pair of gametes having come from the same parent, and is the gametic variance effective number (Wright, 1938). Strictly speaking, formula (3) is valid only if gametes of all parents are identically distributed each generation which is sufficient for the purpose it is to be used here. The transition (3) - (2) produces

$$G^{(t+1)} - G^{(t)} = \frac{1-F^{(t)}}{2N_e} + \left(F^{(t)} - \overline{\theta}^{(t)} \right) \left(\frac{1}{N_e} - \frac{1}{2N-1} \right) \ . \tag{4}$$

If $F = \overline{\theta}$ or $N_e = 2N-1$

$$H = 2NN_e \left(G^{(\infty)} - G^{(0)} \right) \ .$$

$G^{(\infty)}$ will be one if there is not complete splitting of the population as noted by Maruyama (1971). With a single mutant gene initially, and all other alleles identical,

$$G^{(0)} = F^{(0)} = \overline{\theta}^{(0)} = 1 - \frac{1}{N} \ .$$

Consequently, $H = 2N_e$. With two gametes per parent each generation, $N_e = 2N-1$; thus $H = 4N-2$ for any arbitrary structuring or subdivisions as long as there is some migration to avoid permanent sublines. No structuring whatsoever is required for $F = \overline{\theta}$, and then the number of heterozygotes varies according to the gametic variance effective number. In general, subdivisions with reproduction within them cause $F > \overline{\theta}$ and an increase in N_e , but as long as $N_e < 2N-1$ the formulation given by (4) is of little help.

Turn now to an arbitrarily structured population with $N_i^{(t)}$ individuals born in colony i in generation t . $(\sum_i N_i^{(t)} = N)$. The average identity by descent measures are related to those within and among colonies as

$$NF^{(t)} = \sum_i N_i^{(t)} F_i^{(t)} \tag{5}$$

$$N(N-1)\overline{\theta}^{(t)} = \sum_i N_i^{(t)} \left(N_i^{(t)} - 1 \right) \overline{\theta}_{ii}^{(t)} + 2 \sum_i \sum_{i<i'} N_i^{(t)} N_{i'}^{(t)} \overline{\theta}_{ii'}^{(t)}$$

where $F_i^{(t)}$ is the average inbreeding coefficient within colony i , and $\overline{\theta}_{ii}^{(t)}$ and $\overline{\theta}_{ii'}^{(t)}$ is the average coancestry of individuals within colony i and between colonies i and i' , respectively.

We now let these individuals migrate in some arbitrary fashion to form colonies of $N_j^{(t+1)}$ individuals. ($\sum_j N_j^{(t+1)}$ = N) . If ^'s distinguish the identity measures after migration, comparable relationships to those in (5) are

$$N\hat{F}^{(t)} = \sum_j N_j^{(t+1)} \hat{F}_j^{(t)}$$

(6)

$$N(N-1)\hat{\bar{\theta}}^{(t)} = \sum_j N_j^{(t+1)}\left(N_j^{(t+1)} - 1\right)\hat{\theta}_{jj}^{(t)} + 2 \sum_{j<j'} \sum N_j^{(t+1)} N_{j'}^{(t+1)}\hat{\theta}_{jj'}^{(t)} .$$

However, the average identity by descent of genes within and among individuals has not changed with migration, so that $\hat{F}^{(t)} = F^{(t)}$ and $\hat{\bar{\theta}}^{(t)} = \bar{\theta}^{(t)}$ leading to the obvious equalities between (5) and (6).

After migration we assume that reproduction is entirely within each colony, that it follows the rules of a randomly mating monoecious population with equal chance gametic contributions of each parent, and that the same number of offspring is produced as there are parents in a colony. With these specifications, the well-known transition in identity measures is

$$F_j^{(t+1)} = \bar{\theta}_{jj}^{(t+1)} = \frac{1+\hat{F}_j^{(t)}}{2N_j^{(t+1)}} + \left(1 - \frac{1}{N_j^{(t+1)}}\right)\hat{\theta}_{jj}^{(t)} .$$

(7)

The appropriate G in this case is the one used by Maruyama (1971),

$$G^{(t+1)} = \sum_j \sum_{j'} N_j^{(t+1)} N_{j'}^{(t+1)} \bar{\theta}_{jj'}^{(t+1)} / N^2 ,$$

although it has no simple interpretation in terms of probability of identity by descent. Since reproduction is entirely within colonies, the average coancestry of individuals between colonies is the same for the offspring as for their parents, i.e., $\bar{\theta}_{jj'}^{(t+1)} = \hat{\theta}_{jj'}^{(t)}$. Making use of these equalities and of

(7), and then of the equalities between (5) and (6) and of
$F_i^{(t)} = \bar{\theta}_{ii}^{(t)}$,

$$G^{(t+1)} = G^{(t)} + \frac{1-F^{(t)}}{2N} , \tag{8}$$

which is the same result obtained by Maruyama (1971), and

H = 2N .

An alternate derivation, similar to that of Maruyama except it is for the migration of individuals, is given in an Appendix.

Separate sexes provide another kind of subdivision. With unequal numbers, N_m and N_f , of each sex ($N_f + N_m = N$) , equal chance gamete contributions and no other form of subdivision, we may use

$$G^{(t)} = \frac{F^{(t)}}{N_s} + \bar{\theta}_{mf}^{(t)}$$

and

$$G^{(t+1)} = G^{(t)} + \frac{1-F^{(t)}}{2N_s} \searrow$$

where $N_s = 4N_m N_f/N$ and $\bar{\theta}_{mf}^{(t)} \left(= F^{(t+1)} \right)$ is the average co-ancestry between males and females. If the initial mutant occurs in the females, $F^{(0)} = 1 - 1/N$ and $\bar{\theta}_{mf}^{(0)} = 1 - 1/2N_f$ leading to $G^{(0)} = 1/N_s - 1/NN_s + 1 - 1/2N_f$, while $G^{(\infty)} = 1+1/N_s$. Consequently, $H = 4N_m + 2$, and the number of descendent heterozygotes is dependent on the number of individuals in the sex opposite to that in which the mutant occurred. If we consider the mutant to occur in each sex in proportion to the number of individuals and average over these two types of events, $H = 2N_s + 2$, and of course with equal numbers for each sex, $H = 2N+2$. The introduction of subdivisions is unmanageable except for the rather artificial situation where the numbers of males and of females are the same in each colony, and migration is such that the coancestries of males, of females, and of females with males are the same in each colony and for any pair of colonies. Then, $H = 2N+2$.

Separate sexes avoid the mating of 1st degree relatives (selfing). By subdivision or other means one can avoid mating relatives in controlled breeding programs up to a certain degree (Cockerham, 1970). If the population is subdivided into k equal sized colonies which are mated to each other in a fashion to avoid inbreeding as much as possible, with equal chance random mating of individuals between the colonies, it can be shown that

$$H = 2N+2k-2 \quad .$$

This result of course includes separate sexes, $k=2$. With maximum avoidance, $k=N$, and $H=4N-2$. However, this requires two gametes per parent for which we already knew the outcome.

COMMENTS

Kimura and Crow (1963) showed that the total heterozygosity during the course of fixation was the same for randomly mating monoecious and dioecious populations with two gametes per parent each generation. It is for two gametes per parent that the most general type of result obtains: that the total heterozygosity or number of heterozygotes left by a neutral mutant gene is invariant with the system of mating, subdividing, and migrating as long as there is not permanent splitting of the population. The number of descendent heterozygotes is just twice the gametic variance effective number, $2N-1$.

Without any subdivision and random mating, the number of heterozygotes is twice the gametic variance effective number,

N_e . With subdivisions, and other than two gametes per parent, the gametic variance effective number or other effective numbers, inbreeding or gene frequency variance, are not of any help. With k colonies and equal chance gamete formation of twice the number of gametes as are individuals in each colony, N_e = N(2N-1)/(2N-k) . Yet, with random mating within colonies and arbitrary migration, H=2N . Here the rate of inbreeding within colonies is determined by and is inversely related to the rate of migration, and the total heterozygosity remains constant. Completely random migration of individuals to any colony simply leads to panmixia. On the other hand, if super migration is resorted to by crossing k colonies to avoid inbreeding as much as possible, then H = 2N+2k-2 increases as the number of colonies and avoidance of inbreeding increase. Here, genes are forced to migrate through $\log_2 k$ generations before coming together again. At the limit of this super migration or avoidance, two gametes per parent occurs for which H is not affected by subdivision, migration, or avoidance of mating relatives.

Separate sexes avoid self-fertilization and without further subdivisions, H = $2N_s$+2 on the average, where N_s is the well-known modification of N for unequal numbers of each sex. While the further inclusion of arbitrary subdivisions and migration, with random mating within colonies, appears to be an intractable situation, it seems safe to conjecture that the total heterozygosity or number of heterozygotes left by a mutant gene will be altered little if any.

SUMMARY

The number of heterozygotes, H , left by neutral genes as finite populations progress towards fixation was studied as it relates to gametic variance effective number, N_e, subdivision and migration, separate sexes and avoidance of mating relatives. Without subdivisions in a randomly mating monoecious population $H = 2N_e$. With two gametes per parent $N_e = 2N-1$, but for this situation any system of mating, subdividing, and migrating does not affect H as long as there is no permanent splitting of the population. With arbitrary subdivision and migration of individuals among colonies, and random mating within colonies, $H=2N$ as found by Maruyama (1971). If migration is increased beyond that for panmixia by mating individuals of different colonies so as to avoid inbreeding as much as possible, $H = 2N+2k-2$ for k colonies. This includes separate sexes, $k=2$, and with modification of N for unequal numbers in each sex, $H=2N_s+2$.

APPENDIX
Alternate Derivation of G with Migration

Let $n_{ij}^{(t)}$ be the number of individuals that migrate from colony i to colony j to participate in reproduction.

$$\sum_i n_{ij}^{(t)} = N_j^{(t+1)} \quad , \quad \sum_j n_{ij}^{(t)} = N_i^{(t)}$$

Since any individual reproduces in only one colony, the average identity by descent after reproduction between two distinct colonies is

$$\bar{\theta}_{jj'}^{(t+1)} = \sum_{\substack{j \neq j'}} \sum_{i \; i'} \frac{n_{ij} n_{i'j'} \bar{\theta}_{ii'}^{(t)}}{N_j^{(t+1)} N_{j'}^{(t+1)}} \quad .$$

Within a colony, j for example, the probability of a pair of gametes involving parents from the same colony before migration, i for example, is $\left(n_{ij}^{(t)}\right)^2 \Big/ \left(N_j^{(t+1)}\right)^2$. The probability that the gametes are from the same parent is $1/n_{ij}^{(t)}$

and from different parents is $1-1/n_{ij}^{(t)}$, with respective

probabilities of identity by descent of $\left(1+\overline{\theta}_{ii}^{(t)}\right)\!/2$ and

$\overline{\theta}_{ii}^{(t)}$. Consequently,

$$\overline{\theta}_{jj}^{(t+1)} = \frac{\sum\limits_{i}\sum\limits_{i'} n_{ij}^{(t)}\, n_{i'j}^{(t)}\, \overline{\theta}_{ii'}^{(t)}}{\left(N_j^{(t+1)}\right)^2} + \frac{\sum\limits_{i} n_{ij}^{(t)}}{\left(N_j^{(t+1)}\right)^2}\; \frac{1-\overline{\theta}_{ii}^{(t)}}{2} \quad .$$

Forming $G^{(t+1)}$,

$$G^{(t+1)} = \left[\sum_{i}\sum_{i'}\sum_{j}\sum_{j'} n_{ij}^{(t)}\, n_{i'j'}^{(t)}\, \overline{\theta}_{ii'}^{(t)} + \sum_{i}\sum_{j} n_{ij}^{(t)} \left(1-\overline{\theta}_{ii}^{(t)}\right)\!/2\right]\!/N^2$$

$$= \left[\sum_{i}\sum_{i'} N_i^{(t)}\, N_{i'}^{(t)}\, \overline{\theta}_{ii'}^{(t)} + \sum_{i} N_i^{(t)} \left(1-\overline{\theta}_{ii}^{(t)}\right)\!/2\right]\!/N^2$$

$$= G^{(t)} + \frac{1-F^{(t)}}{2N} \quad ,$$

the relationship at (8) is obtained.

REFERENCES

Cockerham, C.C. (1970). In Biomathematics, Vol.1, Mathematical Topics in Population Genetics. K. Kojima Ed., 104-127. Springer-Verlag.

Kimura, M. and J.F. Crow. (1963). Genet. Res. Camb. 4: 399-415.

Maruyama, T. (1971). Genet. Res. Camb. 18: 81-84.

Wright, S. (1938). Science 87: 430-431.

Altruism, Competition and Kin Selection in Populations

I. ESHEL and D. COHEN

INTRODUCTION

The concept of inclusive fitness introduced by Hamilton
(1964) has provided a general basis for the understanding of
selection in populations of genetically related individuals.
As further developed (Hamilton, 1970), it could account for
the evolution of both altruism and spite between individuals
in a population. According to Hamilton, selection will favor
a limited risk-taking by an individual in order to save the
life of an identified relative, when the risk is less than
the excess of genetical relatedness between them above the
mean genetical relatedness in the population. Risk taking
would be expected to be equal between individuals with equal
genetic relatedness. It seems, however, that parental care
and self sacrifice is much stronger and commoner than mutual
help between sibs, inspite of the fact that the genetical
relatedness is identical in both cases. In some extreme
cases there may be spiteful relations between sibs. It is
very likely that the difference in reproductive potential and
in the expected competitivity may account for this difference.

This paper is a further development of the general approach

---

* Department of Statistics, University of Tel-Aviv, Ramat-
Aviv, Israel.

** Department of Botany, Hebrew University, Jerusalem, Israel.

of Hamilton, which takes into account, in addition to genetic relatedness, the inherent competitivity and the reproductive potential of the individuals concerned.

An important new contribution of our model is the consideration of high order reciprocal relationships between individuals with altruistic or spiteful interactions. These reciprocal relations may lead to altruism between unrelated individuals which have a common relative, and in an extreme case, to altruism between unrelated individuals without a common relative, provided that the mutual dependence is strong enough. Reciprocation may also amplify spiteful relations, and in general, allows two or more stable states at different levels of altruism or spite.

THE MODEL

Consider any two individuals i and j with a genetic relatedness a_{ij}, in a group of relatives, $1,\ldots,k,\ldots,n$. Define the competitive effect of j on i,

$$c_{ij} = \frac{W_i - W_{i/j}}{W_{i/j}} \tag{1}$$

or:

$$W_i = W_{i/j}(1 - c_{ij}) \tag{2}$$

where W_i is the expected number of offspring of i in the presence of j, and $W_{i/j}$ is the expected number in the absence of j.

The inclusive fitness of the genes of i in the presence of a relative j and of other relatives is:

$$g_i = W_i + \sum_{k\neq i} a_{ik} W_k .$$

With the death of j, the inclusive fitness of i is:

$$g_{i/j} = W_{i/j} + \sum_{\substack{k \neq i \\ k \neq j}} a_{ik} W_{k/j} \; .$$

Thus, the overall change in the inclusive fitness of $\underline{i}$ when $\underline{j}$ dies is:

$$\Delta g_{i/j} = W_i - W_{i/j} + a_{ij} W_j - \sum_{\substack{k \neq i \\ k \neq j}} a_{ik}(W_{k/j} - W_k) \qquad (3)$$

If an individual $\underline{j}$ is in danger of death, and it can be rescued by $\underline{i}$ with a risk r_{ij} that $\underline{i}$ itself will die, the loss in inclusive fitness of $\underline{i}$ because of the risk is:

$$L_{ij} = r_{ij}[W_i - a_{ij}(W_{j/i} - W_j) - \sum_{\substack{k \neq i \\ k \neq j}} a_{ik}(W_{k/i} - W_k)] \; . \qquad (4)$$

As long as the loss in inclusive fitness of $\underline{i}$ because of the risk is less than the gain in the inclusive fitness of $\underline{i}$ in the presence of $\underline{j}$, there will be selective advantage for taking the risk. The maximal risk will be reached when the loss is equal to the gain.

The maximal risk is:

$$r_{ij}^{*} = \frac{W_i - W_{i/j} + a_{ij} W_j - \sum_{\substack{k \neq i \\ k \neq j}} a_{ik}(W_{k/j} - W_k)}{W_i - a_{ij} W_{j/i} + a_{ij} W_j - \sum_{\substack{k \neq i \\ k \neq j}} a_{ik}(W_{k/j} - W_k)} \qquad (5)$$

r_{ij}^{*} is a decreasing function of $W_{i/j}$ and an increasing function of $W_{j/i}$ and a_{ij}. The quantity under the summation is the decrease in inclusive fitness contributed by $\underline{i}$ or $\underline{j}$ by the competitive effect they have on their relatives. When $r_{ij}^{*} < 0$, it means spite. It is the risk taken by $\underline{i}$ to decrease the survival of $\underline{j}$. r_{ij}^{*} is infinitely high when the loss function is negative, since its inclusive fitness will increase if it commits suicide.

Considering the simplest case of only two related individuals, i and j, (5) reduces to:

$$r^*_{ij} = \frac{W_i + a_{ij}W_j - W_i}{W_i + a_{ij}W_j - a_{ij}W_{j/i}} \qquad (5.1)$$

or, using the definition of C_{ij},

$$r^*_{ij} = \frac{W_i + a_{ij}W_j - \dfrac{W_i}{1 - C_{ij}}}{W_i + a_{ij}W_j - \dfrac{a_{ij}W_j}{1 - C_{ji}}} \; . \qquad (5.2)$$

r^*_{ij} is an increasing function of the reproductive potential of j and of the genetic relatedness between i and j. It is a decreasing function of the reproduction potential of i. r^*_{ij} is a decreasing function of C_{ij} and an increasing function of C_{ji}.

$r^*_{ij} > 0$ when:

$$a_{ij} > \frac{W_{i/j} - W_i}{W_j} \quad \text{or} \quad a_{ij} > \frac{W_i}{W_j}\frac{C_{ij}}{1 - C_{ij}} \qquad (6)$$

when the loss function L_{ij} is positive; otherwise r^*_{ij} is infinitely high.

When possible altruistic or spiteful interaction is with a subset of n individuals out of the whole population, without discrimination of particular individuals, the loss in the expected reproduction of the genes of i through the death of all n individuals is:

$$\Delta g_{i/n} = W_i + \sum_{k \neq i} a_{ik}W_k - W_{i/n} \; .$$

The loss in inclusive fitness because of the risk r_{in} is:

$$L_{in} = r_{in} \left[W_i - \sum_{k \neq i} a_{ik} (W_{k/i} - W_k) \right] .$$

In this case,

$$r_{in}^* = \frac{W_i + \sum_{k \neq i} a_{ik} W_k - W_{i/n}}{W_i + \sum_{k \neq i} a_{ik} W_k - \sum_{k \neq i} a_{ik} W_{k/i}} . \qquad (7)$$

The size of the subset may be determined by the range of a territorial predator, or by the mobility range of the individuals during their feeding and reproductive activities, or by the diffusion of the products of their metabolism.

We may note some special cases of inequality (6).

1. In the symmetrical case $W_i = W_j$, $r_{ij}^* > 0$ when

$$a_{ij} > \frac{C_{ij}}{1 - C_{ij}} .$$ In this case the interaction will be altruistic or spiteful, depending on whether the relatedness is higher or lower than the competitivity.

2. $r_{ij}^* > 0$ when W_i is small or when $C_{ij} = 0$ if all the other parameters are positive. $r_{ij}^* < 0$ when $W_j = 0$, or when $a_{ij} = 0$ when all the other parameters are positive. These extreme cases are quite obvious intuitively.

3. $a_{ij} = 1$ between members of a clone. Altruism between them is of course much more likely than between individuals of lower genetic relatedness. There is, however, the possibility of spite between members of a clone either in the very asymmetrical cases of $W_j \ll W_i$, or in the destructive case of $C_{ij} \sim 1$.

4. In extremely asymmetrical situations it is quite possible to have $r_{ij}^* > 0$ and $r_{ji}^* < 0$. Such, for example, would be the case when $W_i \gg W_j$ or when $C_{ij} \ll C_{ji}$. This would be a situation when one of the two individuals is

nearing the end of its reproductive period, but is still competing for limiting resources with a younger relative with a high reproductive potential. It should be noted, however, that this conclusion is reached without regard to the reciprocal relations of mutual altruism between such individuals, which is discussed below.

In a well mixed closed habitat with N individuals,

$$W_{i/j} = W_i \frac{N}{N-1} \quad .$$

Substituting in (5) and rearranging gives:

$$r^*_{ij} = \frac{a_{ij} - \overline{a}_i}{1 - \overline{a}_i}$$

where $\overline{a}_i = \frac{1}{N} \sum a_{ik}$, including $a_{ii} = 1$. $r^*_{ij} > 0$ when $a_{ij} > \overline{a}_i$; $r^*_{ij} < 0$ when the opposite is the case. This result is similar in principle to that reached by Hamilton (1970).

An interesting case of $r^*_{ij} > 0$ and $r^*_{ji} < 0$ can arise in such a situation when $\overline{a}_j < a_{ij} < \overline{a}_i$, i.e. when $\underline{j}$ is weakly related to $\underline{i}$ and is not related to other individuals, while $\underline{i}$ is strongly related to many other individuals.

When relatedness in the population cannot be identified, and it has some average value $a_{ij} < 1$, or when all individuals are equally related,

$$\overline{a}_i = \frac{1 + (N-1) a_{ij}}{N} \quad ,$$

and we get that $r^*_{ij} = \frac{-1}{N-1}$, which is always negative, i.e. spite, irrespective of the average relatedness. The absolute value of this spite decreases as N increases, becoming negligible at large N .

Real habitats are never perfectly closed. If we assume that there is an average probability ϵ for each individual

to send an offspring to other habitats, then $r_{ij}^* > 0$ if
$(N-1) \in a_{ij} > 1 - a_{ij}$.

THE EFFECT OF RECIPROCATION ON THE
SELECTIVE ADVANTAGE OF ALTRUISM AND SPITE

It is intuitively clear that altruism increases and spite decreases the contribution which one individual has on the inclusive fitness of another individual and thus increase or decrease the optimal risk. We call these mutual effects between interacting individuals reciprocation. We do not consider reciprocation in the sense of Trivers (1971) in which reciprocal altruism is the result of a binding contract between individuals to repay in kind for help received.

Considering the simpler case of only two related individuals, we introduce an increasing function $f_{ji}(r_{ji})$ which is the factor by which W_i is multiplied when a risk r_{ji} is taken by j ; $f(0) = 1$. So in this case,

$$r_{ij}^* = \frac{W_i f_{ji}(r_{ji}^*) + a_{ij} W_j f_{ij}(r_{ij}^*) - W_{i/j}}{W_1 f_{ji}(r_{ji}^*) + a_{ij} W_j f_{ij}(r_{ij}^*) - a_{ij} W_{j/i}} . \qquad (8)$$

In the simplest symmetrical case of two individuals with
$W_i = W_j$, $C_{ij} = C_{ji} = C$, $r_{ij}^* = r_{ji}^* = r$, $a_{ij} = a$, and
$f_{ij} = f_{ji} = 1 + \alpha r$, we get from (8),

$$r = \frac{(1-C)(1+\alpha r)(1+a)-1}{(1-C)(1+\alpha r)(1+a)-a} . \qquad (9)$$

For small values of α the effect of reciprocation is to increase the level of r when it is positive, and to decrease it when it is negative. Reciprocation in this case amplifies the altruistic or spiteful relations expected without it.

From (9) we get the quadratic form

$$\alpha(1-c)(1+a)r^2 + [(1-\alpha)(1-C)(1+a)-a]r+1-(1-C)(1+a) = 0 . \qquad (10)$$

As long as $(1-C)(1+a) > 1$, there are one positive and one negative roots to equation (10). This is the case when competitivity is greater than the genetic relatedness.

When $(1-C)(1+a) < 1$, there are two positive roots when $a > (1-\alpha)(1-C)(1+a)$, which is always true when $\alpha > 1$, and there are two negative roots when $a < (1-\alpha)(1-C)(1+a)$ provided that the discriminant of equation (10) is positive.

A specially interesting case is of two unrelated individuals, i.e. $a = 0$. In this case the discriminant is positive when $\frac{C}{1-C} < \frac{(1-\alpha)^2}{4\alpha}$, which holds when α is very small or quite large. When this is the case, the two roots are positive when $\alpha > 1$, and negative when $\alpha < 1$.

Since genetic relatedness is not required when this condition holds, altruistic relations may become possible between individuals of different species, i.e. symbiosis. It is important to note that no binding contract or gratitude are assumed. The advantage for the helper is that it helps to maintain alive an individual which helps it to survive.

In any of the above cases it seems that any one solution is probably stable over some local neighborhood. Thus, it seems likely that the actual behavior will depend on the initial conditions in the relationship. Very likely, an initially high level of altruism in one individual may select for an increase in altruism in the second individual until a high stable level is reached. Conversely, an initially low level of altruism or a high level of spite may select for further decrease of altruism or for increased spite. This is very likely the explanation for the extremely high aggressiveness found between some insects' larvae (quoted in Hamilton, 1970). It could also explain the great differences in the social cooperation between individuals in some fairly closely related species. Cases of cooperative association between unrelated individuals are known in slime molds, in algal zoospores, and in queen ants (Hamilton, 1964).

DISCUSSION

Slatkin: One thing which bothers me about such reciprocation models is that although the model may explain the social behavior in a given population, I do not see how it can explain the evolution of such a situation.

Eshel: Our model, unlike that of Trivers, has to be understood in genetic terms. Specifically, once a direct kin selection results in fixation of genes which determine a certain optimal level of mutual help among individuals of a given relatedness, then the mutual importance of such relatives in increasing each others' inclusive fitness is also increased. This in turn increases the optimal selected level of mutual help among individuals of the same relatedness and so on. In this case, the limit of the process is given by the smallest positive solution nearest to the Hamilton solution.

Cohen: In general, there will be several solutions to r_{ij} under reciprocation. Selection will increase or decrease the frequencies of altruistic genes according to their frequencies in the population. The final stable level will depend on the initial frequencies. These may depend on some slight differences in the environment, and in small populations may sometimes result from random fluctuations. Thus, the degree of altruism or spite is expected to vary enormously between species with fairly similar ecologies.

Cockerham: What degree of intelligence is required for the evolution of altruism?

Cohen: Fairly complex patterns of behavior are known in organisms with little or no intelligence. It is sufficient that such behavioral patterns are genetically determined. Cooperative interactions are found in very primitive organisms. I am experimenting now with selection in bacterial populations for and against the "altruistic" character of secretion of extracellular enzymes which digest polymeric substances and convert them into soluble substrates for the whole population.

REFERENCES

Hamilton, W.D. (1964). J. Theor. Biol. 7: 1-52.

Hamilton, W.D. (1970). Nature 228: 1218-1220.

Trivers, R.L. (1971). Quart. Rev. Biol. 46: 35-57.

Genetic Modification and Modifier Polymorphisms*

M. W. FELDMAN and J. KRAKAUER

1. INTRODUCTION

For almost forty years after Fisher introduced his theory
of the evolution of dominance in 1928 evolutionary geneticists have understood genetic modification to be the result
of selection for alleles at one or more (modifier) loci which
modify the action of natural selection at other (major) loci.
There are probably two main reasons for the wider interpretation of genetic modification over the past ten years. On
the one hand our knowledge of gene action has developed
greatly. There are well documented and mapped loci which
control recombination and mutation, and which are valuable
experimental tools. On the other hand the interactive theory
of selection and linkage between two or more loci has seen
substantial growth during the same period. These two stimuli
have made it both reasonable and feasible to incorporate the

*
Research supported in part by a grant from the United States-
Israel Bi-National Science Foundation; by Grant GB 37835
from the National Science Foundation and by NIH Grant
USPHS GM 10452-12.

547

greater variety of experimentally validated phenomena into the models of theoretical population genetics.

Much of the recent development in the evolutionary theory of genetic modification has concerned modifier genes which are not under direct natural selection. Such loci are understood to control specific features at major loci, or aspects of the whole phenotype which influence the response to natural selection. This theory owes much to the work of Nei (1967), whose models for the evolution of genes controlling recombination were the point of departure for the neutral modifier theory of Feldman (1972), Balkau and Feldman (1973) and Karlin and McGregor (1972b, 1974).

The study of modifier genes which are under direct selection has also continued. O'Donald (1968) has examined the problem of evolution of dominance when selection is frequency dependent. Prout, Bungaard and Bryant (1974) initiated the study of genes which control the extent to which segregation distortion occurs at other loci. It is worthwhile to return to Fisher's (1930) book and observe that many of the models of modification developed recently are foreshadowed in his discussion. For example, he speculates concerning the evolution of co-adapted gene complexes (i.e., linkage modification) on p.117*; on the evolution of the sex-ratio (p.159*); on the evolution of dispersal (p.142*); and the evolution of mating systems (p.133*). It is doubtful that for these phenomena Fisher was thinking in terms of modifying genes. The impression is that he was rather more concerned with the qualitative outcome of processes than with the biological

---

*
Pages refer to the 1958 edition.

mechanisms by which the processes might occur.

The absence of precise genetic models for many of the processes that interested Fisher led naturally to the discovery of counter-examples to the theory. (A notable case is the evolution of recombination, see e.g., Williams and Mitton, 1973.) It has also produced arguments over whether group selection must be implicated in the evolution of these processes. For example, (Maynard Smith, 1971, and Eshel, 1973) it appears difficult to discuss the evolution of recombination in the manner of Fisher, Muller (1932), Crow and Kimura (1965), Maynard Smith (1968), Eshel and Feldman (1970) and Karlin (1974), without comparing two populations. A mechanism such as a modifier gene, which controls the extent of recombination eliminates this "difficulty" while leaving the process (i.e., the evolution of recombination) essentially unchanged.

We have chosen to incorporate many of the new results we shall report into the text in their appropriate place in the narrative. Section [2C] and [2D] on segregation and sex-ratio distortion summarize results obtained in collaboration with Dr. G. Thomson.

2. MODIFIERS UNDER PRIMARY SELECTION

[2A]   The Evolution of Dominance

The core of the mathematical theory of the evolution of dominance is the model proposed by Wright (1929) and accepted by Fisher (1929) as a fair quantification of Fisher's (1928) ideas. A gene 'A' is understood to have a selective advantage over its allele 'a' while 'a' is held in the population by recurrent mutation of A to a . The mutation rate is μ per generation. The genotypes BB, Bb and bb at a modifying locus influence the fitnesses of the combinations at the primary locus according to the matrix (1)

$$
\begin{array}{cccc}
 & \text{AA} & \text{Aa} & \text{aa} \\
\text{BB} & 1 & 1 & 1-s \\
\text{Bb} & 1 & 1-ks & 1-s \\
\text{bb} & 1 & 1-hs & 1-s
\end{array}
\tag{1}
$$

with $s > 0$ and $0 \leqslant k \leqslant h < 1$. Here the repulsion and coupling double heterozygotes are assumed to have equal fitnesses $1-ks$. Suppose further that the recombination fraction between the B/b and A/a loci is r . Then, apart from the recurrent mutation, the model is a particular case of the general 2-locus 2 allele per locus model.

The modifying locus B/b is usually denoted M/m . We have used the present notation to draw attention to the fact that (1) is a particular case of the usual 2-locus selection matrix W ;

$$
W = ||w_{ij}|| =
\begin{array}{c|cccc}
 & \text{AB} & \text{Ab} & \text{aB} & \text{ab} \\
\hline
\text{AB} & w_{11} & w_{12} & w_{13} & w_{14} \\
\text{Ab} & w_{12} & w_{13} & w_{23} & w_{24} \\
\text{aB} & w_{13} & w_{23} & w_{33} & w_{34} \\
\text{ab} & w_{14} & w_{24} & w_{34} & w_{44}
\end{array}
\tag{2}
$$

Here w_{24} , for example, is the fitness of the genotype Ab/ab . When the coupling and repulsion double heterozygotes are equally fit $(w_{14}=w_{23})$, (2) may be represented in the three dimensional form of which (1) is a special case. We shall refer to (2) often in the remainder of the paper, and always assume $w_{14}=w_{23}$.

Insofar as the evolution of dominance is concerned, we are interested in the progress of B through the population. This process brings the fitness of Aa closer to that of AA

until the two are equally fit, i.e., A has become completely dominant to a .

Since the model (1) was proposed, there has been controversy over the interpretation of the results. Ewens (1965a,b)
(1966), (1967) pointed out that there is an error in Fisher's
and Wright's estimation of the rate of evolution in the final
stages of the process. The case of most interest here is
k=0 . Both Wright and Fisher believed that this would provide an upper limit to the rate at which dominance might
evolve.

The question of initial and final rates of evolution of
dominance was resolved by Feldman and Karlin (1971) using
concepts from two locus theory. If evolution commences when
B is absent then A and a are in a mutation selection balance. If we denote the frequencies of AB, Ab, aB and ab
by x_1, x_2, x_3 and x_4 respectively, then at this initial
stage $x_2 = 1-\mu/hs$ and $x_4 = \mu/hs$. B then increases if
k < h . At the final stages of evolution the frequencies of
AB and aB , x_1 and x_3 are $1-\sqrt{\mu/hs}$ and $\sqrt{\mu/hs}$ respectively.

Initial and final rates can then be calculated from local
linear analysis of the two-locus recursion system near these
two points. For k > 0 the initial and final rates were
shown to be

$$\lambda_1 \approx \frac{1+2\mu r(1-k/h)(1-ks)+\mu ks(1-k/h)}{1-(1-ks)(1-r)} \tag{3}$$

and

$$\lambda_1^* = 1 - k\sqrt{\mu s} - \frac{rk(1-ks)\sqrt{\mu s}}{ks + r(1-ks)} \tag{4}$$

respectively. Thus, evolution commences at the rate 1+bμ
and concludes <u>faster</u> at the rate $1-c\sqrt{\mu}$, where when k > 0 ,
b and c are positive constants. But when k=0 , $\lambda_1^* = 1$ so

that the final rate of evolution of dominance is slower than the initial rate. In fact in this case it is slower than any geometric rate. From (3) and (4) it can also be seen that λ_1 and λ_1^* are increasing and decreasing functions of r respectively. This means that the looser the linkage the faster evolution starts and ends.

It should be emphasized that the results of Feldman and Karlin were obtained in the neighborhoods of the initial and final equilibria. Nothing very useful can be inferred from this analysis as regards the intermediate stages of evolution. Kimura (quoted by Ewens, 1967) has numerically computed B-gene frequency changes, assuming permanent linkage equilibrium which is invalid. From these computations it is clear that when k=0 the final rate of evolution is slow (see also Sved and Mayo, 1970). Ewens pointed this out in his series of papers and his gene frequency treatment of the final stages of evolution is quite comparable to the full 2-locus treatment of Feldman and Karlin.

Still unsolved is the interesting theoretical problem of the evolution of dominance in frequency dependent systems. The models of O'Donald (1968) and O'Donald and Barrett (1973) have not yet been completely analyzed. Detailed analysis of such frequency dependent and density dependent models would contribute significantly to our understanding of the population genetics of Batesian mimicry.

One of the most powerful arguments suggesting that dominance can evolve in a manner at least conceptually similar to Fisher's proposal, is the famous experiment of Clarke and Shepherd (1960) with the African butterfly Papilio dardanus; (see Ford, 1965, pp.35-36). Other examples of such specific modification of fitness have proven difficult to obtain. Those experiments aimed at detecting fitness modifiers (e.g.,

Dobzhansky and Spassky, 1953) have been based on observations of the viabilities of chromosomal rather than genic homozygotes (Lewontin, 1974, pp.52-54).

Karlin and McGregor (1974) discuss a model of genetic modification of fertility. Experiments on this are notoriously difficult. However, segregation distortion is a form of selection which, in the SD system of Drosophila melanogaster, has been amenable to detailed experimental analysis. It has also led to some interesting theory which we proceed to discuss.

[2B] Modification of Segregation Distortion

Prior to the work of Prout, Bungaard and Bryant (1973) there had been no mathematical theory of modifiers of segregation distortion, although there had been biological speculation as to their population dynamics (see e.g., Hiraizumi, Sandler and Crow, 1960 and Zimmering, Sandler and Nicoletti, 1970). The model of Prout et al. has been slightly altered by Hartl (1975) but the changes do not involve increased generality over the original formulation. We therefore stay with this first model. In this section we summarize the results of Prout et al. and in the next section show how these results can be generalized.

Assume that at the distorting locus the alleles are D and d , and that, independent of the modifier locus the fitnesses of DD , Dd and dd respectively are 1 , w_1 and w . The proportions of d-bearing gametes from Dd are assumed to be 1/2 , k_2 and k_1 according to whether the genotype at the modifying locus is mm , Mm , MM . Prout et al. called 'M' a meiotic drive enhancing allele and 'm' a drive suppressing allele. The recombination between the driven locus (D/d) and the driver locus M/m is r .

The simplest model takes $w_1=1$ and $k_1=k_2=1$, so that the intensity of meiotic drive is maximal when it is not suppressed. Let the chromosome frequencies of DM, Dm, dM and dm be x_1, x_2, x_3 and x_4 respectively. Then in this simplest of models Prout et al. showed that there is a polymorphic equilibrium given by

$$\hat{x}_1 = 0 \ , \ \hat{x}_2 = \frac{2r-w}{1-w} \ , \ \hat{x}_3 = \frac{(1-2r)^2}{1-w} \ , \ \hat{x}_4 = \frac{2r(1-2r)}{1-w} \tag{5}$$

which is locally stable when it exists, namely if

$$w < 2r < 1 \ . \tag{6}$$

(Polymorphism here is understood in the sense that both alleles at both loci are present.)

Two interesting theoretical facts emerge from (5) and (6). First, if we define a mean fitness of the population in the usual way, it is an increasing function of the recombination fraction. In tractable two locus selection systems this is a rare occurrence (see [3C]). Second, the magnitude of the linkage disequilibrium, $\hat{x}_1\hat{x}_4 - \hat{x}_2\hat{x}_3$, at equilibrium is not necessarily monotonic in r for all values of w. For small w it increases for small r and then decreases for larger r. Again this is unusual relative to what we know of 2-locus models; (see Karlin, 1975, for more discussion of this point).

Much of the detailed analysis by Prout et al. is restricted, by their choice of a simplifying parameter set, to the face of the frequency simplex with $\hat{x}_1=0$. We next discuss the more general version of the problem as specified at the beginning of this section.

[2C]   A Generalization of the results of Prout et al.

Although many experimental studies report high values of segregation distortion, it is rare that meiotic drive is maximal. It is therefore important to consider cases where, in the above model, k_1 and k_2 are not unity. This has been done by Thomson and Feldman (1975a). The transformation for the chromosome frequencies in this general case, as given by Prout et al. is

(i) $\quad \overline{w}x_1' = x_1^2 + x_1x_2 + 2x_1x_3(1-k_1)w_1 + 2x_1x_4(1-k_2)w_1 - 2rD(1-k_2)w_1$

(ii) $\quad \overline{w}x_2' = x_2^2 + x_1x_2 + x_2x_4w_1 + 2x_2x_3(1-k_2)w_1 + 2r(1-k_2)w_1D$ $\qquad$ (7)

(iii) $\quad \overline{w}x_3 = wx_3^2 + x_3x_4w + 2x_1x_3k_1w_1 + 2x_2x_3k_2w_1 + 2rk_2w_1D$

(iv) $\quad \overline{w}x_4' = x_4^2w + x_3x_4w + x_2x_4w_1 + 2x_1x_4k_2w_1 - 2rk_2w_1D$

where

$$D = x_1x_4 - x_2x_3 \; , \; \overline{w} = 1-2(1-w)p_dp_D - (1-w)p_d^2 \quad \text{with}$$

$$p_d = x_3 + x_4 = 1-p_D \; .$$

$\qquad$ (8)

The equilibrium analysis of this system is quite complicated and, since we shall be reporting the details elsewhere, we shall present only an outline here. There are two isolated boundary equilibrium points:

$$\hat{x}_1 = \hat{x}_3 = 0 \; , \; \hat{x}_2 = \frac{w_1-w}{2w_1-1-w} \; , \; \hat{x}_4 = \frac{w_1-1}{2w_1-1-w}$$

$\qquad$ (9)

$$\hat{x}_1 = \frac{2w_1-w-2k_1w_1}{2w_1-1-w} \; , \; \hat{x}_2 = \hat{x}_4 = 0 \; , \; \hat{x}_3 = \frac{2k_1w_1-1}{2w_1-1-w}$$

$\qquad$ (10)

provided $2w_1-1-w > 0$. (9) and (10) are both unstable in the full frequency simplex although they may be stable in their respective edges, with D and d segregating, under easily obtained conditions. Thus, a new allele at the modifier locus will always be protected from loss when it appears near (9) and (10).

For polymorphic equilibria of (7) it can be shown that either

$$\hat{D} = 0 \qquad\qquad (11a)$$

or

$$\hat{p}_d = \frac{2r(1-2k_2)w_1 + 2w_1 k_2 - 1}{2w_1 - 1 - w} \quad (\text{if} \quad 2w_1 - 1 - w \neq 0) \quad . \qquad (11b)$$

When $k_1 = k_2 = k < 1$ it is easy to see that (11a) is in fact (10). In addition to the boundaries there is the polymorphism

$$\hat{x}_1 = \frac{(1-\hat{p}_d)(1-A)}{2k} \quad , \quad \hat{x}_2 = \frac{(1-\hat{p}_d)(2k-1+A)}{2k}$$

$$\hat{x}_3 = \frac{\hat{p}_d(1-A)}{2(1-k)} \quad , \quad \hat{x}_4 = \frac{\hat{p}_d(1-2k+A)}{2(1-k)} \qquad (12)$$

with $A = \sqrt{1-4k(1-k)(1-2r)}$ and $\hat{p}_d$ as in (11b) with $k_2 = k \neq 0 , 1$. Analysis and numerical computation have both been employed to study the stability of (12). The conditions depend on k, w_1, w and r , and are very complicated.

It has also been possible to explicitly obtain all the polymorphic equilibria in the case $k_1 \neq k_2$. In this case there are three polymorphic solutions, one with $\hat{D} = 0$, and two with $\hat{D} \neq 0$. Again, the conditions for stability are very complicated. The detailed calculations on these models can be found in the paper by Thomson and Feldman. It is clear that the case treated by Prout et al. gives a very simplified picture of the equilibrium behavior of this system. In particular the strange behavior these authors documented on the $\hat{x}_1 = 0$ face of the frequency simplex does not appear to carry over in the general cases.

There are other aspects of segregation distortion which have been subjected to theoretical analysis. The model of Lewontin (1968) for the t-allele system in the house mouse is a case which can be completely analyzed (Karlin, 1972). Another naturally occurring phenomenon, in Drosophila pseudoobscura can also be discussed in terms of segregation distortion

at the chromosomal level. This is the sex-ratio system, which
we proceed to review.

[2D] Models for the evolution of the sex-ratio

Three different approaches have been used in discussing
natural selection and the sex-ratio. The first approach uses
the concept of reproductive value. Fisher (1930, 1958, p.159)
argued that the sex-ratio will so adjust itself, under the influence of natural selection, that the total parental expenditure incurred in respect of children of each sex shall be
equal. Formal mathematical developments of Fisher's approach
were later given by Bodmer and Edwards (1960) and Kolman
(1960).

The second approach has been to consider what happens when
sex is determined by autosomal genes. One set of genotypes is
identified with one sex and the other genotypes with the
second sex. The questions then reduce to the determination of
the relative proportions in the two sex classes at a stable
equilibrium. In many ways these models are formally equivalent to incompatibility models. Such models were discussed by
Finney (1952), Scudo (1964), Karlin and Feldman (1968), with
respect to one locus. For one locus with up to three alleles,
it is usually the case that at a stable isolated equilibrium
the two sexes were equally frequent. That this is not necessarily the case when two loci are considered has been recently
shown by Karlin and Gasko (unpublished). A new autosomal model for the change in the sex ratio has been proposed by Eshel
(1975).

The third approach is to consider that the sex-determining
mechanism is chromosomal and in fact that it is an XX, XY
sex-determining mechanism. Approximate equality of the
frequencies of the sexes is widely accepted. However,

557

careful analysis often reveals wider departures from equality than cursory treatments of data would imply. These departures often appear to be a function of the male, the heterogametic sex in the cases considered (see Zimmering, Sandler and Nicoletti (1970) for a review of the literature) and may be the result of meiotic drive in the heterogametic sex.

In natural populations of Drosophila pseudoobscura, the "sex-ratio" phenomenon occurs quite widely. Males carrying the sex-ratio chromosome, denoted X_r produce mainly, or only, female offspring. Wallace (1948) conducted population cage experiments with Drosophila pseudoobscura, carrying sex-ratio chromosomes and concluded that there were differential viabilities among the karyotypes. Edwards (1961) modeled this system, demonstrating that stable polymorphisms could occur, and calculating the equilibrium using Wallace's viability estimates. Although the resulting equilibrium frequencies did not agree with Wallace's data, reasonable parameter values could be chosen which would produce equilibria compatible with Wallace's data.

Thomson and Feldman (1975b) have extended Edwards' model to incorporate the recent findings of Policansky and Ellison (1970) and Policansky (1974) that Drosophila pseudoobscura sex-ratio (SR) males produce half as many sperm per bundle as do standard males. The augmented model was then used to consider the fate of a new X chromosome, say X_s , introduced while X_r and X were at equilibrium, whose only effect is to alter the proportion of the sex chromosomes in the effective output from the male.

Prior to the introduction of X_s , the model set up is as drawn in Table I. The parameters a,h,b,c and d are relative viabilities of the karyotypes, f_1 is the fertility of a mating involving an X_rY male, and from such a male the gametic output is in the ratio $(1+m_1)X_r : (1-m_1)Y$. Of course the standard X chromosome does not cause any such distortion.

TABLE 1. *Derivation of Genotype Frequencies.*

Genotype / Viability / Mating	XX / a	X_rX / h	X_rX_r / b	XY / c	X_rY / d
XYx XX	$\frac{1}{2}$	0	0	$\frac{1}{2}$	0
XYx XX$_r$	$\frac{1}{4}$	$\frac{1}{4}$	0	$\frac{1}{4}$	$\frac{1}{4}$
XYx X$_r$X$_r$	0	$\frac{1}{2}$	0	0	$\frac{1}{2}$
X$_r$Yx XX	0	$\frac{f_1}{2}(1+m_1)$	0	$\frac{f_1}{2}(1-m_1)$	0
X$_r$Yx XX$_r$	0	$\frac{f_1}{4}(1+m_1)$	$\frac{f_1}{4}(1+m_1)$	$\frac{f_1}{4}(1-m_1)$	$\frac{f_1}{4}(1-m_1)$
X$_r$Yx X$_r$X$_r$	0	0	$\frac{f_1}{2}(1+m_1)$	0	$\frac{f_1}{2}(1-m_1)$

Now assume that from a male X_sY , the gametic output is $(1+m_2)X_s : (1-m_2)Y$, while X_sY males have fertility f_2 . X_s is introduced near the stable equilibrium resulting from the model of Table I. Thomson and Feldman have shown that X_s will then increase if and only if

$$f_1(1+m_1) < f_2(1+m_2) . \qquad (13)$$

Thus, if X_s does not cause a fertility change and if the distortion is in the direction observed in nature, then X_s will become established if it increases this distortion. If the distortion is Y-driven, then a new Y-chromosome will be established if

$$f_2(1-m_2) > f_1(1-m_1) \quad . \tag{14}$$

When $f_1 = f_2$, this is the opposite to (13).

These models have demonstrated that sex-chromosome mediated modification of the sex-ratio can occur, and that polymorphism involving sex-ratio controlling chromosomes is possible. They also explain a limit to the amount of distortion possible through the interaction of m_1, m_2, f_1 and f_2 in (13) and (14). Thus, if $0 < m_1 < m_2$, a loss in fertility due to X_s , i.e. $f_2 < f_1$, may prevent X_s becoming established.

This model of sex-ratio distortion can be viewed in terms of X-linked (or Y-linked for (14)) modification of the sex-ratio. It is conceivable however, that such modification may be mediated by autosomal genes. The following is a very simple model for the control of X-Y segregation by an independently segregating autosomal locus with alleles A and a . We suppose that if the genotype of a male (XY) at the modifier locus is AA then the gametic output is $(1+m_1)X:(1-m_1)Y$. If the genotype at the modifier locus is Aa this ratio is $(1+m_2)X:(1-m_2)Y$ while aa results in $(1+m_3)X:(1-m_3)Y$. Suppose that initially only 'A' is present so that the sexes are in the ratio $1+m_1:1-m_1$ females to males. Then it is easy to see that 'a' increases in frequency if and only if $m_2 > m_1$, provided only that $m_1 > 0$. (If $m_1 = 0$ then the eigenvalue governing the initial increase is unity so that for the result to hold there must be distortion initially.)

Both models of this section lead us to the qualitative conclusion that whether sex-ratio distortion in favor of the homogametic sex (in the form of unequal X and Y proportions in the gametes) is caused by X-linked factors, or by autosomal factors, the amount of distortion will tend to increase. The increase will continue until the genetic agents causing it also cause fertility or viability losses whose effect cancels any advantage due to distortion. It must be remembered that these arguments are purely genetic. Perhaps the most realistic framework for modeling evolution or optimization of the sex ratio would involve a hybrid between the genetic approach and Fisher's original non-genetic argument. This would necessarily involve viability changes for those individuals carrying the new sex-ratio chromosome, which could prevent further distortion.

3. NEUTRAL MODIFIERS AND SECONDARY SELECTION

[3A] Introduction, history and definitions

In Section 2, the overall fitness of a complete genotype depended on its genotype at the modifier locus. Such modifiers are said to be under primary selection. In this section we study loci whose genotypes do not affect the fitnesses of their carriers. These genes influence features of the complete genotype which interact with primary selection at other, major, loci. This selection then feeds back on the modifier locus, and the result is evolution at the modifier locus by what is called secondary selection, even though the modifier is itself selectively neutral.

The study of these neutral modifier loci which influence the frequency of such events as recombination, mutation, and dispersal is relatively recent. As mentioned in the introduction, Fisher speculated on the conditions which might favor increased linkage between loci under selection. Haldane (1931) later criticized Fisher's intuition by providing a model counter-example. The way in which Bodmer and Parsons (1962) addressed the linkage evolution problem is closely related to the later developments. Turner (1967) and Lewontin (1971) also addressed the problem of the evolution of linkage in Fisherian terms. None of these studies included modifier genes.

Nei (1967, 1969) introduced the model which underlies all the subsequent work on neutral modifiers. Here the modifier locus, with alleles M and m influences a parameter θ which in turn interacts with the selection operating on the major loci. The genotypes MM, Mm and mm give rise to values θ_1, θ_2 and θ_3 respectively. In Nei's original model θ was the recombination fraction between the two loci A/a and B/b under selection. The changes in frequency of alleles at the M/m locus determine the changes of θ . We then have a self-contained method of describing the evolution of the biological phenomenon measured by θ .

Following Nei's work the recombination modification problem was adapted by Feldman (1972) into a classical population genetic framework similar to that used for the evolution of dominance in [2A]. The fate of a mutation at the homozygous recombination controlling locus, introduced at a stable equilibrium of the major loci was studied. Balkau and Feldman (1973) followed this with a study on the neutral modification of migration and another on the structure of polymorphic equilibria possible for the system (Feldman and Balkau, 1973). Karlin and McGregor (1972b, 1974) studied the recombination control problem, the migration problem and models of genetic control of the mating system through the proportion of assorting or selfing. They sought a general theoretical principle which would govern the initial increase of a mutant at a homozygous modifier locus introduced at a stable equilibrium of the major loci. In Section [3B] we outline the basic problem in terms of recombination modification which we need in

562

the later section on modifier polymorphism. In [3C] we outline the fitness principle of Karlin and McGregor and summarize the theory of mutation, and migration modification. In
[3D] we present a new general result for modifier polymorphisms and in [3E] we outline a simple model allowing initial
increase of a recombination increasing modifier. Not all results will be reported in complete detail; the purpose is to
present sufficient of the theory that the new result on modifier polymorphisms may be accessible. More details are in
the published papers, or in manuscripts in preparation.

[3B]   Recombination reduction

Nei's original formulation considered selection operating
on two major recombining loci A/a and B/b . If the genotypes at the modifier locus are MM, Mm and mm , the recombination fractions between A/a and B/b are r_1 , r_2 and r_3
respectively. Nei's original analysis assumed nothing about
the type of selection at the major loci, except that the matrix (2) was in force at the A/a - B/b loci irrespective of
the genotype at the M/m locus. His analysis was made in
terms of the gene frequency of M and average recombination
fractions. In addition, the method relied on certain results
about quasi-linkage equilibrium (Kimura, 1965) which were later shown (Feldman and Crow, 1970) not to be applicable to
this situation. Nei's conclusion, however, agreed with that
of Fisher and Bodmer and Parsons, that tighter linkage (reduced recombination) would always be favored as long as there
was additive epistasis in the selection on the major loci.

Feldman (1972) took Nei's three locus formulation but
asked the modification question in the following way. Suppose
the recombination modifying locus is homozygous for MM , and
that the major loci are in a stable polymorphic equilibrium.

A mutation occurs $M \to m$. What are the conditions on M and m and the initial polymorphism which ensure the establishment of the mutant modifier m . This question was answered with reference to the known symmetric equilibria of the two-locus symmetric viability model (Lewontin and Kojima, 1960; Bodmer and Felsenstein, 1967; Karlin and Feldman, 1970). Provided the initial stable equilibrium (with MM at the modifier locus) of the A/a and B/b system is in linkage disequilibrium then m initially increases if and only if $r_2 < r_1$, i.e., it reduces recombination. The procedure here is standard and involves linearization of the full (eight dimensional) 3-locus transformation in the neighborhood of the two locus equilibrium, with M fixed, and determination of the conditions for the leading eigenvalue of the resulting matrix to be larger in absolute value than unity. If the initial equilibrium is in linkage equilibrium (i.e., the fitness matrix (2) is additive, or multiplicative with r , large) then the leading eigenvalue is unity and the modifier cannot advance at a geometric rate. These results on the initial increase of 'm' are independent of the recombination fraction r between M/m and A/a and also of the order of the loci. The amount of interference between crossing over in the two regions M-A and A-B also apparently makes little difference.

From these results two conclusions of importance for evolutionary theory emerge. The first is that in the symmetric viability case where m increased if $r_2 < r_1$, the equilibrium value $\hat{D}(r)$ of the linkage disequilibrium is a decreasing function of r . Second the equilibrium population mean fitness $\hat{w}(r)$ is decreasing in r . The successful mutation at the modifier locus will therefore be one which enhances an adaptively favorable gametic association. This reasoning can

be taken as a theoretical validation of the concept of Dobzhansky (see e.g. 1970, p.145) and others of the co-adapted
gene complex, arising from selectively advantageous linkage
disequilibrium.

Karlin and McGregor (1972b, 1974) used the initial increase
formulation of the neutral modifier theory in attempting to
formulate a set of principles governing the initial increase
of modifiers of various kinds of evolutionarily interesting
parameter. The result was their elegant mean fitness principle.

[3C] The fitness principle

Karlin and McGregor (1974) enunciated a principle, which,
while not a mathematical theorem, has been shown to be valid
in a large number of models of neutral modification with random mating. This fitness principle can be stated as follows.
Suppose that a modifier locus M/m controls a parameter θ
so that the genotypes MM , Mm and mm produce values θ_1 ,
θ_2 , θ_3 respectively. Suppose also that there is a stable
equilibrium frequency vector $\hat{\underline{x}}(\theta)$ (which when M is fixed
is $\hat{\underline{x}}(\theta_1)$) . Let $w(\underline{x})$ be the usual mean fitness of the population and set $F(\theta) = w(\hat{\underline{x}}(\theta))$. Then the new mutant modifier 'm' initially introduced near $\hat{\underline{x}}(\theta_1)$ increases in
frequency if $F(\theta_2) > F(\theta_1)$ and will be lost if $F(\theta_2) <$
$F(\theta_1)$.

In their paper Karlin and McGregor demonstrated the validity of this principle in a number of recombination modification, mutation modification and migration modification
examples. In addition they showed that the principle fails,
in general, for a modifier locus influencing the propensity
of homozygotes to assort in models of Karlin and Scudo (1969)
and Scudo and Karlin (1969). A problem here, which also

arises in all of the segregation distortion models is the appropriate definition of the mean fitness. However, in all cases except the model for the control of the proportion of selfing in a mixed selfing-random mating model, the result of Feldman (1972) for recombination extends: whenever the frequency of 'm' changes at a geometric rate it increases if and only if $\theta_2 < \theta_1$. If $\theta_1 = 0$, or if $\theta_2 = \theta_1$ the controlling eigenvalue is unity, and modification, if it occurs, is a very slow process.

At this point we must introduce the mathematical framework in which the above remarks are valid. For economy we discuss the details of only the recombination and mutation modification cases; the migration modification case is formally analogous and the details can be found in the original papers of Balkau and Feldman (1973) and Karlin and McGregor (1974). The transformation will be written in the notation of Karlin and McGregor.

Consider first the recombination modification problem. Take two major loci A/a and B/b with the general selection matrix W from (2). At the selectively neutral modifier locus the genotypes MM, Mm and mm produce recombination fractions r_1, r_2 and r_3 respectively. The order of the loci is M/m, A/a, B/b with r the recombination fraction between M/m and A/a. No interference is assumed, although in all of our studies this has never appeared to be a restriction. The frequencies of the chromosomes MAB, MAb, MaB and Mab are $\underline{y} = (y_1, y_2, y_3, y_4)$ and those of mAB, mAb, maB and mab are $\underline{x} = (x_1, x_2, x_3, x_4)$. The matrix product of W and the frequency vector x is Wx with $(Wx)_i = \sum_{j=1}^{4} w_{ij} x_j$. The usual scalar product of the two vectors x and y is $(x,y) = (x_1 y_1 + x_2 y_2 + x_3 y_3 + x_4 y_4)$. The

Schur (vector) product of x and y is $x \circ y = (x_1 y_1, x_2 y_2, x_3 y_3, x_4 y_4)$. Define

$$D(x,y) = x_1 y_4 - x_2 y_3 \quad , \quad E(x,y) = x_1 y_2 - y_1 x_2 \quad ,$$

$$F(x,y) = x_3 y_4 - x_4 y_3 \tag{15}$$

Then the full transformation is

$$\overline{W} y' = y \circ W(x+y) + r[x \circ Wy - y \circ Wx] - r_1 D(y,y)\underline{e}$$

$$+ r_2 (1-2r)[E(x,y)\underline{e}_1 + F(x,y)\underline{e}_2] \tag{15a}$$

$$+ r_2 [D(x,y)\underline{e}_4 + D(x,y)\underline{e}_3]$$

$$\overline{W} x' = x \circ W(x+y) + r[y \circ Wx - x \circ Wy] - r_3 D(x,x)\underline{e}$$

$$- r_2 (1-2r)[E(x,y)\underline{e}_1 + F(x,y)\underline{e}_2] \tag{15b}$$

$$+ r_2 [D(x,y)\underline{e}_3 + D(y,x)\underline{e}_4]$$

where

$$\underline{e} = (1,-1,-1,1) \quad , \quad \underline{e}_1 = (1,-1,0,0) \quad ,$$

$$\underline{e}_2 = (0,0,1,-1) \quad , \quad \underline{e}_3 = (r-1,1-r,r,-r) \quad , \tag{16}$$

$$\underline{e}_4 = (-r,r,1-r,r-1)$$

and

$$\overline{W} = \overline{W}(x+y) = (x+y, W(x+y)) = \sum_i \sum_j w_{ij}(x_i+y_i)(x_j+y_j) \quad . \tag{17}$$

Now according to the formulation of Feldman (1972) 'm' is introduced near a stable equilibrium $\hat{y} = (\hat{y}_1, \hat{y}_2, \hat{y}_3, \hat{y}_4)$ of the system

$$\overline{W}(y) \, y' = y \circ Wy - \underline{e} r_1 D(y,y) \tag{18}$$

where only M is present. The fate of 'm' here was des-cribed for the cases where W is a symmetric viability ma-trix by Feldman (1972) and more generally by Karlin and McGregor (1974). The most general results are the following.

For a completely general selection matrix and a stable solution $\hat{y}(r_1)$ of (18) such that $D(\hat{y},\hat{y}) \neq 0$, if r_1 is small enough (i.e., A/a and B/b are initially tightly linked) and r_2 is close enough to r_1, then 'm' increases if and only if $r_2 < r_1$. Further, in a small enough neighborhood of $r=0$, $F(r) = W(\hat{y}(r))$ is a decreasing function of r. Hence 'm' increases if and only if $F(r_1) < F(r_2)$. For the symmetric viability setup these results hold for all r_1 and r_2 throughout the range where $\hat{y}(r_1)$ is a stable symmetric equilibrium.

Remark. The first counter example to the fitness principle for recombination modification was provided by Thomson and Feldman (1974) who considered, as the initial stable equilibrium, the solution (5) to the segregation distortion problem of Prout et al. A third neutral modifier locus was assumed to control the recombination between the driver and the driven loci. These authors showed that although for (5),

$$\hat{w} = 1 - (1-2r_1)^2/(1-w)$$ is increasing in r_1, an allele reducing recombination may be favored. Subsequently Karlin and Carmelli (1975) have discovered numerical examples in regular random mating systems which also violate the principle.

For the mutation modification model, Karlin and McGregor considered a single locus A/a in a mutation selection balance. The fitnesses of AA, Aa and aa are 1,1 and $1-\sigma$ respectively. A mutates to a at the rate μ_1 if the genotype at the neutral, mutation modifying locus is MM, μ_2 with Mm and μ_3 with mm. The recombination fraction between M/m and A/a is r. Let $z = (z_1,z_2,z_3,z_4)$ be the frequencies of MA, Ma, mA and ma respectively. We may write the resulting (four dimensional) transformation in analogy with (17) as

$$N \overset{*}{z}' = (z \circ \overset{*}{S} z) - \underline{er}(1-\mu_2)D(z,z) - \underline{e}_z \circ \underline{\alpha} \qquad (19)$$

where

$$\overset{*}{S} = A \otimes S \quad, \qquad (20)$$

with

$$A = \begin{pmatrix} 1 & 1 \\ 1 & 1 \end{pmatrix} \quad, \qquad S = \begin{pmatrix} 1 & 1 \\ 1 & 1-\sigma \end{pmatrix} \quad,$$

$$\underline{e}_z = (z_1, -z_2, -z_3, z_4) \quad, \qquad \underline{\alpha} = (\alpha_1, \alpha_1, \alpha_2, \alpha_2) \qquad (21)$$

with

$$\alpha_1 = \mu_1 p_M + \mu_2 p_m \quad, \qquad \alpha_2 = \mu_2 p_M + \mu_3 p_m$$

and

$$\overset{*}{N} = (z, \overset{*}{S} z) \quad. \qquad (22)$$

The notation $\otimes$ is used for the usual Kronecker product of two matrices, and $p_M = z_1 + z_2 = 1 - p_m$ is the gene frequency of M . Of course the initial equilibrium MA , Ma is given by the solution $\hat{\nu} = (\hat{\nu}_1, \hat{\nu}_2)$ of the recursion system

$$N\nu' = (\nu \circ S\nu) + \mu_1 \nu_1 \underline{e} \qquad (23)$$

where

$$\underline{e} = (-1, 1)$$

and

$$N(\nu) = (\nu, S\nu) \quad. \qquad (24)$$

The solution is obviously

$$\hat{\nu}_1 = 1 - \sqrt{\mu_1 s} \quad, \qquad \hat{\nu}_2 = \sqrt{\mu_1 s} \quad. \qquad (25)$$

It is then easily shown that 'm' increases near $\hat{\nu}$ if and only if $\mu_2 < \mu_1$ and that $N(\hat{\nu}(\mu_1)) < N(\hat{\nu}(\mu_2))$ if and only if $\mu_2 < \mu_1$. Again the fitness principle is validated.

A number of different models have been studied by Balkau and Feldman, and Karlin and McGregor for migration modification. The essence of these is that two populations, each under selection at a locus A/a exchange a proportion m of their residents every generation. A second selectively neutral locus B/b controls the amount of exchange which goes on. A stable migration-selection polymorphism is established with B fixed and a migration rate m_1. Then 'b' is introduced near this equilibrium. Bb produces an exchange of m_2 per generation and bb, m_3. Karlin and McGregor considered, among others, the case where in both populations AA Aa and aa had viabilities 1, 1-s and 1 respectively. Balkau and Feldman studied the case where in one population the viabilities were 1+s, 1, 1-s, and in the other 1-s, 1, 1+s. They also treated a haploid case where A and a had relative viabilities 1+s:1 in one population and 1:1-s in the other. In all of these cases it has been proven that if b appears near a stable equilibrium whose frequencies are functions of m_1, then b increases if $m_2 < m_1$. Karlin and McGregor again validated the fitness principle in this case.

The qualitative explanation of the results for these selection regimes is as follows. In one of the populations, at the initial equilibrium, A is more frequent due to its selective advantage there while a is more frequent in the other due to its selective advantage there. Most migrants will pass from an environment where they are favored to one where they are at a disadvantage. It is obviously beneficial to both alleles to minimize such movement. The resultant reduction of migration provides an evolutionary basis for the process of geographic isolation without intrinsic barriers to gene flow.

A further interesting aspect of the migration modification problem emerges from the recent work of Christiansen and Feldman (1975). They considered two populations under symmetric viability selection at two loci and exchanging residents.

Here, prior to the introduction of a new migration modifying allele, the populations were shown to have identical gene frequencies. But they did not necessarily have identical chromosome frequencies. In fact if the linkage disequilibria were different in the two populations, decreased migration was again favored. The biological implication here is that differences in chromosome arrangements may be sufficient to spark the process which may eventually result in almost complete separation of two populations while not necessarily causing major differences in gene frequencies.

It is important that in these models of neutral modification there is, in every case, a dominance effect. That is, if $\theta_2 = \theta_1$, whether modification occurs is not discernible from the local linear analysis. But if it occurs, it does so at an extremely slow rate. In fact the rate of modification decreases as θ_2 approaches θ_1. One reason for the genome not congealing could be this retardation of modification of recombination.

Throughout the discussion of the previous two sections we have considered 'm' to be a rare mutation introduced at a stable equilibrium of the major locus or loci, with M fixed; m increased when rare, in the cases of recombination mutation and migration, provided $\theta_2 < \theta_1$. A symmetrical argument can be made that M will increase when rare if $\theta_2 < \theta_3$. If $\theta_2 < \theta_1$ and $\theta_2 < \theta_3$ then there is a 'protected' polymorphism (Prout, 1968) at the modifier locus. In the next section we outline our attempts to describe this polymorphism in more detail.

[3D] A general modifier polymorphism

That a polymorphism for the locus M/m , which controls θ, should exist when $\theta_2 < \theta_1, \theta_3$, was pointed out by Feldman

(1972) in the case of recombination modification and more generally by Karlin and McGregor (1974). Feldman and Balkau
(1973) studied modifier polymorphisms in the recombination
problem. They took the Lewontin-Kojima (1960) and Wright
(1952) versions of the symmetric viability model at the major
locus with MM , Mm and mm producing recombination fractions
r_1, r_2, r_1 respectively. It was shown in these cases that two
classes of polymorphic equilibria exist. For a compact definition, with y and x the frequency vectors as in the previous section, write

$$(y)(x) = (y_1, y_2, y_3, y_4, x_1, x_2, x_3, x_4) \quad ,$$

as the full three locus frequency vector. Suppose $(\hat{y}(r_1))(0)$
is the stable equilibrium on the boundary where M is fixed,
then the first class of equilibria exists when $r_2 < r_1$ and
takes the form

$$(\hat{y}(\frac{r_1+r_2}{2})/2)(\hat{y}(\frac{r_1+r_2}{2})/2) \quad . \tag{26}$$

This equilibrium is stable only for large values of r , i.e.
when M/m is loosely linked to A/a .

The second class of equilibria lacked symmetry and although it could be explicitly obtained, was a complicated
function of r , r_1 and r_2 . A partial stability analysis
indicated that this second class was stable when the first
was not, i.e. for M/m tightly linked to A/a . In fact
when r = 0 , this class of equilibria can be represented in
the form A^*B , A^*b , a^*B , a^*b where $A^* \equiv MA$ and $a^* \equiv ma$
with frequencies $\hat{z} = (z_1, z_2, z_3, z_4) = \hat{y}(r_2)$.

It remains an open problem whether any other polymorphic
equilibria exist, although cursory numerical study suggests
that in the above models they do not.

Feldman and Balkau described the above situation $r_2 < r_1, r_3$

as secondary overdominance. It is natural to ask for other cases of modification whether secondary overdominance produces similar polymorphisms to the recombination balance of Feldman and Balkau. We have recently obtained a result which holds for all the cases of recombination, mutation and migration modification studied so far. An outline of the result follows. The general form and the details of the stability analyses will appear elsewhere.

General recombination balance: For the recombination modifiers defined above, let

$$\hat{p}_M = (r_3 - r_2)/(r_1 + r_3 - 2r_2) = 1 - \hat{p}_m \tag{27}$$

and

$$r^* = \hat{p}_M^2 r_1 + 2\hat{p}_M \hat{p}_m r_2 + \hat{p}_m^2 r_3 = \frac{r_1 r_3 - r_2^2}{r_1 + r_3 - 2r_2} \tag{28}$$

and suppose that $\hat{y}(r^*) = (\hat{y}_1, \hat{y}_2, \hat{y}_3, \hat{y}_4)$ is a solution of (18) with $r = r^*$. Then the Kronecker product

$$(\hat{p}_M, \hat{p}_m) \otimes \hat{y}(r^*) \tag{29}$$

is a polymorphic solution of (17).

Equilibria of the form (29) appear from our preliminary stability analysis to be stable when $\hat{y}(r^*)$ is stable as a solution of (18) and when there is loose linkage between M/m and A/a . These equilibria are in fact generalizations of (26).

For the class of equilibria which can be represented when $r = 0$, in the form $(\hat{z}_1, \hat{z}_2, 0, 0, 0, 0, \hat{z}_3, \hat{z}_4)$ or $(0, 0, \hat{z}_1, \hat{z}_2, \hat{z}_3, \hat{z}_4, 0, 0)$ where $\hat{z} = (\hat{z}_1, \hat{z}_2, \hat{z}_3, \hat{z}_4) = \hat{y}(r_2)$ and $\hat{y}(r_2)$ is a stable solution of (18) with r_2 as recombination parameter, the small parameter theory of Karlin and McGregor (1972a) ensures that for r small and positive, stable interior solutions of (17) exist nearby. Although the linkage of M/m to A/a was irrelevant to the original modification problem (Sections [3C] and [3D]) it is clear that the stability of the modifier polymorphisms is dependent on r .

Mutation modifier balance: For the mutation modifier model defined in [3C], let

$$\hat{p}_M = (\mu_3 - \mu_2)/(\mu_1 + \mu_3 - 2\mu_2) = 1 - \hat{p}_m \tag{30}$$

and

$$\mu^* = \hat{p}_M^2 \mu_1 + 2\hat{p}_M \hat{p}_m \mu_2 + \hat{p}_m^2 \mu_3 = \frac{\mu_1 \mu_3 - \mu_2^2}{\mu_1 + \mu_3 - 2\mu_2} \tag{31}$$

and suppose that $\hat{v}(\mu^*) = (\hat{v}_1, \hat{v}_2)$ is the solution of (25) with $\mu_1 = \mu^*$ then

$$(\hat{p}_M, \hat{p}_m) \otimes \hat{v}(\mu^*) \tag{32}$$

is a stable solution of (19) for all r . A similar result is also true in the migration modification case.

[3E] A model for the initial increase of recombination

Suppose that the three chromosomes MAB , MAb and MaB are in equilibrium under natural selection. The selection matrix is W from (2) with last row and column deleted. The frequencies of the three gametes are $\hat{x}_1$, $\hat{x}_2$ and $\hat{x}_3$ respectively. Suppose that the recombination fraction between A/a and B/b is $r_1 = 0$ in the initial state where only M is present.

Now if a mutation M → m occurs at the M locus, such that Mm has recombination fraction $r_2 > 0$ will 'm' increase? Notice ,that here, the advent of m also entails the appearance of ab and we may assume that selection on the A/a , B/b system is governed by the full W matrix (2). In the sense that we have been considering above the modifier locus is neutral.

To study the fate of m we consider the stability of the

three dimensional equilibrium $(\hat{x}_1, \hat{x}_2, \hat{x}_3)$ as a boundary point of the full 3-locus system. The recursion system (17) is again applicable. Because $r_1 = 0$ the mean fitness $\hat{w}$ of the three dimensional system equals the marginal fitnesses $\hat{w}_1^*$, $\hat{w}_2^*$, $\hat{w}_3^*$ of the three initial gametes. We may then write a simple relation for the frequency p_m of m when it is rare. We have

$$p_m' = p_m + (\frac{w_4^*}{\hat{w}} - 1) \, p_{mab} \qquad (33)$$

where $w_4^* = w_{14}\hat{x}_1 + w_{24}\hat{x}_2 + w_{34}\hat{x}_3$ and p_{mab} is the small frequency of mab. If $w_4^* > \hat{w}$ then m increases initially. It is this initial selective advantage to ab which carries m into the population. But if $w_4^* > \hat{w}$ then, in the absence of recombination, the initial three dimensional equilibrium is unstable as a boundary of the four dimensional simplex. Clearly, m may subsequently disappear. We make no claims about other boundary equilibria to which the population may converge.

4. INVERSIONS AND INVERSION POLYMORPHISMS

Dobzhansky (1970, p.145) suggests that inversion polymorphisms are maintained because the suppression of crossing-over in inversion heterozygotes prevents recombination from breaking up favorable complexes of genes on the same chromosome. The modifier theory in Section 3 provides a simple theoretical framework for modeling the suppression of recombination. Other models have been suggested which are more directly pertinent to the properties of inversions. Haldane (1957), Nei (1967), Deakin (1972), Charlesworth and Charlesworth (1973) and Charlesworth (1974) have all made models specifically aimed at the questions: What are the selection conditions on a population that favor the initial increase of inversions? and, What sort of inversion polymorphisms are possible under a given selection regime? Perhaps the most fruitful approaches have been those of Nei (1967) and Charlesworth (1974). Our purpose here is to summarize the argument of Nei for the problem of initial increase of inversions, and to present a variation on the argument of Charlesworth for the problem of inversion polymorphisms.

The model of Nei (1967), used also by Charlesworth and Charlesworth, begins with the usual two locus recursion system (18). The notation is the same as in (18) with $\hat{y} = (\hat{y}_1, \hat{y}_2, \hat{y}_3, \hat{y}_4)$ the stable equilibrium frequencies of AB, Ab, aB and ab in a usual two locus system with selection matrix W from (2). Now suppose that a new chromosome $\overline{AB}$ is formed from AB with the property that $\overline{AB}$/ab produces no recombinants. Let the frequency of $\overline{AB}$ be u. Then (18) is altered to take the form

(i) $\quad \overline{W}^* u' = u(Wy)^*_1$

(ii) $\quad \overline{W}^* y'_1 = y_1 (Wy)^*_1 - rDw_{14}$

(iii) $\quad \overline{W}^* y'_2 = y_2 (Wy)^*_2 + rDw_{14}$ $\qquad\qquad$ (34)

(iv) $\quad \overline{W}^* y'_3 = y_3 (Wy)^*_3 + rDw_{14}$

(v) $\quad \overline{W}^* y'_4 = y_4 (Wy)^*_4 - rDw_{14}$

where

$$(Wy)^*_i = (Wy)_i + uw_{1i} \quad , \quad \overline{W}^* = \sum_i y_i (Wy)^*_i = (y, Wy) + u(Wy)^*_1$$

with $\quad D = y_1 y_4 - y_2 y_3 \quad , \quad \sum_{i=1}^{4} y_i + u = 1 \quad .$

Clearly for $\overline{AB}$ to increase when rare we require, from

(34)(i)

$$(W\hat{y})^*_1 > \overline{W}^* (\hat{y}) \qquad (35)$$

which is equivalent to $\hat{D} > 0$.

We can therefore state that if the original linkage dis-equilibrium favors the gametic phase in which the inversion occurs, the inversion will initially increase and reinforce the gametic phase imbalance. This is qualitatively similar to the conclusions drawn above for the case of recombination modifiers.

Charlesworth (1974) studied the various polymorphisms which involve the inverted chromosome $\overline{AB}$. Here we discuss one of these, which is the interesting case when all five chromosomes are present. Consider (34) at an equilibrium point $\hat{y}^* = (\hat{u},\hat{y}_1,\hat{y}_2,\hat{y}_3,\hat{y}_4)$ with all coordinates positive. Then, from (34)(i) we have $\overline{W}^* = (Wy)^*_1$. Using this in (34)-(ii) it is clear that $\hat{D} = \hat{y}_1\hat{y}_4 - \hat{y}_2\hat{y}_3 = 0$. It is then easy to see that the frequencies $\hat{u}+\hat{y}_1,\hat{y}_2,\hat{y}_3,\hat{y}_4$ are exactly the four frequencies expected of AB, Ab, aB and ab in the fully polymorphic equilibrium of (18) with $r=0$. Charlesworth called this "the associated $r=0$ 2-locus equilibrium", denoted $\hat{y}_0$. Since $\hat{D}=0$ when $\hat{u} > 0$, the disequilibrium for the associated 2-locus $r=0$ equilibrium must be positive.

The stability of $\hat{y}^*$ is of considerable interest. The following argument demonstrates that for r small enough, $\hat{y}^*$ is stable if the associated $r=0$ two locus equilibrium is stable. As pointed out by Charlesworth, the local stabi-lity matrix for $\hat{y}^*$ takes the form

$$\underset{\sim}{B} = \begin{pmatrix} 1 & \underset{\sim}{c}' \\ \underset{\sim}{b} & \underset{\sim}{A}^* \end{pmatrix} \qquad (36)$$

where

$$\underset{\sim}{b}' = (\beta,-\beta,-\beta,\beta) \; ; \quad \beta = \frac{r\hat{y}_4^* w_{14}}{\overline{w}^*(\hat{y}^*)} \quad ,$$

$$\overline{w}^*(\hat{y}^*)\underset{\sim}{c}' = (w_{11}\hat{u},w_{12}\hat{u},w_{13}\hat{u},w_{14}\hat{u}) \quad ,$$

and $\underset{\sim}{A}^*$ is the four dimensional matrix whose elements are those of the four dimensional matrix $\underset{\sim}{A}$ for the stability of $\hat{y}_0$, plus terms $O(r)$.

This definition of $\underset{\sim}{A}$ as the stability matrix for $\hat{y}_0$ leads to the problem in Charlesworth's argument. The actual local stability matrix for $\hat{y}_0$ is three dimensional, and by considering the four dimensional form a spurious eigenvalue is introduced. When $\hat{y}_0$ is stable this eigenvalue is grea- ter than unity, and when $\hat{y}_0$ is unstable at least two eigen- values of A are larger than unity in absolute value. No- ting this difficulty we can proceed as follows. The charac- teristic polynomial of B , for $r \geqslant 0$, can be expressed in the form

$$g_r(\lambda) = (1-\lambda)A_r^*(\lambda) - \beta M(\lambda) + O(r^2) \tag{37}$$

where $A_r^*(\lambda)$ is the characteristic polynomial of $\underset{\sim}{A}^*$ and $M(\lambda)$ is a polynomial resulting from expansion of $|B-I\lambda|$ by its first column. Since $\beta=0$ when $r=0$, $g_0(\lambda)$ has a root of unity, and by virtue of the previous remark, when $\hat{y}_0$ is stable, $g_0(\lambda)$ has a single root larger than unity and three roots less in absolute value than unity. For r small enough, the stability of $\hat{y}^*$ is determined by the sign of the function $M(\lambda)$ in (37) at $\lambda=1$. A straightforward use of the fact that the determinant of the matrix W must be negative for $\hat{y}_0$ to be stable produces the conclusion that $M(1) < 0$ for r small. Hence for r small enough the four meaningful eigenvalues of B are less than unity in absolute value when $\hat{y}_0$ is stable. Qualitatively, the inversion polymorphism is stable for small r if and only if the associated $r=0$ equilibrium is stable.

5. PERSPECTIVES

The theory of genetic modification we have described can be divided into three aspects; (a) direct modification of factors of natural selection by specific genes, (b) modification of evolutionary parameters other than those of selection by specific genes which do not affect the fitness of their carriers, and (c) chromosomal mechanisms which may alter aspects of selection or other parameters. The evolutionary forces that the theory encompasses include certain types of viability and fertility selection, segregation distortion in a variety of forms, recombination, mutation, migration and aspects of the mating system. There seems every reason to suppose that other problems of considerable biological interest, such as the evolution of diploidy from haploidy might also be studied in terms of modifier theory.

It is often claimed that the equilibrium theory of population genetics is a static theory. In fact it is not necessary to view the equilibrium theory of major loci as the final stage of an evolutionary process and, therefore, only of static interest. Instead it can be viewed, through the medium of genetic modification, as a point of departure for more general evolutionary theory. Testing of such a theory promises to be extremely difficult. The effects of modifiers of recombination, migration and mutation are probably too small to overcome sampling errors in population cage experiments to determine their actual influence on selection.

There are still holes in the theory of genetic modification. We do not know whether there exist classes of 2 locus selection schemes for which recombination between the loci will tend to increase. One suspects that such a class of models would have the property that it allows stable linkage disequilibrium for loose linkage, but even this is a minimally based conjecture. There remain questions concerning mutation modifiers in situations where the selection is not normalizing. There are a number of interesting questions concerning the applicability of the theory to selection in non-constant environment. One aspect of this, currently under study, concerns the problem of canalization, or control of variability in the response of a population to selection. We conceive of a one locus population AA , Aa , aa in a random environment with fitness 1-s , 1 , 1-t where s and t are random (Karlin and Liberman, 1974). Another locus B/b controls the variance of the random variables s and t while not affecting their means. This might be called mean-neutral modification. If the folklore of biology is valid, one would expect selection to minimize the variances of the selection coefficients and in certain special cases we have shown this to be the case. The mean-neutral problem is of some importance for the more general theory of evolution at the phenotypic level.

REFERENCES

Balkau, B. and M.W. Feldman. (1973). Genetics 74: 171-174.

Bodmer, W.F. and A.W.F. Edwards. (1960). Ann. Hum. Genet. 24: 289-294.

Bodmer, W.F. and J. Felsenstein. (1967). Genetics 57: 237-265.

Bodmer, W.F. and P.A. Parsons. (1962). Advan. Genet. II: 1-100.

Charlesworth, B. (1974). Genet. Res. Camb. 23: 259-280.

Charlesworth, B. and D. Charlesworth. (1973). Genet. Res. Camb. 21: 167-183.

Christiansen, F.B. and M.W. Feldman. (1975). Theor. Pop. Biol. 7: 13-38.

Clarke, C.A. and P.M. Shepherd. (1960). Heredity 14: 73-87.

Crow, J.F. and M. Kimura. (1965). Am. Natur. 99: 439-450.

Deakin, M.A.B. (1972). J. Theor. Biol. 35: 191-212.

Dobzhansky, T.H. (1970). Genetics of the Evolutionary Process. Columbia University Press, New York.

Dobzhansky, T.H. and B. Spassky. (1953). Genetics 38: 471-484.

Edwards, A.W.F. (1961). Heredity 16: 291-304.

Eshel, I. (1973). Theor. Pop. Biol. 4: 196-208.

Eshel, I. (1975). Heredity (in press).

Eshel, I. and M.W. Feldman. (1970). Theor. Pop. Biol. 1: 88-100.

Ewens, W.J. (1965a). Ann. Hum. Genet. 29: 85-88.

Ewens, W.J. (1965b). Heredity 20: 443-450.

Ewens, W.J. (1966). Heredity 21: 363-370.

Ewens, W.J. (1967). Am. Natur. 101: 35-40.

Feldman, M.W. (1972). Theor. Pop. Biol. 3: 324-346.

Feldman, M.W. and B. Balkau. (1973). Genetics 74: 713-726.

Feldman, M.W. and J.F. Crow. (1970). Theor, Pop. Biol. 1: 371-391.

Feldman, M.W. and S. Karlin. (171). Theor. Pop. Biol. 2: 482-492.

Finney, D.J. (1952). Genetica 26: 33-64.

Fisher, R.A. (1928). Am. Natur. 62: 115-126.

Fisher, R.A. (1929). Am. Natur. 63: 553-556.

Fisher, R.A. (1930). The Genetical Theory of Natural Selection. (Rev. 2nd Ed. Dover, New York, 1958).

Ford, E.B. (1965). Genetic Polymorphism. Faber and Faber, London.

Haldane, J.B.S. (1931). Proc. Camb. Phil. Soc. 27: 137-142.

Haldane, J.B.S. (1957). J. Genetics 55: 218-225.

Hartl, D. (1975). Theor. Pop. Biol. 7: 168-174.

Hiraizumi, Y., L. Sandler and J.F. Crow. (1960). Evolution 14: 433-444.

Karlin, S. (1972). Am. Math. Mo. 79: 699-739.

Karlin, S. (1974). In Population Dynamics, edited by M. Bartlett and R. Hiorns (1973). Academic Press.

Karlin, S. (1975). Theor. Pop. Biol. 7: 364-398.

Karlin, S. and D. Carmelli. (1975). Theor. Pop. Biol. 7: 399-421.

Karlin, S. and M.W. Feldman. (1968). Genetics 59: 105-116.

Karlin, S. and M.W. Feldman. (1970). Theor. Pop. Biol. 1: 39-71.

Karlin, S. and U. Liberman. (1974). Theor. Pop. Biol. 6: 355-382.

Karlin, S. and J. McGregor. (1972a). Theor. Pop. Biol. 3: 186-209.

Karlin, S. and J. McGregor. (1972b). Proc. Nat. Acad. Sci. 69: 3611-3614.

Karlin, S. and J. McGregor. (1974). Theor. Pop. Biol. 5: 59-103.

Karlin, S. and F.M. Scudo. (1969). Genetics 63: 499-510.

Kimura, M. (1965). Genetics 52: 875-890.

Kolman, W. (1960). Am. Natur. 94: 373-377.

Lewontin, R.C. (1968). Evolution 22: 262-273.

Lewontin, R.C. (1971). Proc. Nat. Acad. Sci. 68: 984-986.

Lewontin, R.C. (1974). The Genetic Basis of Evolutionary Change. Columbia Univ. Press, New York.

Lewontin, R.C. and K. Kojima. (1960). Evolution 14: 458-472.

Maynard Smith, J. (1968). Am. Natur. 102: 469-473.

Maynard Smith, J. (1971). Theor. Pop. Biol. 30: 319-335.

Muller, H.J. (1932). Proc. Sixth Int'l Cong. Genet. I: 213-255.

Nei, M. (1967). Genetics 57: 625-640.

Nei, M. (1969). Genetics 63: 681-699.

O'Donald, P. (1968). Proc. Roy. Soc. Lond. 171: 127-143.

O'Donald, P. and J.A. Barrett. (1973). Theor. Pop. Biol. 4: 173-192.

Policansky, D. (1974). Am. Natur. 108: 75-90.

Policansky, D. and J. Ellison. (1970). Science 169: 888-889.

Prout, T. (1968). Am. Natur. 102: 493-496.

Prout, T., J. Bungaard and S. Bryant. (1973). Theor. Pop. Biol. 4: 446-465.

Scudo, F.M. (1964). La Ricerca Scientifica 34: (II-B): 93-146.

Scudo, F.M. and S. Karlin. (1969). Genetics 63: 479-498.

Sved, J. and O. Mayo. (1970). In Topics in Mathematical Genetics. K. Kojima, Ed., Springer Verlag.

Thomson, G.J. and M.W. Feldman. (1974). Theor. Pop. Biol. 5: 155-162.

Thomson, G.J. and M.W. Feldman. (1975a). Manuscript in preparation.

Thomson, G.J. and M.W. Feldman. (1975b). Theor. Pop. Biol. 8: (in press).

Turner, J. (1967). Evolution 21: 645-656.

Wallace, B. (1948). Evolution 2: 189-217.

Williams, G.C. and J.B. Mitton. (1973). J. Theor. Biol. 39: 545-554.

Wright, S. (1929). Am. Natur. 63: 274-279.

Wright, S. (1952). In Quantitative Inheritance. K. Mather, Ed. 5-41; Her Majesty's Stationary Office, London.

Zimmering, S., L. Sandler and B. Nicoletti. (1970). Ann. Rev. Genet. 4: 409-436.

Travelling Population Fronts

K.P. HADELER

In a large population with random mating and homogeneous genetic background consider a single autosomal locus with two alleles A and a . Let f , 1 , g be the viabilities of the three genotypes AA , Aa , aa . If generations are separated then the state of the population at the beginning of each new generation is determined by the frequency of the gene A . Thus the transition from one generation to the next can be described by a nonlinear mapping of gene frequencies

$$\tilde{p} = \frac{fp^2 + (1-p)\, p}{fp^2 + 2p(1-p) + g(1-p)^2} \quad .\tag{1}$$

In the case of overlapping generations at each moment only the offspring is in Hardy-Weinberg equilibrium, and the appropriate model is a system of differential equations for genotype frequencies. Such models have been discussed by various authors (see e.g. [5], [8] [9], [11], [1]). It can be stated that the equilibrium structure of the population depends very much on the mortalities and the mode of replacement. However, if all mortalities are equal then a population in equilibrium is also in Hardy-Weinberg equilibrium. With this assumption Fisher's equation

$$\dot{p} = p(1-p)(1-g-(2-f-g)p)\tag{2}$$

is a useful approximation at least in the neighborhood of an equilibrium. In [1], [11] the possible deviations from Hardy-Weinberg equilibrium have been investigated.

Now let the population be distributed in a spatial domain (for simplicity one-dimensional). Let $p(s,t)$ be the frequency of the gene A at the point $s \in \mathbb{R}$ at time t. At each point s the population is governed by the law (2), furthermore the individuals migrate with a diffusion rate $D > 0$ independent of the genotype. Thus we are led to a nonlinear diffusion equation [6], [5]

$$p_t = Dp_{ss} + p(1-p)(1-g-(2-f-g)p) \quad , \quad D > 0 \quad . \tag{3}$$

We always assume that A is the advantageous gene. First we consider the case of intermediate heterozygotes, $f \geqslant 1$, $0 \leqslant g < 1$. By an appropriate substitution in t and s we achieve

$$p_t = p_{ss} + p(1-p)(1+\nu p) \quad , \tag{4}$$

where $\nu = -1 + (f-1)/(1-g)$, $-1 \leqslant \nu < +\infty$.

Assume the population starts at $t = 0$ with the distribution

$$p(s,0) = \begin{cases} 1 & \text{for } s \leqslant 0 \\ 0 & \text{for } s > 0 \end{cases} \tag{5}$$

Due to migration and selection the population front will shift to the right for $t > 0$. One can ask whether for large t the front will travel with (asymptotically) constant speed and fixed shape. Mathematically this question amounts to finding a solution of the form $p(s,t) = u(s-ct)$, where u satisfies the ordinary differential equation

$$-c\dot{u} = \ddot{u} + u(1-u)(1+\nu u) \tag{6}$$

and the boundary conditions $u(-\infty) = 1$, $u(+\infty) = 0$.

This question has been posed by Fisher [5], Kolmogorov et al. [17] for the case of complete dominance ($f = 1$ or

$\nu = -1$). The authors [17] investigated a general equation

$$u_t = u_{ss} + F(u) \quad , \tag{7}$$

where $F: [0,1] \to \mathbb{R}$ is continuously differentiable and satisfies

$$F(0) = F(1) = 0 \ , \ F(u) > 0 \text{ for } u \in (0,1) \ , \ F'(0) = \alpha^2 \ , \ \alpha > 0. \tag{8}$$

In [17] the existence of travelling fronts has been established under an additional hypothesis (namely that F' assumes its maximum only for $u = 0$): The minimal possible speed is $c_o = 2\alpha$. To every $c \geq c_o$ corresponds a front with speed c. These results cover Fisher's equation for $\nu \leq 1$ or equivalently for $f+g \leq 2$. In this domain the minimal speed is constant, $c_o = 2$.

For $\nu > 1$ equation (4) does not satisfy the hypothesis of [17]. Franz Rothe and the author [20] have proved the existence of travelling fronts only with the hypothesis (8). A general characterization of the minimal speed could be obtained (see also [1])

$$c_o = \min_{\rho} \ \sup_{0<u<1} \ \left\{ \rho'(u) + \frac{F(u)}{\rho(u)} \right\} ,$$

where $\rho:[0,1] \to \mathbb{R}$ is continuously differentiable and $\rho(0) = 0$, $\rho(u) > 0$ for $u \in (0,1)$, $\rho'(0) > 0$.

With assistance of this extremal principle the minimal speed of a travelling front in the model (4) can be obtained as [20]

$$c_o = \begin{cases} 2 & \text{for } \nu \leq 2 . \\[2mm] \dfrac{\nu+2}{\sqrt{2\nu}} & \text{for } \nu \geq 2 . \end{cases}$$

It can be shown that the initial data (5) give rise to a front with minimal speed c_o , whereas the greater propagation speeds $c > c_o$ are obtained for certain "flattened out" initial data ([13], [20]).

These results can be applied to more general population genetic models, e.g. to models with frequency-dependent fitness coefficients, $p_t = p_{ss} + p(1-p)(h(p) - g(p) - (2h(p) - f(p) - g(p))p)$, where the positive functions f, h, g describe the fertilities of the genotypes AA, Aa, aa, depending on the frequency p . From (8) follows a sufficient condition for the existence of travelling fronts $h(p) - g(p) \geq (2h(p) - f(p) - g(p))p$ for $0 < p < 1$, $h(0) > g(0)$.

Example: Let the fitness parameters of the AA and Aa genotypes be constant, $h(p) \equiv h_o$, $f(p) \equiv f_o \geq h_o$. Then the condition for the aa fitness g is

$$g(p) \leq h_o + (f_o - g_o)p/(1-p) .$$

Thus if $f_o > h_o$ then a travelling front exists (and a is eliminated everywhere) even if the fertility of homozygotes aa is large at low frequencies of a .

Results on the convergence problem have been obtained and will be published elsewhere ([20], see also [13], [1]). Stationary solutions in genetic clines have been investigated in [2].

Existence proofs and the variational principle have been generalized to a much wider class of nonlinear systems of ordinary differential equations depending on a paremter (see [10]). In particular we can treat the diffusion equation (7) when the function F changes sign. If there is a $\mu \in (0,1)$ such that $F(u) < 0$ for $0 < u < \mu$, $F(u) > 0$ for $\mu < u < 1$ then there is a half-line $[c_o, \infty)$ of positive speeds corresponding to monotonely decreasing fronts with $u(-\infty) = 1$, $u(+\infty) = \mu$, a half-line $(-\infty, -\overline{c}_o]$, $\overline{c}_o > 0$, of negative speeds for monotonely decreasing fronts with $u(-\infty) = \mu$, $u(+\infty) = 0$, and finally a single speed c_1 (which may be positive or negative) for a front with $u(-\infty) = 1$, $u(+\infty) = 0$.

These results can be applied to Fisher's model (3) in the case of inferior heterozygotes, $f > 1$, $g > 1$. By a simple substitution we arrive at

$$p_t = p_{ss} + p(1-p)(p-\mu) \quad ,$$

where $\mu = (g-1)/(f+g-2)$, $0 < \mu < 1$.

For this equation we can give a complete description of all types of travelling fronts. In view of the symmetries $s \leftrightarrow -s$ and $f \leftrightarrow g$ we can restrict ourselves to the case $\mu \in (0,1/2)$ and to nonnegative speeds. Let

$$c^* = 2\sqrt{\mu(1-\mu)} \quad ,$$

$$c_o = \begin{cases} (1+\mu)/\sqrt{2} & \text{for} \quad 0 < \mu \leq 1/3 \quad , \\ c^* & \text{for} \quad 1/3 \leq \mu \leq 1/2 \quad , \end{cases}$$

$$c_1 = \frac{1}{\sqrt{2}} - \mu\sqrt{2} \quad .$$

Then:

i) For $c \geq c_o$ there is a monotonely decreasing front with boundary conditions $u(-\infty) = 1$, $u(+\infty) = \mu$.

ii) For $c \geq c^*$ there is a monotonely increasing front with boundary conditions $u(-\infty) = 0$, $u(+\infty) = \mu$.

iii) For $c = c_1$ there is a unique monotone front with boundary conditions $u(-\infty) = 1$, $u(+\infty) = 0$.

In addition there are oscillatory fronts for some $c \in (0,c^*)$, and non-monotone, non-oscillating fronts for some $c \in (c^*,c_o)$. For more details, see [10].

For $\mu > 1/2$, i.e., for $f > g$, the front with speed c_1 travels to the right, for $\mu < 1/2$ to the left.

The aforementioned results on nonlinear systems allow also an application to Kendall's epidemic model [16], [18]. The diffusion equations

$$S_t = -k(I + \rho I_{ss})S \quad , \quad I_t = kS(I + \rho I_{ss}) - \mu I$$

describe the development of a population consisting of susceptibles (S) and the infectious (I). After a suitable normalization the equations assume the form

$$u_t = -u(v+v_{ss}) \quad , \quad v_t = u(v+v_{ss}) - v \quad .$$

The appropriate boundary conditions are

$$u \to \alpha \quad \text{for} \quad s \to -\infty \quad , \quad u \to \beta \quad \text{for} \quad s \to +\infty \quad , \quad 0 < \alpha < \beta \quad . \tag{9}$$

If a travelling epidemic front or wave $u(s-ct)$, $v(s-ct)$ exists, then $c\dot{u} = u(v+\ddot{v})$, $c\dot{v} = -u(v+\ddot{v}) + v$. Since the second order term $\ddot{v}$ occurs in both equations one can reduce to a first order system in two variables, thereby taking care of the boundary conditions

$$\dot{u} = (1 + \frac{1}{c^2})v - (\gamma - u + \log u) \quad , \quad \dot{v} = (\gamma - u + \log u) - v \quad ,$$

where

$$\gamma = \beta - \log \beta = \alpha - \log \alpha \quad . \tag{10}$$

For a given $\beta \neq 1$ there is exactly one $\alpha \neq \beta$ satisfying equation (10). In view of (9) we have $\alpha < 1 < \beta$. Let the density β of susceptibles before infection be prescribed. Then $\beta - \alpha$ measures the local effect of the epidemic (decrease of susceptibles). If α is too large the epidemic slows down, if α is too small then the wave explodes. Thus we have the result:

Epidemic fronts exist iff condition (10) is satisfied. Then there is a half-line $[c_o, \infty)$ of speeds with

$$c_o = 2\sqrt{\beta(\beta-1)} \quad .$$

Noble [19] proposed an epidemic model $S_t = -kIS + S_{ss}$, $I_t = kIS - \mu I + I_{ss}$, in normalized form $u_t = -uv + u_{ss}$, $v_t = uv - v + v_{ss}$ with boundary conditions (9), and compared it with data for the propagation of the black death in the middle ages. Now the ordinary system for a travelling epidemic wave is $\ddot{u}-uv+c\dot{u} = 0$, $\ddot{v}+uv-v+c\dot{v} = 0$. This is a genuinely four-dimensional system. The existence of solutions to boundary conditions (9) cannot be obtained by our methods. However, we can show the necessary conditions

$$\int_{-\infty}^{\infty} v(t)\,dt = c(\beta-\alpha) \quad \text{and} \quad \beta-\log \beta > \alpha - \log \alpha \ .$$

REFERENCES

[1] Aronson, D.G. and H.F. Weinberger.(1975). In Proc. of the Tulane Program in partial differential equations, Lecture Notes in Mathematics. Springer.

[2] Conley, C. J. Math. Biol. (To appear).

[3] Crow, J.F. and M. Kimura. (1970). An introduction to population genetics. Harper and Row, N.Y.

[4] Fife, P.C. and J.B. McLeod. (To appear).

[5] Fisher, R.A. (1930). The genetical theory of natural selection. Oxford Univ. Press.

[6] Fisher, R.A. (1937). Ann. Eugen. 7: 355-369.

[7] Gelfand, I.M. (1959). Uspekhi Math. Nauk (N.S.) 14: 87-158. Am. Math. Soc. Transl. (2), 29: 295-381, (1963).

[8] Hadeler, K.P. (1974). Mathematik für Biologen. Springer Verlag.

[9] Hadeler, K.P. (1974). J. Math. Biol. 1: 51-56.

[10] Hadeler, K.P., and F. Rothe. (To appear).

[11] Hoppensteadt, F.C. (1974). Mathematical theories of populations, demographics, genetics and epidemics. Univ. of West Florida, Pensacola, Fl.

[12] Kanel', Ja.I. (1961). Dokl. Akad. Nauk SSSR 132: 268-271; Soviet Math. Dokl. 1, 533-536.

[13] Kanel' Ja.I. (1961). Dokl. Akad. Nauk SSSR 136: 277-280; Soviet Math. Dokl. 2: 48-51.

[14] Kanel', Ja.I. (1962). Mat. Sbornik (N.S.) 59 (101), suppl. 245-288.

[15] Kanel', Ja.I. (1964). Mat. Sbornik (N.S.) 65 (107) 398-413.

[16] Kendall, D.G. (1965). Mathematical models of the spread of infection, Mathematics and computer science in biology and medicine, Medical Research Council.

[17] Kolmogoroff, A., I. Petrovskij, and N. Piskunov.(1937). Bull. Univ. Moscou, Ser. Internat. Sec. A, 1; 6: 1-25.

[18] Ludwig, D. (1974). In Lecture Notes in Biomathematics 1: Springer Verlag.

[19] Noble, J.V. (1974). Nature 250: 726-728.

[20] Rothe, F. (1975). Dissertation, Univ. Tübingen.

Stochastic Selection and the Maintenance of Genetic Variation

D.L. HARTL and R.D. COOK

INTRODUCTION

Several years ago, motivated largely by anxiety over the assumption commonly made in population genetic models that the selective values of genotypes are constants, an assumption which we felt to be doubtful, we undertook a study of models in which the selection parameters were explicitly assumed to be variable (Hartl and Cook, 1973). Models of spatial variation in selection intensity had been closely studied by Levene (1953, 1967), Prout (1968) and Levins (1968), and the conditions for the maintenance of genetic polymorphism in such models were well known. As a consequence, we restricted our attention to models of temporal variation in selection intensity, and although haploid models of this sort had been stu-

---

* Research supported by NIH grant GM21732, NSF grant GB43209 and Research Career Development Award GM0002301.

died by Dempster (1955) and Jensen and Pollak (1969), there existed no corresponding theory for diploids. The nearest approach was a paper by Haldane and Jayakar (1963), in which they had analyzed the population dynamics of a recessive gene assumed to be deleterious in most generations but favored from time to time.

Temporal variation in fitness is a plausible enough assumption. It was formally invoked by Fisher and Ford (1947) to account for fluctuations in the frequency of the medionigra color morph in a highly isolated population of the moth Panaxia dominula near Oxford, England. Further experimental evidence on the point is scanty, but two recent papers merit special attention. Powell (1971) has studied the amount of isozyme variability maintained in population cages of Drosophila willistoni supplied with several alternative food sources, intended to simulate a spatially heterogeneous environment, and he has observed that the populations maintain a somewhat greater amount of genetic variability when the cages are alternated between 19°C and 25°C than when they are maintained at a constant temperature.

The other recent evidence on temporal variation in fitness is a remarkable paper by Smith (1975), who studied the mating success of two color morphs, chrysippus and dorippus, of the butterfly Danaus chrysippus in a natural population in Tanzania. Smith found that during some months of the year chrysippus males possessed a mating advantage, during other months dorippus males possessed an advantage, and during still other months the mating success of the two color morphs was about equal. The extraordinary finding was that, averaged over the time period of the observations, the mating success of males of both forms was very nearly identical.

Despite the scanty evidence bearing on temporal variation

in selection intensity, the assumption that fitnesses do vary temporally has a certain intuitive appeal. It seems reasonable to assume that fitness depends, at least to some extent, on such factors as the density, growth rate and age structure of the population; on such factors as the prevalence of competitors, parasites, predators and prey; and on such abiotic factors as temperature, humidity, wind velocity, amount of sunlight and so on. It is of course impossible at the present time to specify exactly how fitness depends on all these factors, and a model with all the fitness dependencies and interrelations spelled out would probably be too difficult to comprehend anyway. In our models, as a first approximation, we have assumed that fitness is a random variable. This is tantamount to assuming that the many factors effecting fitness can be treated in much the way as random noise is treated in communications theory.

The striking result of our initial investigation of stochastic selection was that, under certain conditions, purely random selection could maintain genetic polymorphisms even when the expected fitness of each genotype was the same (Hartl and Cook, 1973). Under these conditions, every sample path converges to a single equilibrium value, a phenomenon reminiscent of overdominance in deterministic selection models. These results have been refined and significantly extended by Karlin and Lieberman (1974). In this paper they will be extended still further. The model to be examined is one of stochastic selection in a large population. Using the diffusion approximation for stochastic selection developed by Levikson (1974), we derive the limiting distribution of gene frequencies under the assumption that the arithmetic mean fitness of all genotypes is the same. Various statistical properties of the distribution are then examined. The significance of stochastic selection in the maintenance of genetic polymorphism is then discussed.

THE MODEL

Imagine a locus having two alleles, A and a, in a population of size N, and suppose that the frequency of the A allele is x ($0 \leqslant x \leqslant 1$). Under random mating, the genotype frequencies of AA, Aa and aa will be x^2, $2x(1-x)$ and $(1-x)^2$. Assume that the relative fitnesses of the genotypes in any generation can be written as follows:

Genotype	AA	Aa	aa
Fitness	$1+s_1$	1	$1+s_3$

where (s_1, s_3) is a random sample from some bivariate distribution (S_1, S_3) with $S_1, S_3 \geqslant -1$. Assume further that the fitnesses in successive generations are not autocorrelated.

The model above contains five fitness parameters of interest, namely $E(S_1)$, $E(S_3)$, $E(S_1^2)$, $E(S_3^2)$ and $E(S_1 S_3)$. The expected fitness of AA is $1+E(S_1)$; the expected fitness of aa is $1+E(S_3)$. We assume $E(S_1) = E(S_3) = 0$ so that, on the average, the three genotypes in the population have the same fitness. The model with $E(S_1) = E(S_3) = 0$ is referred to as quasineutrality (Hartl and Cook, 1973). This apparently gratuitous assumption has had its biological plausibility bolstered by Smith's (1975) recent findings concerning mating success of males of Danaus chrysippus. The assumption that the means of S_1 and S_3 are both zero concentrates attention on the consequences of variation in selection intensity itself, apart from other systematic forces attributable to differences in the means.

In this model,

$$\Delta x = \frac{x(1-x)[xS_1-(1-x)S_3]}{1+S_1x^2+S_3(1-x)^2} \ . \tag{1}$$

The appropriate diffusion approximation for this model has drift term $\mu(x) = \lim_{N\to\infty} EN\Delta x$ and diffusion term $\sigma^2(x) = \lim_{N\to\infty} EN(\Delta x)^2$ (Karlin & Levikson, 1974). In the case at hand, using a Taylor series to approximate the denominator of (1),

$$\mu(x) = \lim_{N\to\infty} ENx(1-x)[xS_1-(1-x)S_3][1-S_1x^2-S_3(1-x)^2]$$

$$= x(1-x)[-x^3v_1+(1-x)^3v_3+x(1-x)(2x-1)r] \ ,$$

$$\sigma^2(x) = \lim_{N\to\infty} ENx^2(1-x)^2[xS_1-(1-x)S_3]^2$$

$$= x^2(1-x)^2[x^2v_1+(1-x)^2v_3-2x(1-x)r] \ ,$$

where we have assumed $\lim_{N\to\infty} ENS_1 = \lim_{N\to\infty} ENS_3 = 0$. The parameters are

$$v_1 = \lim_{N\to\infty} ENS_1^2 > 0$$

$$v_3 = \lim_{N\to\infty} ENS_3^2 > 0$$

$$r = \lim_{N\to\infty} ENS_1S_3 \ ,$$

and all higher moments are assumed to be of order $1/N$.

It will be useful in what follows to introduce the parameters

$$\beta = v_3/v_1$$

$$\alpha = r/\sqrt{v_1v_3} \ .$$

The parameter β is therefore the ratio of fitness variances of homozygotes and α is the correlation coefficient.

ANALYSIS

The case $\alpha = +1$ has been extensively analyzed using martingale theory (Hartl and Cook, 1973, Karlin and Lieberman, 1974, Levikson, 1974, Cook and Hartl, 1975). The striking result is that all sample paths converge to

$$\hat{x} = \frac{\sqrt{v_3}}{\sqrt{v_1} + \sqrt{v_3}} = \frac{\sqrt{\beta}}{1 + \sqrt{\beta}} \quad , \tag{2}$$

that is to say, $\text{Prob}\{\lim_{n\to\infty} x_n = \hat{x}\} = 1$. In the first paper on this model (Hartl and Cook, 1973) we parameterized the fitnesses of AA, Aa and aa as $1+K_{11}S$, $1+K_{12}S$ and $1+K_{22}S$ where $E(S) = 0$ and $E(S^2) = 1$. Thus the correspondence is $K_{11} = \sqrt{v_1}$, $K_{12} = 0$ and $K_{22} = \sqrt{v_3}$.

When $\alpha \neq +1$, Levikson (1974) has shown that the model always leads to a globally stable stationary distribution of gene frequency. The probability density of this distribution, $\phi(x)$, satisfies (Ewens, 1969)

$$\phi(x) = \frac{k}{\sigma^2(x)} e^{2\int [\mu(x)/\sigma^2(x)]dx} \quad , \quad 0 \leqslant x \leqslant 1$$

where k is a constant of integration.

Partial fraction expansion of $\mu(x)/\sigma^2(x)$ leads to

$$\frac{\mu(x)}{\sigma^2(x)} = \frac{1}{x} - \frac{1}{1-x} + \frac{(v_3+r) - (v_1+v_3+2r)x}{v_3-2(v_3+r)x+(v_1+v_3+2r)x^2} \quad ,$$

and consequently

$$\phi(x) = \frac{k}{(a+bx+cx^2)^2} \quad , \quad 0 \leqslant x \leqslant 1 \tag{3}$$

where

$$a = v_3 \quad , \quad b = -2(v_3+r) \quad , \quad c = v_1+v_3+2r$$

and the constant of integration, k , is

$$k = 2(v_1 v_3 - r^2)/$$

$$[\frac{v_1+r}{v_1} + \frac{v_3+r}{v_3} + \frac{(v_1+v_3+2r)}{\sqrt{v_1 v_3 - r^2}} \; (\tan^{-1} \frac{v_1+r}{\sqrt{v_1 v_3 - r^2}} + \tan^{-1} \frac{v_3+r}{\sqrt{v_1 v_3 - r^2}})].$$

The density function (3) superficially contains three parameters, v_1, v_3 and r. It is actually a two-parameter density, depending only on the variance ratio β and the correlation coefficient α, although it is convenient for present purposes to write it in the form (3).

CHARACTERIZATION OF THE STATIONARY DENSITY

The density (3) satisfies

$$\frac{d\phi(x)}{dx} = (\frac{-2b - 4cx}{a+bx+cx^2}) \phi(x) \tag{4}$$

where the discriminant of the denominator is $b^2-4ac = -4(v_1 v_3 - r^2) < 0$. Thus $\phi(x)$ is a Pearson Type IV distribution, described by Pearson in 1895 in a paper presciently entitled "Contributions to the mathematical theory of evolution". Pearson's Type IV is defined on $-\infty < x < \infty$, so $\phi(x)$ is more precisely described as a Pearson Type IV doubly truncated at $x=0$ and $x=1$.

From (4) and a minimal amount of straightforward analysis it can be shown that the mode of $\phi(x)$ is given by

$$\text{mode} = \begin{cases} 0 & \text{if } \alpha \leqslant -\sqrt{\beta} \\[2mm] \dfrac{\beta+\alpha\sqrt{\beta}}{1+\beta+2\alpha\sqrt{\beta}} & \text{if } \alpha > \max(-\sqrt{\beta}, -1/\sqrt{\beta} \\[2mm] 1 & \text{if } \alpha \leqslant -1/\sqrt{\beta} . \end{cases}$$

Thus, when $\beta > 1$, the mode is 0 for $\alpha \leqslant -1/\sqrt{\beta}$ and an increasing function of α for $\alpha > -1/\sqrt{\beta}$, the maximum being at $\sqrt{\beta}/(1+\sqrt{\beta})$ when $\alpha = 1$ (compare with equation 2). Conversely, when $\beta < 1$, the mode is 1 for $\alpha \leqslant -\sqrt{\beta}$ and a decreasing function of α for $\alpha > -\sqrt{\beta}$, attaining a minimum of $\sqrt{\beta}/(1+\sqrt{\beta})$ when $\alpha = 1$. When $\beta = 1$, the mode is ½ for all α except $\alpha = -1$, in which case the density has no mode.

The general shape of $\phi(x)$ is shown for several sets of parameters in Figures 1 and 2. Figure 1 has $\beta=1$, and the number that flags each curve is the corresponding value of α . Note that $\alpha = -1$ is the uniform density, and that as α increases from -1 the density becomes increasingly concentrated around ½ . As $\alpha \to 1$ the density becomes degenerate at ½ . Figure 2 exhibits two asymmetrical cases. The left-hand panel has $\beta=2$, the right-hand panel has $\beta=10$. The numbers that mark the curves are again the values of α . In these cases, as $\alpha \to 1$, the density becomes degenerate at $x =\sqrt{\beta}/(1+\sqrt{\beta})$ (equation 2). Note here, especially in the right-hand panel, that the mode is strongly dependent on α . As will be seen later, the mean of the distribution depends only weakly on α .

We have performed simulations of the model to determine how rapidly the gene frequency distribution converges to $\phi(x)$. Five hundred populations evolving independently were periodically grouped into 20 gene frequency classes (0-0.05, 0.05-0.10, 0.10-0.15, etc.), and a χ^2 test was performed for goodness of fit to the limiting density. In these simulations we took $\alpha=0$ and let S_i (i=1,3) be a uniformly distributed pseudorandom number on $(-\sqrt{3v_i},+\sqrt{3v_i})$, so that $ES_i=0$ and $ES_i^2 = v_i$. Taking $v_1 = v_3 = 1/12$ and an initial gene frequency of 0.50 in all populations, we found that the actual distribution of gene frequencies converged to the limiting distribution in an average of about 50 generations, where the criterion of satisfactory fit is a χ^2 value with an associated probability level of 1% or greater. With $v_1 = v_3 = 1/12$ and an initial gene frequency of 0.01 in all populations, convergence was much slower, averaging about 225 generations.

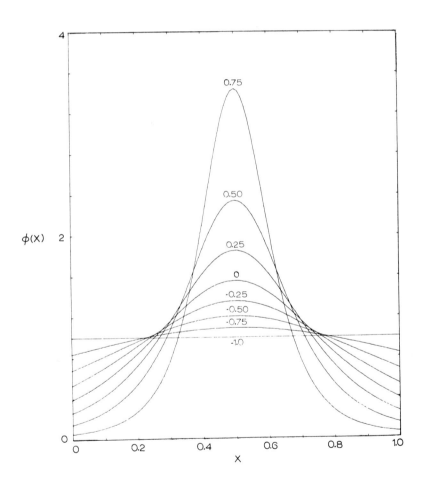

FIGURE 1. The density in equation (3) when β=1 for various values of α . As α → +1 the density becomes degenerate at ½ .

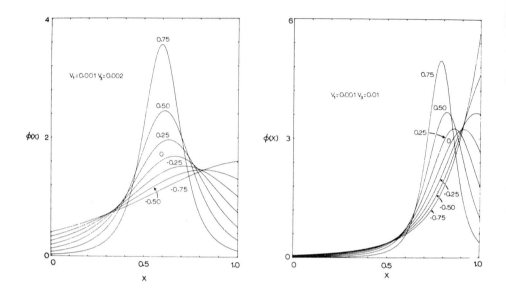

FIGURE 2. *The density in equation (3) when β=2 (left-hand panel) and β=10 (right-hand panel) for various values of α . As α → +1 the densities become degenerate at √β̄/(1+√β̄) . Note the dependence of the mode on α .*

The transformation $z = \sqrt{3}\ c\ (x + \frac{b}{2c})/\sqrt{v_1 v_3 - r^2}$ transforms $\phi(x)$ into a truncated Student's t distribution with 3 degrees of freedom. Letting $k^* = k(v_1 v_3 - r^2)^{-3/2} c/\sqrt{3}$, we have

$$f(z) = \frac{k^*}{(1 + \frac{z^2}{3})^2} \quad , \quad \frac{-(v_3 + r)\sqrt{3}}{\sqrt{v_1 v_3 - r^2}} \leqslant z \leqslant \frac{(v_1 + r)\sqrt{3}}{\sqrt{v_1 v_3 - r^2}} \quad .$$

MOMENTS OF THE STATIONARY DISTRIBUTION

Let $\mu_n = E(x^n)$. From the integration by parts formula

$$\int \frac{x^n}{(a+bx+cx^2)^2} dx = - \frac{x^{n-1}}{(3-n)c(a+bx+cx^2)} - \frac{(2-n)b}{(3-n)c} \int \frac{x^{n-1}}{(a+bx+cx^2)^2} dx +$$

$$\frac{(n-1)a}{(3-n)c} \int \frac{x^{n-2}}{(a+bx+cx^2)^2} dx$$

it is easy to derive the following recursion for the moments
of $\phi(x)$:

$$na\mu_{n-1} + (n-1)b\mu_n + (n-2)c\mu_{n+1} - \frac{k}{v_1} = 0 \quad . \tag{5}$$

Thus, without further integration, $\mu_1 = (k-v_1 b\mu_2)/2av_1$
and $\mu_2 = (av_1-k)/v_1 c$. Therefore

$$\mu_1 = E(x) = \frac{\dfrac{v_1+r}{v_1} + (v_3+r)\Psi}{\dfrac{v_1+r}{v_1} + \dfrac{v_3+r}{v_3} + (v_1+v_3+2r)\Psi} \tag{6}$$

$$\mu_2 = E(x^2) = \frac{\dfrac{r}{v_3} + v_3\Psi}{\dfrac{v_1+r}{v_1} + \dfrac{v_3+r}{v_3} + (v_1+v_3+2r)\Psi} \tag{7}$$

where

$$\Psi = \frac{1}{\sqrt{v_1 v_3 - r^2}} (\tan^{-1}\frac{v_1+r}{\sqrt{v_1 v_3 - r^2}} + \tan^{-1}\frac{v_3+r}{\sqrt{v_1 v_3 - r^2}}) \quad .$$

Recursion (5) is of no use when $n=3$. Evidently the
third moment must be gotten by brute force. After tedious
integration it turns out that $\mu_3 = E(x^3) = A/B$ where

$$A = (v_1+v_3+2r)[2r(v_3+r)-v_3(v_1+r)]+v_1(v_1 v_3-r^2)\log(v_1/v_3)$$

$$+ v_1(v_3+r)(3v_1 v_3+v_3^2+2v_3 r-2r^2)\Psi$$

$$B = v_1(v_1+v_3+2r)^2[\frac{v_1+r}{v_1} + \frac{v_3+r}{v_3} + (v_1+v_3+2r)\Psi] \quad .$$

Higher moments may be obtained recursively from (5).

AN APPROXIMATION TO THE MEAN

It may be shown that $E(x)$ in equation (6) is a decreasing function of α when $\beta > 1$ and an increasing function of α when $\beta < 1$. Thus $E(x)$ for any α is bounded by the mean at the limits $\alpha = -1$ and $\alpha = +1$. That is to say,

$$\frac{\sqrt{\beta}(1+2\sqrt{\beta})}{2(1+\sqrt{\beta}+\beta)} \leq E(x) \leq \frac{\sqrt{\beta}}{1+\sqrt{\beta}} \qquad (8)$$

if $\beta < 1$, and the same expression holds with the inequality signs reversed if $\beta > 1$.

Moreover, the difference between the upper and lower bounds is generally rather small. Denote the left-hand bound of (8) by x^* and the right-hand bound by $\hat{x}$ (as in equation 2). The square of the relative error, $[(x^* - \hat{x})/\hat{x}]^2$, has a local maximum at $\beta = (1+\sqrt{3})^2 \simeq 7.46$, at which point $x^* \simeq 0.79$ and $\hat{x} \simeq 0.73$, an absolute error of 6%. The global maximum of the squared relative error is at $\beta \simeq 0$, in which case $x^* \simeq \sqrt{\beta}/2$ and $\hat{x} \simeq \sqrt{\beta}$, so that the absolute error is merely $\sqrt{\beta}/2$.

The weak dependence of $E(x)$ on α is shown in the left-hand panel of Figure 3. The numbers that identify the curves are the values of $1/\beta$. It will be seen that the curves are nearly linear in α and that the slopes are close to zero. Thus a rough and ready approximation to $E(x)$ for any α may be taken as $\hat{x}$, the value x_n converges to in the case $\alpha = 1$.

The right-hand panel in Figure 3 shows the dependence of $\text{Var}(x)$ on α, calculated from $E(x)$ and $E(x^2)$ in equations (6) and (7). The numbers marking each curve are the corresponding values of $1/\beta$. Note that the variances are

604

identical for any variance ratio $1/\beta$ and its reciprocal β. The variance of gene frequency in the limiting density assumes a maximum at $\alpha = -1$ and decreases monotonically toward 0 as $\alpha \to +1$.

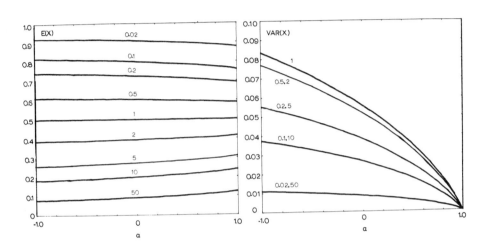

FIGURE 3.   *Left-hand panel: the mean of the stationary den-sity calculated from equation (6) as a function of α for the indicated values of $1/\beta$. Note the rather weak dependence of $E(x)$ on α. Right-hand panel: the variance of the stationary den-sity calculated from equations (6) and (7) as a function of α for the indicated values of $1/\beta$.*

TWO SPECIAL CASES

A.   The case $\alpha = -1$ is of some interest because it con-trasts readily with the opposite extreme of $\alpha = 1$. When $\alpha = -1$,

$$\phi(x) = \frac{k'}{[\sqrt{\beta} + (1-\sqrt{\beta})x]^4} \quad , \quad 0 \leqslant x \leqslant 1$$

where $k' = k/v_1^2$. If $\beta = 1$ the model is one of additive gene effects -- that is, the fitnesses of AA, Aa and aa can be written as $1+S_1$, 1, $1-S_1$. In the case $\beta = 1$, $\phi(x)$ is a uniform density: $\phi(x) = 1$ $(0 \leqslant x \leqslant 1)$, and $E(x) = \frac{1}{2}$, $Var(x) = 1/12$ (Cook and Hartl, 1975). This is worth noting because the uniform density is also the limiting density of a neutral allele in a finite population, conditional on the event that the allele is still segregating (Crow and Kimura, 1970). Thus we have a model of stochastic selection that mimicks in a large population the effects of selective neutrality in small populations, except for the fact that in small populations alleles can be fixed or lost.

In general, when $\alpha = -1$ and $\beta \neq 1$,

$$E(x) = \frac{\sqrt{\beta}(1+2\sqrt{\beta})}{2(1+\sqrt{\beta}+\beta)}$$

$$Var(x) = \frac{3\beta}{4(1+\sqrt{\beta}+\beta)^2} .$$

B. Another case of interest is $v_1 = v_3$ with $\alpha = 0$ because in this case the limiting density becomes considerably simplified. Specifically,

$$\phi(x) = \frac{2}{(2+\pi)[x^2+(1-x)^2]^2} , \quad 0 \leqslant x \leqslant 1$$

and $E(x) = \frac{1}{2}$, $Var(x) = (\pi-2)/4(\pi+2) \simeq 0.05$.

DISCUSSION

The consequences of stochastic selection are often complex, and intuition alone is usually an unreliable guide, often uncertain, at times mistaken. In the case of quasineutrality when $\alpha = +1$, for example, it is an extraordinary and unexpected result that all sample paths converge to one unique equilibrium. It is clear enough, of course, intuitively, that

there will be some kind of selection tending to maintain variation, because the geometric mean fitness of homozygous
genotypes is less than the geometric mean fitness of heterozygous genotypes. That much was perceived by Haldane and
Jayakar in 1963; they saw rather clearly that the geometric
mean fitnesses of the genotypes are critical parameters whenever there is temporal variation in selection intensity. But
that quasineutrality when $\alpha = +1$ should lead to a degenerate
equilibrium distribution of gene frequencies with zero variance could not have been predicted from examination of the
geometric means alone.

Or consider the case of quasineutrality when the variance
in fitness of homozygous genotypes is equal, and $\alpha = -1$.
Relying on comparisons of geometric mean fitness, one could
reasonably expect intuitively that the limiting distribution
of gene frequencies might have a mode at $\frac{1}{2}$; but in fact the
limiting distribution of gene frequencies is uniform. Even
if intuition had not led to the expectation of a mode at $\frac{1}{2}$,
it would not likely have pointed unambiguously to a uniform
density at the stationary state as opposed to, for example,
any one of an infinite variety of U-shaped densities. One
obejctive of rigorous analysis of models of stochastic selection is therefore to sharpen the intuition to deal effectively
with such models.

Stochastic selection bears on the question of the maintenance of genetic polymorphisms. On the one hand, stochastic
selection is clearly a kind of selectionist hypothesis; on
the other hand, models of quasineutrality are, in a certain
sense, kinds of neutralist hypotheses because the expected
fitness of all genotypes is the same. Thus the models are
from one perspective selectionist and from another perspective neutralist. Models of stochastic selection are, in this

sense, a synthesis of neutralist and selectionist models. The synthetic nature of these models is not superficial. In models of neutral alleles there is only one force in the population, a dispersive force resulting from the variance in the sampling of gametes each generation. In models of deterministic overdominance in large populations, there is again only one force, a systematic force tending to move gene frequencies along trajectories leading to the equilibrium. In models of stochastic selection, by contrast, two forces are at work simultaneously. On the one hand, there is a systematic force tending to keep gene frequencies at intermediate values; this force is caused by the geometric mean fitness of homozygotes being less than the geometric mean fitness of heterozygotes, and its magnitude is proportional to the variance in fitness of homozygotes. On the other hand, there is also a dispersive force brought about because a genotype that is favored by selection in one generation may be disfavored in a subsequent one, and the magnitude of this force is also proportional to the variance in fitness of homozygotes. Thus stochastic selection has both a systematic force, analogous to the systematic force in models of deterministic overdominance, and also and simultaneously a dispersive force, analogous to the dispersive force of genetic drift in finite populations. In this way the most important features of neutrality and overdominance are brought together in a single model.

Because models of stochastic selection have both systematic and dispersive effects on gene frequency, their consequences are often somewhat intermediate between models of deterministic overdominance and models of neutrality (Cook and Hartl, 1975). Exceptions include cases like $\alpha = +1$, which behaves very much like deterministic overdominance, and the symmetric case when $\alpha = -1$, which behaves very much like the pure drift process in finite populations that are still segregating neutral alleles. Thus a collection of data from loci subjected to stochastic selection may, upon analysis, have certain features indicative of neutrality and at the same time exhibit other features indicative of deterministic selection in favor of heterozygotes. The data from such loci would therefore be ambiguous if the hypotheses being compared were neutrality versus deterministic selection, but the ambiguities result only becuase the hypotheses are not mutually exclusive; stochastic selection combines elements of both simpler models. Stochastic selection can most straightforwardly be detected by performing long-term, periodic observations of a natural population, such as in Fisher and Ford (1947) and Smith (1975), and these kinds of studies are badly needed.

The fact that the most important features of neutrality and overdominance can be combined in a single model does not, of course, mean that a substantial proportion of genetic variation in natural populations is maintained by stochastic selection. Only observation and experiment can decide that issue. But models of stochastic selection have unique consequences that ought to stimulate a careful evaluation of the potential role of stochastic selection in the maintenance of genetic variation. Stochastic selection does, at the very least, provide a fresh conceptual framework in which to think about the problem of the maintenance of genetic variation, and, like Lewontin (1974), we believe that the theoretical framework exemplified by the neutral theory and deterministic overdominance may be unnecessarily narrow and perhaps even misleading.

ACKNOWLEDGEMENT

Our special thanks to Professor Benny Levikson, Department of Mathematics, Purdue University, for his unflagging help and encouragement.

REFERENCES

Cook, R.D. and D.L. Hartl. (1975). Theor. Pop. Biol. 7: 55-63.

Crow, J.F. and M. Kimura. (1970). An Introduction to Population Genetics Theory. Harper and Row, New York.

Dempster, E.R. (1955). Cold Spring Harbor Symp. Quant. Biol. 20: 25-32.

Ewens, W.J. (1969). Population Genetics. Methuen, London.

Fisher, R.A. and E.B. Ford. (1947). Heredity 1: 143-174.

Haldane, J.B.S. and S.D. Jayakar. (1963). J. Genet. 58: 237-242.

Hartl, D.L. and R.D. Cook. (1973). Theor. Pop. Biol. 4: 163-172.

Jensen, L. and E. Pollak. (1969). J. Appl. Prob. 6: 19-37.

Karlin, S. and U. Lieberman. (1974). Theor. Pop. Biol. 6: 355-382.

Karlin, S. and B. Levikson. (1974). Theor. Pop. Biol. 6: 383-412.

Levene, H. (1953). Am. Natur. 87: 131-133.

Levene, H. (1967). Proc. Fifth Berkeley Symp. Math. Stat. Prob. 4: 305-316.

Levikson, B. (1974). Ph.D. Thesis, Tel-Aviv Univ., Israel.

Levins, R. (1968). Evolution in Changing Environments. Princeton University Press, Princeton, N.J.

Lewontin, R.C. (1974). The Genetic Basis of Evolutionary Change. Columbia Univ. Press, New York.

Pearson, K. (1895). Phil. Trans. Roy. Soc. Ser.A 186: 343-414.

Powell, J.R. (1971). Science 174: 1035-1036.

Prout, T. (1968). Am. Natur. 102: 493–496.

Smith, D.A.S. (1975). Science 187: 664–665.

COMMENTS BY M. NEI

I would like to make two comments on this paper. First, the assumption of an infinitely large population is never fulfilled in real populations. If we remove this assumption, the gene frequency distribution becomes no longer stable, and it would be meaningless to compute the equilibrium distribution without the effect of mutation.

Second, even if we consider a hypothetical population of infinite size, Hartl and Cook's conclusion is highly model-dependent, and a slight change of the model would lead to an entirely different result. This can be seen by the following simple examples: First consider a special case of genic selection in which the fitnesses of A_1A_1, A_1A_2 and A_2A_2 are given by $W_1 = 1-s$, $W_2 = 1$, and $W_3 = 1+s$, respectively, and assume that s is a random variable. It can then be shown that the mean change of gene frequency per generation $(M_{\delta x})$ in diffusion approximations is $V_s x(1-x)(1-2x)$ when the mean of s is 0, where V_s is the variance of s and x is the frequency of gene A_2. However, if we change the model slightly so that $W_1 = 1$, $W_2 = 1+s$, and $W_3 = 1+2s$, then $M_{\delta x}$ becomes $-2V_s x^2(1-x)$. On the other hand, if we assume $W_1 = 1-2s$, $W_2 = 1-s$, and $W_3 = 1$, then $M_{\delta x}$ is $2V_s x(1-x)^2$. Note that all these examples refer to genic selection. Furthermore, if we use the competitive selection model described in my paper (Nei, M., Genetics $\underline{68}$: 169, 1971), $M_{\delta x}$ becomes 0 under genic selection. Therefore, the conclusion about the effect of random fluctuation of selection intensity will be entirely different for different models. Note that the above four models are similar enough to give the same result in diffusion approximations when s is not a random variable but a constant. In genetic literature they

612

are often used interchangeably.

In practice, we do not know which model is most realistic. Probably none of them is really close to the real process. But the above examples show that we cannot make a general inference about nature from a study of a specific mathematical model. In the present case we are studying the second-order effect of selection coefficient, so that a little change in assumptions makes a big difference in results. My own view about the effect of random fluctuation of selection intensity is the same as Fisher and Ford's (Heredity $\underline{1}$: 143, 1947) original idea that it is essentially "random noise" and similar to the effect of random genetic drift due to finite population size.

REPLY BY D. HARTL

Professor Nei criticizes my paper with R.D. Cook on two counts. Nei's first criticism centers on the fact that real populations are of finite size. Yes, they are. However -- and quite apart from the fact that certain populations of viruses, bacteria or fungi are sufficiently large that the assumptions of an effectively infinite population size is probably okay -- the infinite population case serves as a standard against which more realistic models can be compared. The assumption of an infinite population size is no more meaningless than the physicist's undeniably useful assumptions of complete vacuums, absolute zero and frictionless pulleys.

Nei's second criticism concerns an apparent paradox, that the rate of gene frequency change seems to depend on which genotype is chosen as the standard upon which to calculate relative fitnesses. But the manipulations leading to this paradox are incorrect even in the case of deterministic

selection. To be specific, let A_1A_1, A_1A_2 and A_2A_2 have
fitnesses $(W_1, W_2, W_3) = (1-s, 1, 1+s)$, where s is a fixed
nonzero constant. Then $\Delta x = x(1-x)s/[1-(1-2x)s]$, x being
the frequency of A_2 . The fitnesses can be equivalently
written as $[(1-s)/(1-s), 1/(1-s), (1+s)/(1-s)]$ or
$(1, 1+t, 1+2t)$ where then $t = s/(1-s)$. This is Nei's second
case, and note that t cannot equal s . When you set t=s ,
as he does incorrectly, you obtain $\Delta x = x(1-x)s/(1+2xs)$.
The proper correspondence $t = s/(1-s)$ leads to the proper
answer. Writing the fitnesses as $(1-2u, 1-u, 1)$ requires
that $u = s/(1+s)$.

    In the stochastic case, Nei first writes the fitnesses as
$(1-s, 1, 1+s)$ and assumes that the mean of s is zero.
Since $t = s/(1-s)$ is required to transform the fitnesses to
those of his second model, the mean of t cannot be zero. To
the order of the diffusion approximation, $E(t) = V_s$ and
$V_t = V_s$. Similarly, since $u = s/(1+s)$ is required to
obtain the third model, the mean of u cannot be zero. Using
the correct mean and variance of t and u , the expression
for $M_{\delta x}$ turns out to be exactly $V_s x(1-x)(1-2x)$ in all
three models. This resolves the paradox.

    The calculations above show that when the fitnesses of
A_1A_1, A_1A_2 and A_2A_2 are written as $(1-2s, 1-s, 1)$ or
$(1-s, 1, 1+s)$ or $(1, 1+s, 1+2s)$, these are not the same
model, hence they can hardly be expected to lead to exactly
the same results. Nei, however, regards the differences between the models as "slight" and expects the outcome of
selection to be the same in all three instances. And, indeed,
when s is a fixed, positive constant, his intuition is un-
erring; all three modes of selection lead to fixation of A_2 .
However, when s is a random variable with mean zero, then
fitnesses of $(1-2s, 1-s, 1)$ lead to fixation of A_2 ; fit-

nesses of (1-s, 1, 1+s) lead to a uniform stationary density; and fitnesses of (1, 1+s, 1+2s) lead to fixation of A_1. Here the intuition is more subtle. What he chooses to regard as slight changes in the model turn out, on deeper analysis, to be essential changes.

Experience has shown that the appropriate way to think about stochastic selection is in terms of the geometric means of the fitnesses. The geometric means of (1-2s, 1-s, 1) are, to the order of the diffusion approximation, $[\exp(-2V_s), \exp(-V_s/2), 1]$, hence the A_2 allele is certainly favored. Fitnesses of (1-s, 1, 1+s) have geometric means of $[\exp(-V_s/2), 1, \exp(-V_s/2)]$, indicative of overdominance. And fitnesses of (1, 1+s, 1+2s) have geometric means of $[1, \exp(-V_s/2), \exp(-2V_s)]$, so A_1 is clearly favored. From this perspective the behavior of the stochastic models is straightforward.

As for the idea that stochastic selection is similar to the effect of random genetic drift due to finite population size, it is, but only insofar as there is random fluctuation in gene frequency from generation to generation.

I have tried repeatedly to make the point that stochastic selection also creates <u>directed</u> changes in gene frequency when $M_{\delta x} \neq 0$. In this respect stochastic selection is unlike random drift and more like deterministic selection. In random drift, $M_{\delta x} = 0$ and $V_{\delta x} \neq 0$. In deterministic selection, $M_{\delta x} \neq 0$ and $V_{\delta x} = 0$. In stochastic selection, $M_{\delta x} \neq 0$ and $V_{\delta x} \neq 0$.

Population Subdivision and Selection Migration Interaction*

S. KARLIN

INTRODUCTION

Environmental and/or geographical variation in selection patterns and its coupling with gene flow are considered vital ingredients in speciation and differentiation. Recent literature has witnessed increasing emphasis on the formulation and analysis of a hierarchy of models with aim to understand in more precise terms the interaction between spatial and temporal selection variation and population structure.

In the case of finite populations, numerous authors investigated the effects of some forms of population subdivision and migration patterns <u>without selection</u>, with respect to rates of allelic substitution, rates of approach to homozygosity and correlations in gene frequency maintained by linear external pressures. Notable contributors in this vein include Wright (1943); Malécot (1948), (1951), (1959), (1967); Moran (1962); Kimura and Weiss (1964); Karlin (1968, Chap.2); Bodmer and Cavalli-Sforza (1968); Maruyama (1970), (1972); and others. A number of special deterministic migration models coupled with local differential viability forces were set forth by Levene (1953), Prout (1968), J. Maynard Smith (1970), Strobeck (1974), Christiansen (1974), Deakin (1966), (1968) among others. We have cited theoretical analysis, but needless to say, there is

*Research supported in part by National Institutes of Health Grant USPHS 10452-12 and NSF Grant No. MPS71-02905 A03.

voluminous descriptive and taxonometric studies relevant to the above theme.

Several concepts and measures concerning environmental heterogeneity and the degree of migration mixing in relation to the existence of "protected polymorphisms" will be examined in this work. We will report a number of general findings pertaining to a geographical population genetics structure involving almost no restrictions on the parameters of the model.

1. A MULTI DEME POPULATION MODEL SUBJECT TO MIGRATION AND SELECTION FORCES

A population is distributed over a finite region generally composed of separate breeding demes (e.g., geographical or ecological habitats or niches) $\wp_1, \wp_2, \ldots, \wp_n$. Successive generations in the population are discrete and non-overlapping. It is assumed throughout this work that each subpopulation $\wp_i$ is of large size so that the effects of genetic drift are inconsequential. We focus principally on a trait with two possible alleles labelled A and a . The action of selection, migration and mating can be coupled in a variety of forms. A number of the concepts and structures pertinent here are now highlighted and refined.

(i) Spatial selection gradients

We assume that viability selection operates independently in each deme. The transformation of gene frequency under local selection in deme $\wp_i$ is determined by the relation

$$\tilde{\xi} = f_i(\xi) \tag{1.1}^*$$

such that if ξ is the A-frequency in $\wp_i$ then after the

---

*For a multi allele system the transformation would be described by a vector function $\underline{f}_i(\underline{\xi})$ where $\underline{\xi}$ is the vector allelic frequency state in locality $\wp_i$.

action of natural selection, the resulting A-frequency is $\tilde{\xi}$.

Generally, $f_i(\xi)$ is continuous and monotone increasing.

Also, we stipulate throughout this work that $f_i(0) = 0$ and $f_i(1) = 1$ signifying that selection forces maintain a pure population composition. Thus, in this formulation mutation events are ignored, i.e., new mutant forms arising in the time frame under consideration cannot be established.

An important choice for $f_i(x)$ arising from the classical diploid one-locus two-allele viability model has the form

$$f_i(x) = \frac{(1+\sigma_i)x^2 + x(1-x)}{1+\sigma_i x^2 + s_i(1-x)^2} \tag{1.2}$$

where the viability parameters of the genotypes are as listed

AA	Aa	aa
$1+\sigma_i$	1	$1+s_i$

In the corresponding haploid situation, we would take $f_i(x) = \dfrac{\sigma_i x}{\sigma_i x + s_i(1-x)}$. It is generally unnecessary to spell out explicitly the mating system operating in each deme (locality) as the consequences of mating and selection are implicitly coupled and summarized by the local selection functions $f_i(x)$. The choice (1.2) for $f_i(x)$ would, of course, come about from local random mating with standard viability selection in a diploid setting. Other determinations for $f_i(x)$ can be generated by superimposing forms of frequency dependent selection, or selection induced on a single locus when part of a multi locus system or other combined mating and selection forms.

The environmental or geographical selection gradient $\mathcal{E}$ is characterized by the array $\mathcal{E} \approx \{f_1(x),\ldots,f_n(x)\}$.

The extent of environmental heterogeneity is reflected by

the differences existing among the components of $\mathcal{E}$. Where all $f_i(x)$ are identical then unambiguously we speak of a homogeneous environmental selection background. It is still largely unknown how to relate spatial and temporal ecological parameters and selection gradients. Our investigation is mainly in terms of fitness values and we concentrate on achieving qualitative conclusions for different forms of fitness arrays. Specifically, our discussion focuses on concepts involving comparisons of degrees and quality of environmental fitness heterogeneity interrelated with the migration structure.

A general tenet commonly stated is that environmental variability or heterogeneity in the selection gradient is substantially correlated with the proliferation of polymorphism and this is claimed to be the case largely independent of the rate of migration. There are some who contest this as a universal dictum. The above theme is too general and several terms need clarification. By what criteria is a prescribed environmental selection gradient considered more "heterogeneous" or "variable" than another environmental selection form? What are meaningful means for measuring degrees of variability in both ecological and genetic (fitness) terms? The appropriate concepts must take proper account of the migration structure coupled to the spatial selection gradient. We address these questions in Section 2.

(ii) Local relative population sizes

We assume that the individual demes have a characteristic population size at an appropriate stage. Various possibilities have been proposed of which we indicate two :

(a) The relative numbers of offspring contributed from deme

i to the total population is c_i , $(c_i > 0$, $\sum_{i=1}^{n} c_i = 1)$ constant over successive generations.

(b) The relative size c_i reflects the proportion among the whole adult population located in deme $\mathcal{P}_i$ after migration.

The c_i can be construed as a constant expression of "interdeme selection" not altered by the specific genetic composition or local selection forces.

(iii) Hard and soft selection

In a multi deme population there are two principal opposite models relating the interaction between selection and local population size, those of hard and soft selection; this distinction was introduced by Wallace (1968). See also Dempster (1955).

Soft selection stipulates that local viability selection does not change the relative proportions of the deme populations in passing from the offspring to the adult stage. This is the most commonly applied model where each subpopulation carries a constant characteristic fraction of adult individuals in every generation. On the other extreme, hard selection stipulates that each local population after mating includes a characteristic fraction, independent of the generation time, of the total population. With the operation of selection at deme $\mathcal{P}_i$ we postulate the existence of $W_i(x_i)$, a function of the A-gene frequency, x_i , such that $c_i W_i(x_i)$ measures the relative population size resulting from the effects of local differential selection. This conversion can be viewed as a local density regulating factor in the process.

For the choice of (1.2) a common determination has $W_i(x) =$

the mean fitness function in ρ_i , viz., $W_i(x)=1+\sigma_i x^2+s_i(1-x)^2$ where x is the A-gene frequency.

(iv) <u>Migration structure</u>

A basic ingredient in the migration pattern is the prescription of the <u>forward migration matrix</u>

$$\Gamma = ||\mu_{ij}||^n_{i,j=1} \qquad (1.3)$$

where μ_{ij} is the a-priori probability per generation that an individual of deme i will migrate to deme j . Of course

$$\mu_{ij} > 0 \quad \text{and} \quad \sum_{j=1}^n \mu_{ij} = 1 \quad , \quad i = 1,2,\ldots,n \quad .$$

It is worthwhile to highlight a number of old and new examples of relevant migration patterns.

(a) <u>Levene Population Sub-Division Model.</u> In the early literature, two main dispersal and migration patterns were considered. The <u>Island Model</u> introduced by S. Wright divides the population into panmictic units each receiving an equal proportion of the total population. The Levene population subdivision model (1953) slightly generalizes the Wright model: a population after mating at random distributes itself into n separate patches, a fraction c_i going into the i-th patch. Then selection occurs according to the state of the environment in each patch. Notice after migration the subpopulations involve the same mixture of the whole population for each generation. (It has been suggested that this formulation may be appropriate for a species whose numbers are regulated within each of the separate patches but not on the whole population.) For this case

$$\mu_{ij} = c_j \quad \text{independent of} \quad i \quad . \qquad (1.4)$$

622

(b) **Stepping Stone Mode.** A second class of classical migration patterns are based on the principle of isolation by distance where the degree of migration diminishes with the "distance" from a given deme. An extreme, widely applied case is the stepping stone mode. Here the demes occur in an ordered (linear) series. In each generation, a fraction m $(m \leq \frac{1}{2})$ of each deme is exchanged with each contiguous deme as depicted.

$$(1.5)$$

The stepping stone mode of migration including two and higher dimensional versions has been widely used in the study of geographical genetic models without selection by Malécot [1948], [1951], [1959]; Kimura and Weiss [1964]; Fleming and Su [1974]; Maruyama [1972], and others.

(c) **Non-homogeneous Stepping Stone Mode.** Implicit in the stepping stone mode of migration with a constant rate m is the assumption that the demes have essentially equal sizes. Where the relative sizes of the demes differ then the rates of gene flow between neighboring demes are generally not equal or they may have intrinsic unequal rates of migration in reciprocal directions. In this setting the appropriate analog of (1.5) involves general non-constant local migration parameters μ_i , μ_i' such that

$$\mu_{i,i-1} = \mu_i' \quad , \quad \mu_{i,i+1} = \mu_i \quad \text{and} \quad \mu_{ii} = 1-\mu_i-\mu_i' \quad ,$$

$$i = 2,3,\dots,n-1$$

$$(1.6)$$

$$\mu_{1,1} = 1-\mu_1 \quad , \quad \mu_{1,2} = \mu_1 \; ; \; \mu_{n,n-1} = \mu_n' \quad , \quad \mu_{nn} = 1-\mu_n' \quad .$$

We will refer to the migration pattern (1.6) as a "Non-homogeneous stepping stone migration mode".

Other specifications of migration rates depending on distance were considered by Malécot [1959], for models with no differential selection and in practical contexts by Jain and Bradshaw [1966].

As noted by Wright himself [1943], the island model is rather unlikely to be realized in nature whereas the isolation by distance model is more realistic and likely to be interesting.

(d) <u>Circulant Model.</u>   If the demes occur in a circular pattern rather than linear (like around the base of a central mountain or along the shores of a lake), then the homogeneous stepping stone migration mode has the pictorial form

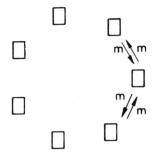

A general circulant isolation by distance migration matrix has the form $\mu_{ij} = a_{|i-j|}$.

(e) <u>A Homogeneous Homing Model.</u>   An appealing extension of the island model was put forth by Deakin [1966], [1968], [1972] and studied further by Christiansen [1974]. The migration matrix is

$$\mu_{ij} = \alpha c_j \qquad\qquad ; \quad i \neq j$$

$$\mu_{ii} = 1-\alpha+\alpha c_i \qquad ; \quad i,j = 1,\ldots,n \ . \tag{1.7}$$

The components in $(c_1, c_2, \ldots, c_n)$ constitute the usual relative deme sizes while the parameter α can be regarded as
a measure of the dispersal rate of organisms in a local deme;
Maynard Smith [1970]. Other meaningful interpretations of α
are possible.

(f) <u>Non-homogeneous Homing Model.</u> We will develop in
Karlin [1976] a number of results for a migration pattern
generalizing (1.7) to the form

$$\mu_{ij} = \alpha_i c_j \qquad\qquad ; \quad i \neq j$$

$$\mu_{ii} = 1-\alpha_i+\alpha_i c_i \tag{1.8}$$

where the rate of homing differs over the respective demes.
Already certain results inferred by Christiansen [1974] on
the basis of the model (1.7) do not apply for (1.8) indicating
that the interactions of migration and selection are more re-
condite in the presence of a non-uniform homing rate.

(g) <u>A Hybrid Island and Isolation by Distance Model.</u>
Another migration pattern in the spirit of (1.8) amenable to
analysis has the form

$$\mu_{ii} = 1-\alpha_i \ , \qquad i = 1,2,\ldots,n$$

$$\mu_{ij} = \begin{cases} \alpha_i c_j \ , \quad 1 \leq i \leq K \ , \ K+1 < j \leq n \quad \sum\limits_{j=K+1}^{n} c_j=1 \ , \ c_j > 0 \\[4mm] \alpha_i d_j \ , \quad K+1 \leq i \leq n \ , \ 1 \leq j \leq K \quad \sum\limits_{j=1}^{K} d_j=1 \ , \ d_j > 0 \end{cases} \tag{1.9}$$

$\mu_{ij} = 0$ otherwise.

Thus, the demes divide into two groupings, $\mathcal{G}_1 = \{\mathcal{P}_1, \mathcal{P}_2, \ldots, \mathcal{P}_K\}$
and $\mathcal{G}_2 = \{\mathcal{P}_{K+1}, \ldots, \mathcal{P}_n\}$ such that an organism either does

not move or when migrating it passes from its group to the other group with a fixed probability of relocation in a specified deme independent of its birthplace. This model and its extensions to r groupings is developed in Karlin [1976]. These can be regarded as hybrid patterns of island and isolation by distance models.

(h) <u>Atoll Migration Pattern.</u> Another migration scheme of some interest has the form

$$
\Gamma = \begin{pmatrix}
1-\mu_1 & , & \mu_1 & 0 \ldots 0 \\
1-\mu_2 & , & 0 & \mu_2 \\
\vdots & & & \\
1-\mu_{n-1} & , & 0 \ldots 0 & \mu_{n-1} \\
1 & , & 0 \ldots 0 & 0
\end{pmatrix}
\tag{1.10}
$$

The motivation conforms to the picture of a sequence of islands resembling an atoll. There is a main central deme $\mathcal{P}_1$ and subsequently smaller demes such that either immigration from a specified deme occurs to the next deme on the right or in the other direction immigration entails return to the central deme. Other interpretations are also possible. The analyses of this migration pattern is quite tractable even allowing variable local homing rates.

(v) <u>Backward Migration Matrices.</u> In order to write the appropriate transformation relations connecting gene frequencies in successive generations and to take proper account of a conglomeration of factors including variable deme sizes, the effect of differential viability selection on deme size and gene flow, the concept of the backward migration matrix is indispensable. The elements of the <u>backward migration</u>

matrix $M = ||m_{ij}||$ after selection and migration specify

$$m_{ij} = \text{the fraction in the i-th deme originating} \qquad (1.11)$$
$$\text{from the j-th deme in a given generation.}$$

We indicate the calculation of (1.11) for the model where selection precedes migration and following Christiansen [1974], we do this separately in the circumstances of <u>soft</u> and <u>hard</u> selection. As pointed out in paragraph (iii) selection converts the relative population sizes into

$$c_i \longrightarrow c_i^* = c_i \qquad \text{(soft selection)}$$
$$\qquad (1.12)$$
$$c_i \longrightarrow c_i^* = \frac{c_i W_i(x_i)}{\sum\limits_{k=1}^{n} c_k W_k(x_k)} \qquad \text{(hard selection)}$$

$$i = 1, 2, \ldots, n ,$$

where $\{W_i(x)\}$ usually stand for the local fitness functions. An elementary calculation involving conditional probabilities gives

$$m_{ij} = \frac{c_j^* \mu_{ji}}{\sum\limits_{k=1}^{n} c_k^* \mu_{ki}} , \qquad i,j = 1, \ldots, n . \qquad (1.13)$$

It is important to emphasize that with hard selection the backward migration matrix depends on the specific genic population composition at hand while in the situation of soft selection $M = ||m_{ij}||$ is independent of the gene frequency configuration. Equivalently, in hard selection differential viability directly influences the migration structure but not with soft selection.

Where all demes are of equal size and $\Gamma = ||\mu_{ij}||$ is symmetric (as in the homogeneous stepping stone model) then for soft selection $M = \Gamma$ showing in this case that the backward and forward matrices coincide. In this case μ_{ij}

reflects the proportion of population exchange between demes
i and j .

(vi) The Influence of the Timing of Migration and Selection Forces. The discussion of paragraph (v) was predicated
on the operational order of the genetic forces in each generation being

$$\text{mating and selection} \xrightarrow{\text{(followed by)}} \text{migration.} \qquad (1.14)$$

Effectively, migration occurs at the adult stage but prior to
mating in the next generation.

Another formulation also relevant in the workings of certain natural populations would have the order of application
of selection and migration reversed; viz.,

$$\text{migration} \longrightarrow \text{selection and mating.} \qquad (1.15)$$

For the model of (1.15) the offspring (infant) rather than the
the adult population migrates (e.g., as in seed and pollen
dispersal) and subsequently differential viability is in
force.

Selection generally has two major components reflecting
fertility and viability effects so that for some natural
populations, either model (1.14) or (1.15), or a mixed model
involving possibly two stages of migration, may be appropriate.

The inherent differences to the timing of migration and
selection effects are well contrasted by writing out the
transformation equations relating gene frequencies in two
successive generations.

(vii) Transformation Equations of the Frequency States.
Let x_i denote the frequency of type A in deme P_i at the
start of a generation and x_i' the frequency for the next
generation. Consider first the model of (1.14). The stan-

dard global transformation equations connecting $\underline{x} =$ $(x_1,\ldots,x_n)$ to $\underline{x}' = (x_1',\ldots,x_n')$ over two successive generations is given by

$$x_i' = \sum_{j=1}^{n} m_{ij} f_j(x_j) \quad , \quad i = 1,2,\ldots,n \qquad (1.16)$$

where $||m_{ij}||$ is the backward migration matrix computed as in (1.13). Recall that in the hard selection model m_{ij} also depends on the frequency state $\underline{x}$.

Where the timing of migration and selection operate in reverse order as in (1.15), the transformation equations read as

$$x_i' = f_i \left(\sum_{j=1}^{n} m_{ij} x_j \right) \quad , \quad i = 1,2,\ldots,n . \qquad (1.17)$$

For the problem concerning the existence of a protected polymorphism the models (1.16) and (1.17) are equivalent, see Bulmer [1972]. Pertaining to the characterization of the actual established equilibria, the timing has a significant influence, e.g., see Karlin and Richter-Dyn [1976].

2. OBJECTIVES, COMPARISONS AND SOME RESULTS FOR SELECTION MIGRATION INTERACTIONS

In the previous section a number of the key concepts and structures underlying a broad class of multi deme population models of n demes subject to local selection forces and migration flow were delineated. The main factors are the following:

(I) The environmental selection gradient described by the collection of local selection functions

$$\{f_1(x),\ldots,f_n(x)\} \qquad (2.1)$$

obeying the conditions of paragraph (i), Section 1.

(II)   The migration pattern characterized by the parameters of the forward and backward migration matrices, respectively

$$\Gamma = ||\mu_{ij}|| \;, \quad M = ||m_{ij}|| \quad \text{(see (1.3) and (1.11))}. \qquad (2.2)$$

(III)   The relative deme sizes given by the vector

$$\underline{c} = (c_1, c_2, \ldots, c_n) \;, \quad c_i > 0 \;, \quad \Sigma c_i = 1 \;. \qquad (2.3)$$

In general terms, the desired objective is to evaluate qualitatively and quantitatively the influence of the factors (I), (II) and (III) separately and in combination on the evolutionary dynamics and equilibrium behavior of a multi deme population obeying the transformation law (1.16). We will consider two categories of problems bearing on the existence of polymorphisms and variability in populations:

(a)   In Section 3 quite precise conditions in terms of the parameters (2.1), (2.2) and (2.3) are set forth guaranteeing the persistence of the alleles A and/or a . The property of persistence of allele A (not going ultimately extinct) even when initially rare is now commonly called protection of the A-allele or A-protection. This is intimately connected to the ascertainment of the initial increase of a new allele. These approaches helping in the study of certain population genetics models is now quite classical and widely used.

The maintenance of a protected polymorphism is more than the existence of a "stable polymorphic equilibrium" since fixation of any allele is precluded as a realizable event from any starting frequency state, assuming of course all types are initially present (i.e., protection holds under all initial conditions). With a protected polymorphism there may be several stable equilibria states (this is already the case even for the Levene population subdivision model) or conceivably oscillatory behavior between several polymorphic states is induced.

In order to make quantitative comparisons we have introduced (in Section 3) a number of measures of environmental
heterogeneity for relating alternative spatial selection
gradients. We will also propose two modes for classifying
intensity of migration mixing (and/or degrees of relative
isolation) among the breeding units. The implications for
"A-protection" of the interactive effects between selection
and migration forces are discussed in a variety of contexts
throughout this paper.

(b) A more difficult problem is the determination and
     characterization of the possible stable polymorphic
states in the multi-deme framework. A description of the
qualitative dependence of all the stable equilibria on the
environmental components embodied in (2.1)-(2.3) would be of
relevance in the explication of variability in natural populations.

We have achieved such characterization in a number of
cline models with the results reported in a series of papers
by Karlin and Richter-Dyn [1976a,b,c]. Several of these
findings are discussed in the following paper (this volume).
We will also present in Karlin [1976] various characterizations
of the possible polymorphic states in the Levene subdivision
model and for some of its extensions.

A. Conditions for protected polymorphisms with general
    environmental parameters

Consider a multi deme population system involving a general
migration structure, selection gradient and distribution of
deme sizes as prescribed in (2.1)-(2.3). Let $M = ||m_{ij}||$
be the backward migration matrix constructed as in (1.13)
focusing on the soft selection model (an analysis of the hard

selection formulation is contained in Karlin [1976]). The familiar analytic criterion assuring protection of the A-allele (e.g., see Bulmer [1972]) is the validity of the inequality

the spectral radius of MD = ρ(MD) > 1   (2.4)

where D is a diagonal matrix involving the values $d_i = f_i'(0)$, $(d_1, d_2, \ldots, d_n)$ such that, $d_i = \dfrac{1}{1+s_i}$ for the example (1.2). Where ρ(MD) < 1 then A goes extinct when its initial frequency is small. Therefore, apart from the non-generic possibility ρ(MD) = 1 , the condition (2.4) is necessary and sufficient for protection of the A-allele. The inequality ρ(MD) > 1 assures protection of the A-allele but information concerning the nature of the ultimate equilibrium state is undetermined. However, it is suggestive that with increasing ρ(MD) the more repellent the state 0 becomes, concomitantly the established A-frequency is expected to be more substantial in at least one deme.

We will now highlight two categories of sufficient conditions of wide scope bearing on the existence of a protected polymorphism. Actually we present the results in terms of protection of the A-allele. An analogous condition pertains to protection of allele a and these together imply the existence of a protected polymorphism.

Let $\underline{v} = (v_1, v_2, \ldots, v_n)$ be the unique left eigenvector corresponding to eigenvalue 1 for the backward migration matrix M obeying the normalization

$$\sum_{i=1}^{n} v_i = 1 \quad , \quad v_i = \sum_{k=1}^{n} v_k m_{ki} \quad , \quad i = 1, 2, \ldots, n \ . \quad (2.5)$$

The following result applies to any migration structure and entails no limitations relating to the vicissitudes of any special examples.

A sufficient condition guaranteeing protection of the A-
allele is the validity of the inequality

$$\prod_{i=1}^{n} \left(\frac{1}{1+s_i} \right)^{v_i} > 1 . \qquad (2.6)$$

Here, the influence of the migration pattern and distribution
of deme sizes is reflected by the components of the left
eigenvector $\underline{v} = (v_1, v_2, \ldots, v_n)$ of M.

Where the relative deme sizes are equal and population
exchanges among demes are such that M is symmetric (as in
the homogeneous stepping stone migration mode with equal deme
sizes) then $v_i = \frac{1}{n}$, $i = 1, 2, \ldots, n$, and the condition (2.6)
reduces to

$$\left(\prod_{i=1}^{n} \frac{1}{1+s_i} \right)^{1/n} > 1 \qquad (2.7)$$

which is always satisfied if the aggregate selection coefficient of the aa-genotype

$$S = \sum_{i=1}^{n} s_i \text{ is non-positive.}$$

The condition (2.6) applies to any migration structure.
It is a sharp inequality since for the particular circulant
permutation migration pattern, example (d) of Section 1, the
equality

$$\rho(MD) = \left(\prod_{i=1}^{n} \frac{1}{1+s_i} \right)^{1/n}$$

holds.

For $\sum_{i=1}^{n} |s_i|$ small the inequality (2.7) is essentially
equivalent to

$$\frac{\sigma^2}{2} + \frac{S^2}{2} \left(\frac{1}{n} + \frac{1}{n^2} \right) > \frac{S}{n} \quad \text{where} \quad \sigma^2 = \frac{1}{n} \sum_{i=1}^{n} (s_i - \frac{S}{n})^2 \qquad (2.8)$$

or $\quad \dfrac{\sigma^2}{2} > \dfrac{S}{n} \quad$ when $\quad S \quad$ is small.

By restricting slightly the class of migration matrices, we can achieve a substantial refinement of the result of (2.6). More specifically, suppose M is positive definite (such is the case for the homogeneous stepping stone migration pattern with equal deme sizes, provided the migration rate $m \leqslant \frac{1}{4}$), then protection of the A-allele is assured provided

$$\frac{1}{n} \sum_{i=1}^{n} \frac{1}{1+s_i} \geqslant 1 \quad . \qquad (2.9)$$

For $\sum_{i=1}^{n} |s_i|$ sufficiently small, such that the cumulative selection effects is of small magnitude, then the condition (2.9) is essentially equivalent to

$$\sigma^2 > \frac{S}{n} \qquad (2.10)$$

Thus a sufficient variance of the spatial selection coeffi-cients can override even a slight cumulative aa-selection advantage and protect the A-allele even when it is initially rare.

Notice that (2.10) does not have the factor $\frac{1}{2}$ entering into (2.8).

Suppose the backward migration matrix M admits the representation

$$M = E_1 K E_2 \qquad (2.11)$$

where E_1 and E_2 are positive diagonal matrices and K is positive definite, then the analog of (2.9) is as follows:

A sufficient condition for A-protection provided M has the form (2.11) is

$$\sum_{i=1}^{n} \frac{v_i}{1+s_i} \geq 1 \quad , \quad (\underline{v} = (v_1, \ldots, v_n) \text{ defined in (2.5))}. \quad (2.12)$$

Examples where (2.11) holds include among others:

(a) Levene population subdivision model

$$M = ||e_i c_j||_{i,j=1}^{n} \quad , \quad \underline{e} = (1,1,\ldots,1) \quad , \quad \underline{c} = (c_1,\ldots,c_n)$$

(b) Most cases of the non-homogeneous stepping stone model.

(c) The non-homogeneous homing model of (1.7).

The circulant migration pattern (example (d), Section 1), does not admit the representation (2.11).

Elaborations and proofs of (2.6) and (2.12) are found in Karlin [1976] and Friedland and Karlin [1975].

We proceed to a concrete application of (2.12) for the homogeneous stepping stone migration mode (1.5) with deme sizes described by the array $\underline{c} = (c_1,\ldots,c_n)$. The calculation (1.13) produces

$$M = \begin{pmatrix} \dfrac{(1-m)c_1}{\gamma_1} & \dfrac{mc_2}{\gamma_1} & 0 & \cdots & & 0 \\[2ex] \dfrac{mc_1}{\gamma_2} & \dfrac{(1-2m)c_2}{\gamma_2} & \dfrac{mc_3}{\gamma_2} & & & \\[2ex] & \ddots & \ddots & \ddots & & \\[2ex] & & \dfrac{mc_{n-2}}{\gamma_{n-1}} & \dfrac{(1-2m)c_{n-1}}{\gamma_{n-1}} & \dfrac{mc_n}{\gamma_{n-1}} \\[2ex] & & & \dfrac{mc_{n-1}}{\gamma_n} & \dfrac{(1-m)c_n}{\gamma_n} \end{pmatrix} \qquad (2.13)$$

with

$$\gamma_i = mc_{i-1} + (1-2m)c_i + mc_{i+1} \quad , \quad i = 2,\ldots,n-1$$

and

$$\gamma_1 = (1-m)c_1 + mc_2 \quad , \quad \gamma_n = mc_{n-1} + (1-m)c_n \quad .$$

A standard determination of the left eigenvector $\underline{v}$ for M in (2.13) leads to

$$v_i = \frac{c_i\gamma_i}{\sum_{k=1}^{n} c_k\gamma_k} \quad , \quad i = 1,2,\ldots,n \quad .$$

Now for $m \leqslant \frac{1}{4}$ so that the rate of population exchange between neighboring demes does not exceed 50% of their inhabitants (a condition undoubtedly always satisfied in practice), the matrix M possesses the representation (2.11). The criterion of (2.12) then asserts protection of the A-allele subject to the inequality

$$\frac{1}{\sum_{i=1}^{n} c_i\gamma_i} \sum_{i=1}^{n} \frac{c_i\gamma_i}{1+s_i} \geqslant 1 \quad .$$

For extensions of this last example allowing unequal local rates of gene flow, consult Karlin and Richter-Dyn [1975a].

B. A method of comparing environmental heterogeneity for classes of selection gradients.

There is a tendency to measure diversity (or heterogeneity) of an environment usually by a single index. Common choices include the variance of selection values (or of an associated ecological parameter), cumulative deviations of selection values (absolute or relative), the inter quartile range of selection values, information index (entropy) for a selection gradient or other indices correlated with those above. A

real valued index for measuring heterogeneity compels essentially a single scaling over all environments. Intrinsically,
an environment is complex and should not and cannot be summarized in a single value. It should also be evident that not
all environments are comparable. We now propose two concepts
for ascertaining that an environment $\mathcal{E}$ is regarded more
heterogeneous than a second environment $\mathcal{E}'$.

Consider an environmental selection regime $\mathcal{E}$ characterized by the local selection functions
$\{f_1(x), f_2(x), \ldots, f_n(x)\}$ and suppose, for definiteness

$$f_i(x) = \frac{x + \sigma_i x^2}{1 + \sigma_1 x^2 + s_i (1-x)^2}$$

associated with the viability parameters

AA	Aa	aa	
$1 + \sigma_i$	1	$1 + s_i$	, $i = 1, 2, \ldots, n$.

In this model the environment is determined by the array of
selection coefficients

$$\underline{s} = \{s_1, s_2, \ldots, s_n\} \quad \text{and} \quad \underline{\sigma} = \{\sigma_1, \sigma_2, \ldots, \sigma_n\} \; . \tag{2.14}$$

Definition 1. We say that the selection regime $(\underline{s}, \underline{\sigma})$ is
more heterogeneous than the selection regime induced by the
parameters

$$\underline{s}' = (s_1', s_2', \ldots, s_n') \quad \text{and} \quad \underline{\sigma}' = (\sigma_1', \sigma_2', \ldots, \sigma_n') \tag{2.15}$$

if $\underline{s}'$ is "an average" of $\underline{s}$ and $\underline{\sigma}'$ is also "an average"
of $\underline{\sigma}$.

We make precise now the notion of "averaging" applied to
vectors. A matrix $A = ||a_{ij}||_1^n$ is said to be doubly sto-
chastic if

$$a_{ij} \geq 0 \ , \quad \sum_{j=1}^{n} a_{ij} = \sum_{i=1}^{n} a_{ij} = 1, \ i,j = 1,2,\ldots,n \ . \quad (2.16)$$

(all the row and column sums are 1) .

The collection of all doubly stochastic matrices is denoted by $\mathcal{a}$.

Now we stipulate

$\underline{s}'$ is an average of $\underline{s}$ $\overset{\text{Definition}}{\Longleftrightarrow}$ provided there exists a matrix A in $\mathcal{a}$ such that

$$\underline{s}' = A\underline{s} \quad \text{that is} \quad s'_i = \sum_{j=1}^{n} a_{ij}s_j \ , \ i = 1,2,\ldots,n \ . \quad (2.17)$$

The averaging operation preserves the aggregate selection effects, viz.,

$$\sum_{i=1}^{n} s'_i = \sum_{i=1}^{n} s_i = S \ . \quad (2.18)$$

Moreover, the relationship (2.17) tends to reduce the variation of the s_i values. In particular, the variance of the $\underline{s}'$ vector is diminished:

$$\sum_{i=1}^{n} (s'_i)^2 \leq \sum_{i=1}^{n} (s_i)^2 \ .$$

More generally, for any convex function, $\phi(\xi)$, we have

$$\sum_{i=1}^{n} \phi(s'_i) \leq \sum_{i=1}^{n} \phi(s_i) \ . \quad (2.19)$$

The relation (2.17) also entails the inequality

$$\left(\prod_{i=1}^{n} s_i \right)^{1/n} < \left(\prod_{i=1}^{n} s'_i \right)^{1/n} \ .$$

The specific averaging matrix having $a_{ij} = \dfrac{1}{n}$ for all i,j converts $\underline{s}$ into the constant

(homogeneous) environmental selection pattern   (2.20)

with $s_i' = \dfrac{S}{n}$, $i = 1,2,\ldots,n$.

To reiterate, we say, the environmental selection gradient

$(\underline{s}',\underline{\sigma}') = \mathcal{E}'$ is more homogeneous than the environmental

selection gradient $(\underline{s},\underline{\sigma}) = \mathcal{E}$ if

s' is an average of $\underline{s}$ and σ' is an average
                                                          (2.21)
of σ in the sense of (2.17).

Formally, (2.21) is equivalent to the existence of A and B

in $\mathcal{a}$ (not necessarily the same), such that

$$\underline{s}' = A\underline{s} \quad \text{and} \quad \underline{\sigma}' = B\underline{\sigma} \quad .$$ (2.22)

We can introduce greater flexibility in the concept (2.21)

by allowing the possibility that $\mathcal{E}$ is more heterogeneous

than $\mathcal{E}'$ with respect to selection on the AA-genotype while

$\mathcal{E}'$ is more heterogeneous than $\mathcal{E}$ with reference to selec-

tion expressed at the aa-genotype. We will not pursue these

ramifications in this work.

With the specification (2.20) we find that for a prescribed

aggregate level of selection coefficients S and Σ for the

aa and AA-genotypes, respectively, then the constant selec-

tion gradient characterized by the constant selection coeffi-

cients

$$s_i = \frac{S}{n} \quad \text{and} \quad \sigma_i = \frac{\Sigma}{n} \quad , \quad i = 1,2,\ldots,n$$

is more homogeneous than any other environmental gradient with

a selection array having the same cumulative selection effects

S and Σ .

The following question is natural.

How does increased heterogeneity of the environmental

selection gradient correlate with the realization of $\underline{A}$ and

$\underline{a}$ protection and the maintenance of polymorphism?

639

Definition 1 provides a framework for dealing with this problem. The averaging concept is appealing but appears to be unnatural. It is not a correct fact that the existence of a protected polymorphism is more likely in a more hetero- geneous environment (taken in the sense of Definition 1). The weakness is that Definition 1 refers only to selection gra- dients and does not take account of the nature and inter- action of selection with gene flow.

We now extend the idea for comparing selection gradients in a manner to mesh it better with the underlying migration structure. Let M be a fixed backward migration matrix having eigenvectors

$$\underline{v}M = \underline{v} \ , \ M\underline{e} = \underline{e} \ , \ \underline{e} = (1,1,\ldots,1) \ , \ \underline{v} = (v_1,\ldots,v_n) \qquad (2.23)$$

and $\underline{v}$ normalized to satisfy $\sum_{i=1}^{n} v_i = 1$.

Let $\mathcal{Q}(\underline{v},\underline{e})$ consist of the collection of all non-negative matrices A with the properties (2.23). $\mathcal{Q}(\underline{v},\underline{e})$ constitutes a convex closed set of matrices containing M and with each A all its powers. The rank one matrix $J = ||e_i v_j||$, $(e_i \equiv 1)$ is also a member of $\mathcal{Q}(\underline{v},\underline{e})$. When $\underline{v} = \underline{e}$, plainly $\mathcal{Q}(\underline{e},\underline{e})$ coincides with the collection of all doubly stochastic matrices.

Definition 2. Consider two arrays of selection coefficients $\underline{s} = (s_1,s_2,\ldots,s_n)$ and $\underline{s}' = (s_1',\ldots,s_n')$ reflecting two different environmental selection gradients $\mathcal{E}$ and $\mathcal{E}'$ [#],

---

[#]To ease the exposition we have focused on comparing sets of aa-genotype selection coefficients. The extension to selec- tion functions is obvious.

respectively. We say that $\mathcal{E}'$ is less heterogeneous than $\mathcal{E}$ with respect to the migration structure M if the relation

$$\underline{s}' = A\underline{s} \quad \text{holds for some} \quad A \in \mathcal{a}\,(\underline{v},\underline{e}) \quad . \tag{2.24}$$

The least heterogeneous environment in the hierarchy implicit to the above definition is the constant vector

$$\underline{\tilde{s}} = (\tilde{s}_1, \tilde{s}_2, \ldots, \tilde{s}_n) \quad \text{with} \quad \tilde{s}_i = \sum_{i=1}^{n} s_i v_i = S_{\underline{v}} \quad \text{for all} \quad i \; .$$

For $\underline{s}'$ determined as in (2.24) the analog of (2.18) is

$$\sum_{i=1}^{n} s_i' v_i = \sum_{i=1}^{n} s_i v_i \quad . \tag{2.25}$$

It also follows that

$$\sum_{i=1}^{n} v_i (s_i' - S_{\underline{v}})^2 \leqslant \sum_{i=1}^{n} v_i (s_i - S_{\underline{v}})^2$$

showing that the environmental selection variance (<u>weighting</u> <u>subpopulation</u> i <u>by the factor</u> v_i) is smaller for environment $\mathcal{E}'$ than for environment $\mathcal{E}$.

The following general result holds in many circumstances:

<u>Principle I. Let</u> M <u>be a backward migration matrix of the</u> <u>structure (2.11). Let</u> $\mathcal{E}$ <u>and</u> $\mathcal{E}'$ <u>be two environmental</u> <u>selection gradients such that</u> $\mathcal{E}'$ <u>is less heterogeneous than</u> $\mathcal{E}$ <u>with respect to the migration structure</u> M <u>in the sense</u> <u>of Definition 2. Symbolically, we write</u> $\mathcal{E}' \prec \mathcal{E}$. <u>Define</u> D' <u>to be diagonal selection matrix engendered by</u> $\mathcal{E}'$ <u>i.e.,</u>

$$D' = \text{diag}\,(\frac{1}{1+s_i'}\,,\,\frac{1}{1+s_2'}\,,\ldots,\,\frac{1}{1+s_n'}) \quad \underline{\text{and}} \quad D \quad \underline{\text{analogously}}$$

<u>determined from the selection coefficients</u> $(s_1, s_2, \ldots, s_n)$.
<u>Then</u>

$$\rho\,(MD) \geqslant \rho\,(MD') \quad . \tag{2.26}$$

<u>Accordingly, protection of the A-allele is more likely in the</u> <u>more heterogeneous environment</u> $\mathcal{E}$ <u>over that of</u> $\mathcal{E}'$.

For the extreme case $\underline{s}' = (S_v, S_v, \ldots, S_v)$, $S_v = \sum\limits_{i=1}^{n} s_i v_i$ then $\mathcal{E}' < \mathcal{E} = \{s_1, s_2, \ldots, s_n\}$ and indeed (2.26) holds by virtue of the analysis leading to (2.12).

Comparison of the models of hard and soft selection with reference to the existence of protected polymorphism reduces to an important case of Principle I. It can be proved that the environments of soft selection $\mathcal{E}^{(S)}$ is more heterogeneous than the environment of hard selection ($\mathcal{E}^H$) in the guise of Definition 2.

We would expect from Principle I that the phenomenon of a protected polymorphism is more fascile with soft selection over that of hard selection: Where local fitnesses also influence the migration flow, the resulting environmental structure amalgamates to a more homogeneous population behavior entailing increased possibilities for total fixation.

The validity of Principle I is established in a number of examples including the stepping stone migration pattern for a monotone cline model, see Karlin and Richter-Dyn [1976a], and in the Deakin migration form and other cases, see Karlin [1976]. This fact for the Deakin case was discovered first by Christiansen [1975]. Principle I appears not to be correct in complete generality without imposing some restrictions on the migration structure.

C. Protection for different degrees of isolation and mixing in migration structures

When does one migration pattern entail more mixing than a second migration pattern? We will introduce two criteria to deal with this question and discuss their implications with reference to the manifestation of protected polymorphisms.

642

(i) Two stage versus one stage migration flow

We start with the following definition.

__Definition 3. A backward migration matrix__ M_1 __is said to be__
__more mixing than the backward matrix__ M_2 __provided__ M_1 __has__
__the form__

$M_1 = M_3 M_2$ with M_2 and M_3 commuting

(i.e., $M_2 M_3 = M_3 M_2$)

(2.27)

__and where__ M_3 __is also a migration matrix.__

Thus the extent of migration involved in M_1 is effectively the outcome of two stages of exchange (and/or) immigration with one stage corresponding to M_2 .

It is generally anticipated that two operations of gene
flow spread the effects of the local selection forces engen-
dering the workings of a more homogeneous population. This
is not a valid general conclusion. Where M_2 and M_3 entail
excessive movement possibly cancelling each other then M_1
can reflect less movement than M_2 or M_3 separately.
Indeed, by Definition 3, M^2 is more mixing than M , but
the extreme example

$$M = \begin{pmatrix} 0 & 1 \\ 1 & 0 \end{pmatrix} \quad \text{and} \quad M^2 = \begin{pmatrix} 1 & 0 \\ 0 & 1 \end{pmatrix}$$

shows that M^2 may involve no exchange of population while
M entails a total exchange. Accordingly, there are essential
limitations on the amount of mobility ascribed to M_2 and
M_3 in order that M_1 reflect genuinely more mixing than
M_2 . The exact requirements on M_2 and M_3 are embodied in
the condition (2.28) below which imposes a constraint on the
magnitude of movement and mixing satisfied in many biologically reasonable contexts. These include cases of stepping
stone migration, the examples of (1.8) and others.

We have established a precise result enabling us to compare

the influence of different levels of mixing with respect to the existence of protected polymorphisms and this is now stated formally. Further interpretations and implications are set forth in Section 3.

Result I. Suppose each M_i in (2.27) admits the representation

$$M_i = F_i P_i G_i \quad , \quad i = 1,2,3 \tag{2.28}$$

where F_i and G_i are positive definite diagonal matrices with P_i a positive semi definite matrix (cf. the discussion of (2.11)). If M_1 is more mixing than M_2 such that (2.28) holds, then for any selection matrix D we have

$$\rho(M_1 D) \leq \rho(M_2 D) \quad (\text{see } (2.4)). \tag{2.29}$$

Thus, where the multi deme population determined by the migration selection parameter set $\{M_1, D\}$ entails A-protection then with the migration pattern M_2 (which is less mixing than M_1 in the sense of Definition 3) and the same selection structure of D , protection of the A-allele is, a fortiori, assured.

In particular, if M possesses the representation (2.28) then for each integer k , we have

$$\rho(M^{k+1} D) \leq \rho(M^k D) \quad . \tag{2.30}$$

It is important to underscore the fact that the relation (2.29) is not universally correct with respect to any two comparable migration patterns. In fact, consider a system of 2-subpopulations having equal deme sizes with homogeneous migration matrix

$$M = \begin{pmatrix} 1-\gamma, \gamma \\ \gamma, 1-\gamma \end{pmatrix} \quad .$$

It is elementary to check that M_{γ_1} is more mixing than M_{γ_2}

644

in the sense of Definition 3 if and only if $\gamma_1 > \gamma_2$. However, for any $D = \begin{pmatrix} d_1 & 0 \\ 0 & d_2 \end{pmatrix}$, $\rho(M_\gamma D)$ decreases to a minimum attained when $\gamma = \frac{1}{2}$ and afterwards increases. Of course, M_γ for $\gamma > \frac{1}{2}$ does not fulfill the requirement of (2.28). The significance of these examples is tantamount to the phenomenon that where the migration structure induces excessive oscillatory mixing then the possibilities for a protected polymorphism are diminished.

The hypotheses underlying Result I are satisfied for the homogeneous stepping stone forward migration matrix of any number of demes provided $m \leqslant \frac{1}{4}$ allowing for a general prescription of deme sizes.

(ii) Rates of homing

The following criterion for comparison of two migration patterns seems natural.

Definition 4. Let $M^{(1)}$ and $M^{(2)}$ be two (backward) migration matrices. If for each i

$$m_{ij}^{(2)} \geqslant m_{ij}^{(1)} \quad \text{for all } j \neq i \tag{2.31}$$

then it is suggestive to say that $M^{(2)}$ is more mobile than $M^{(1)}$.

The relation (2.31) tells us that after migration the number of inhabitants in locality $\wp_i$ originating from any other locality other than $\wp_i$ is larger for the migration mode $M^{(2)}$ as against $M^{(1)}$ and this property holds for all i .

A set of matrices comparable in the sense of (2.31) incorporates the one parameter family

$$M^{(\alpha)} = (1-\alpha)I + \alpha M \quad \text{(M is a fixed stochastic matrix).} \tag{2.32}$$

The Deakin migration pattern (1.7) is a very special

example of (2.32) with $M = ||e_i c_j||$, $(e_i \equiv 1)$. We can interpret $1-\alpha$ as the innate propensity of an organism to actively home, independent of selection and deme sizes. A proportion α of the population follows the migration pattern M . <u>When</u> $\alpha = 0$ all demes are strictly isolated and when $\alpha = 1$ the migration behavior of the total population per generation is summarized by M .

It is trivial to check that $M^{(\alpha_1)}$ is more mobile (in the sense of Definition 4) than $M^{(\alpha_2)}$ if and only if $\alpha_1 > \alpha_2$.

Allowing for dispersal rates varying with the deme origin we obtain an n-parameter family of matrices

$$m_{ij}^{(\alpha)} = (1-\alpha_i)\delta_{ij} + \alpha_i m_{ij} \quad , \quad i,j = 1,\ldots,n \tag{2.33}$$

$$(M = ||m_{ij}|| \ , \ \underline{\alpha} = (\alpha_1,\ldots,\alpha_n)).$$

Obviously the matrix $M^{(\underline{\alpha})}$ is more mobile than $M^{(\underline{\beta})}$ constructed with dispersal parameter sets $\underline{\alpha} = (\alpha_1,\alpha_2,\ldots,\alpha_n)$ and $\underline{\beta} = (\beta_1,\ldots,\beta_n)$, respectively, if $\alpha_i \geq \beta_i$ for every i .

To what extent does "more mobility" enhance the maintenance of a protected polymorphism? Comparison of the migration structure $M^{(\underline{\alpha})}$ and $M^{(\underline{\beta})}$ with n genuine parameters is formidable and does not point to a coherent relationship. In fact, decreasing only the first component α_1 <u>need not</u> ameliorate the occurrence of protected polymorphisms.

For the case of a uniform dispersal rate (the model of (2.32)), we find in substantial generality, independent of the selection gradient, that the likelihood in favor of a protected polymorphism becomes stronger as the degree of mobility diminishes (α decreases).

The following general result is correct.

Result II. Consider the one parameter family of migration matrices (2.32) where M has the form (2.28). Let D be a diagonal matrix with positive terms on the diagonal induced by the spatial array of aa-selection coefficients (see (2.4)). Then

$$\rho(M_\alpha D) = \rho(\alpha)$$

is a decreasing function of α .

It follows that if a protected polymorphism exists for a level of homing $1-\alpha_0$, and migration structure $M^{(\alpha)}$ then a protected polymorphism is assured for any higher level of homing. This finding is consistent with the small parameter theory of Karlin and McGregor [1972].

3. DISCUSSION

In explaining polymorphisms and clines, emphasis is usually given to changes in selective factors between and within environments. It is also widely recognized that in many natural situations migration may play an important, even a dominant, role. The following theme recurs in many works concerned with population genetics: The ongoing process of evolution probably requires adjustment to a constantly varying environment and to the combination of characteristics that survive from the different populations. Important sources of variability in natural population can be genes and gene complexes transferred from other populations. Also widely recognized is that differentiated populations retain the ability for exchange of genetic material. Put in a more descriptive language, spatial and temporal variation in environment are considered to be highly involved in the main-tenance of genetic variation in populations (Darlington [1957], Wright [1968], Dobzhansky [1967]). The results reported in

647

Section 2 bear a variety of implications pertaining to the
theme cited above.

Through a series of mathematical models the representations
of genetic variability in a subdivided population acted on by
migration selection forces has been recently studied by a
number of authors including Deakin [1966], [1972], Prout
[1968], Maynard-Smith [1970], Christiansen [1974], [1975],
Strobeck [1974] and others. The existence of "stable polymorphic equilibria" has been mostly confirmed by showing that
each allele is protected against disappearance. (This method
used for the confirmation of polymorphism is reliable only in
the context of two alleles.) All the above theoretical works
confined attention to very special models mostly variations
on the Wright Island model. The discussion of special
examples are undoubtedly of some separate interest and may fit
some natural situations. But even here, complete exact
results on protection for the important stepping stone cline
models are as yet unavailable. (In this connection see
Karlin and Richter-Dyn [1976].)

1. It is commonly stated that a spatial and temporal
environmental variation and increased population subdivision
enhance the occurrence of polymorphism. The theory expounded
in this work and its detailed development in Karlin [1976]
circumscribes somewhat the scope and validity of this con-
tention. For this purpose it is essential to delimit carefully the concept when two environmental selection gradients
can be compared with reference to their degrees of heterogeneity (this is not always well defined). Such comparisons
must take proper account of the migration structure coupled
to the spatial selection gradient.

Various authors have emphasized that "average heterozygosity seems to increase with increasing environmental

variability". Most averages are usually computed by weighting equally likely over space and/or time. We have determined in Section 2 that it is unnatural when constructing the average to improperly scale the effects of deme sizes, differences in local migration rates and the spectrum of selection influences. Formulas (2.12) and (2.6) indicate possible appropriate weightings.

2. A precise sense in which more heterogeneous selection gradients engenders more polymorphism is the intent of Principle I, now restated (see Section 2 for its detailed formulation). If the environment $\mathcal{E}$ (characterized by the aa-genotype spatial selection coefficient array $\{s_i\}_1^n$) is more heterogeneous than environment $\mathcal{E}' \approx \{s_i'\}_1^n$ with respect to the migration structure M in accordance with Definition 2, then protection of the A-allele is more likely with $\mathcal{E}$ over $\mathcal{E}'$.

The above assertion appears to be true in substantial generality. We have accomplished its validation for several important models but we do not have a complete classification. The comparison of soft versus hard selection fits perfectly the framework of Definition 2. We have established in a number of cases, including the migration selection cline setting, that Principle I applies with hard selection corresponding to a less heterogeneous environment vis a vis soft selection. However, the conclusion that protection for hard selection entails protection for soft selection is not universally correct. Some restrictions on the nature of the migration structure are essential.

3. Recent theoretical studies show that with temporally fluctuating selection intensities the extent of polymorphism

increases, see Gillespie [1973], [1974], Hartl and Cook
[1973], Karlin and Lieberman [1974]. Bryant [1974] has
reviewed some of the literature on temporal and spatial
selection heterogeneity related to natural enzyme polymorphisms. He dwells on the relative roles of spatial and temporal environmental variation and claims on the basis of
mathematical work of Haldane and Jayakar [1963] and some work
of Charlesworth and Giesel [1972] and Giesel [1972] that the
conditions for polymorphic stability in the presence of temporal variation are more stringent than for spatial variation.
Bryant goes on to conclude that "the major trend of genetic
variation seems intimately associated with temporal variation
in the environment, while the remaining trends in some cases
may be related to other parameters, including spatial heterogeneity".

This conclusion is not concordant with our findings reported in Section 2. With population subdivision and moderate
migration flow a sufficient condition for the existence of,
say, protection of the A-allele is

$$\sum_{i=1}^{n} v_i \frac{1}{1+s_i} \geq 1 \tag{*}$$

where $\{s_i\}_{i=1}^{n}$ consists of the spatial array of aa-genotype
selection coefficients among the n localities and the components of $(v_1, v_2, \ldots, v_n)$ reflect the influence of migration
and varying deme sizes (see (2.23) and (2.12)).

For a cyclically (e.g. seasonal) varying set of selection
effects $\{s_i\}_1^{n}$ of period length n a sufficient condition
for protection is

$$\prod_{i=1}^{n} \left(\frac{1}{1+s_i} \right)^{v_i} > 1 \tag{**}$$

where v_i now relates to the variable population sizes over successive generations.

The generalized arithmetic geometric mean inequality

$$\sum_{i=1}^{n} v_i \frac{1}{1+s_i} > \prod_{i=1}^{n} \left(\frac{1}{1+s_i}\right)^{v_i}$$

shows that <u>protection of the</u> A-<u>allele is more easily maintained with spatial as against temporal variation</u> in selection coefficients. This suggests that <u>spatial rather than temporal heterogeneity of the environments is a more powerful force for polymorphism.</u> For temporal heterogeneity the determining factor is a generalized geometric mean of fitness values while in spatial heterogeneity a generalized arithmetic mean of fitness values is critical. It should be emphasized that we are comparing the same average levels of selection in the two cases.

The contrast is more manifest with small cumulative selection effects, ($\sum |s_i|$ small), then (*) is essentially equivalent to

$$\sigma^2 > S_v \quad \text{with} \quad \sigma^2 = \sum_{i=1}^{n} v_i (s_i - S_v)^2 \quad \text{and} \tag{†}$$

$$S_v = \sum_{i=1}^{n} s_i v_i$$

while (**) reduces to

$$\frac{\sigma^2}{2} > S_v \quad . \tag{‡}$$

Thus with temporal fluctuating selection intensities the inequality (‡) (by a factor $\frac{1}{2}$) brings less likelihood of protection.

Hartl suggested an intuitive argument for the above conclusion. In the circumstance of cyclic temporal selection variation, once fixation occurs in a generation then fixation persists thereafter. Whereas in the presence of spatial selection variation even with fixation in one locality for a generation, still the alternative type can be reintroduced by migration from other localities.

4. An accurate assessment of the significance and descriptions of the degree of homogeneity or heterogeneity in gene frequency patterns correlated to the environmental selection gradient and population structure could only come from a determination (qualitative or explicit) of all the stable equilibria, their domains of attraction and the dynamic behavior of the process. This is undoubtedly a formidable analytic task. We had some success on this objective for the cline stepping stone model (Karlin and Richter-Dyn [1976]). The evaluation of $\rho(DM)$ (see (2.4) for the definition) does give some information concerning the gene frequency patterns that are possible in the general case. To wit, if $\rho(DM)$ is substantially larger than 1 then certainly in some locality, at least one, we could expect a significant frequency of the A-allele. If $\rho(DM)$ is close to 1 but still exceeding 1 then the A-allele is protected but generally represented throughout the population in small frequency. The early discussion of Section 2 gives quite good lower estimates of ρ for several important cases of migration patterns.

5. It is also of interest to contrast migration structures as to their degrees of mixing and isolation. Two such concepts were introduced in part C of Section 2 and analyzed. The influence of migration structure on the maintenance of a protected polymorphism and its characteristics can be divided into four categories according to the extent of migration flow; very small, small to moderate, moderate to uniform

mixing, and strongly oscillatory movement. In each case, based on the analysis of Section 2, a number of qualitative inferences are highlighted and discussed.

(i) Very small migration flow. In this circumstance, the degree of environmental heterogeneity coupled to the initial frequency state plays a decisive role in the evolutionary development of the population:

(a) With selection forces favoring different genotypes in different niches (demes), a preponderance of one or other alleles predominate in each deme. The average level of heterozygosity is low but the level of polymorphism is large. The emerging gene frequency arrays are considerably hetero-geneous. The exclusive contingency of avoiding polymorphism for any sets of initial conditions is that a single allele has selective advantage throughout the population range (cf., Karlin and McGregor [1972a], [1972b]).

(b) With a homogeneous selection gradient involving local heterozygote advantage, a relatively homogeneous poly-morphic frequency state is achieved expressing a high average heterozygosity.

(c) A mixture of underdominance, directional and overdominant spatially varying selection expression can produce a wide variety of stable polymorphic and/or fixation states and the actual equilibrium established depends sensitively on the initial frequency state.

(ii) Small to moderate outbreeding or mobility rates. Result II (Section 2) tells us that the strength of a pro-tected polymorphism increases with the extent of isolation of demes. It is important to caution that this result applies in general form only if the rate of outbreeding is diminished uniformly independent of the deme sites. A decrease of dispersal at a particular deme while the other dispersal rates are kept constant, can engender the opposite effect making fixation more likely.

Increasing strength of protection means that the fixation states are more repellant and that each allelic frequency is represented with substantial frequency in at least one deme. There appears to be no relationship between the strength of a protected polymorphism and the form of the polymorphic equilibrium. With low migration rates we would expect considerable heterogeneity in gene frequency. For moderate migration, more monomorphic outcomes are revealed unless substantial heterozygote advantage is operating in each deme.

(iii) <u>Moderate to uniform mixing migration rates.</u> The contribution of the demes substantially blend in all respects. The outcomes now depend in a complex manner on all parameters of the model producing both fixation and polymorphic possibilities with fixation occurrences usually more frequent unless other forces are involved. With local heterozygote advantage a usually unique global polymorphism is maintained independent of the nature of gene flow.

(iv) <u>Strongly oscillating migration patterns.</u> Protection is now again more likely than with uniform mixing. There appears to be a threshhold level of medium migration flow such that the maintenance of a stable polymorphism is minimal at that rate of migration.

6. Several authors have recently appealed to Levins [1968] to help explicate the influence of fine versus coarse grain environmental expression pertinent to genetic variability. Although this theory is regarded as mathematically based, it is principally graphical and descriptive in

654

character. One commentary of this theory is that a very mobile organism experiences many different conditions, the average of which is similar for all members of a population. The effective environment is accordingly fine grained signifying little uncertainty and therefore the organism may well fix on a given genotype adaptive to the bulk of its experiences. For relatively immobile population, the environment experiences is likely to be uncertain and therefore the adaptive strategy of the population is to maintain substantial variability with different alleles predominant over appropriate ranges of the population.

The tenuous contact of these concepts with our work is that a migration pattern with substantial flow has indeed decreased opportunities for polymorphism. More precisely, Result II provides an analytic assertion that with increasing outbreeding (or mobility), the manifestation of multiple phenotypes and genotypes is reduced. There are restrictions on the validity of Result II. The reduced mobility must apply essentially uniformly over the whole range of species.

A different approach to the evaluation of degrees of mixing is the substance of Result I in paragraph C of Section 2.

REFERENCES

Bodmer, W.F. and L.L. Cavalli-Sforza. (1968). Genetics 59: 565-592.

Bryant, E.H. (1974). Am. Natur. 108: 1-19.

Bulmer, M.G. (1972). Am. Natur. 106: 254-257.

Charlesworth, B. and J.T. Giesel. (1972). Am. Natur. 106: 388-401.

Christiansen, F. (1974). Am. Natur. 108: 157-166.

Christiansen, F. (1975). Am. Natur. 109: 11-16.

Darlington, P.J. (1957). Zoogeography: The geographical distribution of animals. John Wiley, New York.

Deakin, M.A.B. (1966). Am. Natur. 100: 690-692.

Deakin, M.A.B. (1968). Aust. J. Biol. Sci. 21: 165-168.

Deakin, M.A.B. (1972). Aust. J. Biol. Sci. 25: 213-214.

Dempster, E. (1955). Quant. Biol. 20: 25-32.

Dobzhansky, Th. (1967). Proc. Fifth Berkeley Symp. Math. Stat. Prob. 4: 295-304.

Endler, J.A. (1973). Science 179: 243-250.

Fleming, W.H. and C.H. Su. (1974). Theor. Pop. Biol. 5: 431-449.

Friedland, S. and S. Karlin. (1975). Duke Math. J. 42: 365-387.

Giesel, J.A. (1972). Am. Natur. 106: 412-414.

Gillespie, J.H. (1973). Theor. Pop. Biol. 4: (2) 193-195.

Gillespie, J.H. (1974). Am. Natur. 108: No.964, 831-836.

Haldane, J.B.S. and S.D. Jayakar. (1963). Genetics 58: 232-242.

Hanson, W.A. (1966). Biometrics 22: 453-468.

Hartl, D.L. and R.D. Cook. (1973). Theor. Pop. Biol. 4: 163-172.

Jain, S.K. and A.D. Bradshaw. (1966). Heredity 21: 407-441.

Karlin, S. (1968). J. Appl. Prob. 5: 231-313, 487-566.

Karlin, S. and J. McGregor. (1972a). Theor. Pop. Biol. 3: 210-238.

Karlin, S. and J. McGregor. (1972b). Theor. Pop. Biol. 3: 186-209.

Karlin, S. and U. Lieberman. (1974). Theor. Pop. Biol. 6: 355-382.

Karlin, S. (1975). Theor. Pop. Biol. 7: 364-398.

Karlin, S. and N. Richter-Dyn. (1976). (This volume).

Karlin, S. and N. Richter-Dyn. (1976a,b,c,d). On the theory of clines. (To appear).

Karlin, S. (1976-77). Topics in mathematical genetics. Academic Press, New York. (To appear).

Kimura, M. and G.H. Weiss. (1964). Genetics 49: 561-576.

Kimura, M. and T. Maruyama. (1971). Genet. Res. 18: 125-131.

Levene, H. (1953). Am. Natur. 87: 331-333.

Levins, R. (1965). Genetics 52: 891-904.

Levins, R. (1968). Evolution in changing environments. Princeton University Press, Princeton, N.J. 120 pp.

Lewontin, R.C. (1974). The genetic basis of evolutionary change. Columbia University Press, New York and London.

Malécot, G. (1948, 1970). Les Mathématiques de L'Hérédité. Masson et Cie., Paris. Revised English translation: Freeman, San Francisco.

Malécot, G. (1951). Ann. Univ. Lyon Sciences, Sec.A, 14: 79-118.

Malécot, G. (1959). Publ. Inst. Stat. Univ. Paris 8: 173-210.

Malécot, G. (1967). In Proc. Fifth Berkeley Symp. Math. Stat. Prob. IV: 317-332. Univ. of California Press, Berkeley.

Maruyama, T. (1970). Adv. Appl. Prob. 2: 229-258.

Maruyama T. (1972). Math. Biosciences 14: 325-335.

Maynard Smith, J. (1970). Am. Natur. 104: 487-490.

Moran, P.A.P. (1962). The statistical processes of evolutionary theory. The Clarendon Press, Oxford.

Prout, T. (1968). Am. Natur. 102: 493-496.

Strobeck, C. (1974). Am. Natur. 108: 166-172.

Wallace, B. (1968). Topics in population genetics. Norton.

Wright, S. (1943). Genetics 23: 114-138.

Wright, S. (1968). Evolution and the genetics population, Vol. 1. Genetic and Biometric Foundation. Univ. of Chicago Press, Chicago.

Some Theoretical Analyses of Migration Selection Interaction in a Cline: A Generalized Two Range Environment

S. KARLIN and N. RICHTER-DYN

1. INTRODUCTION

A frequency cline refers generally to a gradient pattern in a character with respect to a <u>linear series of demes</u> (sub-populations of discrete breeding units) subject to some inter-deme migration. The slope of the cline between regions (and/or demes) has partly been taken to be indicative of the degree of differentiation among the different sub-populations and used in estimating the strength of selection acting to maintain a cline.

The phenomena of clinal variation in gene frequency is voluminously documented both in experimental and natural populations. Classic field studies of clines include the works of Kettlewell [1961] and Kettlewell and Berry [1961], [1969] on melanism and the extensive analyses of Jain and Bradshaw [1966] which highlight 5 clines of grasses and

* Department of Mathematics, The Weizmann Institute of Science, Rehovot, Israel - and

Department of Mathematics, Stanford University, Stanford, California, U.S.A.

**Department of Mathematics, Tel-Aviv University, Ramat-Aviv, Israel.

plants along certain transects. Frequency clines have long been described for morphological, physiological, behavioral and chromosomal characters. Endler [1973] contains a substantial set of references reporting cases of observed morphotones (clines). In recent years enzyme polymorphic clines have been increasingly uncovered, e.g., Koehn [1969], Beardmore [1970], Marshall and Allard [1972], Hamrick and Allard [1972], Merritt [1972], Rockwood Slus et al. [1973], Frydenberg and Christiansen [1974], Bryant [1974], Nevo and Bar [1976] and others.

A commonly held belief has been that clinal variations in gene frequency originates in response or is correlated to one or more ecological and/or genetic environmental parameters; e.g., temperature and humidity gradients over space and time, variation in substrate availabilities, characteristics of soil and water composition, the distribution of related predators, factors of cryptography and coloration background, etc. It is a trite comment in the analyses of clinal patterns that genotypic differentiation is promulgated by joint effects of local selection forces and gene flow. Important contributing factors determining the equilibrium and dynamic structure of clines would include the rate and form of gene flow, the extent of population subdivision or multiplicity of demes, the variation in local population density, the variegated intensity of selection forces operating over the range of the species, and similarly.

It is often assumed that near a sharp discontinuity in gene frequency there necessarily exists some environmental factor change. Arguments have been put forth by Haldane [1948], Hanson [1966], and others, indicating that small selection differences can already engender marked local genotypic differentiation but the amount of migration is critical.

This sensitivity to the magnitude of gene flow is questioned by Endler [1973]. There are alternative conclusions pertaining to the balance necessary between migration and selection entailing the establishment of clinal variation. For example Wright [1943] has stated: "diversity in degree and direction of selection among localities... may bring about great differentiation if not overbalanced by migration". On the other hand, Jain and Bradshaw infer on the basis of their simulation studies, "Selection can cause very localized patterns of microgeographical variation despite migration through pollen dispersal". "Populations of plants closely adjacent can diverge very remarkably even if there is only a small degree of geographical isolation". Of course both sets of pronouncements may be valid under appropriate circumstances.

A number of mathematical models incorporating some forms of gene flow and a geographical selection gradient were introduced by Haldane [1948], Fisher [1950] and others, most recently Slatkin [1973]. The above authors all dealt with an infinite linear array of demes (actually a continuum of subpopulations) where the gene flow in each generation occurs only to adjacent populations. The formulation of the analogous finite discrete model is as follows. A population is distributed on a finite region composed of separate breeding demes (localities), e.g., geographical or ecological habitats, $\wp_1, \wp_2, \ldots, \wp_N$ arranged in linear order as indicated. Successive generations in the population are discrete and non-overlapping. Each sub-population $\wp_i$ is of large size so that genetic drift effects are not pertinent. Consider a trait with two possible alleles labeled A and a . Mating and natural selection operate independently in each deme before some individuals disperse to the neighboring demes.

Thus after selection, a fraction m (m ≤ ½) of each deme is exchanged with each contiguous deme as depicted below.

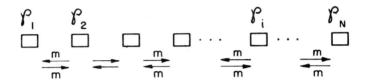

The migration pattern with movement allowed only to neighboring demes over successive generations, is commonly dubbed the homogeneous stepping stone migration mode and was studied extensively by Malécot [1948], [1951], [1959], later by Kimura and Weiss [1964], Weiss and Kimura [1965], and recently by Fleming and Su [1974]. These and other authors including Latter [1973], Maruyama [1970], [1972], concentrated mainly on finite population size effects with no selection differences among genotypes and demes.

Slatkin [1973] ostensibly considered more general migration patterns but the approximations he employed, in essence, reduce the model to that of a continuous version of the stepping stone model. In a later study we discuss forms of dispersal other than the "continuous movements" implicit to the stepping stone flux.

The two-range selection regime

Haldane, Fisher and Slatkin treat mainly the case when the range of the species is clearly divided into two sections. Fisher [1950] writes "The interesting case arises in which a gene enjoys a selective advantage in one part of a species range while in the remainder it is at a selective disadvantage. On the boundary between these regions, selection is neutral between two alleleomorphic genes."

It is useful to represent such a geographical selection gradient schematically as follows:

$$\begin{array}{cccccccc} & 1 & 2 & \ell_1 & \ell_1+1 & \ell_2 & \ell_2+1 & N \\ \text{Demes} & \square & \square & \cdots \square & \square & \cdots \square & \square & \cdots \square \\ & & \underbrace{\qquad}_{-} & & \underbrace{\qquad}_{0} & & \underbrace{\qquad}_{+} & \end{array} \qquad (1.1)$$

where $-$ signifies that A is <u>disadvantageous</u> at the localities $1 \leqslant i \leqslant \ell_1$ (meaning that where such a deme is isolated, reproduction and selection would bring the local population to fixation of the a-allele); $+$ connotes that A is <u>advantageous</u> at the positions specified and in the region of the 0 designation the A and a alleles are understood to be neutral (that is, no selection differences exist among genotypes). Thus, allele A is advantageous in the right portion of the range and allele a is advantageous in the left portion. The neutral zone may or may not exist. We can view the model of (1.1) as a generalized two range selection regime, each environment (range) favoring a different alleleomorph. The intensity of selection may vary in an arbitrary manner between demes consistent only with the $-$, 0 , $+$ arrangement of (1.1).

Fisher examined a subcase of (1.1) where the heterozygote carries intermediate fitness values corresponding to additive allelic effects. Haldane concentrated on an autosomal dominant gene and its alleleomorph. Kettlewell and Berry [1961] strove to fit their data into the Haldane setting. In the case of melanism, the Tingwell valley appears to be a dividing boundary between the two generalized environments. Kettlewell and Berry did subsidiary experiments which gave reasonable evidence for the existence of selection, namely by the introduction of organisms from one part of the cline into another part of the cline and studying their subsequent evolution. Jain and Bradshaw [1966] simulated a model involving two geographic environments with selection favoring different alleles in each region. The latter authors considered a migration structure of the kind consonant to isolation by distance where the extent of individual migration is described by a probability distribution. In their computer runs, they take some special choices considered appropriate for describing pollen and seed dispersal patterns. Thus migration per generation is not limited to small movements. Furthermore, a significant element in the grass clines of Jain and Bradshaw is the fact of unequal rates of gene flow in reciprocal directions (mediated partially by wind conditions). These authors found that strong local selection was more potent than the mitigating effects of even high gene flow. The concept of grades of local selection intensity will be discussed later in this work.

A number of examples of studies of clines in the human population are reviewed in the book of Cavalli and Bodmer [1971, p.483-490]. This discussion includes an application of the Fisher and Haldane model to a cline in A B O gene frequencies running latitudinally through the Islands of Japan.

The selection gradient

The transformation of gene frequency accountable to the mating structure and selection forces in deme ρ_i is reflected by the relation

$$x' = f_i(x) \tag{1.2}$$

such that if x is the A-frequency in ρ_i at the start of a generation, then after the action of mating and natural

selection the resulting A frequency prior to dispersal is
x' . The local selection functions satisfy $0 \leqslant f_i(x) \leqslant 1$,
$0 \leqslant x \leqslant 1$. Generally, $f_i(x)$ is continuous and monotone
increasing. Also we stipulate in this work that $f_i(0) = 0$
and $f_i(1) = 1$ indicating that the selection forces maintain
a pure population composition (or equivalently, new mutant
types in the time frame under consideration cannot be
established so that effectively mutation events are ignored).

An important choice for $f_i(x)$ arising from the classical
diploid one-locus two-allele viability model has the form

$$f_i(x) = \frac{(1+\sigma_i)x^2 + x(1-x)}{1+\sigma_i x^2 + s_i(1-x)^2} \qquad (1.3)$$

when the viability parameters of the genotypes are as listed

AA	Aa	aa	
$1+\sigma_i$	1	$1+s_i$	(1.4)

In a haploid situation we would take

$$f_i(x) = \frac{\sigma_i x}{\sigma_i x + s_i(1-x)} \qquad . \qquad (1.5)$$

Other determinations for $f_i(x)$ can be generated by superimposing various forms of frequency dependent selection and
other selection functions can be constructed induced on a
single locus when part of a complex genome. The cline models
discussed in this work also apply in some cases of selection
functions based on considerations of competition, prey predator, host pathogen or other ecological contexts.

The statement that A is _advantageous_ in deme i is
equivalent to the formal relation

$$f_i(x) > x \quad \text{for all} \quad 0 < x < 1 \quad , \qquad (1.6)$$

disadvantageous if

$$f_i(x) < x \quad \text{for all} \quad 0 < x < 1 \quad , \qquad (1.7)$$

665

and <u>neutral</u> if $f_i(x) \equiv x$.

For directional selection favoring allele A with additive fitnesses, $\sigma = -s > 0$ in (1.3). Where A is dominant and advantageous then $\sigma = 0$ and $s = -t$ $(0 < t \leqslant 1)$. The case of heterozygote advantage corresponds to $\sigma = -\gamma$, $s = -\delta$ $(0 < \gamma , \delta < 1)$.

A <u>selection function</u> $f(x)$ is said to be <u>more favorable</u> (stronger) <u>than</u> $\tilde{f}(x)$ with respect to the a-allele (in symbols $f \prec \tilde{f}$) , if

$$f(x) < \tilde{f}(x) \quad \text{for all} \quad x \quad (0 < x < 1) . \qquad (1.8)$$

It is easy to check that for $f_{\sigma,s}(x)$ defined in (1.3),

$$f_{\sigma,s} \prec f_{\sigma_1,s_1} \quad \text{provided} \quad \sigma \leqslant \sigma_1 , \text{ and } s_1 \leqslant s .$$

<u>The environmental (geographical) selection gradient is</u> <u>characterized by the complete prescription of the local</u> <u>selection functions</u> $f_i(x)$, $i = 1, 2, \ldots, N$.

In the circumstance where the heterozygote has additive fitness values, each local selection function is determined by a single selection coefficient σ_i , $-1 \leqslant \sigma_i \leqslant 1$, viz.,

	AA	Aa	aa		
fitnesses	$1+\sigma_i$	1	$1-\sigma_i$	in deme $\mathcal{P}_i$.	(1.9)

Where allele A (a) is advantageous, $\sigma_i > 0$ $(\sigma_i < 0)$ holds.*

---

*
Gillespie and Langley [1974] survey biochemical evidence pointing to wide validity of the assumption of heterozygote intermediacy. Thus the postulates of Section 2 of additive allelic fitness effects is possibly an important case in the study of allozyme polymorphisms; in this connection, see also Latter [this volume].

With local selection functions of the form (1.3) and fitnesses (1.9), the schematization of the selection gradient
(1.1) in terms of the parameters $\{\sigma_1, \sigma_2, \ldots, \sigma_N\}$ can have a
general form as below

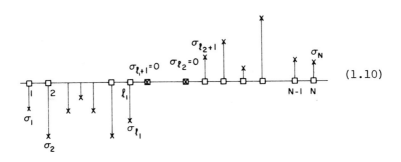

$$(1.10)$$

A more regular pattern has σ_i monotone separately in each
environmental range. We describe three cases whose equilibrium behavior will be contrasted in our later discussion.

In Case I, selection becomes stronger with distance from
the neutral zone (or from the demarcation point):

Condition I

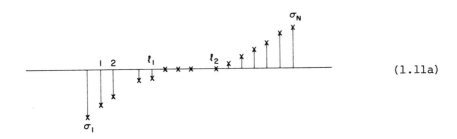

$$(1.11a)$$

In Case II, the selection coefficients are constant in each
environmental range:

Condition II

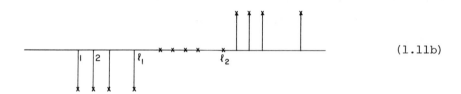

(1.11b)

In Case III, the selection intensity decreases with distance from the neutral region:

Condition III

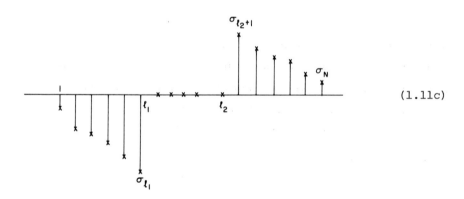

(1.11c)

To have a meaningful comparison, it is stipulated that the selection regimes I-III share the value

$$S_L = \sum_{i=1}^{\ell_1} \sigma_i \qquad (1.12a)$$

i.e., the aggregate selection coefficients over the left range where selection favors allele a coincide for Conditions I, II and III.

Similarly,

$$S_R = \sum_{i=\ell_2+1}^{N} \sigma_i \qquad (1.12b)$$

is the common cumulative selection value in the right range
for models I-III. Conditions II and III may possibly correspond to abrupt environmental changes, induced by a geological
fault or a sharp change in two contiguous soil compositions
while Condition I may be relevant to a more gradual climate
variation, background coloration, etc.

Formulation of objectives and problems

The analysis of migration selection interaction in multi
deme models can be implemented on two levels:

(i) It is possible to obtain rather precise conditions
guaranteeing the existence of a protected polymorphism. For
substantial studies of this problem in the general context of
geographical population models we refer to Karlin [1976]. For
more refined results concerning protection in cline systems,
see Karlin and Richter-Dyn [1976a].

(ii) The characterizations and comparisons of the equilibria under different conditions on selection and gene flow is
in general, a formidable undertaking. In this paper we
report a number of results describing the nature of polymorphic states in the selection cline model with stepping
stone migration.

The specific objectives of this presentation and the subsequent detailed works (Karlin and Richter-Dyn [1976b,c,d]
seek to clarify the interrelations between patterns of geographically differentiated gene frequency and the background
selection and gene flow structure.

Concentrating on a general two-environmental selection
gradient of the form (1.1) we investigate the following
questions:

669

(1) For a prescribed environmental or geographical selection gradient (determined by the $f_i(x)$ that reflect the local selection forces) and a migration rate m, we wish to ascertain qualitatively and quantitatively the form of the stable equilibrium configurations (possible clines) that can evolve.

In particular it is of value to compare and contrast the shape of a steady state (equilibrium) cline with the shape of the background environmental selection pattern. Particular pertinent inquiries include:

(2) To what extent does increasing gene flow swamp out differences in local selection expressions?

(3) Where does the maximum slope of the cline occur and what is its significance?

(4) Is a sharp local change in selection intensity between two adjacent demes reflected in a corresponding marked change in gene frequency nearby?

(5) It is of interest to evaluate the effects on the shape of the cline in lengthening the neutral region.

(6) With prescribed aggregate levels of selection coefficients favoring allele A in the right range and allele a in the left range, how does the equilibrium cline differ in the contrasting cases of the selection patterns I, II and III described in (1.11)?

(7) How would the character of the cline change if the demes of the central (interface) zone involve local heterozygote advantage instead of each bearing no selection differences among genotypes?

(8) What would be the influence if each deme of the central range expresses underdominance?

(9) What are the effects of unequal rates of migration, a non-local migration range, different relative deme sizes and

multiplicity of demes, the influence of boundary conditions, a selection gradient involving more than two environments, timing (or order) of the operation of selection and migration forces, multi alleles, multi loci, dominance relations, etc.

Several of our findings on these problems are reported in Sections 2 and 3. Some of their implications and limitations are described in the concluding Section 4. In a series of papers we will elaborate the validations of the results high-lighted in the later sections plus other aspects of selection clines.

We close this first section with some comments on other recent literature on this topic. Endler [1973] reported simulation and computer studies of the equilibrium state for some examples of clines involving 15 sub-populations. Speci-fically, Endler focused primary attention on the model of (1.1) for several regular selection patterns where each f_i is determined by the formula (1.3) with fitness parameters $w_i(AA)$, $w_i(Aa)$, $w_i(aa)$ for deme i of the genotypes AA , Aa and aa , respectively. We list the choices he made:

(α) $w_i(Aa) = h > 1$, $w_i(AA) = 1 - \frac{i-1}{15}$, $w_i(aa) = \frac{i}{15}$,

i = 1,...,15 (heterozygote advantage throughout the range).

(β) $w_i(Aa) = 1$, $w_i(AA) = 1 + \frac{i-8}{15}$, $w_i(aa) = 1 + \frac{8-i}{15}$,

(additive fitnesses).   (1.13)

(γ) $w_i(Aa) = \frac{|i-8|}{15} + h$, $w_i(AA) = 1 + \frac{i-8}{15}$,

$w_i(aa) = 1 + \frac{8-i}{15}$

(local heterozygote advantage).

The effective population size in each deme was so large that the influences of genetic drift is considered inconsequential.

671

The cases (1.13α,β) were designed to reflect a phenomenon where a particular genotype fitness increased uniformly with position along a transect while the other two genotype fitnesses decreased at different <u>constant</u> rates along the same transect. His simulation runs yielded monotone clines in all cases which exhibited a marked steepening in the neighborhood of the center deme $i=8$ in case (β). The outcome for case (γ) was close to that of (α). But in contrast the heterozygous advantage cline (α) produced a roughly linear monotone cline. He observed some boundary effects in the cline as revealed in the computer runs.

Hastings [preprint] conducted some stochastic simulations of the stepping-stone model with 15 demes of "small" population sizes 10, 20 and 40 along a selection gradient of the kinds treated by Endler and showed that gene flow had several important effects. As gene flow increased, the stochastic effects were reduced and after 50 generations the yield approached the deterministic solutions, thus paralleling the effects of increasing size for each deme.

It is convenient, for later comparisons, to have available a number of Endler's qualitative claims and evaluations which we present by direct citation from his paper.

"There are many possible spatial patterns of selection and gene flow that can produce a given cline structure."

"Irregularities in environmental gradients increase the sensitivity of clines to the effects of gene flow in proportion to the increase in the differences in gene frequencies between the emigrants and the demes receiving the immigrants."

"Any asymmetry in gene flow does not lead to dedifferentiation if the environmental gradient is smooth; it merely shifts the position of the transition zone between the differentiated areas from that which would be expected if there were no asymmetry. Abrupt geographic differences in gene, genotype, or morph frequencies should not, therefore, be interpreted as evidence for environmental changes in the immediate vicinity of the steepest part of the cline."

"Gene flow may be unimportant in the differentiation of populations along environmental gradients."

We can test a number of Endler's assertions on the basis of our exact analysis of his and our more general models. A variety of new insights emerge.

Slatkin [1973] contends that there is a single critical parameter which mainly determines the shape of the cline at least near the barrier where the character of selection sharply changes from favoring allele A to favoring allele a . This conclusion was correct for the special model he considered but not typical for the general description of the interaction between gene flow and the selection gradient structure as will be more apparent from the discussion and results reported in this paper.

2. RESULTS AND COMPARISONS OF FREQUENCY CLINES

We will describe a series of results bearing on the objectives and problems set forth in Section 1.

Consider a geographical cline displaying a two range selection regime having the schematic form

$$\text{Demes } \underbrace{\rho_1, \ldots, \rho_{\ell_1}}_{-} \quad \underbrace{\rho_{\ell_1+1}, \ldots, \rho_{\ell_2}}_{0} \quad \underbrace{\rho_{\ell_2+1}, \ldots, \rho_N}_{+} \tag{2.1}$$

$$(1 < \ell_1 \leqslant \ell_2 < N)$$

where in the + demes, A is advantageous, and in the -

demes the a-type is advantageous, while in the central zone there are no selection differences (A and a are mutually neutral) (see (1.6) and (1.7) earlier).

Apart from the general structure of (2.1) there are no further requirements imposed on the detailed nature of the local selection functions. In particular we stipulate no regularity or pattern for the strength or degree of advantage for A in the respective demes of the + region and similarly for the - region. Endler [1973], Slatkin [1973] examined some special regular patterns fitting (2.1) by numerical means.

The associated migration form is assumed at first to be uniform gene flow between contiguous demes, with exchange rate m . The assumption tacitly implies that the relative deme sizes are equal. (Later we will discuss the model of unequal local rates of migration and the consequences of unequal relative deme sizes, paragraph E.)

Where selection operates first and migration occurs in the adult stage, the transformation equations relating the A-frequency distribution $\underline{x} = (x_1, \ldots, x_N)$ and $\underline{x}' = (x_1', \ldots, x_N')$ over two successive generations, x_i representing the proportion of the A-allele in deme i , are given by

$$x_1' = (1-m) f_1(x_1) + m f_2(x_2)$$

$$x_i' = m f_{i-1}(x_{i-1}) + (1-2m) f_i(x_i) + m f_{i+1}(x_{i+1}) \, ,$$

$$i = 2, \ldots, N-1 \tag{2.2}$$

$$x_N' = m f_{N-1}(x_{N-1}) + (1-m) f_N(x_N) \quad .$$

Several criteria for the ascertainment of local instability of the fixation states $\underline{0} = (0, \ldots, 0)$ and $\underline{1} = (1, \ldots, 1)$, corresponding to fixation of the a-allele and A-allele, respectively, in the context of an ordered geographical structure, are developed in Karlin and Richter-Dyn [1976a], and for general multi deme population systems in Karlin [1976]. This is the problem of "protection" in a cline.

An equilibrium

$$\underline{\hat{x}} = (\hat{x}_1, \hat{x}_2, \ldots, \hat{x}_N) \tag{2.3}$$

of the system (2.2) is called <u>polymorphic</u> if $0 < \hat{x}_i < 1$ for all i. <u>The equilibrium</u> $\hat{x}$ <u>is said to be monotone</u> (a <u>morphotone</u>) if

$$\hat{x}_i < \hat{x}_{i+1} \quad \text{for all} \quad i. \tag{2.4}$$

The <u>slope of the cline</u> for the equilibrium state $\underline{\hat{x}}$ is described by the vector

$$\underline{\hat{\delta}} = (\hat{\delta}_1, \hat{\delta}_2, \ldots, \hat{\delta}_{N-1}) \quad \text{where} \quad \hat{\delta}_i = \hat{x}_{i+1} - \hat{x}_i. \tag{2.5}$$

A. <u>The nature of polymorphic equilibria.</u>

The result reported next is rather remarkable in that it is independent of any specific form of the selection functions except that the general pattern of (2.1) prevails.

<u>Result I.</u> (<u>Every polymorphic equilibrium describes a mono-</u>
<u>tone cline</u>)

<u>Consider a two range selection gradient as in</u> (2.1). <u>Any</u> <u>polymorphic equilibrium</u> $\hat{x}$ <u>is</u> <u>monotone.</u> <u>Moreover, the slope</u> <u>changes have the form</u>

$$\hat{\delta}_1 < \hat{\delta}_2 < \ldots < \hat{\delta}_{\ell_1} < \hat{\delta}_{\ell_1+1} = \ldots = \hat{\delta}_{\ell_2-1}$$
$$> \hat{\delta}_{\ell_2} > \hat{\delta}_{\ell_2+1} > \ldots > \hat{\delta}_{N-1}. \tag{2.6}$$

A pictorial view of a polymorphic cline concomitant with the above result is now drawn:

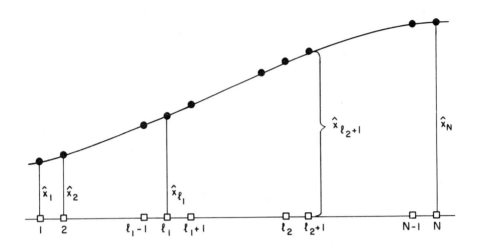

showing that the maximum slope occurs in the neutral region.
We stress again that every polymorphism is necessarily monotone and (2.6) holds independently of the variations in intensity and strength expressed in the local selection functions
provided merely the geographical selection gradient displays
the two range form of (2.1). A formal proof of this and the
later results are elaborated in Karlin and Richter-Dyn [1976b].

Result II. (Uniqueness of a polymorphic cline)

If each local selection component function of
$\{f_1(x),\ldots,f_N(x)\}$ corresponds to additive fitness values as
in (1.9) and subject to a mild restriction limiting the maximum extent of migration exchange then there exists at most
one polymorphic equilibrium.

A formal sufficient condition for the validity of Result II
is that the magnitude of migration rate relative to the selection coefficients obey the inequalities

$$m \leq \frac{1}{|s_i|} \left\{ 1 - \sqrt{1-s_i^2} \right\} \qquad i = 1,2,\ldots,N \quad .$$

$$\approx |s_i|/2 \quad \text{for} \quad s_i \quad \text{small}$$

In many circumstances the above restriction can be much relaxed. Thus for an environmental selection pattern involving a central locality such that the selection expression is the same but directed in favor of different alleleomorphs on opposite sides of the center then the conclusion of Result II is valid for all levels of migration flow.

Under the conditions of Result II we secure the following classification pertaining to the dynamic behavior of the cline realization.

(a)   If a-fixation is a locally stable equilibrium state and the a-allele is protected then the a-allele will be fixed from any internal initial state.

(a')   If the A-allele is protected and $\underline{1}$ is locally stable then fixation of the A-type is established independent of the initial population makeup.

(b)   Where a protected polymorphism exists then the population evolves at a geometric rate to a unique monotone polymorphic cline whose slope variation satisfies (2.6).

Cases (a) and (a') can occur only for m large and (b) is certainly assured for m small. In a symmetric model (Section 3), (b) holds for all migration rates.

Where the hypotheses of Result II are omitted then a selection gradient consistent with (2.1) can be constructed without a unique polymorphic equilibria, in fact including at least 3 polymorphisms. (According to Result I all such polymorphisms represent morphotones.) Examples also exist where both fixation states and at least one further polymorphic monotone cline are simultaneously stable. For general selection

677

functions there can exist cases with any number of polymorphisms. Henceforth unless stated otherwise, we assume Result
II in force.

B.   Comparison of two selection gradients the second having
     a longer range where allele A is favored.

In this context we seek to determine the relative change in
the morphotone that emerges where the region in which A is
advantageous is extended. The selection patterns to be compared are as indicated

$$
\begin{array}{lllllll}
\underline{\text{Population 1}} & 1 & \ell_1 & \ell_2 & N & \text{Demes} \\
& \Box \cdots \Box \cdots \Box \cdots \Box & & & & \\
& f_1 \quad\; f_{\ell_1} \quad\; f_{\ell_2} \quad\; f_N & & & & \text{associated selection} \\
& & & & & \text{functions}
\end{array}
$$

(2.7)

$$
\begin{array}{llllllll}
\underline{\text{Population 2}} & & & & & N+1 & \tilde{N} & \\
& \Box \cdots \Box \cdots \Box \cdots \Box & & & & \Box \cdots \Box & \tilde{N} > N \\
& \overline{f}_1 \;\; \overline{f}_{\ell_1} \;\; \overline{f}_{\ell_2} \;\; \tilde{f}_N & & & & \tilde{f}_{N+1} \;\; \tilde{f}_{\tilde{N}}
\end{array}
$$

The demes numbered $1,\ldots,N$ are identical for the two versions and accordingly $f_i(x) \equiv \tilde{f}_i(x)$, $i = 1,2,\ldots,N$. However, there appear in population 2 new demes connected to
deme N such that A is advantageous also in demes
$N+1,\ldots,\tilde{N}$. In other words, population 2 involves a longer
segment (region) in favor of the A-type.

Result III.   (Enlarged A-favorable region)

   Let $\hat{x} = (\hat{x}_1,\ldots,\hat{x}_N)$ be a polymorphic equilibrium for the
selection regime of population 1. For population 2 either
only A-fixation is possible or if there exists a polymorphic
cline
$$\underline{\tilde{x}} = (\tilde{x}_1,\ldots,\tilde{x}_N,\tilde{x}_{N+1} \cdots \tilde{x}_{\tilde{N}})$$
then
$$\hat{x}_i < \tilde{x}_i \text{ and } \hat{\delta}_i < \tilde{\delta}_i , \quad i = 1,2,\ldots,N .$$

(2.8)

Thus, <u>the extended cline is steeper</u> (<u>exhibiting more gene</u> <u>frequency differentiation</u>) <u>over the common region of the</u> <u>population distribution provided its polymorphic character is</u> <u>retained. Where</u> $\tilde{x}$ <u>does not exist then necessarily A-</u> <u>fixation is the exclusive stable outcome</u> evolving from the selection regime of population 2.

With any morphotone, it is obvious that the slope between localities at the neutral region cannot exceed $\frac{1}{k+1}$ where k is the number of demes composing the neutral zone.

COMPARISON OF CLINES IN CASE OF AN EXTENDED A-FAVOURABLE RANGE

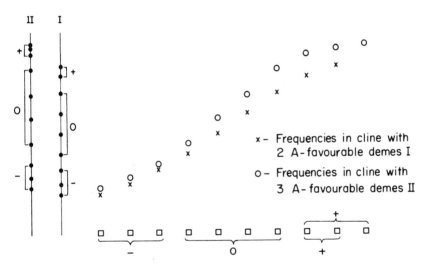

x – Frequencies in cline with
2 A-favourable demes I

o – Frequencies in cline with
3 A-favourable demes II

C.   The effects of the length of the neutral zone.

Another comparison of interest concerns the following set-up.   Consider two versions of environmental selection gradients as now depicted.

(2.9)

$$\tilde{\ell}_1 = \ell_1 \quad , \quad N+r = \tilde{N} \quad , \quad r = \tilde{\ell}_2 - \ell_2 > 0$$

where the selection functions are related according to

$$f_i(x) \equiv \tilde{f}_i(x) \quad , \quad i = 1, \ldots, \ell_1$$

$$f_{N-j}(x) \equiv \tilde{f}_{\tilde{N}-j}(x) \quad , \quad j = 0, \ldots, N-\ell_2-1$$

(2.10a)

while

$$f_\mu(x) = \tilde{f}_\nu(x) = x \quad , \quad \ell_1+1 \leqslant \mu \leqslant \ell_2 \quad , \quad \tilde{\ell}_1+1 \leqslant \nu \leqslant \tilde{\ell}_2 . \quad (2.10b)$$

Thus the neutral zone is longer under the condition (ii) and otherwise the selection patterns coincide. The next assertion shows how the equilibrium cline is affected by the extent of the neutral region.

Result IV.   Consider the model of (2.9) and suppose the polymorphism

$$\hat{\underline{x}} = (\hat{x}_1, \ldots, \hat{x}_N) \quad \text{for condition (i)} \quad \text{and}$$

$$\tilde{\underline{x}} = (\tilde{x}_1, \ldots, \tilde{x}_{\tilde{N}}) \quad \text{for condition (ii)}$$

exist.   Then

$$\hat{x}_i > \tilde{x}_i \quad \text{and} \quad \hat{\delta}_i > \tilde{\delta}_i \quad \text{for} \quad i = 1, 2, \ldots, \ell_2 \quad\quad (2.11)$$

and

$$\tilde{x}_{\tilde{N}-j} > \hat{x}_{N-j} \quad \text{and} \quad \tilde{\delta}_{\tilde{N}-j} < \hat{\delta}_{N-j} \quad \text{for}$$

$$j = 0, 1, \ldots, N-\ell_1-1 .$$

(2.12)

Thus, a more prolonged neutral zone tends to <u>accent more</u> the <u>selection forces operating on the outer regions of the population distribution.</u> This finding is perhaps unintuitive; its validity applies independent of the detailed nature of the selection functions operating in the $-$ and $+$ range of the population. It is worth emphasis that over the neutral zone the frequency cline is linear bearing a more pronounced flat but discernible slope with increasing length of this portion. In another work we will highlight several more sensitive distinguishing characteristics of a morphotone corresponding to a two range environment with a long neutral region vis a vis the form of a cline associated with an underlying selection regime expressing a regular monotone pattern of heterozygote advantage over the range of the species.

By combining the estimates (2.11) and (2.12) we extract the following bounds on the equilibrium frequencies in the neutral zone

$$\tilde{x}_i < \hat{x}_i < \tilde{x}_{i+r} \quad (r = \tilde{\ell}_2 - \ell_2) \quad i = \ell_1 + 1, \ldots, \ell_2 \quad . \qquad (2.13)$$

The comparison of (2.11) and (2.12) can be graphically displayed

COMPARISON OF CLINES IN CASE OF AN EXTENDED NEUTRAL REGION

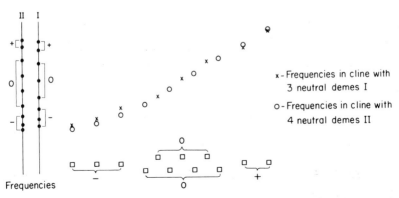

x - Frequencies in cline with 3 neutral demes I

o - Frequencies in cline with 4 neutral demes II

The following precise stability fact is of interest.

Result V. Consider the same conditions as in IV above. Where under Condition (i) there exists a polymorphism, then under Condition (ii) (involving an enlarged neutral region) a globally stable polymorphism necessarily occurs.

The above finding asserts that with an underlying monotone ecotone, (a term used by Endler [1973] for a two range selection regime), the possibilities of a stable polymorphism are significantly increased as the neutral zone is extended while preserving the same selection pattern in the non-neutral locations.

D.   The influence of the magnitude of the migration rate.

Consider an ecotone selection pattern where the localities $\mathscr{P}_1, \ldots, \mathscr{P}_{\ell_1}$ have the a-allele advantageous, $\mathscr{P}_{\ell_2+1}, \ldots, \mathscr{P}_N$ confer advantage to the A-allele and the demes of the center region $\mathscr{P}_{\ell_1+1}, \ldots, \mathscr{P}_{\ell_2}$ $(\ell_2 \geq \ell_1+1)$ bear no selection differences. Suppose there exists a polymorphic equilibrium $\hat{x} = (\hat{x}_1, \ldots, \hat{x}_N)$. We know on the basis of Result I that $\hat{x}$ describes a monotone cline (morphotone) endowed with the properties (2.4) and (2.6). The next result concerns the changes of $\hat{x}(m)$ as m increases (showing its dependence on the migration rate).

Result VI. Suppose the existence of equilibrium frequency clines $\hat{x}(m)$ and $\hat{x}(m')$ for the migration rates m and m' , respectively with m < m' . Then there exists a deme site k_o such that

(i) $\hat{x}_i(m) < \hat{x}_i(m')$, $i = 1,2,\ldots,k_o-1$

(ii) $\hat{x}_j(m') < \hat{x}_j(m)$, $j = k_o+1,\ldots,N$.

(2.14)

In the symmetric model k_o corresponds to the center deme (see Section 3).

The interpretation of (2.14) is the natural one:
Increased gene flow either causes fixation or attenuates the
frequency differentiation between demes to the extent that in
both portions of the population distribution the attained
gene frequencies appear more flat.

E.   Unequal local migration rates

Consider a two environment selection regime as in (2.1)
with a stepping stone migration flow but allowing unequal
migration exchanges between demes (and also not necessarily
equal rates in reciprocal directions at a locality).   Even
with a uniform individual migration propensity but where the
deme sizes are unequal: for example, let the relative deme
sizes at $\mathcal{P}_1, \ldots, \mathcal{P}_N$ be $c_1, c_2, \ldots, c_N$, respectively, $c_i > 0$,
$\Sigma_i c_i = 1$, then the associated backward migration matrix
$M = ||m_{ij}||_1^N$ [#] is for the case at hand

$$M = ||m_{ij}|| =$$

$$\begin{pmatrix} \dfrac{(1-m)c_1}{\gamma_1} & , & \dfrac{mc_2}{\gamma_1} & , & 0 & , & 0, \ldots, 0 \\[2ex] \dfrac{mc_1}{\gamma_2} & , & \dfrac{(1-2m)c_2}{\gamma_2} & , & \dfrac{mc_3}{\gamma_2} & , & 0 \\[2ex] 0 & & \dfrac{mc_2}{\gamma_3} & , & \dfrac{(1-2m)c_3}{\gamma_3} & , & \dfrac{mc_4}{\gamma_3} \\[2ex] & & & & & & \\ & & 0 & , & \dfrac{mc_{N-1}}{\gamma_N} & , & \dfrac{(1-m)c_N}{\gamma_N} \end{pmatrix} \quad (2.15)$$

[#] (m_{ij} = fraction of individuals of deme j contributed to
deme i after selection and migration).

where $\gamma_i = mc_{i-1}+(1-2m)c_i+mc_{i+1}$, $2 \leqslant i \leqslant N-1$,
$\gamma_1 = (1-m)c_1+mc_2$, $\gamma_N = mc_{N-1}+(1-m)c_N$.

Where the migration flow between neighboring demes can also vary, then the backward migration matrix $M = ||m_{ij}||$ attains the general form

$$
M = \begin{pmatrix}
r_1 & m_1' & 0 & 0 \\
m_2 & r_2 & m_2' & 0 \\
0 & m_3 & r_3 & m_3' \\
& & \ddots & \ddots & \ddots \\
& & & m_N & r_N
\end{pmatrix}
\qquad
\begin{array}{c}
m_i+m_i'+r_i=1 \\
i=1,2,\ldots,N
\end{array}
\qquad (2.16)
$$

The transformation equations version of (2.2) in this general clinal framework becomes

$$x_1' = r_1 f_1(x_1) + m_1' f_2(x_2)$$

$$x_i' = m_i f_{i-1}(x_{i-1})+r_i f_i(x_i)+m_i' f_{i+1}(x_{i+1}) \ , \quad (2.17)$$

$$2 \leqslant i \leqslant N-1$$

$$x_N' = m_N f_{N-1}(x_{N-1})+r_N f_N(x_N) \ .$$

Conditions for the existence of a protected polymorphism for this non-homogeneous stepping stone model (2.16) are developed in Karlin and Richter-Dyn [1976a]. We highlight now the analog of Result I for this case of unequal local migration rates.

Result VII. Suppose $\hat{x} = (\hat{x}_1,\hat{x}_2,\ldots,\hat{x}_N)$ is a polymorphic equilibrium for the general stepping stone migration mode (2.16) involving a two range selection gradient of the character (2.1). Then $\hat{x}$ describes a morphotone, i.e.,

684

$$\hat{x}_1 < \hat{x}_2 < \ldots < \hat{x}_N \quad (\hat{x}_i \text{ is increasing}). \tag{2.18}$$

Let $\hat{\delta}_i = \hat{x}_{i+1} - \hat{x}_i$, $i = 1, 2, \ldots, N-1$ be the slope of the morphotone measuring local gene frequency differentiation and <u>define</u> $a_1 = 1$, $a_k = \prod\limits_{i=1}^{k-1} \left(\dfrac{m_i'}{m_{i+1}'} \right)$, $k = 2, \ldots, N$. If demes $\wp_1$ to $\wp_{\ell_1}$ constitute the $-$ region (cf. (2.1)), $\wp_{\ell_1+1}$ to $\wp_{\ell_2}$ comprise the neutral zone and demes $\wp_{\ell_2+1}$ to $\wp_N$ the $+$ environmental portion <u>then</u>

$$m_1' a_1 \hat{\delta}_1 < m_2' a_2 \hat{\delta}_2 < \ldots < m_{\ell_1}' a_{\ell_1} \hat{\delta}_{\ell_1} < m_{\ell_1+1}' a_{\ell_1+1} \hat{\delta}_{\ell_1+1} = \ldots =$$

$$m_{\ell_2-1}' a_{\ell_2-1} \hat{\delta}_{\ell_2-1} > m_{\ell_2}' a_{\ell_2} \hat{\delta}_{\ell_2} > m_{\ell_2+1}' a_{\ell_2+1} \hat{\delta}_{\ell_2+1} > \ldots > \tag{2.19}$$

$$m_{N-1}' a_{N-1} \hat{\delta}_{N-1} \quad .$$

<u>The relations of (2.19) are equivalent to</u>

$$m_i' \hat{\delta}_{i-1} < m_i' \hat{\delta}_i \quad \text{in the} \quad - \quad \text{region}$$

$$= \quad \text{in the} \quad 0 \quad \text{region} \tag{2.20}$$

$$> \quad \text{in the} \quad + \quad \text{region}$$

It is manifest in the present case, that the actual maximum gene frequency differentiation, the position where $\max \hat{\delta}_i$ is attained, <u>need not occur in the neutral zone</u> as in the uniform flow situation.

Unlike the smooth convex-linear concave shape attendant to a polymorphism with homogeneous migration rates with any background of selection intensities consistent with (2.1), for variable local migration rates it is now possible to have kinks in the clinal shape as depicted

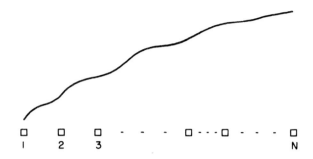

but persistently a morphotone (the property of increasing gene frequency) appears.

For the special case of (2.15) we have $m_i'/m_i = c_{i+1}/c_{i-1}$. Then, where relative population size decreases from both ends of the species range toward the central zone, the maximum gene frequency differentiation (max $\hat{\delta}_i$) occurs again in the neutral region. More precisely

$$\frac{\hat{\delta}_{i-1}}{\hat{\delta}_i} \begin{cases} < & \text{in the } - \text{ region} \\ = \dfrac{c_{i+1}}{c_{i-1}} & \text{in the } 0 \text{ region} \\ > & \text{in the } + \text{ region} \end{cases}$$

On the basis of the relations of (2.19) and (2.20) and observations of the gene frequency clinal patterns, it may be possible to estimate local migration intensities.

It is worth pointing out that all the qualitative comparisons of paragraphs B-D apply in the general stepping stone context with variable local migration rates. Result VI requires a reformulation which will be set forth elsewhere.

F. The influence of the order of migration and selection pressures on the morphotone

The previous results were derived for the model where the genetic forces operating in each generation had the timing

local selection $\longrightarrow$ migration (at adult stage).
followed by

$$(2.21)$$

We discuss briefly the model where the order is reversed in (2.21). Accordingly, consider now the order to be

$$\text{followed by}$$
migration (as gametes) $\longrightarrow$ selection. $\qquad$ (2.22)

The analog of (2.2) for the model (2.22) with homogeneous migration flow follows the transformation equations

$$y_1' = f_1((1-m)y_1 + my_2)$$
$$y_i' = f_i(my_{i-1} + (1-2m)y_i + my_{i+1}) \ , \ i = 2,3,\ldots,N-1 \qquad (2.23)$$
$$y_N' = f_N(my_{N-1} + (1-m)y_N)$$

where $\underline{y} = (y_1,\ldots,y_N)$ and $\underline{y}' = (y_1',\ldots,y_N')$ denote the frequency state in two successive generations.

The analog of Result I carries over completely:

Result I'. Suppose $\hat{\underline{y}} = (\hat{y}_1,\ldots,\hat{y}_N)$ is a polymorphic equilibrium for (2.23) with underlying selection regime as in (2.1).
Then necessarily $\hat{\underline{y}}$ describes a morphotone; viz.,

$$\hat{y}_1 < \hat{y}_2 < \ldots < \hat{y}_N \qquad (2.24)$$

and the slope vector $\hat{n}_i = \hat{y}_{i+1} - \hat{y}_i$ satisfies

$$\hat{n}_1 < \hat{n}_2 < \ldots < \hat{n}_{\ell_1} = \hat{n}_{\ell_1+1} = \ldots = \hat{n}_{\ell_2} > \hat{n}_{\ell_2+1} > \ldots > \hat{n}_{N-1}$$

$$(\text{cf. (2.6).}) \qquad\qquad (2.25)$$

The uniqueness version of Result II carries over and all the comparisons of Results III-VI are valid in the setting of (2.22).

What is of interest is to compare the morphotone (2.24) for the model (2.22) with the morphotone $\hat{\underline{x}}$ emerging in the formulation of (2.21).

687

<u>Result VIII.</u> <u>Suppose</u> $\hat{x}$ <u>and</u> $\hat{y}$ <u>are morphotones for the</u>
<u>models of</u> (2.21) <u>and</u> (2.22), <u>respectively acted on by the same</u>
<u>selection gradient and homogeneous migration flow.</u> (Here
only the order of applications of the forces are interchanged.)
<u>Then</u> $\hat{y}_i = f_i(\hat{x}_i)$ <u>and therefore</u>

$$\hat{y}_i < \hat{x}_i \quad , \quad i = 1,2,\ldots,\ell_1 \qquad \text{(over the } - \text{ region)}$$

$$\hat{y}_i = \hat{x}_i \qquad i = \ell_1+1,\ldots,\ell_2 \qquad \text{(over the neutral region)}$$

$$\hat{y}_i > \hat{x}_i \qquad i = \ell_2+1,\ldots,N \qquad \text{(over the } + \text{ region).}$$

$$(2.26)$$

The deduction of (2.26) is perhaps intuitive after the fact.
When selection follows migration the effects due to selection
are more pronounced in each environmental range. Equivalently,
migration flow acting after selection will smoothen the
selection impression operating in the two extremes of the
range.

3. RESULTS FOR SYMMETRIC TWO RANGE SELECTION REGIMES

When there is a natural symmetrical center in the distribution of demes, e.g., a position at which selection changes
from favoring one type to favoring the other type, it is more
convenient to label the demes in the sequence

$$\wp_{-K} \; \wp_{-K+1}, \ldots, \wp_{-1}, \wp_0, \wp_1, \ldots, \wp_K \qquad {}^* \qquad (3.1)$$

with $\wp_0$ as the deme in the center of the region and $\wp_{-K}$
and $\wp_K$ the end demes. The associated selection functions

---

*The actual deme $\wp_0$ may or may not be present but position
0 is a demarcation point.

reflecting the environmental selection gradient are described with the same notation

$$\{f_{-K}(x), \ldots, f_0(x), \ldots, f_K(x)\} \quad . \tag{3.2}$$

Definition. A selection pattern is said to be <u>symmetrical</u> (it may be more logical to call it asymmetrical or odd) <u>with respect to the center deme locality</u> called for brevity, "symmetric", if

$$f_i(x) \equiv 1 - f_{-i}(1-x) \quad , \quad 0 \leq x \leq 1 \; , \; i = 0,1,2,\ldots,K \; . \tag{3.3}$$

One can envision the symmetric model as depicting a selection gradient <u>anti-symmetrical</u> as a function of the distance from $\mathcal{P}_0$. Throughout this section we assume uniform gene flow (with stepping stone migration mode) so that $m_i = m$ for all i .

A symmetrical two range ecotone has the following structure

$$\tag{3.4}$$

where in the + (-) demes the A (a)-allele is advantageous while the 0 demes (those numbered $i = 0, \pm 1, \ldots, \pm \ell$) are neutral in selection expression.

The pattern of (3.4) coupled with (3.3) reflects a picture where the ecological selection expression in one direction varies in intensity, not necessarily in any regular way but persists in favor of the A-allele, and in a symmetrical fashion involving the same intensities the a-allele is favored for the localities in the opposite direction from the center position.

<u>Many of the results reported in this section are valid for geographical linear population structures approximately</u>

symmetric.

Result IX.   For a symmetrical monotone selection gradient, there always exists a polymorphic state $\underline{x}^* =$ $(x_{-K}^*, \ldots, x_0^*, \ldots, x_K^*)$ with the properties

$$x_i^* = 1 - x_{-i}^* \quad \text{and} \quad x_0^* = \frac{1}{2} \; . \tag{3.5}$$

We call such an equilibrium a symmetric polymorphism.

For the case where each local selection function reflects additive fitnesses (not necessarily with the same selection coefficients) then, for each migration rate, $\underline{x}^*$ is the unique globally stable polymorphic equilibrium.

A.   The changes in the frequency cline when extending the ecotones.

We will make some comparisons similar to the Results III and IV with the proviso here of preserving the symmetry of the selection ecotone. Consider the situation of (3.4) (referred to as Condition (i)) vis a vis the enlarged geographical symmetric ecotone involving $2L+1 > 2K+1$ demes

Condition (ii) $L > K$ (3.6)

stipulating that the clines (3.4) and (3.6) coincide in the positions from $-K$ to K .

Thus we have in (3.6) extra A-advantageous demes at the right end of the geographical region and by symmetry extra a-advantageous demes present for the left region.

Let

$$\underline{x}^* = (x_{-K}^*, \ldots, x_{-1}^*, x_0^*, x_1^*, \ldots, x_{-K}^*) \tag{3.7}$$

and

$$\underset{\sim}{x} = (\underset{\sim}{x}_{-L}, \ldots, \underset{\sim}{x}_{-K}, \ldots, \underset{\sim}{x}_{0}, \ldots, \underset{\sim}{x}_{K}, \ldots, \underset{\sim}{x}_{L}) \tag{3.8}$$

be the symmetric equilibrium frequency clines present in conditions (i) and (ii), respectively. The analogue of Result III is stated below in Result X. Notice that unlike the setup in III where the conditions (i) and (ii) differed only in the addition of new advantageous A-type positions, the condition of (ii) now involves simultaneously in a symmetrical setting new localities favorable for the A and a-alleles.

<u>Result X.</u> <u>For the population conditions</u> (i) <u>and</u> (ii), <u>we have</u>

$$\underset{\sim}{x}_{i} > x_{i}^{*} \ , \quad i = 1,2,\ldots,K \ , \quad \underset{\sim}{x}_{0} = x_{0}^{*} = \frac{1}{2} \tag{3.9}$$

<u>and</u>

$$\max_{i \geqslant 0} \underset{\sim}{\delta}_{i} = \underset{\sim}{\delta}_{0} = \underset{\sim}{x}_{1} - \frac{1}{2} > x_{1}^{*} - \frac{1}{2} = \delta_{0}^{*} = \max_{i \geqslant 0} \delta_{i}^{*} \tag{3.10}$$

$$(\delta_{i} = x_{i+1} - x_{i}) \ .$$

<u>Moreover,</u> $\underset{\sim}{\delta}_{i} > \delta_{i}^{*}$, $i = 1,2,\ldots,K$.

Thus, where the advantageous A and a regions are extended preserving the symmetric ecotone structure, the maximal frequency differentiation occurring at the center deme is sharpened. In more picturesque language the whole cline tilts more steeply about the null position as the cline is lengthened symmetrically.

A graphical display of the above statements is now given.

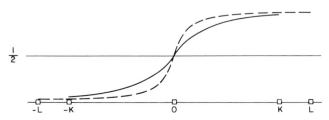

Slope at center increases if new demes are added (in a symmetrical manner) to both ranges .

The conclusion of (3.9) is correct for any magnitude of migration flow.

B.   Very long species range.

Consider a doubly infinite monotone symmetric cline C with deme location (for notational convenience) at all the integers (positive and negative). The setting is that of an infinite or far-reaching horizon (in both directions) for the distribution of the population. Let C_K be the truncated environment involving the $2K+1$ localities embracing the population demes P_i , $-K \le i \le K$. We denote by

$$\hat{x}^{(K)} = (\hat{x}_{-K}^{(K)}, \ldots, \hat{x}_0^{(K)}, \ldots, \hat{x}_K^{(K)}) \tag{3.11}$$

the unique symmetric polymorphic frequency cline (necessarily monotone) for a population confined to C_K .

Result XI.   Over a very long species range, we have

$$\lim_{K \to \infty} \hat{x}_i^{(K)} = \hat{x}_i^{(\infty)} \quad , \quad i = 0, \pm 1, \ldots \tag{3.12}$$

$$\frac{1}{2} = \hat{x}_0^{(\infty)} < \hat{x}_1^{(\infty)} < \hat{x}_2^{(\infty)} < \ldots < \ldots \to 1$$

and

$x_i^{(\infty)} = 1 - x_{-i}^{(\infty)}$ <u>is an equilibrium cline for the extended</u>

<u>population.</u>

In practice, already for $K \geqslant 5$, $\hat{x}_i^{(K)}$ is quite close to its limit at localities $i = 1$ and 2. What is, perhaps, surprising is that the frequency gradient at 0 does not become vertical. Thus, the slope is not as steep as might be anticipated due to the prolonged opposite selection pressures operating on the two sides of the null position.

C. <u>The shape of the frequency cline as a function of</u>
<u>migration rate.</u>

Consider an ecotone as in (3.4) having the symmetric polymorphic equilibrium frequency array

$$\hat{x}(m) = (\hat{x}_{-K}(m), \ldots, \hat{x}_{-1}(m), \frac{1}{2}, \hat{x}_1(m), \ldots, \hat{x}_K(m)) \tag{3.13}$$

where we display its dependence on the migration rate m.

<u>Result XII.</u> <u>As</u> m <u>increases</u> (larger migration rate) $\hat{x}_i(m)$, $i \geqslant 1$ <u>decreases.</u> More specifically for $m < m'$

$$\frac{1}{2} < \hat{x}_i(m') < \hat{x}_i(m) \quad , \quad i = 1, 2, \ldots, K \quad . \tag{3.14}$$

This finding confirms the intuitive property that with any background symmetric monotone selection gradient, increased gene flow levels the local selection effects yielding a more constant less differentiated gene frequency pattern. Endler [1973] claims on the basis of computer calculations performed on some special regular selection patterns that the influence of the magnitude of migration rate is only slight. Our analysis shows that there is a definitive trend independent of the individual selection components such that increasing migration rates cause more equalized gene frequencies provided only the monotone character of the ecotone is maintained. Further analysis suggests that the swamping effect of increased gene flow is more pronounced in the presence of a population distribution involving a few demes than in the circumstances of many numbers of demes.

D.   The case of hybrid (heterotic) central demes.

We alter the model (3.4) where in place of the central neutral zone we assume the local selection functions $f_i(x)$, $-\ell \leq i \leq \ell$, express heterozygote advantage symmetric in the A and a genes. This situation is certainly of common occurrence where the interface between the areas favoring the A and a-alleles provides a region of coadaptability of both genes in that the heterozygote has superior fitness.

We will refer to the environmental selection gradient now depicted.

$$(3.15)$$

where the demes with symbol (+ -) have selection functions expressing heterozygote advantage, as the mixed heterotic ecotone. (The opposite model where heterozygote disadvantage is manifested in the center region will be discussed in the following paragraph.)

In Karlin and Richter-Dyn [1976c] we will treat models with local selection functions manifesting some regular patterns of heterozygote advantage at each deme. We will compare and contrast the resulting symmetric equilibrium frequency cline in these cases with the partially heterotic hybrid model associated to the ecotone (3.15).

Our principal finding on the mixed heterotic monotone cline is as follows

Result XIII. Consider a selection regime of a mixed heterotic ecotone. There exists a symmetric polymorphic monotone cline

$$\underset{\sim}{x} = (\tilde{x}_{-K}, \tilde{x}_{-K+1}, \ldots, \tilde{x}_{-1}, \tilde{x}_0, \tilde{x}_1, \ldots, \tilde{x}_K) \ . \tag{3.16}$$

Every symmetric polymorphic cline has the following properties

$$\frac{1}{2} = \tilde{x}_0 < \tilde{x}_1 < \tilde{x}_2 < \ldots < \tilde{x}_K \ , \ \tilde{x}_{-i} = 1 - \tilde{x}_i \ , \ i = 1, \ldots, K \tag{3.17}$$

and $\tilde{\delta}_i = \tilde{x}_{i+1} - \tilde{x}_i$ satisfies

$$0 < \tilde{\delta}_1 < \tilde{\delta}_2 < \ldots < \tilde{\delta}_\ell \ , \ \tilde{\delta}_{\ell+1} > \ldots > \tilde{\delta}_{K-1} \tag{3.18}$$

and, of course, $\tilde{\delta}_{-i} = -\tilde{\delta}_i$.

If $\underset{\sim}{x}^*$ is the unique symmetric polymorphism for the cline (3.4) then

$$\tilde{x}_i < x_i^* \ , \ i \geqslant 1 \ . \ \tilde{\delta}_i > \delta_i^* \ , \ i = \ell+1, \ldots, K-1 \ . \tag{3.19}$$

In view of (3.18), the steepest slope in the cline of (3.16) does not occur at the null position (the central deme) but at the boundary demes ℓ and $-\ell$ where the local selection expression changes pronouncedly from that of overdominance to that of directional selection. The picture is as follows.

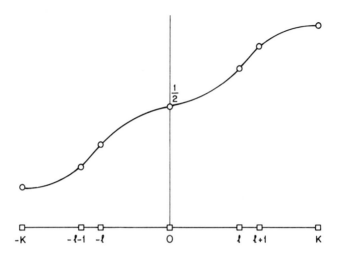

The relation (3.19) shows that with a central overdominant region the gene frequency array tends to be more flat at the center than with a dividing neutral zone.

As $m \downarrow 0$ the equilibrium $\tilde{x}(m)$ converges to $\tilde{x}(0)$ with coordinates

Deme # $-K$ $-\ell-1, -\ell$ ℓ $\ell+1$ K

$$\underline{x} = (0,\ldots,0\ ,\ \frac{1}{2}\ ,\ \frac{1}{2}\ ,\ldots,\ \frac{1}{2}\ ,\ 1\ ,\ldots,1)$$

where the transition from 0 to $\frac{1}{2}$ occurs between demes $-\ell$ and $-\ell-1$ and from $\frac{1}{2}$ to 1 at the location of demes ℓ to $\ell+1$.

E.   The case of underdominant central demes.

The model now differs from (3.15) in that the local selection functions for the central demes reflect <u>heterozygote disadvantage</u> (underdominance) symmetric in the two alleles. We call this model an <u>ecotone with underdominant central zone.</u> Here in the interface zone both homozygotes carry superior fitness to the heterozygote. The hybrid is less adapted to the central environmental habitats. The equilibrium frequency cline now takes the following shape

Result XIV. There exists a symmetric polymorphic morphotone

$$\tilde{\underline{x}} = (\tilde{x}_{-K}, \dots, \tilde{x}_{-1}, \tilde{x}_0, \tilde{x}_1, \dots, \tilde{x}_K) \quad , \quad \tilde{x}_0 = \frac{1}{2}$$

$$\tilde{x}_i > x_i^* \quad , \quad 1 \le i \le K \quad , \quad \tilde{\delta}_i < \delta_i^* \quad , \quad \ell+1 \le i \le K \quad ,$$

which satisfies

$$\tilde{\delta}_1 > \tilde{\delta}_2 > \dots > \tilde{\delta}_{K-1} > 0 \quad (\tilde{\delta}_i = \tilde{x}_i - \tilde{x}_{i-1}) \quad . \tag{3.20}$$

The existence of the underdominant demes in the center tends to sharpen the gene frequency differentiation manifested at the center deme over that of a plain neutral zone. It should be cautioned that in this underdominant model for slight migration rate there occur non-monotone stable poly- morphisms (see Karlin and McGregor [1972]). However, for moderate migration flow $m \ge .05$ these non-monotone clines apparently do not persist provided the selection coefficients are of moderate magnitude.

Symmetric environmental selection gradient with different central zone

Case 1 : Central zone is neutral.

Case 2 : Central zone has heterozygote disadvantage at demes $(P_{-\ell}, \dots, P_\ell)$
Both homozygotes have superior fitness to heterozygote.

Case 3 : Central zone has heterozygote advantage, symmetric fitnesses with respect to alleles A and a at demes $(P_{-\ell}, \dots, P_\ell)$.

Equilibrium frequency clines

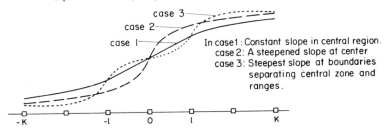

case 3
case 2
case 1

In case 1 : Constant slope in central region.
case 2 : A steepened slope at center
case 3 : Steepest slope at boundaries separating central zone and ranges.

-K   -ℓ   0   ℓ   K

F.   Comparisons of morphotones for increased population sub-
      division with a prescribed aggregate level of selection.

We concentrate on the symmetric model of this section with
local additive allelic fitness effects. For definiteness and
ease of exposition we consider the two environmental selection
gradient as in (3.4) with a single neutral deme at the center.
Let S be the cumultative selection coefficient in the right
portion of the species range corresponding to the demes of
positive index (see 1.12a) and suppose S is divided equally
among the K demes so that the AA-genotype selection coefficient per deme is S/K . In the negative range symmetry
entails that the aa-genotype selection coefficient per deme
is -S/K . Let $x^*_{(K)}$ denote the equilibrium frequency value
in deme 1 .

It can be proved that the <u>maximum slope of the cline</u>

$$\delta^*_{(K)} = x^*_{(K)} - \frac{1}{2} \tag{3.21}$$

<u>decreases as</u> K <u>increases. Thus, when preserving the</u>
<u>total selection effects, the resulting morphotone is less</u>
<u>steep with increasing population subdivision.</u>

G.   <u>The nature of the morphotone with three comparable</u>
      <u>environments.</u>

We displayed in (1.11) three environments I, II, III, each
more heterogeneous than the preceding in the sense of Definition 2 of Karlin [1976] having the same <u>aggregate</u> selection
intensity in the right and left population ranges.

We assume the environmental selection gradients of I-III
are each symmetric with respect to the center deme.

<u>Result XV.</u> <u>Let</u> δ_I , δ_{II} <u>and</u> δ_{III} <u>be the maximum slopes of</u>
<u>the morphotones for the symmetric selection regimes</u> I, II <u>and</u>

III respectively (given in (1.11)). Also let $\{x_i^{I}\}_{-K}^{K}$,
$\{x_i^{II}\}_{-K}^{K}$ and $\{x_i^{III}\}_{-K}^{K}$ be the unique stable symmetric cline

associated with the respective gradients. Then for all
migration rates, $0 \leqslant m \leqslant \dfrac{1}{4}$,

$$\delta_I < \delta_{II} . \tag{3.22}$$

Moreover, there exists a deme position k , $1 \leqslant k \leqslant K$
such that

$$x_i^{I} < x_i^{II} , \quad i = 1,2,\ldots,k$$
$$x_i^{I} > x_i^{II} , \quad i = k+1,\ldots,K . \tag{3.23}$$

It appears that

$$\delta_{II} < \delta_{III} . \tag{3.24}$$

The foregoing result is, perhaps, not surprising. It
states that with the intensity of selection differences,
strongest in the neighborhood of the central deme, a more
pronounced gene frequency change is established there.

4. SUMMARY AND DISCUSSION

A number of problems were formulated in Section 1 with
objective to help explicate and discriminate the abundantly
observed clinal patterns of gene frequency variation. A
hierarchy of mathematical cline models incorporating some
forms of selection migration interaction have been studied by
analytic means or with the aid of computer runs by Haldane
[1948], Fisher [1950], Hansen [1966], Slatkin [1973], Endler
[1973] and others. Following these authors we concentrate in
this work on a two-range geographical selection gradient
where allele A is favored in the right range of the population and the alternative allele a is favored in the left

range. This kind of selection background, referred to as an ecotone, is prominently present in a variety of biogeographic situations. The gene flow pattern is stipulated to be that of the stepping stone migration mode. (Ehrlich and Raven [1969] cite many examples of animal and plant populations where gene flow is considerably localized.)

The detailed model is set forth in Section 1. We have reported in Sections 2 and 3 a series of results character- izing and contrasting the gene frequency arrays that are established with different forms of underlying ecotones. We presently summarize several of the salient findings and note briefly some of their implications with reference particularly to the questions of Section 1 :

(i) For an ecotone selection gradient, it is established that any polymorphic equilibrium describes a morphotone (monotone cline) that is, the A-allele frequency increases along the population transect going from left to right. The fact that every equilibrium is a morphotone is exceptionally robust prevailing with:

(a) unequal local rates of exchange between neighboring demes,

(b) in the presence of variable deme sizes, and

(c) allowing irregularities in the local selection intensity provided only the ecotone property is maintained. (See Results I, I', VII of Section 2.)

With this outcome, in order to correlate the selection gradient to the associated frequency array, it is essential to plot and examine in detail the changes in gene frequency (the slope function). By these means it may be possible to evaluate (and sometimes estimate) the influence of unequal local migration rates (see especially Result VII of Section 2).

(ii) Where gene flow is <u>homogeneous</u> over the linear population range, the resulting A frequency morphotone has a precise monotone shape exhibiting exactly one inflection point.
The graph is therefore of the form (from left to right) first
convex increasing over the region where allele <u>a</u> is advantageous, linear on the neutral range and concave increasing
over the region where A is favored. These descriptions
apply independently of the detailed nature of the local
selection functions.

A very kinky morphotone cannot be attributable to an
inherent uniform migration flow even in the presence of substantial spatial variation in selection intensities.

Observed frequency clines, not of this precise convex-
linear - concave shape, would entail that either the model
is not relevant because the underlying selection gradient is
not an ecotone or the migration flow is not homogeneous or
the local deme sizes fluctuate substantially.

(iii) The question is frequently asked: Does a marked
local change in gene frequency correspond to a sharp local
change in selection intensity? Under the conditions of (ii),
the maximum gene frequency differentiation (the maximum slope
of the cline) in the presence of uniform local gene flow is
attained at the neutral zone. If the neutral zone is extensive, then the change in gene frequency may still be significant without involving nearby any selection differences
among the genotypes.

Extending the range on which allele A is advantageous,
balanced on the other extreme by more demes advantageous to
allele a , steepens the slope of the cline throughout the
range (Results X, XI).

(iv) A more prolonged neutral zone tends to <u>accent</u> the
selection forces operating on the outer regions of the popu-

lation distribution but flattens the slope over the population range (see Result IV).

The findings of (iii) and (iv) provide a possible explanation for the phenomenon of "area effects" (Goodhardt [1969]; Clarke [1966]).

(v)   If the zone separating the regions favoring the A and a alleles has a selection expression of heterozygote advantage rather than being neutral, the corresponding cline is more flat in the center but steeper at the boundary between demes expressing directional selection as against overdominance.   (See Result XIII on mixed heterotic ecotones.) On the other hand, underdominance in the central zone sharpens the cline slope at the center (Result XIV).

An implication attendant to (iii)-(v) is that a marked gene frequency change can be related to the following factors:

(a)   an ecotone with underdominant selection expression at the deme and neighboring demes at hand;

(b)   a long ecotone with the deme in question at the interface between the regions favoring A and a alleles respectively.

(c)   sharp differences in migration flow in reciprocal directions.

On the other hand, local irregularities in selection intensity within the confines of an ecotone is not a decisive factor in locating sharp morph differentiation.

(vi)   The influence of increased migration flow is unambiguous and substantial.   Indeed Result VI affirms that with two different rates of local migration exchange the spatial gene frequency pattern is always more homogeneous with larger migration or some state of fixation occurs.   The different possibilities between very small as against moderate to uniform mixing migration rates is pronounced.   In fact, for very

small migration flow a cline is established manifesting generally a steep slope between the two ranges of the population
independent of the detailed character of the ecotone. For
moderate up to uniform mixing inter deme exchange rates, the
evolutionary outcome of the population becomes sensitive to
the aggregate selection coefficients on the two extended
environmental ranges of the A-favorable as against the a-
favorable portions. However, for the special circumstance
where the selection pattern is completely symmetric with respect to a center position, then an increased migration rate
can never lead to fixation. Endler [1973] in his computer
analysis of clines unfortunately restricted attention primarily to specific symmetric models and on this basis hastened
to conclude that the rate of migration was of minor significance. Even in this case the cline systematically flattens
with increasing migration levels.

(vii) When migration occurs at the infant stage (as in
pollen and seed dispersal) vis a vis the adult stage, the
effects due to selection are more pronounced in each environmental range (Result VIII).

(viii) Spreading an aggregate selection effect over more
demes (or equivalently increased population subdivision) tends
to bring about a flattened morphotone, that is, yielding more
spatial homogeneity in gene frequency.

(ix) We have contrasted three different environmental
ecotones with prescribed aggregate selection coefficients for
each range (see paragraph G of Section 3), the steepest slope
apparently occurs for environment III where the intensity of
selection decreases with distance from the center, that is,
where the change in selection between the - and + region
is the most abrupt.

All the above results will be elaborated and extended in

703

Karlin and Richter-Dyn [1976b,c,d] to a background selection gradient involving more than two environments. We will also contrast the nature of the morphotones with an underlying regular selection gradient of heterozygote advantage.

The properties and characterizations described above and elaborated more precisely in the body of Sections 2 and 3 afford some means both qualitatively and quantitatively of discriminating better among cline structure relating to spatial patterns of selection and gene flow and help also to understand more thoroughly the effects of environmental parameters.

REFERENCES

Beardmore, J. (1970). In: Essays in evolution and genetics, pp. 293-314. M.K. Hecht and W.C. Steere (eds.) Appleton-Century-Crofts, New York.

Bryant, E.H. (1974). Am. Natur. 108: 1-19.

Cain, A.J. and J.D. Curray. (1963). Phil. Trans. B. 246: 1-81.

Cavalli-Sforza, L. and W. Bodmer. (1971). Human Population Genetics. Freeman, S.F.

Christiansen, F. and O. Frydenberg. (1974). Genetics 77: 765-770.

Clarke, B.C. (1956). Am. Natur. 100: 389-402.

Ehrlich, P.R. and P.H. Raven. (1969). Science 165: 1228-1232.

Endler, J.A. (1973). Science 179: 243-250.

Fisher, R.A. (1950). Biometrics 6: 353-361.

Fleming, W.H. and C.H. Su. (1974). Theor. Pop. Biol. 5: 431-449.

Gillespie, J.H. and C.H. Langley. (1974). Genetics 76: 837-848.

Goodhart, C.B. (1969). Heredity 18: 459-465.

Haldane, J.B.S. (1948). Genetics 48: 277-284.

Hamrick, J.L. and R.W. Allard. (1972). Proc. Nat. Acad. Sci. 69: 2100-2104.

Hanson, W.A. (1966). Biometrics 22: 453-468.

Jain, S.K. and A.D. Bradshaw. (1966). Heredity 21: 407-441.

Karlin, S. (1972a). Am. Math. Monthly 79: 699-739.

Karlin, S. and J. McGregor. (1972b). Theor. Pop. Biol. 3: 186-208.

Karlin, S. (1976). This volume.

Karlin, S. and N. Richter-Dyn. (1976a,b,c,d). To appear.

Kettlewell, H.B.D. (1961). Heredity 21: 407-441.

Kettlewell, H.B.D. and R.J. Berry. (1961). Heredity 16: 403-414.

Kettlewell, H.B.D. and R.J. Berry. (1969). Heredity 24: 1.

Kimura, M. and G.H. Weiss. (1964). Genetics 49: 561-576.

Koehn, R.K. (1969). Science 163: 943-944.

Kojima, K., P. Smouse, S. Yang, P. Nair and D. Brncic. (1972). Genetics 72: 721-731.

Latter, B. (1973). Genetics 73: 147-157.

Malécot, G. (1948, 1970). Les Mathématiques de L'Hérédité. Masson et Cie., Paris. Revised English translation. Freeman, S.F.

Malécot, G. (1950). Ann. Univ. Lyon Sciences, Sec.A 13: 37-60.

Malécot, G. (1951). Ann. Univ. Lyon Sciences, Sec.A 14: 79-118.

Malécot, G. (1959). Publ. Inst. Stat. Univ. Paris, 8: 173-210.

Malécot, G. (1967). Proc. Fifth Berkeley Symp. Math. Stat. Prob. IV: 317-332. Univ. of Calif. Press, Berkeley.

Marshall, D.R. and R.W. Allard. (1972). Genetics 72: 721-731.

Maruyama, T. (1972). Math. Biosciences 14: 325-335.

Merritt, R.B. (1972). Am. Natur. 106: 173-185.

Nevo, E. and Z. Bar. (1976). This volume.

Rockwood-Sluss, E.S., J.S. Johnson and W.B. Heed. (1973). Genetics 73: 135-146.

705

Slatkin, M. (1973). *Genetics* 75: 733-756.

Weiss, G.K. and M. Kimura. (1965). *J. Appl. Prob.* 2: 129-149.

Wright, S. (1943). *Genetics* 23: 114-138.

Models of Density-Frequency Dependent Selection for the Exploitation of Resources I: Intraspecific Competition

C. MATESSI and S.D. JAYAKAR

INTRODUCTION

The study of competition for resources at the purely interspecific level forms a major part of ecological theory.   Only
recently, however, have models been presented which consider
the effects of this ecological factor on the genetic evolution of a species.   An important concept which emerged from
this approach is the distinction between r-selection and K-
selection (MacArthur and Wilson, 1967).   The constant fitness
of classical population genetics theory is interpreted by this
approach as a measure of the intrinsic rate of increase of a
genotype, and genetic differences in this quantity result in
r-selection.   A different type of selective effect (K-selec-
tion) comes into play in a crowded environment where differ-

On leave from: Laboratorio di Genetica Biochimica ed Evoluzionistica, C.N.R., 27100, Pavia, Italy.

ences in competitive ability between genotypes becomes the dominant selective factor. In one set of papers (MacArthur, 1962; Anderson, 1971; Charlesworth, 1971; Roughgarden, 1971), the way in which $\underline{K}$-selection is formulated has the effect of making the fitness of any given genotype dependent on population size alone. Clarke (1972), on the other hand, has presented a more general formulation such that the fitness of a given genotype depends on the size as well as the genotypic composition of the population. León (1974) has extended this approach to the case of two competing species, both of which show genetic variability, with fitness depending on the population sizes of the two species.

This combination of population genetics and ecology is essential in order to study the evolution of the ecological properties of one or more species in a given ecosystem. In this paper we wish to exploit this approach in order to examine the role of competition for resources in the control of the degree of ecological specialization of a species, or in other words, the effects of competition on the niche width of the species. Roughgarden (1972) has shown that in an asexually reproducing organism the effect of intrapopulation competition is to expand the niche occupied by the population. He also argues that a sexually reproducing species is not as efficient in expanding its niche.

There is undoubtedly a large number of factors which are responsible for the evolution of the niche-shape of a species, the most important of which are intra- and inter-specific competition, the availability and the quality of resources in the habitat, predators and parasites. We have limited ourselves to studying intra- and inter-specific competition alone. The results of the analysis of inter-specific competition will be published elsewhere (Jayakar and Matessi, 1975).

708

1. COMPETITION BETWEEN GENOTYPES OF A SINGLE SPECIES

The model most frequently used in mathematical ecology to describe competition between species is that due to Lotka and Volterra in which the growth rate of each species decreases linearly with the population sizes of the members of the community. Several authors have extended this to competition between genotypes of a single species (see e.g., Roughgarden, 1972), and we shall do the same in this paper. A continuous time analysis leads to the complication that even at equilibrium, the genotypes are not in Hardy-Weinberg proportions, and this is due simply to the fact that there is a continuous carry-over of selected genotypes from previous generations. We have decided therefore to concentrate on the discrete-time, separate generations formulation. Competition has in general, effects both on fertility and on survival. For convenience, we consider here only the latter, and the fertility is therefore assumed to be independent of the numbers of all the genotypes. Further, we assume that the relevant genetic variability is controlled by a single locus.

Let us then assume that there are $\underline{m}$ alleles at an autosomal locus. The subscripts i,j etc. will denote alleles, and the subscripts α, β etc. will denote genotypes, that is α corresponds to any pair (i,j) and so on. In a given generation N_α individuals of the genotype α are born, and the total population size at this stage is $N = \sum_\alpha N_\alpha$. Only N_α^* of these survive to reproduction, i.e., after the effects of competition. From the Lotka-Volterra model we have:

$$N_\alpha^* = (b - \sum_\beta a_{\alpha\beta} N_\beta) N_\alpha \quad , \quad (a_{\alpha\beta} \geq 0) \quad . \tag{1.1}$$

The total number of survivors is $N^* = \sum_\alpha N_\alpha^* = bN - \sum_{\alpha\beta} a_{\alpha\beta} N_\alpha N_\beta$. In the absence of competition a larger proportion, $\underline{b}$, of individuals would survive; $\underline{b}$ is then the intrinsic survival rate of an individual. Notice that we are assuming $\underline{b}$ independent of the genotype. The array $\{a_{\alpha\beta}\}$ is the competition matrix. A particular case is that treated by Mac-Arthur (1962), Anderson (1971) and Roughgarden (1971), where $a_{\alpha\beta} = a_{\alpha\alpha}$ for all β.

The surviving population then undergoes random mating to produce the next generation. Each individual on the average produces $\underline{r}$ offspring. Again we assume that this fertility does not depend on the genotype. Thus the population size in the next generation is

$$N' = rN^* = r(bN - \sum_{\alpha\beta} a_{\alpha\beta}N_\alpha N_\beta) \quad . \tag{1.2}$$

The number of individuals of genotype α born in the next generation is,

$$N'_\alpha = \frac{1}{N^*} (\sum_{\beta\gamma} N^*_\beta N^*_\gamma \lambda^{(\alpha)}_{\beta\gamma})$$

where $\lambda^{(\alpha)}_{\beta\gamma}$ is the proportion of offspring of genotype α expected from a pair $(\beta \times \gamma)$. The set $\{N_\alpha\}$ is a redundant description of the system since $\{N'_\alpha\}$ depends only on the allelic frequencies among $\{N^*_\alpha\}$ and the population size N. The fitness of the genotype α is given by:

$$\frac{N^*_\alpha}{N_\alpha} = b - \sum_\beta a_{\alpha\beta}N_\beta = b - N \sum_\beta a_{\alpha\beta}x_\beta$$

where $\underline{x}_\beta$ is the frequency of genotype β among the newborn. If we denote by $\underline{p} = \{p_i\}^m_{i=1}$ the vector of allelic frequencies, we have:

$$x_{ii} = p_i^2 \quad ; \quad x_{ij} = 2p_ip_j \quad ,$$

and the evolution of the population is therefore described by the following system of recurrence equations:

$$p_i' = p_i \frac{b - N \sum\limits_{jkr} a_{ijkr}\, p_j p_k p_r}{b - N \sum\limits_{ljkr} a_{ljkr}\, p_l p_j p_k p_r} \quad , \quad i = 1, \ldots, m$$

$$N' = r\, N\, (b - N \sum\limits_{ijkl} a_{ijkl}\, p_i p_j p_k p_l) \quad .$$

It is convenient to use the notation

$$V_i(\underline{p}) = \sum\limits_{jkl} a_{ijkl}\, p_j p_k p_l \qquad (1.3)$$

$$V(\underline{p}) = \sum\limits_{ijkl} a_{ijkl}\, p_i p_j p_k p_l \quad . \qquad (1.4)$$

The functions $V_i(\underline{p})$ and $V(\underline{p})$ are defined on the set

$$0 \leqslant p_i \leqslant 1 \,, \quad i = 1, \ldots, m \,; \quad \sum\limits_i p_i = 1 \quad .$$

$V(\underline{p})$ is the average competition in the population and $V_i(\underline{p})$ is, so to speak, the competition against the i-th allele. The recurrence equations are then

$$p_i' = p_i \frac{b - N V_i(\underline{p})}{b - N V(\underline{p})} \quad , \quad i = 1, \ldots, m \quad , \qquad (1.5)$$

$$N' = r\, N\{b - N V(\underline{p})\} \quad . \qquad (1.6)$$

Notice that b is a redundant parameter since we can replace r by R = r b , and measure the $a_{\alpha\beta}$ relative to b . The use of the Lotka-Volterra formulation, in the present discrete generation context introduces the inconvenience that perfectly meaningful values of p and N can give negative, and there-fore meaningless, values of either of these quantities in the following generation. There are at least two ways out of these difficulties. We can either redefine (1.5), (1.6) such that $p_i' = 0$ or $N' = 0$ whenever these values become nega-tive. Our results, which deal only with existence and local stability of the equilibria of the system (1.5), (1.6), retain their validity even under this more rigorous version of the model. Alternatively, we can restrict the parameter space in such a way that negative values of p_i and N can never occur. For instance this requisite is satisfied for all successive generations if

$$R < 4 \frac{\min(a_{\alpha\beta})}{\max(a_{\alpha\beta})}$$

and the population size is small enough in the initial generation.

Of this general model we will discuss only two special cases:

(i) a symmetric competition matrix $||\underline{a}_{\alpha\beta}||$ with $\underline{m}$ alleles;

(ii) a general competition matrix with only two alleles and complete dominance.

Symmetric competition matrix.

In most of the ecological literature only competition matrices $||\underline{a}_{ij}||$ of the following form are used:

$$a_{ij} = \frac{c_{ij}}{\varepsilon_i} \quad ,$$

where $||\underline{c}_{ij}||$ is symmetric, and ε_i is regarded as a measure of the abundance of resources relative to the efficiency of exploitation of the $\underline{i}$-th type. Since, in assuming $\underline{b}$ to be equal for all types, we have implied equality of efficiencies, the only additional implication of the symmetry of $||\underline{a}_{ij}||$ is the uniform distribution of resources over the habitat of the population.

The principal results which can be proved assuming symmetry of the competition matrix are:

(1) At any equilibrium the average competition $V(\underline{p})$ is at a stationary point (i.e., its first derivatives are zero) with respect to all non-zero p_i's, under the constraint

712

$\Sigma p_i = 1$. All genotypes present have the same fitness.

(2) Any equilibrium is stable only if $1 < R < U$, where U depends on the competition matrix $||a_{\alpha\beta}||$ and is bounded by the interval $[\frac{5}{3} , 3]$.

(3) If $\underline{R}$ is in the desired range, any equilibrium $(\hat{\underline{N}}, \hat{\underline{p}})$ is stable if and only if the average competition $V(\underline{p})$ has a local minimum at $\hat{\underline{p}}$ with respect to all non-zero p_i's , and $V_i(\hat{\underline{p}}) > V(\hat{\underline{p}})$ for all $\underline{i}$ such that $\hat{p}_i = 0$. Observe that if the average competition $V(\hat{\underline{p}})$ is a local minimum, then the population size at this equilibrium is at a local maximum, in the sense that $N(\hat{\underline{p}})$ is larger than the "equilibrium" size that the population would reach if the gene frequencies were frozen at any point $\underline{p}$ in the neighborhood of $\hat{\underline{p}}$.

The maximization of population size mentioned in (3) is analogous to the result obtained by Charlesworth (1971) who treated a 2 allele model with density-dependent but not frequency-dependent fitnesses. The demonstration of the results in (1) and (3) is similar to the analysis in the classical models of constant selection at an autosomal locus. The function $V(\underline{p})$ plays here the same role as the average fitness in that case. The difference between the two analyses lies in the fact that $V(\underline{p})$ is a fourth degree polynomial whereas the average fitness in the classical model is a quadratic.

Two alleles with dominance.

In this case, there are only two ecologically distinct phenotypes. This reduces the number of distinct competition parameters to four. These will be called a_{11}, a_{12}, a_{21} and a_{22} where the subscripts now refer to phenotypes, 1 corres-

ponding to the dominant and 2 to the recessive. Also, p
will refer to the frequency of the dominant allele. Further,
it is convenient to define:

$$\rho = 1 - \frac{1}{R} \; .$$

With this model, there is at most one polymorphic equilibrium. The necessary and sufficient condition for its existence is

$$(a_{11} - a_{21})(a_{22} - a_{12}) > 0 \; .$$

The equilibrium is given by

$$\hat{N} = \frac{\rho(a_{11}-a_{12}-a_{21}+a_{22})}{a_{11}\,a_{22}-a_{12}\,a_{21}}$$

and $\hat{p} = 1 - \left\{ \dfrac{a_{11}-a_{21}}{a_{11}-a_{12}-a_{21}+a_{22}} \right\}^{1/2} \; .$

This equilibrium is stable if and only if

$$a_{11} > a_{21} \; , \quad a_{22} > a_{12}$$

and $1 < R < \min(3, u)$

where

$$u = 1 + \frac{(a_{11}a_{22}-a_{12}a_{21})\{(a_{11}-a_{12}-a_{21}+a_{22})^{\frac{1}{2}}+(a_{11}-a_{21})^{\frac{1}{2}}\}(a_{11}-a_{21})^{\frac{1}{2}}}{(a_{11}-a_{21})^2(a_{22}-a_{12})} \; .$$

We have not been able to determine in general the conditions for u to be greater than 3 , but at least when
$a_{11}a_{12} = a_{22}a_{21}$, this is always so.

When a stable polymorphic equilibrium exists, the population size at this equilibrium is not necessarily larger than
those at each of the monomorphic equilibria. This is only
true if

$$a_{11} \geqslant a_{22} > a_{21} \quad \text{or} \quad a_{22} \geqslant a_{11} > a_{12}$$

and is therefore true in the case of a symmetric competition matrix. It is of course true that if no intermediate equilibrium exists the population size is largest at the stable monomorphic equilibrium.

2. THE NICHE OF A SPECIES WITHOUT COMPETITORS

The result stated earlier that the average competition at a stable equilibrium is minimum is biologically interesting though we are uncertain as to how robust it is with respect to the assumption of symmetry of the competition matrix. In an asymmetric situation the stable equilibria do not occur at minima of $V(\underline{p})$ but if deviations from symmetry are small, stable equilibria will occur not far from these points, hence the population will move away from points of high competition to points of lower competition. In a general asymmetric situation with two alleles, we have found conditions under which the competition is less at the stable polymorphic equilibrium than at both the trivial equilibria; this occurs whenever competition between individuals of different ecological types is less than between those of the same type, which should be a fairly common situation. When a polymorphic equilibrium does not exist, the population is fixed for the type with the least competition. It would seem therefore that though intrapopulation competition is not strictly minimized in asymmetric situations, natural selection does however tend to reduce it.

An ecologically meaningful interpretation of this result is that the niche of the population becomes wider. This will become more convincing if we make use of the theory developed by MacArthur (for reviews see MacArthur, 1972, and May, 1973) and based on the concept of the resource utilization function of a species. Let us suppose that the set of resources utilized by a population can be represented by a single conti-

nuous variable $\underline{Y}$. The axis along which $\underline{Y}$ is measured is
called the resource axis. A particular population is charac-
terized by a function $f(\underline{y})$ which gives the probability that
a resource at a position $\underline{y}$ on the axis is utilized by the
population; $f(\underline{y})$ is called the utilization function of the
population. Some index of the dispersion of $f(\underline{y})$, for
example its variance, is taken as a measure of the niche
width. Following MacArthur, the competition among members of
the population, V , can be interpreted as

$$V = \int f^2(y) \, dy \quad .$$

Let us suppose that at two successive stages of its evolution,
the population has utilization functions $f_0(\underline{y})$ and $f_1(\underline{y})$,
and competition V_0 and V_1 respectively. Suppose also that
$f_0(\underline{y})$ and $f_1(\underline{y})$ have the same form and differ only by a
scale factor, that is,

$$f_1(y) = k \, f_0(ky) \quad .$$

(Multiplication by $\underline{k}$ is necessary for normalization.) We
see immediately that $V_1 = \underline{k} \, V_0$, thus if V_1 is smaller
than V_0 , $\underline{k} < 1$, that is f_1 has larger dispersion than
f_0 . In other words, the niche has become wider. This simple argument thus supports our conjecture but only approximately since the shape of the utilization function need not
remain constant during the process analyzed in Section 1.
Notice however that the argument is not affected if there is
change only in the location along the resource axis.

This theme can be developed further, in order to examine
the extent of the validity of our conjecture on niche expansion, by looking at the detailed structure of $f(\underline{y})$, the
overall utilization function of the population, in terms of
the corresponding functions of the different genotypes. In
order to do this we shall look at two particular rather

extremely opposed situations.

In the first situation we assume that the different types in the population have unimodal utilization functions differing only by a scale factor. Let α and β be the two types with utilization functions $f(\underline{y})$ and $g(\underline{y}) = \underline{k}f(\underline{k}\underline{y})$ respectively with $\underline{k} > 1$, which means that α has a wider niche than β. Then it can be easily shown that

$$a_{\alpha\alpha} < a_{\alpha\beta} < a_{\beta\beta} \quad ,$$

and further that $\underline{a}_{\alpha\beta}$ is an increasing function of $\underline{k}$. Therefore if γ is any other type with a niche even narrower than that of β, $\underline{a}_{\alpha\gamma} > \underline{a}_{\alpha\beta}$. If α, β, γ refer to the three genotypes at a di-allelic locus, β being the heterozygote, it is an immediate consequence of our analysis that the only stable equilibrium is that where only the homozygote genotype α is present.

In the general case of $\underline{m}$ alleles, let us suppose that

(a) each heterozygote has a niche intermediate in width to those of the corresponding homozygotes, and

(b) for any three alleles $\underline{i}$, $\underline{j}$, $\underline{k}$ such that the homozygote $(\underline{i},\underline{i})$ has a wider niche than $(\underline{j},\underline{j})$, the niche of $(\underline{i},\underline{k})$ is wider than that of $(\underline{j},\underline{k})$.

It can then be shown by applying results of the previous section that the only stable equilibrium is the monomorphic one involving the homozygote with the widest niche. Even without the assumption (b) above, it could be argued, in view of the 2-allele result, that if mutations of this kind occur rarely in a population, there is a progressive widening of the niche through successive replacement of one allele by another with a wider niche. It is perhaps of interest that in these hypothetical situations, the niche is widened through a reduction in genetic variability.

In the second situation we assume that the different types differ only in the location of their utilization functions along the resource axis. For a wide class of utilization functions, the competition between two types decreases with the distance between their locations. In these circumstances, in a population with two alleles and the heterozygote intermediate in location, the only stable equilibria are polymorphic ones. Both with complete dominance and with additivity with respect to location there is just one non-trivial equilibrium. We do not know whether in other cases there can be more equilibria. In any case there can be at most 3 non-trivial equilibria two of which will be stable.

In a population with three alleles, if we make the simplifying assumptions that

(a) the alleles are additive in their effects on location, and

(b) the distances between the locations of adjacent homozygotes are equal,

it can be shown that the two alleles whose homozygotes have the extreme locations are always maintained in the population, while the central one may or may not be eliminated, depending on the degree of packing. There are however certain shapes of utilization functions for which the central one is never eliminated no matter how tight the packing. This is true for example when the competition between two types decays exponentially with the distance between their locations. Field data analyzed by Roughgarden (1972) fit this kind of exponential decay. May (1974) however, has derived the utilization function which leads to this kind of competition and showed that it has an "unnatural" discontinuity.

All these results lend credibility to our conjecture that there is a tendency for a population which has to cope with

718

intraspecific competition alone to expand its niche in a
variety of situations. This does not necessarily imply a
maximization of genetic variability since even in the case of
genotypes which differ in their locations along a resource
axis, the intermediate ones can be eliminated.

3. DISCUSSION

A phenomenon which is frequently observed in a species
which is successfully introduced in an environment with an
impoverished fauna, as is often the case on islands, is that
which is referred to in the literature as "ecological release".
This consists of an expansion of the niche of the species in
comparison to that in its original habitat. In animals with
considerable behavioral plasticity, there will be a substantial non-genetic component involved in this process. Ecological release can also be correlated however with certain morphological changes, such as the increase in the variability of
certain bill measurements in birds (see Van Valen, 1965, and
for a general discussion of this argument, MacArthur and Wilson, 1967). Very likely, then, ecological release is the result of a remoulding of the genetic structure of the population. Our results concerning a single species provide a
possible mathematical description of this process. We have
considered in some detail two extreme modes of accomplishing
niche expansion. The first of these is by selection for single genotypes which are most generalized in their resource
utilization. The second mode is through the establishment of
a polymorphism with each morph being rather specialized. In
the former case there is no increase in genetic variability.
These two are not alternative modes, and which of these is
more relevant depends on the type of genetic variability which

719

is available to the population. It would be desirable to investigate a mixed situation, where genetic variability for
both location and dispersion along the resource axis is available. We have the impression that an increase in niche width
is not necessarily accompanied by an increase in genetic
variability. There have been several attempts to find a correlation between the genetic variability in a population and
the width of its niche. The results are far from clear cut
(e.g., Powell, 1971; Sabath, 1974, Shugart and Blaylok, 1973;
Soulé and Stewart, 1970).

4. SUMMARY

In order to study the effects of intra-specific competition
on the genetic structure of a population, at an autosomal
locus, we analyze a mathematical model in which competition
is represented by the classical Lotka-Volterra approach. This
gives rise to selective pressures which are both density and
frequency dependent.

In a population without competitors of a different species,
the amount of intra-specific competition is minimized by natural selection, whenever the competition matrix between the
various genotypes for an arbitrary number of alleles is symmetrical. Competition is not minimized if the competition
matrix is asymmetrical but, at a stable equilibrium, the
amount of competition may be smaller than at unstable ones.
These results are interpreted as a process of niche expansion
and a theoretical description of "ecological release".

REFERENCES

Anderson, W.W. (1971). Am. Natur. 105: 489-498.

Charlesworth, B. (1971). Ecology 52: 469-474.

Clarke, B. (1972). Am. Natur. 106: 1-13.

Jayakar, S.D. and C. Matessi. (1975). Theor. Pop. Biol. (To appear).

León, J.A. (1974). Am. Natur. 108: 739-757.

MacArthur, R.M. (1962). Proc. Nat. Acad. Sci. 48: 1893-1897.

MacArthur, R.M. (1972). Geographical Ecology. Harper & Row, New York.

MacArthur, R.M. and E.O. Wilson. (1967). The Theory of Island Biogeography. Princeton Univ. Press, Princeton, N.J.

May, R.M. (1973). Stability and Complexity in Model Ecosystems. Princeton Univ. Press, Princeton, N.J.

May, R.M. (1974). Theor. Pop. Biol. 5: 297-332.

Powell, J.R. (1971). Science 174: 1035-1036.

Roughgarden, J. (1971). Ecology 52: 453-468.

Roughgarden, J. (1972). Am. Natur. 106: 683-718.

Sabath, M.D. (1974). Am. Natur. 108: 533-540.

Shugart, H.H. and G.B. Blaylok. (1973). Am. Natur. 107: 575-579.

Soulé, M.R. and B.R. Stewart. (1970). Am. Natur. 104: 85-97.

Van Valen, L. (1965). Am. Natur. 99: 377:390.

ACKNOWLEDGMENTS

We wish to thank Professor S. Karlin for reviewing the manuscript and pointing out inaccuracies in the text. Part of this research has been supported by Grant AEC AT(04-3)326-#33.

Mathematical Models of Speciation and Genetic Distance*

M. NEI

INTRODUCTION

Until recently, population genetics was concerned mainly
with short-term changes of the genetic structure of populations. Long-term changes such as speciation and genetic divergence between species have been simply conjectured as a
continuation of short-term changes. This is because, before
the development of molecular biology, geneticists were not
aware how genes change in the evolutionary process. In the
past ten years this situation has changed drastically, and
thanks to the development of molecular techniques such as
amino acid sequencing and electrophoresis, we can now study
not only the proportion of common genes shared by a pair of
species but also the number of gene substitutions that occurred
in the process of speciation. In the study of speciation
electrophoresis has been very useful, since by this method
gene differences between populations can be studied quickly
for a large number of loci. After the pioneering work by
Hubby and Throckmorton (1965, 1968), a large amount of gene

---

*Supported by Public Health Service Research Grant GM 20293.

frequency data in related species has been accumulated, notably by the effort of Selander (Selander and Johnson, 1973)
and Ayala (Ayala et al., 1974).

In the past few years I have been interested in the interpretation of these data in terms of mathematical models.
First, I tried to estimate the accumulated number of gene
substitutions or the number of codon differences per locus
between two species by using Hubby and Throckmorton's (1965,
1968) data (Nei,1971a). Although I made a number of assumptions, the result seemed to be satisfactory as a first approximation. Later, taking into account the polymorphism
within populations, I proposed a model of population differentiation and presented a new measure of genetic distance,
the unit of which is the accumulated number of gene substitutions (codon differences) per locus (Nei, 1971b, 1972). Recently, in collaboration with my colleagues, I have made further studies on the properties of this model and extended it
to various biological situations. I have also gotten interested in modeling the development of reproductive isolation
under geographical isolation. Reproductive isolation seems
to be the crux of speciation but few mathematical studies
have been done. In this paper I would like to discuss some
of our latest studies, first on the latter problem and then
the former.

MODELS OF REPRODUCTIVE ISOLATION

Speciation consists of two distinct genetic processes, i.e.,
establishment of reproductive isolation and genetic divergence of populations after reproductive isolation. Reproductive isolation may be attained by either premating or postmating isolation mechanisms or both. According to Dobzhansky
(1970), the former includes ecological isolation, seasonal

isolation, sexual or ethological isolation, mechanical isolation, isolation by different pollinators, and gametic selection, while the latter includes hybrid inviability, hybrid sterility, and hybrid breakdown. The evolutionary scheme of reproductive isolation would vary considerably with different isolation mechanisms. Ecological and seasonal isolation mechanisms may be developed by a single gene substitution, though generally more than one gene substitution would be involved. Similarly, isolation by different pollinators may evolve by a single gene substitution in the host plant. For the evolution of ethological, mechanical, and gametic isolation, however, more than two gene substitutions seem to be required. Similarly, more than two gene substitutions appear to be involved in the evolution of postmating isolation mechanisms.

(1) Postmating isolation

In the evolution of postmating isolation mechanisms several gene loci for viability and fertility seem to be involved, though it is not impossible to produce reproductive isolation by means of repeated gene substitution at a single locus. Dobzhansky (1937) and Muller (1939, 1942) suggested the following scheme. Consider two loci (or two sets of loci) which control some type of postmating reproductive isolation, and let $A_0A_0B_0B_0$ be the genotype for these loci of the foundation stock from which populations 1 and 2 are derived. If these two populations are geographically isolated, it is possible that in population 1 A_0 mutates to A_1 and this mutant gene may be fixed in the population, provided that $A_0A_1B_0B_0$ and $A_1A_1B_0B_0$ are as viable and fertile as $A_0A_0B_0B_0$. Similarly, in population 2 mutation may occur at the B locus and genotype $A_0A_0B_0B_0$ may be replaced by $A_0A_0B_2B_2$ without loss of viability and fertility (see Figure 1).

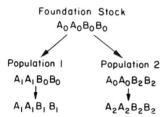

Foundation Stock
$A_0 A_0 B_0 B_0$

Population I
$A_1 A_1 B_0 B_0$
↓
$A_1 A_1 B_1 B_1$

Population 2
$A_0 A_0 B_2 B_2$
↓
$A_2 A_2 B_2 B_2$

Figure 1. Dobzhansky-Muller scheme of evolution of postmating reproductive isolation.

However, if there is gene interaction such that any combination of mutant genes A_1 and B_2 results in inviability or sterility, the hybrids $(A_0 A_1 B_0 B_2)$ between the two populations will be inviable or infertile (see Table 1).

TABLE 1. *Genotype fitnesses for mutant genes controlling hybrid sterility or inviability.*

(1) Diploid model

	$A_0 A_0$	$A_0 A_1$	$A_1 A_1$
$B_0 B_0$	1	$1 + s_1$	$1 + s_2$
$B_0 B_2$	$1 + s_1$	$1 - t_1$	$1 - t_2$
$B_2 B_2$	$1 + s_2$	$1 - t_2$	$1 - t_3$

(2) Haploid model

		A_0	A_1
B_0	Fitness	1	$1 + s_1$
	Frequency	$(1-x)(1-y)$	$x(1-y)$
B_2	Fitness	$1 + s_2$	$1 - t$
	Frequency	$(1-x)y$	xy

726

If A_1 and B_2 reduce viability only when they are both homozygous, the F_1 hybrids will show a complete viability but there will occur hybrid sterility or F_2 breakdown. In practice, of course, more than one gene substitution may be necessary in each population. Thus, for hybrid sterility to be complete the B_0 gene in population 1 may have to be replaced by B_1, while in population 2 the A_0 gene may have to be replaced by A_2. The genotypes of populations 1 and 2 then become $A_1A_1B_1B_1$ and $A_2A_2B_2B_2$, respectively. A possible scheme of this type of gene interaction at the molecular level has recently been discussed (Nei, 1975).

It should be noted here that neither Dobzhansky nor Muller seriously considered how the mutant gene A_1 or B_2 increases in frequency and reaches fixation. They both stated, however, that fixation of such mutant genes is possible if it affects some other character and increases the fitness of its carrier. This problem can be treated easily if we use the haploid model and assume that the A and B loci are in linkage equilibrium. In practice, the assumption of linkage equilibrium does not hold in this case, but this assumption is unlikely to affect our qualitative conclusion. Theoretically, this problem can be studied without the assumption of linkage equilibrium but it makes the mathematical treatment very complicated. We note that the following method is similar to that used by Crow and Kimura (1965) in their study of evolution of "coadapted genes". At any rate, let x and y be the frequencies of mutant genes A_1 and B_2, respectively. If the fitnesses of genotypes A_0A_0, A_1B_0, A_0B_2 and A_1B_2 are 1, $1+s_1$, $1+s_2$ and $1-t$, respectively, then the amounts of gene frequency change per generation are given by

$$\Delta x = x(1-x)[s_1 - (s_1+s_2+t)y]/\overline{W} \ , \tag{1}$$

727

$$\Delta y = y(1-y)[s_2 - (s_1+s_2+t)x]/\bar{W} \quad , \tag{2}$$

where $\bar{W} = 1+s_1x(1-y) + s_2(1-x)y - txy$. Therefore, x increases if y is smaller than $\hat{y} = s_1/(s_1+s_2+t)$, while it decreases if y is larger than $\hat{y}$. Similarly, y increases if x is smaller than $\hat{x} = s_2/(s_1+s_2+t)$ but decreases if x is larger than $\hat{x}$. If mutant gene A_1 is adapted in the environment of population 1 but gene B_2 is not, then s_2 may be 0 or even negative. In this case B_2 never spreads through the population as long as A_1 is present. If mutation A_1 occurs earlier than B_2 , it will easily be incorporated into the population. However, if mutation B_2 occurs earlier than A_1 and $s_2 = 0$, then it may increase in frequency by random genetic drift. If mutant gene A_1 occurs at this stage, its frequency increases only when $y < \hat{y}$. It is now clear that the same situation may happen in the incorporation of the B_2 gene in population 2, if it is adapted in the environment of this population.

The above treatment indicates that the Dobzhansky-Muller statement is roughly correct as a possibility, though we need some qualifications. Unfortunately, however, we do not have much evidence that the genes for hybrid sterility or inviability are actually selected for through their pleiotropic effects on other characters. At least in some cases (e.g., gamete sterility genes in rice; Oka, 1974), the mutant genes do not appear to have any noticeable phenotypic effect. Can postmating reproductive isolation evolve without any pleiotropic effect of the genes concerned? The answer to this question is, of course, "yes" in relatively small populations, as noted by Muller (1942). In small populations the effect of random genetic drift becomes important, and mutant genes A_1 and B_2 may be fixed in the population without the aid of natural selection. This is particularly so when mutation occurs irreversibly from A_0 to A_1 or B_0 to B_2. The assumption of irreversible mutation may be justified when a large number of loci are concerned with reproductive isolation and any mutation at these loci behaves as an A_1 or B_2 allele. The probability of fixation of such mutant genes is higher when they are recessive (t_1 and t_2 in the diploid model in Table 1 are 0). In this case the problem becomes identical with that of fixation of nonfunctional genes at duplicate loci (Nei and Roychoudhury, 1973). At any rate, if evolution of postmating isolation occurs in this way, it is expected to evolve faster in small populations than in large populations.

(2) Ethological isolation

In bisexual organisms ethological isolation is one of the important components of reproductive isolation. There are two different hypotheses about the evolution of ethological isolation, though they are not mutually exclusive. Muller (1939, 1942) postulated that ethological isolation is a byproduct of genetic divergence of populations and evolves in the same fashion as that of hybrid inviability or sterility. On the other hand, Fisher (1930) and Dobzhansky (1940, 1970) emphasized the importance of natural selection in developing ethological isolation. For this type of natural selection to operate, however, there must be already a considerable amount of genetic differentiation between the populations, so that the hybrids between them show a reduced viability or fertility. As indicated by Dobzhansky, if there is any hybrid inviability or infertility and the two populations meet in the same geographical area, males or females which tend to mate with the individuals in their own population will have selective advantage over those which tend to mate with the individuals from the other population. Therefore, if there is genetic variability with respect to mating preference, ethological isolation will gradually be developed.

There have been a number of experiments to intensify the ethological isolation between two genetic groups or species by artificial selection (e.g., Koopman, 1950; Knight et al. 1956; Fukatami and Moriwaki, 1970). The results obtained are quite inconsistent, selection being sometimes effective but sometimes not. Ineffectiveness of selection, of course, does not refute the Fisher-Dobzhansky hypothesis. It may simply be a result of no genetic variability in the initial population.

While the Fisher-Dobzhansky scheme of evolution of ethological isolation should work in the presence of genetic variability of mating behavior, it is not a necessary condition for the development of ethological isolation. Even in the absence of selection, ethological isolation may evolve according to the scheme envisaged by Muller. Particularly, if the two diverging populations never meet after geographical isolation, there is no chance for selection to operate. In this case, ethological isolation must evolve in Muller's scheme. In the following, I present a mathematical model which follows Muller's argument to some extent.

In many organisms such as Drosophila and birds, females choose their mates, while males do not have any mate preference. Suppose that loci A and B control male-limited and female-limited morphological, physiological or behavior characters, respectively, and that the original genotype of a population is aabb . Mutant gene A changes the male character, while mutant gene B changes the female character. Because of these changed characters, Bb or BB females may prefer Aa or AA males. This is a kind of assortative mating, but unless A and B are the same allele controlled by the same locus, the consequence of this mating is quite different from that of ordinary assortative mating. A simplest mathematical model of this type of mating is given in Table 2. In this model BB and Bb females choose AA and Aa males more frequently than aa males, the relative probability of choice being 1 and $1-m_a$, respectively. On the other hand, bb females choose aa males more frequently than AA and Aa males with relative probabilities of 1 and $1-m_A$. Here we have assumed, for simplicity, that alleles A and B are dominant over a and b , respectively, but it can easily be extended to the case of an arbitrary degree of dominance. This model is an extension of O'Donald's (1963) sexual selection, and if $m_a = 1$ and $m_A = 0$, it becomes identical to his. Of course, O'Donald used this type of model to study the evolution of secondary sex characters rather than the evolution of ethological isolation. At any rate, he has shown that if $m_a = 1$ and $m_A = 0$, the frequencies of A and B increase with increasing generation though the initial rate of increase is very low. If $m_A \neq 0$, however, this model creates unstable equilibria, and if gene frequencies are less than certain critical values, they decrease rather than increase.

TABLE 2. *Mating preferences among genotypes. Only females are assumed to have mating preference.*

Females \ Males	AA, Aa(A)	aa(a)	Normalizing factor
	p	1-p	
BB, Bb (B) q	1	$1-m_a$	$1-(1-p)m_a$
bb (b) 1-q	$1-m_A$	1	$1-pm_A$

The exact treatment of this problem with the diploid model is very complicated, but it can be seen easily if we use the haploid model and assume that the gene frequencies are the same for males and females and the two loci are in linkage equilibrium.

Let p and q be the gene frequencies of A and B, respectively. It can be shown from Table 2 that the frequency of A in the next generation is

$$p' = \frac{pq}{1 - (1-p)m_a} + \frac{p(1-q)(1-m_A)}{1-pm_A} .$$ (3)

Thus, the amount of gene frequency change per generation is

$$\Delta p = p(1-p) \left[\frac{qm_a}{1 - (1-p)m_a} - \frac{(1-q)m_A}{1-pm_A} \right]$$

$$= \frac{p(1-p)\left[qm_a - pm_A m_a - (1-q)m_A(1-m_a)\right]}{\left[1 - (1-p)m_a\right](1-pm_A)} .$$ (4)

On the other hand, the frequency of B remains constant, as it should be. The above formula shows that if m_a is 1 and $m_A > 0$, the change of p depends on the value of $q-pm_A$. Namely, p increases only when q is larger than pm_A. If m_a is smaller than 1, the situation is worse; only when q is very large, p increases. Since q must be very small in the early generations, this model suggests that ethological isolation cannot evolve in large populations unless m_A is small compared with m_a. If m_A is 0, the evolution of ethological isolation becomes the same as the evolution of secondary sex characters studied by O'Donald, as mentioned earlier. In this case O'Donald (1967) has shown that the frequencies of both A and B genes increase particularly when A is recessive rather than dominant. The increase of the frequency of B is caused by the association of alleles A

and B due to nonrandom mating, and if the A_1 allele is fixed, the frequency of B ceases to increase in large populations.

In small populations, however, the frequency of allele B may increase owing to random genetic drift. This is true even if m_A is not 0 . Namely, before any mutation occurs at the A locus, the mutant gene B may arise and increase in frequency substantially by chance. If mutant gene A arises after the frequency of B becomes larger than a certain critical value, the frequency of A is expected to increase from the beginning (see formula (4)). The increase of the frequency of A genes will cause the nonrandom association of alleles A and B and increase the frequency of B as well. It is also noted that in the absence of A genes, gene B may be fixed in the population purely by random genetic drift. Once mutant gene B is fixed in the population, gene A will have a large selective advantage over a . If we note that population size often goes through bottlenecks in the process of speciation, the effect of random genetic drift must be large. In fact, Carson (1970) and Prakash (1972) presented possible examples of rapid speciation in Drosophila, where a population went through a bottleneck.

TIME REQUIRED FOR DEVELOPING REPRODUCTIVE ISOLATION

It is important to know how fast reproductive isolation evolves. Obviously, the time required for developing reproductive isolation will depend on a number of factors. First, it will depend on the type of organism. Recently, Wilson et al. (1974) and Prager and Wilson (1974) have shown that reproductive isolation as measured by hybrid inviability develops much faster in mammals than in frogs and birds. By using the method of immunological dating, they estimate that in mammals the average divergence time between hybridizable species is about 2 million years, whereas in frogs and birds it is about 20 million years. They ascribe this difference to the probable difference in the rate of "regulatory mutation". We note, however, that this difference can also be explained by the difference in postmating reproductive system. Namely, the reproductive system after fertilization is more complex in mammals than in frogs and birds. If the reproductive system is complex, it would require a large number of genes for the system to function properly, and reproductive isolation (hybrid inviability) may be produced by mutations at any of these loci. Thus, it is expected that reproductive isolation evolves faster in organisms with a complex reproductive system than in those with a simpler system.

Second, even in the same type or organisms the evolutionary time of reproductive isolation will depend on whether there are any mutant genes for reproductive isolation in the initial population and, if not, how fast such mutations arise. It also depends on whether the mutant genes are fixed by selection or by genetic drift. We have seen that in the development of reproductive isolation mutation and genetic drift are very important, though there must be some sort of epistatic selection. If we ignore the effect of selection, the evolutionary time of reproductive isolation can be studied by using Crow and Kimura's (1970) gene frequency distribution under irreversible mutation.

Although generally a large number of loci are involved in the evolution of reproductive isolation, we consider a single locus for simplicity. We designate the original and mutant alleles by b and B , respectively. At the molecular level, each mutant allele may be unique, but as long as they affect

reproductive isolation, they can be pooled together and treated as a single allele. We assume that mutation occurs irreversibly from b to B at a rate of v per generation and
all mutations are selectively neutral. This assumption is
satisfied in the evolution of "female character" in ethological isolation considered above in the absence of mutant gene
A . At any rate, under this assumption the probability that
the mutant gene B is fixed by the t-th generation in a population of effective size N is given by

$$f(p,1;t) = 1 - (1-p) \sum_{i=1}^{\infty} (-1)^{i-1} \frac{\Gamma(i-1+M)(M+2i-1)}{\Gamma(M)i!}$$

$$\times F(1-i, i+M, M, p)\exp[-i(v + \frac{i-1}{4N})t] \quad , \tag{5}$$

where M = 4Nv , p is the initial gene frequency and
F(·,·,·,·) is the hypergeometric function (Crow and Kimura,
1970). Therefore, the probability density that the mutant
gene is fixed at the t-th generation is

$$g(p,t) = \frac{\partial f(p,1;t)}{\partial t}$$

$$= (1-p) \sum_{i=1}^{\infty} (-1)^{i-1} \frac{\Gamma(i-1+M)(M+2i-1)}{\Gamma(M)i!}$$

$$\times F(1-i,i+M,M,p)\lambda_i e^{-\lambda_i t} \quad , \tag{6}$$

where λ_i = i[v+(i-1)/(4N)] . If the initial population does
not have any mutant genes, then

$$g(t) = \lim_{p\to 0} g(p,t)$$

$$= \sum_{i=1}^{\infty} (-1)^{i-1} \frac{\Gamma(i-1+M)(M+2i-1)}{\Gamma(M)i!} \lambda_i e^{-\lambda_i t} . \tag{7}$$

The n-th moment of fixation time is obtained by μ_n' =
$\int_0^{\infty} t^n g(p,t)dt$ and becomes

$$\mu_n' = (1-p)n! \sum_{i=1}^{\infty} (-1)^{i-1} \frac{\Gamma(i-1+M)(M+2i-1)}{\Gamma(M)i!(\lambda_i)^n}$$

$$\times F(1-i,i+M,M,p) \qquad . \qquad (8)$$

When $p \to 0$, this reduces to

$$\mu_n' = (4N)^n n! \sum_{i=1}^{\infty} (-1)^{i-1} \frac{\Gamma(i-1+M)(M+2i-1)}{\Gamma(M)i![i(M+i-1)]^n} \qquad . \qquad (9)$$

If $4Nv$ is much smaller than 1, only the first term in (9) is important. In this case the first four moments are approximately given by

$$\mu_1' = (M+1)/v = 4N+1/v \quad , \qquad (10a)$$

$$\mu_2' = 2(M+1)/v^2 \quad , \qquad (10b)$$

$$\mu_3' = 6(M+1)/v^3 \quad , \qquad (10c)$$

$$\mu_4' = 24(M+1)/v^4 \quad . \qquad (10d)$$

The mean fixation time is, therefore, of the order of recipro-cal of mutation rate. On the other hand, the standard devia-tion is $(1-M^2)^{\frac{1}{2}}/v$, while the skewness and kurtosis are gi-ven by $\gamma_1 = \mu_3/\sigma^3 \approx 2$ and $\gamma_2 = \mu_4/\sigma^4 - 3 \approx 6$ approximately, respectively. Figure 2 shows the probability distribution of fixation time for the case of $4Nv = 0.1$. It is seen that fixation time has a very wide distribution. This distribution is more skewed and more leptokertic than that of conditional fixation time of a single mutant gene (Kimura, 1970), and the absolute time is much longer in the present case. If $4Nv$ is larger than about 0.5, the approximate formulae (10) do not work, but the above conclusion remains qualitatively the same unless $4Nv$ is very large.

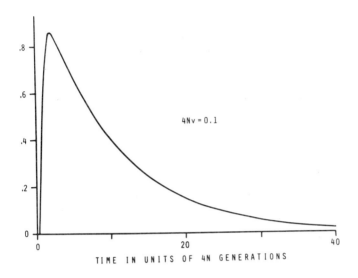

FIGURE 2. Distribution of fixation time of mutant genes
under irreversible mutation pressure.

In the above computation we have assumed that all mutant
genes are selectively neutral. In the case of reproductive
isolation, mutant genes may have some selective advantage.
Selection decreases fixation time considerably for a given
value of 4Nv when there is no dominance and no epistasis
(Li and Nei, unpublished). Some of the results of our studies on this problem are presented in Table 3. It is clear
that mutation and selection are almost equally important in
determining the fixation time of mutant genes. In practice,
of course, selection for reproductive isolation involves
epistasis, so that the exact treatment of fixation time is
quite complicated. On the other hand, if the initial population contains mutations, the fixation time can be very

737

short but still has a large variance.

TABLE 3. *Average fixation times of mutant genes under re-*
current mutation pressure in units of 4N generations.
These fixation times were computed by using exact formulae.
The initial population is assumed to have no mutant genes.
Selection is additive. S = 4Ns , where s is the selective
advantage of mutant genes over the original alleles (Li and
Nei, unpublished).

Selection	$4N\upsilon$			
intensity	1	0.1	0.01	0.001
S = 0	1.65	10.94	101.00	1001.0
S = 10	0.544	1.53	10.42	99.07
S = 100	0.104	0.197	1.05	9.61

These studies suggest that evolutionary time of reproductive isolation varies greatly in individual cases. This is
supported by the experimental results recently obtained by
Wilson et al. (1974) and Prager and Wilson (1974). In frogs,
for example, they have shown that the immunological distance
between hybridizable species varies from 0 to 91 , the
latter value corresponding to 55 million years of divergence
time if we accept their immunological dating. A possible
example of extremely rapid evolution of reproductive isolation has been reported by Prakash (1972). He presents evidence showing that the partial sterility barrier between the
United States and Bogota populations of Drosophila pseudoobscura evolved possibly within ten years by means of the
bottleneck effect. If this is true, there must have been
hybrid-sterility genes segregating in the foundation stock of
the Bogota population.

THE WRIGHT MODEL OF POPULATION DIFFERENTIATION

The mathematical model of genetic differentiation among isolated populations was first studied by Wright (1943). In his model a population is split into a large number of populations of equal size N, and the genetic differentiation of populations <u>relative to the total population</u> is studied in terms of F_{ST}, which is defined as $\sigma_x^2[\bar{x}(1-\bar{x})]$, where $\bar{x}$ and σ_x^2 are the mean and variance of gene frequency among the populations. He showed that in the absence of mutation, selection, and migration, the F_{ST} in the t-th generation is given by

$$F_{ST} = 1 - (1 - \frac{1}{2N})^t$$

$$\approx 1 - e^{-t/2N} \quad . \tag{11}$$

It is worthwhile to note that this formula is independent of the initial gene frequency.

In applying this formula to natural populations, however, there arise some problems even if selection and migration are negligible. First, the number of isolated populations is often very small. This is particularly so if we consider a group of populations which are derived at the same evolutionary time. In this case $1 - (1 - 1/2N)^t$ is not the correct expectation of $\sigma_x^2/[\bar{x}(1-\bar{x})]$. Second, in the present model no mutation is assumed, so that all populations eventually become homozygous. In practice, of course, mutation always occurs so that most natural populations contain a certain amount of genetic variability at any moment of evolutionary time. Third, in the presence of mutation F_{ST} is not a good measure of the genetic differentiation of populations. This is because at the molecular level new mutations are almost always different from the pre-existing alleles, and a locus may contain more than two alleles.

In view of these difficulties I have recently proposed a new measure of genetic differentiation called the coefficient of gene differentiation (Nei, 1973a). The definition of this parameter is dependent on the analysis of gene diversity (heterozygosity) in the total population into its components, i.e. the gene diversity within (H_S) and between (D_{ST}) populations. The coefficient of gene differentiation is then defined as the proportion of the interpopulational gene diversity, i.e.

$$G_{ST} = D_{ST}/H_T \quad . \tag{12}$$

In the special case of two alleles at a locus this becomes identical with Wright's F_{ST} . In this connection, however, it should be noted that G_{ST} was primarily designed to be applied to a large number of randomly chosen loci in order to measure the gene differentiation of the whole genome, though it can be applied to any number of loci in practice.

The theoretical value of G_{ST} under mutation and genetic drift can be obtained by evaluating the expectations of H_S , D_{ST} , and H_T . We assume that a population is split into s isolated populations of effective size N . Let x_{ik} be the frequency of the k-th allele at a locus in the i-th population. The gene diversity in this population is then defined as $H_i = 1 - \Sigma_k x_{ik}^2$. The quantity $\Sigma_k x_{ik}^2$ is called homozygosity or gene identity and denoted by J_i (Nei, 1973a). We denote the average of H_i over all subpopulations by H_S and $1 - H_S$ by J_S . The gene diversity in the total population is given by $H_T \equiv 1 - J_T = 1 - \Sigma_k x_k^2$, where $x_k = \Sigma_i x_{ik}/s$. The difference $D_{ST} = H_T - H_S$ is the gene diversity due to interpopulational gene differences. It can be shown that

$$D_{ST} = (1-1/s) \sum_{i \neq j} D_{ij}/[s(s-1)] \quad , \tag{13}$$

where $D_{ij} = H_{ij} - (H_i + H_j)/2$, in which $H_{ij} = 1 - \Sigma_k x_{ik} x_{jk}$.

Let us now derive the expectation of G_{ST} defined as $E(D_{ST})/E(H_T)$ under the effects of mutation and genetic drift. If we assume that new mutations are different from the pre-existing ones in the entire population, the expectations of J_i and $J_{ij} \equiv 1 - H_{ij}$ are easily obtained. It is clear that $E(J_i)$ is the same for all i and $E(J_{ij})$ is the same for all i and j $(j \neq i)$ in the present case. Nei and Feldman (1972) have shown that

$$E(J_i^{(t)}) = J^{(\infty)} + (J^{(0)} - J^{(\infty)}) e^{-(2v + \frac{1}{2N})t} \qquad , \qquad (14)$$

$$E(J_{ij}^{(t)}) = J_{ij}^{(0)} e^{-2vt} \qquad (15)$$

approximately, where $J^{(\infty)} = (4Nv+1)^{-1}$. The formula for $E(J_i^{(t)})$ is essentially the same as Malécot's (1948). Therefore, using Nei and Feldman's results we obtain

$$G_{ST}^{(t)} = E(D_{ST}^{(t)})/E(H_T^{(t)}) \qquad (16)$$

where

$$E(D_{ST}^{(t)}) = (1-1/s)[J_S^{(\infty)} - J_S^{(0)} e^{-2vt} - (J_S^{(\infty)} - J_S^{(0)}) e^{-(2v+1/2N)t}]$$

and

$$E(H_T^{(t)}) = H_S^{(\infty)} + (1-1/s)(J_S^{(\infty)} - J_S^{(0)} e^{-2vt})$$

$$+ s^{-1}(J_S^{(\infty)} - J_S^{(0)}) e^{-(2v+1/2N)t} \qquad ,$$

where $J_S^{(\infty)} \equiv 1 - H_S^{(\infty)} = (4Nv+1)^{-1}$.

In most natural populations average heterozygosity or gene diversity for a large number of loci apparently remains more or less constant (Nei, 1975). In this case we may assume $J_S^{(0)} = J_S^{(\infty)}$. Then,

$$G_{ST}^{(t)} = \frac{(1-1/s)J_S^{(\infty)}(1-e^{-2vt})}{H_S^{(\infty)}+(1-1/s)J_S^{(\infty)}(1-e^{-2vt})} \quad . \tag{17}$$

It is now clear that G_{ST} is not a simple function of t and depends on N, v and s as well. This indicates that G_{ST} is not a good parameter to estimate evolutionary time, though it is a useful measure of the degree of gene differentiation relative to the total population. When $t \ll 2N$, however, $G_{ST}^{(t)}$ can be written as

$$G_{ST}^{(t)} = \frac{s-1}{s}\frac{t}{2N} \tag{18}$$

approximately. Therefore, it increases linearly in the initial stage.

Incidentally, if small populations are derived from a large foundation stock, then the gene differentiation occurs mostly by genetic drift and new mutation may be neglected. In this case $4Nv \ll 1$ and $J_0^{(\infty)} = 1$ approximately. Therefore,

$$G_{ST}^{(t)} = \frac{(1-1/s)(1-e^{-t/2N})}{1-(1-e^{-t/2N})/s} \quad . \tag{19}$$

When $s \to \infty$, $G_{ST}^{(t)}$ becomes $1-\exp(-t/2N)$, agreeing to Wright's result, as expected. On the other hand, if $t \ll 2N$, (19) can be approximated by (18). Namely, as long as $t \ll 2N$, the same formula applies whether there is a balance between mutation and genetic drift or not.

GENE DIFFERENTIATION BETWEEN TWO POPULATIONS
AND GENETIC DISTANCE

As noted earlier, F_{ST} or G_{ST} measures the degree of genetic differentiation <u>relative to the total population</u>.

Therefore, their value varies according to how many populations are included or which populations are included in the computation. Furthermore, splitting of natural populations generally occurs not simultaneously as assumed above but sequentially one after another. These make it difficult to relate the value of F_{ST} or G_{ST} to the evolutionary time. The difficulties are, however, removed if we consider just two populations and measure the genetic differentiation in terms of absolute gene or gene frequency differences. To my knowledge, the first serious attempt of modeling the process of genetic differentiation between two populations was done by Cavalli-Sforza and Edwards (1967), though the statistical properties of their model had been studied earlier by Bhattacharyya (1946). At a locus with multiple alleles, they represented two populations on the hypersphere in an Euclidean space in terms of Bhattacharyya's angular transformation, and conceived the genetic differentiation as a result of the independent Brownian motion of the two populations on the hypersphere. They assumed that no mutation and no random loss or fixation of genes occur during the process and that gene frequencies are not close to 0 or 1 .

This assumption seems to be satisfactory as long as the evolutionary time considered is short. In practice, however, we generally do not know the evolutionary time. In applying this theory to actual data only polymorphic loci with intermediate gene frequencies at the present time are used, but if the actual evolutionary time is long, some of them might have had gene frequencies close to 0 or 1 when the two populations were separated. Furthermore, the distance measure originally suggested by these authors is not linear with evolutionary time, although the alternative measure suggested by Cavalli-Sforza (1969) is approximately linear with time in the early generations, as will be discussed later.

In the study of speciation, which generally requires a long evolutionary time, it is important to consider both genetic drift and mutation even if selection and migration are negligible. Under the effects of mutation and genetic drift the turnover of genes is always occurring in a population. Therefore, a particular locus may be polymorphic at an evolutionary time but completely monomorphic at another time. Yet, if we consider the whole genome of an organism, the average heterozygosity per locus or even the distribution of gene frequencies may remain constant. Consideration of this possibility led me to propose a new model of genetic differentiation of populations and a measure of genetic distance generated by this model (Nei, 1971a, 1972). In this model I assumed that at an evolutionary time a population, which is in equilibrium with respect to the effects of mutation, selection, and genetic drift, is split into two of the same size as that of the parental population, and thereafter each population undergoes independent evolution. New mutations were assumed to be always different from the preexisting alleles in either population. I would like to emphasize that in this model, the gene frequencies at the time of separation vary from locus to locus, and if all mutations are neutral and mutation rate is the same for all loci, the distribution of gene frequencies is given by Kimura and Crow's (1964) formula $\Phi(x) = 4Nv(1-x)^{4Nv-1}x^{-1}$. Note that if $4Nv \leqslant 0.1$, a majority of loci are monomorphic. Essentially the same model was used by Latter (1972) in his computer simulation of genetic divergence of populations.

In the present model we have assumed that a population is split suddenly at an evolutionary time. This assumption is certainly unrealistic, since in the early stage of population differentiation there usually occurs some migration between populations. In reality, however, no substantial gene differentiation occurs as long as there is migration with a rate appreciably higher than mutation rate (Nei and Feldman, 1972), so that the model seems to be useful for practical purposes. In fact, as will be discussed later, the effect of migration in the early generations is not so important in the study of long-term evolution.

A number of authors have defined genetic distance as the geometric distance between two populations represented in a multidimensional space, the number of dimensions often being equal to the number of alleles at a locus. This definition, however, does not have much biological meaning. Furthermore, in long-term evolution it has mathematical problems. Namely, if a new mutation occurs, we must add one dimension, whereas if an allele is lost we must subtract one dimension. Theoretically, this problem can be solved if we equate the number of dimensions to the possible number of alleles at a locus. In practice, however, this number is almost infinite, so that the mathematical treatment becomes very complicated. In measuring genetic distance between populations, a large number of loci must be used, since a single locus does not provide much information. In this case spatial representation of populations is terribly complicated.

In my opinion spatial representation of populations is not required in the study of evolution, and the best measure of genetic distance is to use the accumulated number of gene substitutions per locus. Since gene substitution is the most basic process of evolution, the genetic distance thus defined will reflect the evolutionary history of populations. If the rate of gene substitution per unit length of time is constant, as is roughly the case with some proteins, the genetic distance is proportional to evolutionary time. In the model of population differentiation I have described above this genetic distance can be estimated by

$$D = -\log_e I , \qquad (20)$$

where I is the normalized identity of genes (Nei, 1971a, 1972). The I value is computed by

$$I = J_{XY}/\sqrt{J_X J_Y} , \qquad (21)$$

in which J_X and J_Y are the average homozygosity per locus in populations X and Y , respectively, and J_{XY} is the average identity of genes between X and Y .

It has been shown that if the rate of gene substitution per locus per generation is α , the I value in generation t after divergence of the two populations is given by $e^{-2\alpha t}$. Therefore, the genetic distance defined by (20) is

$$D = 2\alpha t , \qquad (22)$$

which is linear with divergence time, if α remains constant. It is also clear that D measures the accumulated number of gene substitutions per locus. Since each gene substitution generally produces one codon difference, D can also be interpreted as the number of codon differences per locus between the two populations. Of course, this interpretation is dependent on the model we have set up. However, the minimum estimate of codon differences per locus (minimum genetic distance) can be obtained in any circumstance. It is given by

$$D_m = \sum_{j=1}^{n} d_{m_j}/n , \qquad (23)$$

where $d_{m_j} = (\Sigma x_{ij}^2 + \Sigma y_{ij}^2)/2 - \Sigma x_{ij} y_{ij} = \Sigma (x_{ij} - y_{ij})^2/2$, in which x_{ij} and y_{ij} are the frequencies of the i-th allele at the j-th locus in populations X and Y , respectively (Nei and Roychoudhury, 1972; Nei, 1973b). In contrast to this minimum

genetic distance, I have called D the <u>standard genetic dis-</u>
<u>tance.</u> In the absence of selection α is identical with the
mutation rate v , and the expectation of D_m is given by
$J(1-e^{-2vt})$. We note that Latter's (1972) coefficient of
genetic divergence (γ) is D_m/J .

One of the purposes of computing genetic distance between
populations is to construct a phylogenetic tree and estimate
the time after divergence between each pair of populations or
species. In this case the measure of genetic distance should
be linear with divergence time. Our standard genetic distance meets this requirement, but how about other distance
measures?

This problem was studied to some extent by Latter (1972,
1973) by means of computer simulation. He showed that his
measure of genetic divergence (γ) increases almost linearly
with time for the first 5N generations when 4Nv = 0.2 ,
while Cavalli-Sforza and Edwards' (1967) distance measure (d)
rapidly increases in the early generations but the rate of
increase quickly declines in the later generations. In his
1973 paper he compared Balakrishnan and Sanghvi's (1968) measure G_S^2 , Cavalli-Sforza's (1969) f_θ , and Yasuda's (1968)
ϕ_Y with Malécot's (1948) formula $\phi(t) = (1 + 4Nv)^{-1} \times$
$\times [1-\exp\{-(2v+1/2N)t\}]$. He showed that while G_S^2 and ϕ_Y
are far from the value of $\phi(t)$, Cavalli-Sforza's f_θ is
close to $\phi(t)$ in the early generations when there is natural
selection. Latter claimed a good agreement between f_θ and
$\phi(t)$ even in the absence of selection. However, a close
look at his figure indicates that f_θ is somewhat smaller
than $\phi(t)$ in the early generations but later it exceeds the
value of $\phi(t)$. Of course, Malécot's formula refers to the
<u>expected homozygosity within populations</u> $[E(J_i^{(t)})$ in (14)],

and in the model of infinite possible alleles as used by
Latter it is not the measure of population differentiation.
Therefore, the agreement between a distance measure and $\phi(t)$
neither proves nor disproves its suitability as a measure of
genetic divergence. Furthermore, the agreement between f_θ
and $\phi(t)$ in the presence of selection must be coincidental,
since $\phi(t)$ is based on the assumption of no selection.

With this in mind, I started a computer simulation work
with the help of Yoshio Tateno. We were particularly interested in the linear relationship between distance and divergence time. Since this linearity did not hold for Balakrishnan and Sanghvi's G_S^2 and Yasuda's ϕ_Y in Latter's
simulation, we excluded these from our studies. The distance
measures we studied are Cavalli-Sforza's f_θ , Roger's (1972)
coefficient of dissimilarity (D_R) and Hedrick's (1971) dissimilarity (D_H) . They are defined as follows:

$$f_\theta \equiv D_C = 4\Sigma_j[1-\Sigma_i(x_{ij}y_{ij})^{1/2}]/\Sigma_j(n_j-1) \quad ,$$

$$D_R = \frac{1}{r}\Sigma_j[\frac{1}{2}\Sigma_i(x_{ij}-y_{ij})^2]^{1/2} \quad ,$$

$$D_H = 1 - \frac{1}{r}\Sigma_j \frac{\Sigma_i x_{ij}Y_{ij}}{\frac{1}{2}(\Sigma_i x_{ij}^2 + \Sigma_i Y_{ij}^2)} \quad ,$$

where n_j is the number of alleles at the j-th locus, r is
the number of loci used, and X_{ij} and Y_{ij} are the i-th
genotype at the j-th locus in populations X and Y , respectively. In our simulation X_{ij} and Y_{ij} were obtained
from gene frequencies under the assumption of Hardy-Weinberg
equilibrium. The method of simulation was briefly as follows:
In many natural populations the average heterozygosity per
locus is about 10 percent (Lewontin, 1974; Nei, 1975). This
suggests that $4Nv$ is roughly 0.1 . The mutation rate for

electrophoretically detectable alleles has been estimated to be 10^{-7} per locus _per year_ (not per generation) (Kimura and Ohta, 1971). In Drosophila there are about ten generations in a year, so that the mutation rate per generation in this organism is estimated to be about 10^{-8}. However, since the mean and variance of genetic distances D and D_m are almost exclusively determined by $M = 4Nv$ and $T = t/N$ (Li and Nei, 1975), we used $N = 50$ and $v = 0.0005$ with $4Nv = 0.1$ to save computer time. We assumed that a population which is in equilibrium with respect to the effects of mutation and genetic drift is split into two populations in a generation and thereafter each population evolves independently. Mutation was introduced according to the Poisson distribution with mean $2Nv = 0.05$ in every generation. The number of loci (replications) studied was 200. (The details of the procedure will be published elsewhere.) Using the gene frequency data for these 200 loci, the values of D, f_θ, D_R and D_H were computed in a number of specified generations.

The results obtained are presented in Figure 3. As expected, the D value increases linearly with time, but f_θ and D_R increase curvilinearly when a long period of evolutionary time is considered. The distance D_H closely followed D_R, so that it is not given in this figure. For constructing a phylogenetic tree, therefore, D is a better measure of distance than f_θ, D_R and D_H.

For a short period of time (up to $2N$ generations), however, f_θ increases almost linearly. This property is also observed in Latter's simulation. Therefore, as long as the evolutionary time is short, as was originally assumed by Cavalli-Sforza, f_θ is a satisfactory distance for constructing a phylogenetic tree. Nevertheless, it is not clear how the absolute value of f_θ is related to the evolutionary time except in the case, where there are only two alleles segregating at a locus and the effect of new mutation is negligible.

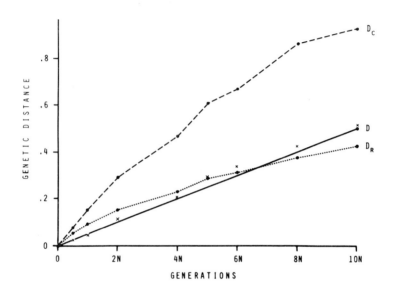

FIGURE 3. *Relationships between genetic distance measures and evolutionary time (generations). D_C = Cavalli-Sforza's distance (f_θ) . D_R = Rogers' dissimilarity. D = Nei's standard genetic distance; the straight line is the theoretical value and crossmark × represents the observed value. Theoretical values for D_C and D_R are not known.*

As mentioned earlier, Latter claimed that the value of f_θ was close to Malécot's $\phi(t)$. The values of these quantities in our simulation are given in Table 4.

TABLE 4. *Comparisons of Cavalli-Sforza's* f_θ *with* $\phi(t) =$ $(4Nv+1)^{-1}[1-exp(2v+1/2N)t]$ *and* $f(t) = 1-exp(-t/2N)$.

Generation	f_θ	$\phi(t)$	$f(t)$
0.5 N	.0789 ± .0108	.2186	.2122
N	.1554 ± .0222	.3846	.3935
2N	.2914 ± .0304	.6065	.6321
4N	.4683 ± .0501	.8084	.8647
6N	.6695 ± .0722	.8756	.9502
8N	.8665 ± .0804	.8979	.9817
ION	.9314 ± .0866	.9054	.9933

It is clear that, unlike Latter's case, the agreement between f_θ and $\phi(t)$ is very poor. Cavalli-Sforza (1969) stated that the expectation of f_θ is approximately equal to $f(t) = 1-exp(-t/2N)$. Table 4, however, shows that the agreement between f_θ and $f(t)$ is also poor.

In a study of the genetic distances for individual loci among human races, Nei and Roychoudhury (1974a) noted a large variation of d_m among different loci. A large variation of genetic distance among loci was also observed by Ayala and Tracey (1974) and Ayala et al.(1974) in Drosophila. Recently Li and Nei (1975) studied the expected variance of genetic distance due to mutation and genetic drift, and showed that the ratio of the standard deviation of genetic distance to the mean is often larger than 2 . This indicates that in order to study the genetic relationship of populations, a large number of loci must be used. Compared with this inter-locus variance, the sampling variance of genetic distance at the time of gene frequency survey is generally small unless the sample size is very small (Nei and Roychoudhury, 1974b).

So far we have assumed that the sizes of the two descendant populations are equal to each other. In many circumstances this assumption may not hold. However, Chakraborty and Nei (1974) have shown that the genetic distance D is quite robust and even if one population is 100 times larger than the other, the relationship D = 2vt is roughly correct when long-term evolution is considered.

Our theory is also dependent on the assumption that the effective size (N) of the descendant populations is the same as that (N_0) of the foundation stock. If this assumption does not hold, the rate of gene substitution may change temporally. This can be seen in the following way. When the time after divergence is relatively short, the minimum genetic distance may be written as

$$D_m = H_0(t/2N) \tag{24}$$

approximately, where H_0 is the average heterozygosity at the time of divergence and given by $4N_0 v/(4N_0 v+1)$ (Li and Nei, 1975). Therefore,

$$D_m = \frac{2v(N_0/N)t}{4N_0 v + 1} . \tag{25}$$

When $N = N_0$, the rate of gene substitution per generation is 2v approximately and remains constant. On the other hand, if $N < N_0$, the rate is accelerated temporarily, while if $N > N_0$, it is decelerated temporarily. This indicates that if population size fluctates, the rate of gene substitution is expected to vary even if there is no selection. Of course, if we consider a long-term evolution, this kind of fluctuation will be averaged out.

One might also object to our assumption that as soon as two populations are separated, there occurs no migration. In practice, it is probable that in the early generations migration occurs, but the rate of migration gradually declines. Recently, the effect of this type of migration on genetic distance was studied by Wen-Hsiung Li (unpublished). His results indicate that if a short period of evolutionary time is considered the effect of migration is very large, but as far as concerned with an evolutionary time which would encompass the formation of new species migration in the early stage has little effect on the final value of genetic distance.

Natural selection is expected to have a profound effect on genetic distance. It is, however, difficult to do a precise mathematical study on this effect except when all loci are subject to a more or less similar type of selection. In this case the rate of gene substitution per generation is given by $\alpha = 2Nvu$, where u is the probability of fixation of a new mutation (Kimura and Ohta, 1971). For example, if all mutant genes have selective advantage s in heterozygoes and $2s$ in homozygotes compared with the original wild type homozygotes, $u = 2s$ in large populations (Haldane, 1927). Therefore, $\alpha = 4Nvs$. As long as α remains constant, our formula (22) holds true. Latter (1972) confirmed this by computer simulation.

DATA ANALYSIS

With the help of Dr. A. K. Roychoudhury, I have computed the interracial and interspecific genetic distances in various organisms by using published data on gene frequencies (Nei, 1973b; 1975). The distance measure used was the standard genetic distance. The results obtained are presented in Table 5 in a condensed form. In this computation only those organisms in which gene frequency data for a relatively large number of loci were available were used.

753

TABLE 5. *Estimates of genetic distance from electrophoretic data.*

Taxa	No. of taxa	No. of loci	$D = -\log_e I$
A. Local races			
Man	3	35	.011 ~ .019
Rodents	13	18 ~ 41	.000 ~ .058
Drosophila	12	11 ~ 24	.001 ~ .010
B. Subspecies			
Rodents	16	27 ~ 41	.004 ~ .262
Lizards	4	23	.335 ~ .351
Fish	9	17	.062 ~ .218
Drosophila	11	12 ~ 25	.028 ~ .234
C. Species			
Mammals	7	14 ~ 27	.12 ~ .63
Lizards	4	23	1.32 ~ 1.75
Drosophila	45	13 ~ 28	.05 ~ 2.54
D. Genera			
Fish	5	16	1.1 ~ 2.8 (∞)
E. Man-chimp (Families)		42	.62
F. Man-horse (Orders)			(18)*

*This was estimated from amino acid sequence data (Nei,1975).

Yet the standard error of genetic distance was quite large, amounting to 1/4 ~ 1/2 of the distance itself (see Nei and Roychoudhury, 1974b, for the computation of standard errors). The genetic distance estimates are classified into five categories according to the rank of the taxa compared, i.e., local races, subspecies, species, genera, and families. The distinction between local races and subspecies was not always easy. I generally followed the classification by the authors who published gene frequency data, but when there is evidence that no or little migration occurs between a given pair of taxa, I classified them as subspecies. The gene frequency data used in this study were all examined by electrophoresis. Therefore, our estimate of genetic distance refers to the number of electrophoretically detectable codon differences per locus.

The genetic distance between races is generally very small and always less than a few percent. A relatively large value (0.058) was obtained between Arizona and Texas populations in kangaroo rats. This organism, however, has a short migration distance and the two populations may be reproductively isolated (Johnson and Selander, 1971). In most other cases the distance was less than 0.02. This result is in agreement with the theoretical expectation that genetic distance cannot be very large as long as there is migration (Nei and Feldman, 1972). It is of interest to note that the genetic distances among three major races of man, Caucasoids, Negoids and Mongoloids, are of the same order of magnitude as those of local races in other organisms.

Estimates of genetic distances between subspecies are generally much larger than those between races, while the distances between species are still larger. The distribution of the former is, however, overlapped with that of the latter. This is, of course, expected, since the definition of species largely depends on whether the two taxa in question are reproductively isolated or not and the evolutionary time for reproductive isolation may vary considerably independent of the genetic divergence at average loci, as discussed earlier. The genetic distance between species is generally larger than 0.2 and smaller than 2. But in some pairs of sibling species such as Drosophila pseudoobscura and D. persimilis it is as small as 0.05. In general nonsibling species are genetically more divergent than sibling species (Hubby and Throckmorton, 1968; Nei, 1971a).

It is of great interest to know how many gene substitutions occur when a new species is formed. A rough estimate of the minimum number of gene substitutions that occur in speciation can be obtained from genetic distances between species. The smallest value of interspecific genetic distance so far obtained is 0.05 between D. pseudoobscura and D. persimilis. The next smallest value is 0.06 between D. victoria and D. lebanonensis (see Nei, 1975). Therefore, if electrophoresis detects only a quarter of codon differences, the actual number of codon differences is estimated to be about 0.2 per locus, neglecting synonymous codons. If the Drosophila genome has 5,000 structural genes, this is equivalent to 1,000 codon differences per genome. If both species compared experienced an equal number of gene substitutions during speciation, about 500 gene substitutions must have occurred in each species. In practice, of course, the number of gene substitutions accompanying speciation should vary greatly with species, since the definition of species largely depends on reproductive isolation. In fact, some pairs of subspecies have a genetic distance of as large as 0.35 (e.g. the U.S. mainland and the Bimini Island populations of the lizard Anolis carolinensis). This would correspond to 1.0 codon differences per locus. Yet, they have not developed a complete reproductive isolation.

Gene differences between different genera have been studied only in a few organisms. The data in the Sciaenidae in fish indicate that intergeneric genetic distance is still larger than interspecific distance. In all cases examined, the D value was larger than 1 . In one of the twelve intergeneric comparisons studied no common proteins were shared by the two genera.

It is now clear that genetic distance generally increases as the rank of taxa compared becomes higher, as expected. The genetic distance between man and chimpanzee is, however, inconsistent with this general trend. They belong to different families, but the distance is only 0.62 (King and Wilson, 1975), which corresponds to the interspecific genetic distance in other organisms. There are three possible explanations for this relatively small value of D . First, primates have been considerably oversplit relative to other groups as a simple result of anthropocentrism. Second, morphological differences between species in other taxa are not as easily distinguishable as differences between primates. Third, for a given amount of change at the gene level there has been more morphological and behavioral change between man and chimpanzee than between species in other organisms. Arguing that the actual morphological differences between man and chimpanzee are much larger than those between species of mouse, lizards, and Drosophila, King and Wilson prefer the third explanation.

As mentioned earlier, the genetic distance D is expected to increase linearly with time, i.e. $D = 2\alpha t$. Therefore, if we know the rate of gene substitution, α , then we can estimate the divergence time from D . From the rate of amino acid substitution in some proteins in evolution, α has been estimated to be 10^{-7} per locus per year for electrophoretically detectable alleles (Kimura and Ohta, 1971; Nei, 1975). Thus, the time after divergence between two taxa can be estimated by

$$t = 5 \times 10^6 \, D \text{ years} \tag{26}$$

approximately. Application of this formula for the evolution

of human major races and of the cave fish <u>Astyanax</u> <u>mexicanus</u>
has given rather reasonable divergence times which are consistent with geological evidence (Nei and Roychoudhury, 1974a;
Chakraborty and Nei, 1974). A similar conclusion was also
obtained by Nevo <u>et</u> <u>al</u>. (1974) in gophers. Nevertheless, it
should be emphasized that our estimate of α depends on a
number of assumptions and these assumptions have to be
examined carefully in the future.

It should also be noted that formula (26) tends to give an
underestimate of t when D is large, say, larger than 0.5.
There are two reasons for this. First, as the number of
mutations increase, the difference in net charge of a protein
between the taxa, which is induced by a certain amino acid
substitution in one of the two taxa, may be cancelled out by
a second amino acid substitution occurring in the same species
or the other. A correction for this factor can be made by
the method given by Nei and Chakraborty (1973). Second, if
the rate of amino acid substitution varies with protein, formula (26) is no longer correct. If the mean and variance of
2α are $\overline{2\alpha}$ and $V_{\overline{2\alpha}}$, then $D \equiv -\log_e I = \overline{2\alpha} - \log_e (1 + V_{\overline{2\alpha}}/2)$
approximately (Nei, 1971a). Unfortunately, the magnitude of
$V_{\overline{2\alpha}}$ is not known at the present time.

Genetic distance is useful for constructing a phylogenetic
tree (Cavalli-Sforza and Edwards, 1964). For this purpose,
the standard distance seems to be most appropriate, since it
is proportional to divergence time. I have used this distance for constructing a phylogenetic tree in the <u>virilis</u>
group of Drosophila species (Nei, 1971a). The results obtained were consistent with the cytological phylogeny by
Stone <u>et</u> <u>al</u>. (1960). Yang <u>et</u> <u>al</u>. (1974) also used this distance for constructing a tree in the <u>Anolis</u> <u>roquet</u> group of

lizard species. The results obtained were quite consistent
with the geological history of the formation of the Lesser
Antilles in the Caribbean Sea, where the lizards live.
Dr. G. C. Gorman at the University of California at Los Angeles has informed me that the tree produced from the D value
is virtually the same as that produced by using Sarich and
Wilson's (1967) immunological distance.

Making a phylogenetic tree is the first step to study the
evolutionary process of related species. If we know this, we
can then study what kinds of genetic changes were important
in creating a new species or a new group of species. We will
also be able to estimate the rate at which a certain morphological and physiological character has evolved. In this
area, however, both biological and statistical techniques are
far from complete. For a group of distantly related species,
data on amino acid sequences in some proteins and immunological distance are very useful, while in closely related
species electrophoretic data seem to be better. But electrophoresis detects only those amino acid substitutions that
change the net charge of a protein. Therefore, estimates of
the number of gene substitutions based on electrophoretic
data are not very accurate. It is hoped that in the future a
more refined method such as heat denaturation in combination
with electrophoresis (Bernstein et al.,1973) will be developed
and widely used. The current statistical technique of constructing a tree is also not very satisfactory in several
aspects, as pointed out by a number of authors. For example,
for a given matrix of genetic distances we do not know the
best way to construct a tree. Clearly, more intensive
studies have to be made in this area.

SUMMARY

(1) Mathematical models for the evolution of postmating reproductive isolation and ethological isolation are developed. The models developed support the Dobzhansky-Muller hypothesis that postmating reproductive isolation evolves by fixation of mutant genes affecting hybrid inviability or sterility through their pleiotropic effects on morphological and physiological characters. Of course, this is not proof of the Dobzhansky-Muller hypothesis. It may also evolve by random fixation of mutant genes in small populations. While evolution of ethological isolation may be aided by natural selection through hybrid inviability and sterility, as argued by Fisher and Dobzhansky, it is not a requirement. It may well be developed without this type of selection. Here random events (mutation and genetic drift) seem to play an important role.

(2) Formulae for the distribution and moments of fixation time of mutant genes under recurrent mutation pressure are derived. These formulae suggest that the mean fixation time required for the development of reproductive isolation is of the order of reciprocal of mutation rate, if the genes concerned are selectively neutral and the original population does not have any mutant genes. If the mutant genes have selective advantage through their pleiotropic effects, the mean fixation time decreases considerably. Yet the effect of mutation rate remains very important. The distribution of fixation time has a wide variation. The wide variation of evolutionary time between hybridizable pairs of frog species recently observed by Wilson et al. (1974) is consistent with this theoretical expectation.

(3) The degree of gene differentiation among subpopulations relative to the total population may be measured by Nei's coefficient of gene differentiation (G_{ST}) , an extension of Wright's F_{ST} to the case of multiple alleles and a finite number of subpopulations. The theoretical expectation of this coefficient under the effect of mutation and random genetic drift is derived. While it is a good measure of the degree of gene differentiation among subpopulations, it is not a simple function of evolutionary time except in the early generations.

(4) Mathematical models of gene differentiation between two isolated populations are discussed, taking into account the effects of mutation, genetic drift, and selection. It is argued that genetic distance should be defined as a measure of the gene differences between populations rather than a geometric distance between populations represented in an Euclidean space. If we measure genetic distance in terms of the number of gene substitutions (or codon differences) per locus, it can be directly related to the evolutionary process. For a measure of genetic distance to be useful for constructing a phylogenetic tree, it should be linearly related to evolutionary time. From this point of view, the relationships between evolutionary time and a number of distance measures were studied by computer simulation, taking into account the effects of mutation and genetic drift. It is also indicated that in the estimation of genetic distance a large number of loci which are ideally a random sample of the genome should be used, since the interlocus variance of genetic distance is very large.

(5) Interracial and interspecific genetic distances in various organisms were computed by using published data on

gene frequencies for protein loci. The genetic distance obtained (Nei's standard distance) is generally 0.00 ~ 0.05 between races, 0.02 ~ 0.20 between subspecies, 0.1 ~ 2.0 between species, and more than 1 between genera. Thus, genetic distance increases as the rank of taxa compared becomes higher, as expected. The minimum number of gene substitutions accompanying speciation was estimated to be about 500 in Drosophila. Problems arising in the estimation of evolutionary time and reconstruction of phylogenetic trees from genetic distance estimates are discussed.

REFERENCES

Ayala, F.J. and M.L. Tracey. (1974). Proc. Nat. Acad. Sci. U.S. 71: 999-1003.

Ayala, F.J., M.L. Tracey, L.G. Barr, J.F. McDonald and S. Perez-Salas. (1974). Genetics 77: 343-384.

Balakrishnan, V. and L.D. Sanghvi. (1968). Biometrics 24: 859-865.

Bernstein, S.C., L.H. Throckmorton and J.L. Hubby. (1973). Proc. Nat. Acad. Sci. U.S. 70: 3928-3931.

Bhattacharyya, A. (1946). Sankhya 7: 401-406.

Carson, H.L. (1970). Science 168: 1414-1418.

Cavalli-Sforza, L.L. (1969). Proc. 12th Interl. Cong. Genet. (Tokyo), 3: 405-416.

Cavalli-Sforza, L.L. and A.W.F. Edwards. (1964). In: Genetics Today, Proc. 11th Interl. Cong. Genet. (The Hague), Pergamon Press, Oxford, 923-933.

Cavalli-Sforza, L.L. and A.W.F. Edwards. (1967). Am. J. Human Genet. 19: 233-257.

Chakraborty, R. and M. Nei. (1974). Theor. Pop. Biol. 5: 460-469.

Crow, J.F. and M. Kimura. (1965). Am. Natur. 99: 439-450.

Crow, J.F. and M. Kimura. (1970). An Introduction to Population Genetics Theory. Harper, New York.

Dobzhansky, Th. (1937). Genetics and the Origin of Species. Columbia Univ. Press, New York.

Dobzhansky, Th. (1940). Am. Natur. 74: 312-321.

Dobzhansky, Th. (1970). Genetics of the Evolutionary Process. Columbia Univ. Press, New York.

Fisher, R.A. (1930). The Genetical Theory of Natural Selection. Clarendon Press, Oxford.

Fukatami, A. and D. Moriwaki. (1970). Japan. J. Genet. 45: 193-204.

Haldane, J.B.S. (1927). Proc. Cambridge Philos. Soc. 23: 838-844.

Hedrick, P.W. (1971). Evolution 25: 276-280.

Hubby, J.L. and L.H. Throckmorton. (1965). Genetics 52: 203-215.

Hubby, J.L. and L.H. Throckmorton. (1968). Am. Natur. 102: 193-205.

Johnson, W.E. and R.K. Selander. (1971). Systemat. Zool. 20: 377-405.

Kimura, M. (1970). Genet. Res. 15: 131-133.

Kimura, M. and J.F. Crow. (1964). Genetics 49: 725-738.

Kimura, M. and T. Ohta. (1971). Nature 229: 467-469.

King, M. and A.C. Wilson. (1975). Science 188: 107-116.

Knight, G.R., A. Robertson and C.H. Waddington. (1956). Evolution 10: 14-22.

Koopman, K.F. (1950). Evolution 4: 135-148.

Latter, B.D.H. (1972). Genetics 70: 475-490.

Latter, B.D.H. (1973). Am. J. Human Genet. 25: 247-261.

Lewontin, R.C. (1974). The Genetic Basis of Evolutionary Change. Columbia Univ. Press, New York.

Li, W.H. and M. Nei. (1975). Genet. Res. (In press).

Malécot, G. (1948). Les Mathématiques de l'hérédité. Masson et Cie, Paris.

Muller, H.J. (1939). Biol. Rev. Camb. Phil. Soc. 14: 261-280.

Muller, H.J. (1942). Biol. Symp. 6: 71-125.

Nei, M. (1971a). Am. Natur. 105: 385-398.

Nei, M. (1971b). Genetics 68: s47.

Nei, M. (1972). Am. Natur. 106: 283-292.

Nei, M. (1973a). Proc. Nat. Acad. Sci. U.S. 70: 3321-3323.

Nei, M. (1973b). In: Genetic Structure of Populations.
N.E. Morton, Ed. Univ. Press of Hawaii, Honolulu, 45-54.

Nei, M. (1975). Molecular Population Genetics and Evolution.
North-Holland, Amsterdam.

Nei, M. and R. Chakraborty. (1973). J. Molec. Evol. 2:
323-328.

Nei, M. and M.W. Feldman. (1972). Theor. Pop. Biol. 3:
460-465.

Nei, M. and A.K. Roychoudhury. (1972). Science 177: 434-436.

Nei, M. and A.K. Roychoudhury. (1973). Am. Natur. 107:
362-372.

Nei, M. and A.K. Roychoudhury. (1974a). Am. J. Human Genet.
26: 421-443.

Nei, M. and A.K. Roychoudhury. (1974b). Genetics 76:
379-390.

Nevo, E., Y.J. Kim, C.R. Shaw and C.S. Thaeler, Jr. (1974).
Evolution 28: 1-23.

O'Donald, P. (1963). Heredity 18: 451-457.

O'Donald, P. (1967). Heredity 22: 499-518.

Oka, H. (1974). Genetics 77: 521-534.

Prager, E.M. and A.C. Wilson. (1974). Proc. Nat. Acad. Sci.
U.S. 72: 200-204.

Prakash, S. (1972). Genetics 72: 143-155.

Rogers, J.S. (1972). In Studies in Genetics VII. (Univ. of
Texas Publ. No. 7213): 145-153.

Sarich, V.M. and A.C. Wilson. (1967). Science 158:
1200-1203.

Selander, R.K. and W.E. Johnson. (1973). Ann. Rev. Ecol.
Systemat. 4: 75-91.

Stone, W.S., W.C. Guest and F.D. Wilson. (1960). Proc. Nat.
Acad. Sci. U.S. 46: 350-361.

Wilson, A.C., L.R. Maxson and V.M. Sarich. (1974). _Proc. Nat. Acad. Sci. U.S._ 71: 2843-2847.

Wright, S. (1943). _Genetics_ 28: 114-138.

Yang, S.Y., M. Soulé and G.C. Gorman. (1974). _Systemat. Zool._ 23: 387-399.

Yasuda, N. (1968). _Am. J. Human Genet._ 20: 1-23.

The Rate of Spread of an Advantageous Allele in a Subdivided Population

M. SLATKIN

The rate of spread of an advantageous trait through a geographically structured population will determine the extent
to which the population can be expected to be homogeneous in
its genetical properties. If the rate is high then a trait
in one part of the population's range would usually be found
throughout the range unless it is selected against in some
areas. However, if the rate is low and of the same order of
magnitude as ecological and geological changes affecting the
population, then there are likely to be local differences
even though the population would be uniform if the equilibrium state were reached. Those genetic differences could
provide the basis for speciation or adaptation to local conditions. I will investigate here the consequences of a simple model which predicts the rate of spread of an advantageous allele through a subdivided population. The model
takes into account the fact that when a new allele arrives in
a population it has only a small chance of ultimately becoming fixed even though it has some selective advantage. I
will show that, under many circumstances, the rate of spread
of an advantageous allele due to local migration alone can be
very small but can be greatly increased by a small amount of

long distance migration.

Consider a diploid species with non-overlapping generations and a single locus with two possible alleles, A and a . If there are two isolated, panmictic populations exchanging m individuals each generation and if A has a selective advantage in both of the populations, then the only stable equilibrium state for the locus is complete fixation of the A allele. The quantity of interest is the average time until the equilibrium is reached. While an exact calculation of the rate of approach to equilibrium would involve the solution to a complex diffusion equation and would be quite difficult, some useful information can be obtained by making some simple approximations.

Assume that in one of the two populations A starts with an initial frequency, p_0 , and that in the other it is absent. Initially, $p_0 = 1/2N$, where N is the population size (assumed to be the same for both populations), when A is a new mutant or when A has just arrived. If there is no dominance at the locus and if A has a selective advantage s over a , then the probability that A will ultimately be fixed in the first population is 2s (Fisher, 1958; Kimura, 1962) when 2Ns >> 1 and 2s << 1 . We shall assume that N is large enough that if A is ultimately fixed, its time course is according to the differential equation

$$\frac{dp(t)}{dt} = sp(t)(1-p(t)) \tag{1}$$

where $p(t)$ is the frequency of A at t given that $p(0) = 1/2N$. That is, the effects of genetic drift are ignored and the only effect of the finiteness of the population is the determination of the initial frequency. This approximation greatly simplifies the analysis; its validity will be discussed below. Solving (1), we get

$$p(t) = \frac{1}{(1+(2N-1)e^{-st})} \quad . \tag{2}$$

As $\underline{A}$ increases in frequency in the first population, it will become more likely that it will arrive at the second. However, not every arriving $\underline{A}$ allele will ultimately be fixed. We would like the distribution of arrival times of the first $\underline{A}$ allele which will ultimately go to fixation. We can compute this distribution approximately by making several simplifying assumptions which reduce the problem to one which is analytically tractable. I will then test the validity of this approach by comparing the approximate results with those obtained by simulating the same model on a computer. As I will show, the results obtained with the approximation technique are close to those from the simulation for the range of parameters for which the approximations might reasonably be expected to be valid.

If each individual in the first population has a probability $\underline{m/N}$ of migrating to the second, then the distribution of the number of arriving migrants is a Poisson distribution with mean $\underline{m}$

$$\text{prob(i migrants)} = \frac{m^i}{i!} e^{-m} \quad . \tag{3}$$

Given that there are $\underline{i}$ migrants, there are $2i$ alleles, each with a probability of $\underline{p(t)}$ of being an $\underline{A}$. Therefore, the distribution of the number of arriving $\underline{A}$ alleles given that there are $\underline{i}$ migrants is

$$\text{prob(}\underline{j}\ \underline{A}\text{ alleles} \mid \underline{i}\text{ migrants)} = \binom{2i}{j} p(t)^j (1-p(t))^{2i-j}. \tag{4}$$

Finally, if we assume that each of the arriving $\underline{A}$ alleles will be fixed or disappear independently of each other or independently of the $\underline{A}$'s already present, then each has a probability of approximately $2\underline{s}$ of ultimately being fixed and

769

(1-2$\underline{s}$) of disappearing (Fisher, 1958; Kimura, 1962). This approximation is valid if $\underline{m}$ is not large and if there is no building up of $\underline{A}$'s which would eventually disappear but have not yet done so. Kimura and Ohta (1969) have shown that the expected time to fixation of an advantageous allele which will not be fixed ($\bar{t}_0(1/2\underline{N})$ in their notation) is on the order of a few generations. Thus only a few $\underline{A}$'s would be present at any one time before fixation begins. The assumption of the non-independence of the newly arriving $\underline{A}$ alleles would be violated if there were a significant number of $\underline{AA}$ homozygotes formed where one of the $\underline{A}$'s was newly arrived and the other was present already. These homozygotes would increase the likelihood of fixation of a new allele from $2\underline{s}$. Clearly the frequency of these homozygotes would decrease with total population size as long as $\bar{t}_0(1/2\underline{N})$ increases less rapidly than linearly with $\underline{N}$, as is implied by the diffusion result of Kimura and Ohta (1969). Thus we would expect this approximation to be best when $\underline{m}$ is on the order of 1 or 2 or less and $\underline{N}$ is large.

If each of the $\underline{j}$ newly arrived $\underline{A}$ alleles has a probability of 1-2$\underline{s}$ of ultimately not being fixed, then the probability, $\underline{q}$, that fixation will not begin in generation t is

$$q(t) = (1-2s)^j \; . \tag{5}$$

The expectation of $\underline{q}$ is found by averaging over $\underline{j}$ using (4) to get

$$E[q(t)|i \text{ migrants}] = (1-2sp(t))^{2i} \; ,$$

and then averaging over $\underline{i}$ using (3) we find

$$E[q(t)] = \exp(-4msp(t)+4ms^2p(t)) \; . \tag{6}$$

Since we will consider only those cases in which $\underline{s}$ is small, we can ignore the second term in the exponent of (6).

In order for fixation to begin in generation $\underline{t}$, it must have not begun in any of the previous $t-1$ generations. Thus $\underline{F(t)}$, the probability that fixation of $\underline{A}$ begins in the second population in generation $\underline{t}$ is approximately

$$E[q(0)]\cdot E[q(1)] \ \ldots \ E[q(t-1)]\cdot (1-E[q(t)])$$

$$= (1-e^{-4msp(t)})\exp\ (-\sum_{t'=0}^{t-1} 4msp(t')) \tag{7}$$

or

$$F(t) = 4msp(t)\exp(-\int_{0}^{t} 4msp(t')dt') \tag{8}$$

where the sum can be replaced by the integral when s is small.

If we denote the first and second moments of $\underline{F(t)}$ by $\overline{t}$ and $\overline{t^2}$ then,

$$\overline{t} = \int_{0}^{\infty} tF(t)dt = \int_{0}^{\infty} e^{-4ms \int_{0}^{t} p(t')dt'} \, dt \tag{9}$$

and

$$\overline{t^2} = \int_{0}^{\infty} t^2 F(t)dt = 2 \int_{0}^{\infty} te^{-4ms \int_{0}^{t} p(t')dt'} \, dt \tag{10}$$

after integrating by parts. Substituting for $\underline{p(t)}$ from (2), (9), and (10) reduce to

$$\overline{t} = \int_{0}^{\infty} (p(0)e^{st}+1-p(0)^{-4m}dt \tag{11}$$

and

$$\overline{t^2} = 2 \int_{0}^{\infty} t(p(0)e^{st}+1-p(0))^{-4m}dt \ . \tag{12}$$

While I could find no simple reduction of (11) and (12), after a change of variables the denominator of the integrand in each can be expanded in a series to yield

$$\overline{t} = \frac{1}{s} \sum_{i=0}^{\infty} \frac{(1-p(0))^i}{(n+4m)} \tag{13}$$

and

$$\overline{t^2} = \frac{1}{s^2} \sum_{i=0}^{\infty} \sum_{j=0}^{\infty} \frac{(1-p(0))^{i+j}}{(i+4m)(i+j+4m)} \ . \quad (14)$$

The sums in (13) and (14) in general have to be evaluated on a computer, but there is some information which can be obtained directly. If $4\underline{m}$ is an integer, then (13) can be summed to give

$$\overline{t} = \frac{1}{s}[\frac{1}{4m} + \frac{1}{(1-p(0))} \ 4m \ (\ln \frac{1}{p(0)} - \sum_{i=1}^{4m} (1-p(0))^i)] \quad (15)$$

(Mangulis, 1965, p.73). For small $\underline{m}$, at least, we can see that $\overline{t}$ varies roughly inversely with $\underline{m}$ and with the logarithm of $2\underline{N}$. Values for $\overline{t}$ obtained by using (13) are shown in Table 1, part A, for different values of $\underline{m}$ and $\underline{N}$.

The variance (σ^2) in the time until fixation begins can be computed by subtracting the square of (13) from (14) to obtain

$$\sigma^2 = \frac{1}{s^2} \sum_{i=0}^{\infty} \sum_{j=0}^{\infty} \frac{(1-p(0))^{i+j}}{(i+4m)(j+4m)} \ \frac{4m+i-j}{4m+i+j} \quad (16)$$

Each term in the above sum is the same as the corresponding term in the sum for $\overline{t^2}$ except that it is multiplied by

$$\frac{4m+i-j}{4m+i+j} \quad (17)$$

which is always less than one. Therefore, the coefficient of variation $(\sigma/\overline{t})$ is always less than one, and it is reasonable to use the values of $\overline{t}$ shown in Table 1 as estimate of actual pattern of change in gene frequencies. For the values of $\underline{m}$ and $\underline{N}$ shown in Table 1, part A, most of the coefficients of variation were about .7 to .8 with a few smaller.

The results presented in Table 1, part A, were based partly on the assumption that once $\underline{A}$ starts to increase in the first population $(\underline{t}=0)$ it will increase deterministically

772

according to equation (2). We can test that assumption by comparing the times to fixation using (2) with those found by Kimura and Ohta (1969) who used a diffusion approximation and a Monte Carlo simulation to solve the same problem. The time to fixation using (2), t_f , is found by solving (2) for t with $p(t_f) = 1-1/2N$. Thus

$$t_f = \frac{1}{s} \ln\left(\left(\frac{1}{p(t_f)} -1\right) \frac{p(0)}{1-p(0)}\right) \quad . \tag{18}$$

Kimura and Ohta (1969, Figure 2) show their results for the expected time until fixation of an advantageous allele given that it will ultimately be fixed $(\bar{t}_1(p))$. A comparison of their results (obtained by estimated values from their graph) with those of equation (15) is shown in Table 2. Their selection coefficient was divided by 2 to be comparable to the case here because of a difference in the specification of the model.

From Table 2, we infer that when $Ns \gg 1$, the case considered here, equation (18) provides an underestimate of the times until fixation. If that is the case, then the estimate of the time lag between the beginning of fixation in populations 1 and 2 using equation (2) is also an underestimate. Ewens (1963, 1964) and Karlin and Levikson (1974) show that the diffusion approximation overestimates the expected time until fixation but they do not consider the expected time until fixation conditioned on fixation of the advantageous allele. In the absence of further information on the extent to which the diffusion technique overestimates $\bar{t}_1(1/2N)$ we must rely on Kimura and Ohta's (1969) results.

I carried out a computer simulation study of the model of migration between two populations to determine the accuracy of the approximations used here. The steps in the simulation were as follows:

(1) The $\underline{A}$ allele increased in the first population according to equation (2).

(2) If $\underline{m} > 1$, m migrants were chosen from population 1 every generation. If $\underline{m} < 1$, then a random number in $(0,1)$, $\underline{z}$, was generated. If $\underline{z} > \underline{m}$, there was no migrant that generation, otherwise there was one.

(3) The genotypes of both the migrant to population 2 and the individual it replaced was determined by comparing random numbers in $(0,1)$ with the appropriate genotypic frequencies.

(4) The gametic frequencies in population 2 were calculated by counting the genotypes after migration. Individuals to make up the next generation were generated one at a time by sampling with replacement from the gametic pool. If the individual generated was an $\underline{AA}$ genotype, it was assumed to reach the adult stage. If it was $\underline{Aa}$ or $\underline{aa}$, then a new random number was generated and compared with $1-\underline{s}$ or $1-2\underline{s}$ to see if it should be kept or discarded, thus modelling the proper selection scheme.

The results from the simulation are shown in Table 1, part B, for some of the values of the parameters. The time of beginning of fixation was estimated by recording the time of the first crossing of some threshold of gene frequencies. The values shown are for a threshold of 0.1 but other values for the threshold did not produce very different sorts of results. For the simulation results in the coefficients of variation were always on the order of .35 or less.

Comparing the two parts of Table 1, we see that the approximate results obtained from (13) are not too far from those of the simulation, particularly when Ns > 1 . The simulations indicate that there is greater dependence on population size than implied by equation (15). However, the simulations confirm the fact that there usually is a long delay before fixation begins in the second population. Another way of looking at the delay is to compare the average time until fixation begins with the time at which the frequency in population 1 is .5 . This time can also be found from (18) to be $\ln(2N-1)/s$ or 459, 529, 621, 691 for N=50, 100, 250 and 500 and s = 0.01 . Especially when Ns > 1 , we can see that fixation does not begin until p(t) is almost .5 in population 1.

So far we have considered only a pair of populations but we can extend the results to a one or two dimensional array. For a one dimensional stepping stone model, if we assume that there is exchange of individuals only between adjacent populations, then the average time between the start of fixation of A in one population and in another population n steps apart is n t̄ . The advantageous allele would be progressing through the collection of populations in a wave-like fashion with the wave "velocity" equal to the interpopulation distance divided by t̄ . In this case, when there is only local gene flow, there would always be a sharp gradient between the regions where A is and is not fixed.

We can compare this result with that obtained by Fisher (1937) and Hadeler (1975) from a continuum model of the same problem. Fisher also found that a relatively sharp gradient would exist. However, the wave velocity found with the present model is considerably slower than in Fisher's model, as long as the interpopulation distance in the present model is of the same order of magnitude as the mean migration distance used by Fisher. For example, Fisher (1937, p.365) finds that if the average migration distance is 100 yards and s = 0.01 , the wave velocity is approximately 14 yards per generation. With the present model, if we assume that the interpopulation distance is 500 yards, $m=1$, and $N=500$, then from Table 1 we find that the velocity is roughly one yard per generation. If, as mentioned before, the assumption that the increase in A in the first population is approximately deterministic results in an overestimate in A's rate of increase, then the actual velocity would be smaller. Of course, if this results in an underestimate then the velocity would be correspondingly larger. Clearly the results here are quite sensitive to the actual time course of A once it begins to increase in frequency and a more careful analysis is required.

The extremely slow rate of progress of an advantageous allele found above is partly a result of the fact that there is assumed to be migration only between adjacent populations. If long distance migration is also considered, the results can be quite different. For example, with $m=1$, $N=500$ and $s=0.01$, the advantageous allele would take slightly more than 10,000 generations to reach a population 20 steps from the one in which it originated. If the frequency of migrant individuals reaching a population 20 steps away is 1% of those reaching adjacent populations ($m=0.01$) , then the advantageous allele would reach a population 20 steps away in roughly 3,200 generations as a result of this long distance migration. With other choices of the parameters, the differences in effect between long and short distance migrants can be even greater. An occasional long distance migrant individual can be very much more effective in distributing an advantageous allele than a larger number of short distance migrants.

The potentially greater importance of long distance migrants in this model is in contrast to Fisher's (1937) result, in which only the mean migration distance was significant. It is also in contrast to the results found for the equilibrium shape of a cline maintained by migration and spatially varying selection (Slatkin, 1973). This result is unfortunate in some ways since the frequency of rare, successful, long distance migrants is extremely difficult to measure. Some estimate of the importance of long distance migrants might be obtained by measuring the steepness of the cline at the advancing wave front. The less steep the cline is, presumably the more common are the longer distance migrants, although a quantitative analysis of the problem would be very complex.

In a two dimensional array of populations, the rate of spread of an advantageous allele would be somewhat increased because there would be more pathways that the allele could take between two populations. That is consistent with results obtained from other models which compare the effect of migration in one and two dimensions (Nagylaki, 1975). However, it is reasonable to assume that rate of spread of the advantageous allele is determined primarily by the lag time associated with the movement of the allele between adjacent populations, $\bar{t}$. Therefore, the relative importance of local and long distance migrants would not be greatly different from the one dimensional case.

The conclusion that can be drawn from the model presented here is that the rate of approach to equilibrium in a subdivided population can be very long, on the order of thousands or tens of thousands of generations. This is true even when nearby populations exchange one or more migrants per generation. Therefore the possibilities for genetic divergence at the extremes of the population's range can be great even if the population would be genetically uniform at equilibrium. The equilibrium may never be reached because of new advantageous alleles possibly serving the same function arising in different parts of the range. Occasional, long distance migrants can have a great effect in reducing the chance for divergence.

A problem of interest in population genetics is the patterns in geographic distribution of selectively neutral and advantageous alleles. This problem is greatly complicated by the fact that many of the populations for which allele frequencies can be measured may not be at equilibrium. At equilibrium, spatial differences in selection pressures can produce a sharp gradient or cline in allele frequencies. The results here suggest that relatively sharp gradient would also result before an advantageous allele is fixed throughout a population's range. This is in contrast to the results found for the geographic patterns in the distribution of neutral alleles. Kimura and Maruyama (1971) and Maruyama and Kimura (1974) found that, under a wide variety of conditions, the loci with neutral alleles will be approximately uniform through a population's range. In general, no steep gradients would exist. Slatkin and Maruyama (1975) find a similar pattern for the rate of accumulation of genetic differences due to genetic drift. Widely separated populations do not diverge much more rapidly than do adjacent populations. Therefore, during the approach to equilibrium, sharp gradients in allele frequency would not be expected.

Thus, if the present results are supported by more exact models, then there may be some hope of identifying some characteristic "signature" of geographic patterns resulting from selective and non-selective mechanisms. In many species the pattern of local migration can be determined by mark-release-recapture studies. These data could then be used to put limits on the gradients in gene frequencies expected under different assumptions, possibly leading to the elimination of some of the explanations for the observed patterns.

ACKNOWLEDGEMENTS

I wish to thank R. May for a helpful discussion of this topic, J.D. Cowan for assistance with the analysis, and various participants at the Conference for valuable comments on the material described here. The research was supported by AEC Contract No. AT(11-1)-2467.

REFERENCES

Ewens, W.J. (1963). Biometrica 50: 241-249.

Ewens, W.J. (1964). J. Appl. Prob. 1: 141-156.

Fisher, R.A. (1937). Ann. Eugenics 7: 355-369.

Fisher, R.A. (1958). The Genetical Theory of Natural Selection. Second Edition. Dover Press, New York.

Karlin, S. and B. Levikson. (1974). Theor. Pop. Biol. 6: 383-412.

Kimura, M. (1962). Genetics 47: 713-719.

Kimura, M. and T. Maruyama. (1971). Genet. Res. 18: 125-131.

Kimura, M. and T. Ohta. (1969). Genetics 61: 763-771.

Mangulis, V. (1965). Handbook of Series for Scientists and Engineers. Academic Press, New York.

Maruyama, T. and M. Kimura. (1974). Nature 249: 30-32.

Nagylaki, T. (1975). Genetics (in press).

Slatkin, M. (1973). Genetics 75: 733-756.

Slatkin, M. and T. Maruyama. (1975). Amer. Natur. (in press).

TABLE 1. *Relative times until fixation of an advantageous allele begins. Actual times are obtained by multiplying entry by 1/s .*

m	N 500	250	100	50
Part A.	Approximate Results Obtained by Computing Equation (9).			
(Underlined values are the ones for which simulations were run.)				
.001	256.6	256.1	255.3	254.6
.005	56.84	56.16	55.26	54.57
.01	31.81	31.13	30.23	29.54
.02	19.25	18.57	17.67	16.98
.05	11.61	10.92	10.01	9.327
.1	8.885	8.199	7.292	6.609
.2	7.292	6.608	5.706	5.030
.5	5.911	5.233	4.345	3.687
1.0	5.088	4.417	3.548	2.916
2.0	4.347	3.689	2.850	2.254
Part B.	Results From a Simulation Study of the Same Model as Described in the Text. s=0.01			
0.1	9.181	6.982	5.029	3.431
0.5		5.126	3.038	2.323
1.0	5.678	4.778	2.904	1.546
2.0	5.176	4.328	2.342	1.116

TABLE 2. *The average time to fixation of an allele with selective advantage s using (1) Equation (18) in text, (2) Diffusion equation approximations and (3) Monte Carlo simulations both from Kimura and Ohta (1969, Fig. 2, p.768). $N = 10$, $p_o = 0.1$.*

	(1)	(2)	(3)
s=0.05	102.8	36	32
0.1	51.4	32	34
0.15	34.3	27	28
0.2	25.7	25	26
0.25	20.6	20	22
0.3	17.1	18	20

The Three Locus Model with Multiplicative Fitness Values: The Crystallization of the Genome

C. STROBECK

1. INTRODUCTION

One of the frequently made criticisms of the theory of population genetics is that it considers only a few genes instead of the total genome. In order to get around this objection, Monte-Carlo simulations have been made using a relatively large number of genes. Such studies can be used to gain insight into the dynamical and equilibrium behaviour of a large genetical system.

In their simulations of a large number of genes in a small length of chromosome, Franklin and Lewontin (1970) observed that first there appears "a localized increase in linkage disequilibrium which then spreads to all loci in a block". Thus the genome appears to "crystallize" around a "nucleus" of high linkage disequilibrium. The explanation they suggested for this phenomenon was that two loci in linkage disequilibrium will "interact (more) strongly with other adjacent loci" than two loci in linkage equilibrium.

In this paper the three locus model with multiplicative fitness values is used to study this "crystallization" phenomenon. This is done by comparing the stability of the equi-

781

librium point with all loci in linkage equilibrium with the stability of the equilibrium point with two loci in linkage disequilibrium and the third locus in linkage equilibrium with the other two.

2. THEORY

The recurrence equations for the general three locus model with two alleles at each locus are used. The two alleles at the three loci are denoted by A_1 and A_0, B_1 and B_0, and C_1 and C_0. The frequency of the gamete $A_iB_jC_k$ is denoted by X_{ijk} and the relative fitness of the genotype $A_iA_jB_kB_lC_mC_n$ by $a_{i+j,k+l,m+n}$. For the multiplicative fitness model

$$a_{ijk} = w_i^A w_j^B w_k^C$$

where, for instance, the relative fitnesses of the genotypes A_1A_1, A_1A_0, and A_0A_0 are w_2^A, w_1^A, and w_0^A.

The recombination value between A and B is defined to be r_1; between B and C, r_2; and between A and C, r_3.

Instead of using the gametic frequencies, which are subject to the constraint that they add to one, it is convenient to use the seven independent variables:

$$P_1 = X_{111} + X_{110} + X_{101} + X_{100}$$

$$P_2 = \frac{X_{111} + X_{110}}{X_{111} + X_{110} + X_{101} + X_{100}}$$

$$P_3 = \frac{X_{011} + X_{010}}{X_{011} + X_{010} + X_{001} + X_{000}}$$

$$P_4 = \frac{X_{111}}{X_{111} + X_{110}} \qquad P_5 = \frac{X_{101}}{X_{101} + X_{100}}$$

$$P_6 = \frac{X_{011}}{X_{011} + X_{010}} \qquad P_7 = \frac{X_{001}}{X_{001} + X_{000}}$$

For these variables the gametic frequencies are

$$X_{111} = p_1 p_2 p_4 \qquad X_{011} = (1-p_1)p_3 p_6$$
$$X_{110} = p_1 p_2 (1-p_4) \qquad X_{010} = (1-p_1)p_3(1-p_6)$$
$$X_{101} = p_1(1-p_2)p_5 \qquad X_{001} = (1-p_1)(1-p_3)p_7$$
$$X_{100} = p_1(1-p_2)(1-p_5) \qquad X_{000} = (1-p_1)(1-p_3)(1-p_7) \ .$$

(1) Linkage equilibrium between all loci

The equilibrium point with linkage equilibrium between all loci is

$$\hat{p}_1 = \frac{b_A}{a_A+b_A} \qquad \hat{p}_2 = \hat{p}_3 = \frac{b_B}{a_B+b_B}$$

$$\hat{p}_4 = \hat{p}_5 = \hat{p}_6 = \hat{p}_7 = \frac{b_C}{a_C+b_C}$$

where $a = w_2-w_1$ and $b = w_0-w_1$.

The eigenvalues of the Jacobian evaluated at this equilibrium point are

$$\lambda_1 = 1 + \frac{K_A}{1+K_A} \qquad \lambda_2 = 1 + \frac{K_B}{1+K_B} \qquad \lambda_3 = 1 + \frac{K_C}{1+K_C}$$

$$\lambda_4 = 1 + \frac{K_A K_B K_B - r_1(\frac12+K_C) - r_2(\frac12+K_A) - r_3(\frac12+K_B)}{(1+K_A)(1+K_B)(1+K_C)}$$

$$\lambda_5 = 1 + \frac{K_A K_B - r_1}{(1+K_A)(1+K_B)}$$

$$\lambda_6 = 1 + \frac{K_B K_C - r_2}{(1+K_B)(1+K_C)}$$

$$\lambda_7 = 1 + \frac{K_A K_C - r_3}{(1+K_A)(1+K_C)}$$

where

$$K = \frac{ab}{a+b} \ .$$

These eigenvalues are obtained as in Strobeck (1973) or in Roux (1974). If there is a heterozygotic advantage at all loci then λ_1, λ_2, λ_3 and λ_4 are less than one in absolute value. λ_5, λ_6 and λ_7 are the eigenvalues which would be obtained if each pair of loci were considered as an isolated two locus system (Bodmer and Felsenstein, 1967).

(2) Linkage disequilibrium between two loci

The equilibria with linkage disequilibrium cannot be determined for the two locus model with general multiplicative fitness values. However, they have been found for the symmetric model (Lewontin and Kojima, 1960; Karlin and Feldman, 1970). Therefore, it is now assumed that $w_2^A = w_0^A = w$ and $w_2^B = w_0^B = v$. If

$$r_1 < \frac{(w-1)(v-1)}{4} = K_A K_B$$

then there exists two equilibria with linkage disequilibrium with

$$\hat{p}_1 = \tfrac{1}{2} \qquad\qquad \hat{p}_2 = \tfrac{1}{2}+2D \qquad\qquad \hat{p}_3 = \tfrac{1}{2}-2D$$

$$\hat{p}_4 = \hat{p}_5 = \hat{p}_6 = \hat{p}_7 = \frac{b_C}{a_C+b_C}$$

where

$$D = \pm \tfrac{1}{4} \sqrt{1 - \frac{4r_1}{(w-1)(v-1)}} \quad .$$

These equilibria are of the type which Feldman, Franklin and Thomson (1974) designated as class 2.

(a) Stability analysis with $r_1=0$.

If $r_1=0$ the equilibria lie in a boundary. For the purpose of making a comparison it is sufficient to analyze the stability of these equilibria in the boundary. In the boundary the problem reduces to determining the stability for the two locus model with multiplicative fitness values. The fitness values at the 'AB' locus are $wv = w_2^{AB} = w_0^{AB}$ and the recombination value between the 'AB' and C loci is $r = r_2 = r_3$. The eigenvalues of the Jacobian evaluated at this equilibrium are therefore

$$\lambda_1 = 1 + \frac{K_{AB}}{1+K_{AB}} \qquad \lambda_2 = 1 + \frac{K_C}{1+K_C}$$

$$\lambda_3 = 1 + \frac{K_{AB}K_C - r}{(1+K_{AB})(1+K_C)} \quad .$$

(b) Stability analysis with $r_1 \neq 0$.

If $r_1 \neq 0$ the Jacobian evaluated at this equilibrium is $J = $ diagonal (A,B) where A is a 3×3 matrix and B a 4×4 matrix. The eigenvalues of A are

$$\lambda_1 = \frac{\tfrac{1}{2}w + \tfrac{1}{2}v + r_1}{\overline{W}}$$

and the two roots of the quadratic equation

$$\overline{W}^2 \lambda^2 - \overline{W}(wv + \tfrac{1}{2}w + \tfrac{1}{2}v)\lambda + \tfrac{1}{2}(w+v)wv + wvr_1 = 0$$

where $\overline{W} = \tfrac{1}{2}wv + \tfrac{1}{2} - r_1$. These are the same eigenvalues as obtained for the equilibrium with linkage disequilibrium in the two locus symmetric model (Bodmer and Felsenstein, 1967; Karlin and Feldman, 1970). All three eigenvalues are less than one in absolute value. The eigenvalues of B are

$$\lambda_4 = 1 + \frac{K_C}{1+K_C} \quad , \qquad \lambda_5 = \frac{(\tfrac{1}{2}w + \tfrac{1}{2}v + r_1)(1+K_C) - \tfrac{1}{2}r_1 - \tfrac{1}{2}wr_2 - \tfrac{1}{2}vr_3}{(\tfrac{1}{2}wv + \tfrac{1}{2} - r_1)(1+K_C)}$$

and the two roots of a quadratic equation too complicated to be written down here. It can be shown that $0 \leqslant \lambda_4$, $\lambda_5 < 1$ if there is a heterozygotic advantage at each locus and if $r_1 < K_A K_B$.

Although the other two eigenvalues of B are too complicated to be useful, an understanding of their behaviour is gained from the study of four special cases.

Case 1 : For the gene arrangement ABC , if $v=0$ and $r_3 = r_1+r_2-2r_1r_2$ (no interference) then the other two eigenvalues are

$$\lambda_6 = \frac{1-2r_2}{1+K_C} \quad \text{and} \quad \lambda_7 = \frac{\tfrac{1}{2}w(1+K_C-r_2) + \tfrac{1}{2}r_1(1-2r_2)}{(\tfrac{1}{2}-r_1)(1+K_C)} .$$

If there is a heterozygotic advantage at each locus and $r_1 < K_A K_B$ then $0 \leqslant \lambda_7 < 1$.

Case 2 : For the gene arrangement ABC , if $v \neq 0$ and $r_3 = r_1+r_2-2r_1r_2$ then these two eigenvalues must in general be calculated numerically. As a representative example, the larger eigenvalue of B is given in Table 1 for different values of r_1 and r_2 when $w = v = .5$ and $K_C = -.25$.

TABLE 1. *The largest eigenvalue of B for different values of r_1 and r_2 (w = v = .5 and $K_C = -.25$)*

		r_2				
	·00	·02	·04	·06	·08	·10
·01	1·192	1·150	1·108	1·066	1·024	·982
·02	1·183	1·142	1·100	1·059	1·018	·977
·03	1·173	1·132	1·092	1·051	1·011	·970
r_1 ·04	1·160	1·121	1·081	1·042	1·002	·970
·05	1·144	1·106	1·067	1·029	·991	·953
·06	1·120	1·084	1·047	1·011	·975	·938
·0625	1·111	1·076	1·040	1·004	·969	·933

Case 3 : For the gene arrangement ACB , if $w=v$ and $r_2 = r_3 = r$ and there is no interference, then the eigenvalues are

786

$$\frac{\frac{1}{4}(w+1)^2 + \frac{1}{2}(w+1)wK_C - \frac{1}{2}(w+1)r \pm D[(w-1)^2 + 2(w-1)wK_C + 2(w-1)r]}{(\frac{1}{2}w^2 + \frac{1}{2} - r_1)(1+K_C)}$$

Case 4 : If some degree of interference is assumed, then the eigenvalues must in general be calculated numerically. However, if $w=v=0$, then the eigenvalues are

$$\frac{\frac{1}{2}(1-r_2-r_3) \pm \sqrt{\frac{1}{4}(1-r_2-r_3)^2 - 4r_1(\frac{1}{2}-r_2)(\frac{1}{2}-r_3)}}{(1+K_C)(1-2r_1)}$$

3. DISCUSSION

Using the preceding examples it is possible to make some general statements about the process of the 'crystallization' of the genome. Since we are interested in the effect of linkage disequilibrium between the A and B loci on the stability of the C locus, we wish to compare the eigenvalues

$$\lambda_6 = 1 + \frac{K_B K_C - r_2}{(1+K_B)(1+K_C)} \quad \text{and} \quad \lambda_7 = 1 + \frac{K_A K_C - r_3}{(1+K_A)(1+K_C)}$$

when linkage equilibrium is assumed between all loci with the largest eigenvalue when A and B are in linkage disequilibrium and C in linkage equilibrium with A and B . Such a comparison gives information about the relative stability of the C locus in the two cases and also about the relative rates of increase of the deviations away from linkage equilibrium if unstable. (This follows since λ_6 and λ_7 have eigenvectors which only contain non-zero coordinates that involve deviations from the equilibrium at the C locus and they are the only two which can be greater than one whose eigenvectors contain any non-zero coordinates involving the deviations from equilibrium at the C locus.)

787

For the gene arrangement ABC , three statements can be made if there is no interference.

Statement 1: If both homozygotes are lethal at the B locus, (i.e., v=0) , then

$$\lambda = \frac{1 - 2r_2}{1 + K_C}$$

is an eigenvalue for both equilibria and if $\lambda \geqslant 1$ then it is the largest eigenvalue for both equilibria involving the C locus.

Thus the instability at the C locus is the same when the nearest locus is a balanced lethal. This is intuitive since with multiplicative fitness values the most linkage disequilibrium can do is to produce a 'supergene' which is a balanced lethal.

Statement 2: For $v \neq 0$, if $\lambda_6 \geqslant 1$ or $\lambda_7 \geqslant 1$, then the largest eigenvalue for the equilibrium point with linkage disequilibrium between A and B is greater than λ_6 or λ_7 .

Therefore if there is not a balanced lethal at the B locus, then the instability at the C locus is increased by linkage disequilibrium between A and B . Since the rate of increase away from linkage equilibrium is greater near a region of high linkage disequilibrium, one would expect to see a genome 'crystallize around a nucleus' of high linkage disequilibrium.

Since the eigenvalues of a matrix are continuous functions, Statement 3: For $v \neq 0$, the range of recombination values and fitness values which cause instability at the C locus is greater when there is linkage disequilibrium between A and B than when all loci are in linkage equilibrium.

The above statements are not necessarily true if some

degree of interference is assumed. This is seen by using Case 4 with $K_C = -.5$ (i.e., C is a balanced lethal) and $r_1 = .2$ and $r_2 = .24$. If there is complete interference then $r_3 = .44$. The largest eigenvalue involving the C locus if there is linkage equilibrium between all loci is 1.04 . If there is linkage disequilibrium between A and B the largest eigenvalue is .915 . Thus linkage disequilibrium can actually stabilize an equilibrium which would be unstable if there was linkage equilibrium between all loci.

For the gene arrangement ACB and with no interference, we have

Statement 4: If $\lambda_6 \geq 1$ or $\lambda_7 \geq 1$ when there is linkage equilibrium between all loci, then the largest eigenvalue when there is linkage disequilibrium between A and B is greater than λ_6 or λ_7 .

Thus again, the range of recombination and fitness values which cause instability at the C locus is greater when A and B are in linkage disequilibrium than when all loci are in linkage equilibrium.

SUMMARY

The 'crystallization' of a genome around a 'nucleus' of high linkage disequilibrium observed in the simulations of Franklin and Lewontin (1970) is studied using the three locus model with multiplicative fitness values. This is done by comparing the eigenvalues of the equilibrium point with all loci in linkage equilibrium with the eigenvalues of the equilibrium point with two loci in linkage disequilibrium and the third locus in linkage equilibrium with the other two.

It is shown that if there is no interference then the rate of increase of the deviations away from the equilibrium at the third locus is greater when the other two loci are in linkage disequilibrium than when they are in linkage equilibrium (unless the locus nearer the third locus is a balanced lethal). This greater instability of loci near a region of high linkage disequilibrium would cause the genome to appear to 'crystallize' around this region.

This is not necessarily true if there is some degree of interference. In fact, the equilibrium at the third locus can be stabilized by linkage disequilibrium at the other two loci if there is complete interference.

REFERENCES

Bodmer, W.F. and J. Felsenstein. (1967). Genetics 57: 237-265.

Franklin, I. and R.C. Lewontin. (1970). Genetics 65: 701-734.

Karlin, S. and M.W. Feldman. (1970). Theor. Pop. Biol. 1: 39-71.

Lewontin, R.C. and K. Kojima. (1960). Evolution 14: 458-472.

Roux, C.Z. (1974). Theor. Pop. Biol. 5: 393-416.

Strobeck, C. (1973). Genet. Res. 22: 195-200.

PROBLEMS, OBJECTIVES AND COMMENTS ON POPULATION GENETICS

The final session of the Conference was a tape-recorded
discussion of problems, experiments, polemics and concepts of
population theory and practice. The tapes were edited and
then amended by the individual participants. The comments
centered around aspects of the following themes.

1) The merits and demerits of global versus particular
    population genetic studies.

2) Robustness manifested in nature and expressed in
    models.

3) Meaning and uses of statistical analysis and problems
    of interpretation of data.

4) Desiratum and proposals for laboratory, field and
    theoretical work in population genetics.

5) The relationships between ecological and genetic
    variables.

6) The relevance of equilibrium studies vis a vis non-
    equilibrium phenomenon.

7) Aspects of coadaptation and speciation.

8) Problems connected with linkage equilibrium and multi-
    locus systems.

M.W. Feldman: *General and specific population genetic theory*

I want to define two classes of problems as the stuff of
theoretical population genetics. The first class comprises
general, the second specific phenomena. The mathematical
framework of subjects such as multi-locus theory, neutral
mutation-drift theory, the theory of clines, to name a few
examples, is intended to be relevant to the genetics of a
large and general class of evolving biological systems. (This
is not to impute more generality to specific models with restricted parameter sets than is their due. Rather I refer to
the model building framework involved.) These are to be contrasted with the models stimulated by phenomena specific to a
certain limited group of organisms for which special assumptions might be made and tested. Included in this latter class
might be regular inbreeding, mating or incompatibility systems,
or even overdominance applied to the study of laboratory populations of pairs of inversions in Drosophila.

The class of general models is not at present amenable to
testing in the usual statistical or even biological senses.
The preference for a specific interpretation over others is
more a matter of religion than science in these cases. Such
general models might at the present time be consigned to a
category called "evolutionary theory". In such models specific parameter values cannot be very meaningful until more is
known about the robustness of the theories to such nuisances
as random changes in the parameters.

The class of specific models on the other hand might be
amenable to statistical analysis, and might be termed
"population genetics" models. The study of evolutionary
theory might be thought of as circumscribing the possible ways
in which presently observed phenomena might have arisen. On
the other hand, the study of population genetics might involve

the statistical verification that a process is going on and the statistical elucidation of the parameters of the process. An example might be the theory of segregation distortion in Drosophila and Mouse or certain self-incompatibility mating systems in plants.

It certainly seems that few models in theoretical genetics can presently be viewed as having reached the latter stage. This is despite the fact that so much statistics is in fact done in population genetics. One of the principal problems we face is that in the attempt to reconcile the general theory with empirical observation the "scientific method" has not been observed. The sequence (1) specific observation, (2) theoretical prediction for a class of such cases, (3) further observation and validation is not a common one in population genetics. This is due on the one hand to the difficulty of reconciling the general theory with the biology, a point Ewens and I stressed in our paper. On the other hand, it is due to band-wagon biology with unprecise empirical questions behind it.

D.L. Hartl: The use of specific models in population genetics

I want to say a few words in support of looking at specific models in population genetics as opposed to general evolutionary theory. Specific models can be used to stimulate and guide experiments; the results of the experiments feed back upon the model, which is made more realistic; and the new model stimulates still more experiments.

Specific models can be employed to forge a sort of vertical integration of the interrelated phenomena that occur at distinct levels of biological organization, from molecules to ecosystems. This integration is in contrast to the sweeping, horizontal generalizations that evolutionary biologists usually seek, though broad generalizations can be extremely useful.

My greatest familiarity with the use of specific models as guides to experiments and as a framework for vertical integration involves the segregation distorter (SD) system in Drosophila melanogaster. In this case we have a type of second chromosome which causes a very large segregation bias in males, an effect which is counterbalanced by recessive lethality or male sterility of homozygous SD genotypes. The two selective effects are sufficiently large that the intensity of other selective forces on the system are likely to be minor in comparison. I have placed considerable emphasis on understanding the genetic structure of SD chromosomes. The developmental effects of SD on spermatogenesis have also been explored. The segregation bias in the case of SD is attributable to the dysfunction of a large fraction of the non-SD-bearing sperm, and morphological concomitants of the dysfunction can be observed in both the electron microscope and the light microscope. We also have some promising clues about what the dysfunction involves at the molecular level.

Specific models and a knowledge of the system at several levels have been extremely valuable in understanding the evolutionary implications of the segregation distorter system. In one recent instance, knowledge of the genetic structure of SD chromosomes led us to suspect that certain widespread dominant suppressors of SD found in natural populations were not really conventional suppressors at all, but were one of the integral genetic components of the SD system itself. This is certainly what a plausible population genetic model based on the genetic structure suggested. We have recently tested a number of dominant suppressors from natural populations, and the theoretical predictions have been borne out (Hartl and Hartung, _Evolution_, in press).

Moreover, these results have suggested that certain inversion-bearing chromosomes that suppress SD also carry the high-frequency component of the SD system, although their ability to suppress distortion had long been attributed to the inversions themselves. A recent re-analysis of these inversion-bearing chromosomes has revealed that they do indeed carry the high-frequency component of SD (Hartl, _Genetics_, in press).

Thus, in the case of SD, a specific population genetic model based on detailed information about the formal genetics of the system has led to an important discovery about the true nature of one type of suppressor segregating in natural populations. The results of the population studies then caused a reassessment of one aspect of the formal genetics, and we now have shown that chromosome pairing is not essential for distortion. The use of specific models and data from several levels of biological organization is a powerful way of looking at the world.

D. Cohen: _Robustness in nature_

The commonly and repeatedly occurring instances of parallel evolution and of converging evolution of ecological specializations, of behavioral patterns, and of the genetic systems themselves strongly suggest that evolutionary processes are very robust, at least over a limited domain. It is useful, therefore, to search for such general and robust processes in nature and to search for the robustness in the models which underlie them.

Briefly, if the behavior of a model of some natural processes depends very strongly on some rather specific assumptions about the exact form of the functional relationships
between the variables, then this type of model is unlikely to
be a good representation of nature because: (a) in the real
world it is not very likely that the relationships are exactly
those that had been assumed, and (b) random perturbations are
expected to spread the natural relationships over quite a
wide range.

S. Karlin: *Value of theoretical models*

Some comments on the value and uses of theoretical models.
A model is a suitable abstraction of reality preserving the
essential elements in such a way that its analysis affords
insights into both the original concrete situation and other
situations which have the same formal structure. The solution
of the model will generally not be strictly applicable, but
should help provide a qualitative conceptual basis in the
interpretation of experimental findings, as well as in stimu-
lating the intuition and imagination of the naturalist for his
further field studies.

Undoubtedly, a number of special models unambiguously
defined have intrinsic interest and their thorough analysis
is valuable. However, there has been a tendency in theoretical
biological studies to offer sweeping conclusions on the basis
of those particular cases where the solutions can be explicitly
calculated. These inferences at times are misleading and
present a quite limited view of the scope of the theory. It
is worthwhile to proceed, whenever possible, in line with the
modern branch of mathematics (called "global analysis") whose
emphasis is on the deduction of qualitative results for
generic models (involving general parameter values) rather

than explicit evaluations for only particular examples.

In my view, with model building one should attempt a complete classification of the formal structures elucidating functional relations between parameter specifications of the genetic and environmental factors and the resulting frequency patterns. It is while modeling and abstracting that one becomes more aware of implicit assumptions and robustness principles. The analysis of models is in essence an educating process rather than an engineering construction. Accordingly, it is not crucial to be able to estimate well the parameters of the system in order to contrast effects and distinguish qualitative outcomes and relations. I believe generally that it is a mistake to test data or extract quantitative predictions based on formulas derived from specific models. Rather, the qualitative information furnished by the relevant analytic models in conjunction with the intimate experience of the naturalist can probably better suggest suitable empirically oriented indices or scale functions for estimation purposes and the use of appropriate descriptive statistics in the interpretation of the data.

J.R.G. Turner: *Global theories of selection*

The scientific method is not equipped to deal with unique historical phenomena. In this sense it cannot cope with a unique historical process like evolution. However we can isolate certain repeatable phenomena within evolution and analyze them by the scientific method: thus obviously although each species is unique we hope that we can find repeatable phenomena in the process of speciation, and although each mutation is unique or very nearly so we hope we can find repeating processes in the evolution of nucleic acid. How do we deal with this problem when we are examining natural selection in

a sexual outbreeder? How do we isolate events which are
causal in the sense in which we are interested in them, from
events which are random in that sense when all we can observe
is a population of unique individuals, each of which is
uniquely born, suffers and dies? When we are dealing with a
single locus or a few loci this is easy, as we take the
average survival and fertility of each of the genotypes at the
interesting loci and average them across the rest of the
genome to obtain an a posteriori estimate of the fitness of
the genotypes. If we wish to analyze selection on the whole
genome then this becomes impossible and it is difficult to
see what is meant by the fitness of an individual, as it is
impossible to distinguish death or reproduction failure due to
genetic causes from that which is merely accidental. Does
this mean that ultimately there is no global scientific theory
of natural selection and that beanbag genetics or a close
approximation to it dealing with only a few loci is the best
theory that we can hope for?

S.K. Jain: *Of the status of modeling, the role of rough-and-*
ready methods, etc.

The claim that there are many repeating patterns of
variation, population structure, mating and migration systems,
or genetic distances among related taxa, has not been adequately documented. Yes, there are numerous predictions
around on the correlations among body size, generation time,
mobility, amount of variation, degree of ecological specialization, colonizing ability, and so on. It would be nice to
have large comparative sets of data, thoroughly reviewed, to
verify them. However, I find it somewhat premature to talk
about testing families of models against such patterns if
those do indeed occur widely.

Will the facts of natural history, recorded by rough-and-

ready, and impatient tools match up with the theoretician's
requirements for complexity, pace of "advancements" in the
awareness for interactions, etc.? Can we hope to find
intellectual satisfaction and funds for long-term descriptive
field studies to gather such data? Will the present evolutionary theory require any radical changes and new practical
applications with the projected theoretical developments? I
have no answers but we need to pose these questions. I raise
the practicality issue here since there are indeed some important areas such as domestication of plants and animals,
breeding for genetic improvement, biological control, and
nature conservation, which require tools of population biology.
But note an analogy here: most breeding work uses quantitative
genetics only at an elementary level whereas most theoretical
refinements call for either too large an experimental effort,
or suffer from our dilemma about the general or specific
genetic models to be used in modeling.

W. Ewens: *Theory and data analysis*

I would like to say a few words concerning the relationship
between population genetics theory and data analysis. It is
my belief that we are still in the comparatively early days
of population genetics theory, and that we still need more
theoretical development before being confident about analyzing
genetic data successfully. In my view theoreticians should,
in developing this theory, concentrate on general processes,
try to develop general theoretical findings, using (as much
as we can) model-free results, rather than being overly tied
to particular models. I feel that much current data analysis
uses theory inappropriate for the data. All too often we plug
our data into a formula developed by a certain theory, paying
practically no attention to the question of whether this theory

is appropriate for the form of data we are using. The use of
general model-free methods, to which I have just referred, is
intended as a first step in overcoming this problem. This is
one of the important points on the use of the F statistic for
testing neutrality in the conclusions of my paper with Feldman.

W.F. Bodmer: Theory and interpretation of data

In human population genetics you have to take the data that
exists, accept their limitations, and make the most that you
can out of it. You clearly must be aware of the limits in
variance relative to the model postulated. In fitting theory
to population data, obviously, sometimes you make a mistake
because you are not aware of the sorts of alternative theories
that can exist. But nevertheless I think it is very important
to try to take your theory and to consider its properties in
relation to its predictions and to see how it fits data. I
believe there are situations where one can do this successfully, and I like to think the HLA system is one. I still
offer the challenge to come up with a model that is plausible
in terms of what we actually know about the likely human
situation including for example meaningful migration rates,
which would explain the levels of persistent linkage disequilibrium observed for the HLA system in the absence of
selection. It is a blend of the quantitative and the qualitative in attempting to explain the data that one learns from
the theory. I think if we forego that opportunity and possibility, we are really throwing away much of what we do in
population genetics. What we have to ask ourselves is what
aspects of the data we can look at in that way and what critical points exist.

S. Karlin: Statistical methodology

Over the past decade the emphasis in general statistical

practice has been on descriptive analyses as contrasted with
formal procedures of testing hypothesis, estimation and
decision theoretic concepts. Recall that the Neymann Pearson
theory allows only to reject hypothesis at a level of significance and never the proof or acceptance of a hypothesis.
The concept of alternative hypothesis was mainly formalized
by decision theorists presumably to distinguish among tests.
Unfortunately, experience has revealed many of the special
classical tests and goodness of fit procedures are much too
sensitive to distribution assumptions as well as having other
innate difficulties. The use of stepwise regression schemes,
higher order interaction evaluations, and much of the standard
formal multivariate techniques have become suspect (already
largely discarded in the treatment of economic time series
and in other disciplines). These formalisms have been mostly
abandoned in favor of rough and ready methods, jackknifing
principles, index measures, scaling concepts, intuition,
educated judgments, and a general mixture of semiformal ad hoc
techniques. I would suggest that the analysis and interpretation of biological data be done in this vein.

W.G. Hill: *Some complex statistical interpretations*

As experimentalists and field workers study polymorphisms
at more loci, an increasing number of these loci will be
linked and data which gives gene frequencies at n loci also
gives roughly n^2 pair-wise frequencies, together with three
and more locus associations. The theoreticians have to find
what information about individual genes and groups of genes
can be extracted from such data, for if loci are studied only
one at a time clearly much is lost.

There is a need both for better statistical techniques and
for better theories with which the data can be confronted.
Most theory involving two or more loci has been concerned with

numbers and stability of polymorphisms at which there is
linkage disequilibrium, and it has been possible to say that
for recombinations less than some function of selective values
such disequilibrium exists. Yet these are deductions from
infinite populations and in the real world all populations are
finite. How much relevance do these infinite population
results have?

D.L. Hartl: *Welding theory and observation*

Population genetics is a field blessed with a multitude of
interesting parameters and cursed with an inability to measure
them. The problem has been enhanced by, on the one hand, the
development of an increasingly elaborate theory and, on the
other hand, the proliferation of essentially descriptive
studies of natural populations. A major gap in population
genetics is in sampling theory and experimental design. We
ought to know, for example, the optimal sampling procedure
for assessing the amount of genetic variability in a population. We ought to find out, for instance, the optimal
design of an experiment to estimate fitnesses of genotypes in
nature. We ought to decide what kinds of analyses data are
to be subjected to, and what kinds of inferences can validly
be drawn. The theory of population genetics and the observations and experiments that can be carried out in natural
populations should be much more intimately intertwined. It
should be pointed out that certain predictions of population
genetics theory, such as the direction of change of average
fitness under selection, can actually be tested in such
organisms as Drosophila melanogaster. The most important
question is whether real populations of plants and animals
behave genetically as population geneticists think they ought
to.

B. Wallace: Use and interpretations of theories

Like Turner, I would also claim that there may be no general theory of evolution, especially one that includes speciation. More precisely, I suspect that such a theory, if formulated, would rank with Calvin Coolidge's explanation of unemployment (that which occurs if large numbers of persons are unable to find jobs) or J.B.S. Haldane's statement on death ("You, reader, will die of oxygen want."). Surely there are numerous mechanisms that initiate the speciation process; any theorem that would encompass all of these would border on triviality. In my opinion, a useful general theorem of "evolution" must explain the coherence of local populations, the advantages gained by the sharing through time of recombined genomes, and the constraints accompanying the use of DNA as a basis for life. Given this understanding, we shall be able in any instance to identify and explain the incompatibilities which accompany speciation but we shall not be able to predict with accuracy when and where speciation will occur in nature.

Because of the above belief, I am not nearly as pessimistic about our knowledge of the genetics of speciation as some persons appear to be. For those features of genetics which can be seen and for which simple mechanical explanations suffice, we know a great deal; Darlington wrote The Evolution of Genetic Systems nearly thirty years ago. Because of the difficulty in their study, molecular geneticists have yet to explain the structure of gene control regions so that possible incompatibilities of these regions can be understood. Neither do we know how to predict which proteins (or other macro-molecules) can or cannot coexist in individual cells. Despite the lack of both chemical and structural information at the molecular level, I believe that concepts are already waiting to receive and interpret the details as they are provided.

In a sense these comments bear on an earlier discussion
between Karlin and Ewens over the use of statistical hypotheses. I see the Mendelian population as the object to be
studied and, in a manner of speaking, the equivalent of the
null hypothesis; separate species comprise populations that
are not Mendelian populations and for that reason demand
explanation.

E. Nevo: *Statement of some problems*

I will list several problems that are both bothersome and
sufficiently challenging for further discussion and work.
(1) Can the real biological world be explained in terms of
just a few principles or are there many principles intrinsically involved which correspond to basically different
situations? In other words, should we strive chiefly at the
discovery of universals rather than search for differences?
(2) The long debated problem of the relative importance of
non-random and random processes in the genetic structure of
finite populations has now shifted to the selectionist-
neutralist controversy of protein polymorphism and evolution.
Short of biochemical and physiological correlates, a promising
alternative to resolve the problem is to find correlations
among allozymes over populations. Yet the absence of satisfactory genetic maps in most studied organisms makes the
finding of linkage correlations a basic hindrance. We therefore remain with the challenging problem of assessing
associations between unlinked loci over different populations
as a possible potential outlet. (3) The structure of the
environment still remains a major problem. Until a satisfactory niche quantification methodology is established, the
niche-variation model, as well as any other relationship between the ecological background structure and genetic patterns

will remain a vague concept. (4) The relative roles and interaction of spatial and temporal variation in maintaining genetic polymorphisms are vaguely known at present, yet their interaction may possibly be a first order phenomena. (5) The selection-migration structure is still largely unknown in natural populations despite the essential roles they are supposed to play in both genetic differentiation of populations and speciation. (6) The problem of multi-loci versus one locus model is a challenging problem still barely touched experimentally and theoretically. (7) The theory and methodology of genetic distance between populations and species lack robustness yet genetic distance is of primary importance in assessing evolutionary relationship. (8) There is no quantitative theory of the genetics of speciation. At present, the genetic changes involved in speciation are largely unknown. Does the acquisition of reproductive isolation require "genomic revolution" as has long been advocated, or it may be achieved with very few genomic changes as recently suggested? (9) The problem of experimental design and data analysis must be critically reviewed by statisticians and mathematicians. There are available an assortment of multivariate and stepwise multiple regression techniques. We are not always cognizant of their scope of validity and implicit cryptic assumptions embedded in the method. The whole problem of testing of significance of conclusions appears to be a lax procedure. (10) The connection between haploid and diploid models are not well understood. It does not seem proper just to substitute models and study one for the other. (11) Are mathematical models of a "strategic" type meaningful and realistic or rather gross oversimplifications? (12) What is the best way of integrating mathematical and experimental models? And a final remark: testing ideas with field data is essential yet it is far at present from satisfactory implementation.

R. Koehn: *Some problems of characterizations of allelic function*

There is an era which I believe is tremendously important to major questions in population genetics, but which has received no mention in our discussions here. That is, the functional and structural characterizations of alleles from natural populations and laboratory experiments designed to test hypotheses generated from the results of these character- izations. Another way of putting this is, might we not learn more about variability in general, particularly selection mechanisms, by an approach other than large surveys of poly- morphic variation? I would like to see empirical population genetics move away from general surveys of heterozygosity, however unique the habitat of life cycle of an organism. There is great heterogeneity among loci in magnitudes of variation, physiological function of their products, organi- zation of alleles with alleles of other loci, patterns of spatial and/or temporal variation in allele frequencies, degree of interspecific differentiation, etc. To me, hetero- zygosity surveys have been largely uninformative, leading to contradictory conclusions (heterozygosity can be related to environmental stability, heterogeneity, and time). These con- tradictory explanations, I believe, are not unrelated to the diversity among the gene products being considered. Now we hear, that what we really need are more precise measurements of genic heterozygosity, which, when the data are in, will increase the power of certain theoretical models to test various hypotheses. I would suggest, that while we do indeed need accurate estimates of heterozygosity, a more fruitful approach to understanding variation is via a detailed bio- chemical examination of gene products, and not by another round of heterozygosity estimates using different techniques. Detailed functional and structural characterizations of allelic variants will provide a number of important kinds of information, such as structural uniqueness of electrophoretic components, the structural bases for unique electrophoretic behavior, magnitudes of genetic (i.e. amino acid sequence) differences among segregating alleles (both within and among species), and the magnitudes and classes of functional (i.e. phenotypic) diversity among segregating alleles. We desperately need information on the relationship between magnitudes of genetic change and phenotypic diversity among the products of a locus. This knowledge should lead us to a better understanding of the forces that promote and maintain population variation.

D.L. Hartl: Long-term population studies

An understanding of various evolutionary phenomena would be greatly enhanced by long-term, generation-by-generation studies of chromosome frequencies and population size in particular local populations. Ideally, the local populations will be of a species in which the migration rates are exceedingly small, or at least estimable. A prototype of this kind of study is the famous Fisher and Ford (1947, Heredity 1: 143-174) report of the year-by-year frequencies of the medionigra color morph and population sizes in a natural population of Panaxia dominula. Studies of this kind would require considerable commitment, a careful choice of the species, the local populations, the loci to be studied, and a reliable source of long-term funds. Although the monitoring of chromosome frequencies and population size generation after generation after generation might get to be something of a bore, such studies are not markedly more tedious in this respect than long-term artificial selection experiments. Moreover, data of this sort are badly needed, and the importance of the information would well repay the effort. These kinds of studies are perhaps the only ones that will permit an assessment of the variability of selective values and the importance of random drift to be made. The importance of this kind of study can be judged super-ficially by the fact that Fisher and Ford's paper was cited more than 30 times in the literature in the decade 1964-1974. Thus, there seems to be considerable emphasis on their results. The obvious danger in this case is one of putting altogether too much emphasis on a single set of data, a set that may not be representative of what happens in other populations.

J.A. Beardmore: *Some challenging problems in population genetics*

I would like to make one or two comments on some basic things in population genetics, about which we are still remark-ably ignorant. One of these is the spontaneous rate of mutation in terms of both mean and variance in range of life forms. It has been thought by many people that we are well informed on these matters but this is simply not so. Another important problem is the genetic architecture of the gene pool as it relates to quantitative characters. Quite a lot of work of course, has been done on this but in fact our knowledge of the number of loci, the relative effects of loci, and the location of loci, in any precise terms, is not very good. If we are going to talk about selection on phenotypes, and many of these phenotypes have to do with quantitative characters, this ignorance is unfortunate. Another matter is linkage dis-equilibrium in the strict sense, that is to say when it con-cerns loci on the same chromosome pair. There is an astoni-shing variety of opinion as to the real extent of such linkage disequilibrium, and its importance. There is one factor common to the three topics I have mentioned, namely that it is

laborious to find out anything about them in detail. They
are all topics of the sort in which projects have to be thought
of in terms of man-years, and on the whole we are not prepared
to do it, because it's a grinding process. We have therefore,
to be prepared as experimentalists, to sit down and do these
basic but very tedious experiments, or we have to search for
new and more efficient methods of getting these answers, which
are needed for progress in understanding basic evolutionary
processes. Some other more or less important areas in which
more critical basic data are needed are the genetic effects of
stabilizing selection, the role of intra-allelic recombination
in generating new alleles and the extent of cryptic genetic
variation within single electrophoretic allele classes.

S.K. Jain: *Problems of evolution of breeding systems, notion*
of optimality, and role of models

Since very little was said at this conference about the
evolution of inbreeding versus outbreeding, and the notion of
optimal strategies in evolving genetic systems, perhaps this
brief review of an outstanding problem would be of some
interest. Based on the experience of Darwin and many breeders,
earlier evolutionists considered a predominantly inbred species
as "blind alley" with no future for long-term survivial or
phyletic evolution. Progressive loss of variability was predicted under inbreeding (and implied role of random drift and/
or selection for the best homozygous line). K. Mather drew
a distinction between short term fitness and long term flexibility criteria in adaptation.

In recent years, however, an embarrassing richness of
"evolutionary virtues" of inbreeders has accumulated. We are
told that inbreeders may often have the optimum balance between too much and too little recombination within linked gene
complexes, optimum breeding structure in terms of gene flow

among neighborhoods (Wright), high colonizing ability, mechanisms for taking advantage of polyploidy and fixed karyotypic changes, the wider adaptedness under marginal environments, and so on. Evolution of self-fertility is suggested for minimizing migrational load under conditions for localized multi-niche selection. Automatic frequency response of selfing genes such as a gene for cleistogamy or homostyly follows from a model due to R.A. Fisher. Haldane had argued for an advantage of inbreeders in terms of more rapid fixation of a favorable allele (or allelic combination). Several other more recent ideas will appear in a review shortly (Ann. Rev. Ecol. and Systematics, 1976). One may be tempted to ask like John Turner did about linkage: Why aren't all species inbreeders?

Two points should be noted for discussion: (1) Clearly, new models have suggested most of these hypotheses, and not new facts. We now have a wide variety of hypotheses, not always stated precisely, that call for massive biological information. At least the three major unknowns are the population sizes, overdominance component in selection and such alternatives for heterosis as phenotypic plasticity and developmental homeostasis. (2) The notion of optimality, in relation to some measure of population fitness ($\bar{W}$, r_m, net reproductive rate, etc.), is often used in these models. This requires both ecological and genetic interpretations as well as some general agreement on the experimental tests of evolution toward optimality. Breeding systems could indeed provide some excellent opportunities in this context.

B. Wallace: *Limitations of electrophoretic studies*

The comments of Nevo and Jain remind me of Lewontin's statement: "To the present moment no one has succeeded in measuring with any accuracy the net fitnesses of genotypes for

any locus in any species in any environment in nature." In a
discussion of problems for the future, such a statement might
prompt some to believe that Heaven resides in having measured
the net fitnesses of genotypes at all loci in all species in
all environments in nature. To set the second version as one's
goal (or the goal of all experimentalists) would be a disas-
trous waste of time; on the other hand, to set about studying
the net fitness of genotypes at one locus in one species in
one environment in nature would be nearly as ridiculous. If
Lewontin's statement has merit, it must reside at some intermediate level --- some genotypes, some loci, some species and
some environments. But, this is approximately what we already
know and for this reason, I suspect, electrophoretic studies
until now have disappointed many population geneticists:
aside from the precise information on genetic variation which
they provide, these studies have to a large extent merely
swelled data of a sort that is already in plentiful supply.
This is not a blanket criticism of electrophoretic studies; on
the contrary, I believe that these were necessary studies
which under no circumstance could have been avoided once the
technique became available.

J.R.G. Turner: *The study of natural selection*

Natural selection should now be regarded as a proved phenomenon. There seems to be little point in establishing any
more cases of natural selection simply for the sake of con-
firming this phenomenon, in the same way that there has long
ceased to be any point in demonstrating Mendelian inheritance
in yet another organism. For a good use of scientific man-
power studies of natural selection should be confined to
instances where it is interesting to discover the form of
selection for some other particular reason, such as

establishing the form which selection is likely to take on allozymes, or the form which selection takes in a particularly interesting ecological context.

M. Nei: Level of protein polymorphism and mutation rate

I would like to direct our attention to some other problems. We now know that there is a great deal of protein polymorphism in natural populations. In practice, however, the level of heterozygosity for protein loci is generally lower than the level that is expected from the estimated rate of neutral mutations and population size. This suggests that there must be some mechanism working to reduce the extent of genetic polymorphism rather than to increase it. Of course, there are a number of such possible mechanisms. One is the bottleneck effect and another is the random fluctuation of selection intensity.

Another important problem with respect to the mechanism of maintenance of genetic polymorphism is the determination of accurate mutation rate. At the present time, the rate of neutral mutations is estimated from the rate of amino acid substitutions in proteins by using the property that the rate of gene substitution is equal to the mutation rate. This method of estimation gives a mutation rate of 10^{-7} per locus per year rather than per generation for electrophoretically detectable alleles. However, this estimate has not been proven experimentally. Since the approximate constancy of the rate of neutral mutations per year is the crucial point in the neutral mutation theory, it is very important to determine the rate and property of mutation. If the constancy of the rate of neutral mutation per year is false, the neutral mutation theory will be seriously damaged.

J.R.G. Turner: Substitution of dominant alleles

Given that both for stochastic and deterministic reasons
dominant alleles are much more likely to be substituted in the
course of evolution than are recessive alleles, what over a
long period of time is to be the expected proportion of mutants
which are dominant or recessive to their established wild-type
allele, and how is this ratio altered by the hitch-hiking
effect of advantageous recessive alleles which become established in the population because they happen to be linked to
an advantageous dominant which is substituting at the same
time? There seems to be no hard theory to deal with this
extremely important phenomenon in the evolution of the genome.

A. Robertson: Information from molecular biology

We have been talking almost entirely about that part of
the DNA essentially as a single copy set which calls for protein. How much is this of the total; possibly it depends on
the species, in some species a very small amount. It seems
quite clear from the information that is now coming out about
the satelite DNA, in which there are lateral repetitions of
certain sequences, with 20% of the DNA of the guinea pig of
this character. In the case of humans and chimpanzees there
is satelite DNA in humans which has evolved since humans split
from chimpanzees, which is now identified in three of our
chromosomes. We find that the ribosomal RNA is a multi copy
situation. But all species in terms of sequences are apparently identical. There are obviously still a lot of facts
of this sort coming out of molecular biology which are just
not properly coped with by population genetics.

W.F. Bodmer: Molecular biology and population genetics

There are a lot of interesting facts that are coming out of
molecular biology. I am not sure I would put it in the frame-

work that much of the repetitive DNA may not be functional; it's there and poses an interesting problem, although it may not be relevant to the evolution we are interested in. One could draw a parallel to the selection-neutralist argument, between the molecular biologists who believe that a lot of DNA is functional and those who believe that a little is. You have for example, within ribosomal DNA coding sequences, a certain part of the repeated sequence that codes for the actual RNA in the ribosome and so-called spacer which is not even transcribed into RNA, and has no obvious function. The sequences that make the RNA in the ribosome are relatively constant. The spacer sequences are quite variable, even within the individual, and there is no obvious explanation for this. It is this sort of phenomenon which has raised the comment that molecular biologists often subscribe to (and which I don't subscribe to) that sequences which have no obvious function can evolve very quickly. The rate of evolution of sequences like this poses very interesting problems. The satelite DNA, is an extraordinary phenomenon. This is a very large amount of DNA, highly repetitive, that can apparently spread from one chromosome to another, presumable not by conventional genetical means and can evolve very rapidly even though it has no obvious functional significance. Some argue that it might be a relevant factor in hybrid sterility, but I believe this to be very unlikely. Perhaps the closest analogue is a segregation distorter or a similar mechanism. It has an evolution of its own and has an advantage of its own within the species. It certainly is true that there is fascinating data of this sort, that one should take note of which poses very interesting problems.

S. Karlin: *Problems of heterogeneity in environments*

A number of verbal dictums abound the evolutionary literature, concerning the relationships inherent to the environmental parameters and patterns of gene frequency variation.
We paraphrase some of these: (i) Polymorphism will be more
likely in more variable environments while unlikely in constant
environments. (ii) Greater heterogeneity in environmental
selection gradients enhances polymorphism. (iii) Decreased
migration flow in a population entails more cases of polymorphism. (iv) More population division facilitates the
establishment of polymorphism.

In these pronouncements, there are several undefined and
vague concepts. What does it mean for one migration pattern
to involve relatively less migration flow than a second
migration pattern? Is there a consistent scale (or a vector
of indices) by which to assess rate of mobility, or degrees
of isolation? By what criteria is a prescribed environmental
selection gradient judged more heterogeneous than another
environmental structure? Undoubtedly not all environmental
grains are comparable. What are meaningful means for measuring
degrees of heterogeneity in both ecological and genetic
(fitness) terms? A common measure used involves the variance
taken over time or space of some ecological variable. This is
obviously too restrictive.

A number of results on these topics were reported in my
papers of this conference. Several ways of comparing environmental structures and migration patterns were introduced, but
a serious limitation in these concepts is that environmental
comparisons are done in terms of fitness parameters. How to
convert the ecological or geographical parameters or related
environmental profiles into fitness terms remains a challenging
and fundamental problems, (i.e. the problem of translating

ecological into genetic ranges of parameters values). This problem is not the same as that of disentangling the effects of selection acting on phenotypes versus genotypes. Proper listings of ecological parameters by naturalists are needed and more importantly the development of criteria for comparing ecological profiles suitably standardized appears to be essential.

R.J. Berry: *Difficulties with interpretation of clines*

The difficulties experienced by naturalists when they encounter theoreticians can be well illustrated by the cline in melanism found in the moth Amathes glareosa Esp. in the Shetland Islands. A dominantly-inherited morph (f. edda) has a frequency of 97% at the north of the island group, but declines more or less steadily to less than 2% at the southern end, 54 miles away. Initially this seemed to represent a straightforward problem in gene flow and changing selection pressures, and the data from all but one site (this exception may be symptomatic and significant) could be fitted to a simple model proposed by Haldane (Haldane 1948; Kettlewell & Berry, 1961). This simplistic interpretation was criticized by Ford (1964) on several grounds, not least that a mathematical model can suggest only the average advantage along a cline and demands gross over-simplification. However Wallace (1968) defended it since "without a model, it is unimportant what happened to edda on Shetland; this fine work takes on meaning only in relation to other studies and from the generalizations which emerge from them. It has no meaning standing in splendid isolation".

Wallace is undoubtedly right in principle; the difficulty is that Ford was almost certainly correct with his reservations. The genetical structure of A. glareosa in Shetland is affected by habitat selection, behavioral differences, predation, dominance modification, and local adaptation - all distinct from the main edda cline, yet inevitably interacting with it. At least three entirely theoretical analyses have been made of the cline data but the effective fitness factors affecting the frequencies of the edda allele are still virtually unknown. This is a case where theoretical insight might guide study to the benefit of all.

The rift between the "Two Cultures" expounded by C.P. Snow exists to some extent between theoretical and practical population geneticists. It applies to the problems analyzed by each, but perhaps more basically to a lack of communication produced by the differing languages of each. The situation is not unlike two islands with little migration between them. Only a few scientists are equally at home in each 'niche'.

The work of the genetical "naturalist" is to identify and investigate field problems. Some - perhaps all - of these will have a wide relevance which can profit from a generalized approach. A good example of this is the study of the Fair Isle Arctic Skua polymorphism. Clues to the reason why this exists came from a simple observational approach combined with an un-sophisticated analysis of data collected mainly for other purposes (Berry and Davis, Proc. R. Soc. B 175, 1970). We have profited enormously by the more rigorous studies of Peter O'Donald, combined with data specifically collected (O'Donald, this volume). It would undoubtedly benefit many theoreticians to spend time on field work so that they might learn the problems and apply their own insights to particular situations. The littoral zone (particularly of rocky shores) is one eco-system which might profit from joint practical and theoretical studies of the genetical influences acting on and in it.

S.K. Jain: *Relations between ecological and genetic problems*

I wish to refer to the tie between the ecological and genetic aspects of population processes. There are numerous life cycle components that determine the success of a local population to an ecologist whereas there are equally numerous variables of genetic systems and population structure that underlie population genetic predictions of adaptability. The genetic variance in fitness and thereby determining how much and in what stages of life cycle selection occurs constitute one of the most exciting meeting grounds for ecologists and geneticists. As also noted by Karlin, another fundamental problem is how to relate environmental descriptors to the variation in fitness and the kinds of adaptive responses by individuals as well as populations as units (metapopulations, perhaps). We need to collect such information not only as snapshots of a process but in time sequence in order to avoid the assumptions of equilibrium before testing it. Another point to be made by my own field studies is that of finding rather large variances around the estimates of various parameters (e.g. rate of gene flow, population size, outcrossing rate). I would like to think that very often that is the reality and not that our methods are faulty or sloppy. Most specific attempts to fit data into some neat predictions of a model may even suffer in objectivity if we choose to ignore this messy aspect of real world data.

M.W. Feldman: *Genotypes vs. phenotypes*

The question of the relevance of the current models of clines to observations about clines reduces to one of the central themes of Lewontin's book. The observations consist of frequencies of genes, geographic variables and occasionally phenotypic variables. The models are phrased in terms of selection parameters which act on genotypes and which cannot be estimated. The extent of migration cannot be estimated for most organisms. Should the models be structured in terms of phenotypic variables? Should we set about developing new statistical techniques for estimation of population size and migration rates? All of these areas are in unsatisfactory shape at present.

E. Nevo: *Levels of explanation*

There are two levels of explanation to allozymic variation or cline structure. The first level would be to describe the variations in space and time and to relate them to environmental background structure and suggest the possible forces operating to maintain these variations. The second level would be to know the biochemical and physiological correlates of the alleles we are handling, preferably under ecological conditions which are involved in the natural life cycle of the organism. I believe both levels of operation are compatible and should ideally complement each other. Ultimately, we wish to know the relative contributions of all forces to the fitnesses. I would suggest that one promising way should be to study a variety of natural populations of different organisms differing in environmental background complexities, history, and population structure and dynamics, where the critical experiments are defined cooperatively by naturalists and theoreticians. Finally, since models involve hidden assumptions it is imperative to make these explicit to the naturalist so he can judge the applicability of a model to a real situation.

G. Thomson: *Data is non-equilibrium*

I would like to question the use of equilibrium information for non-equilibrium situations. We must in reality ask the question if populations ever are at equilibrium? Because everything, all the parameters we are looking at, are generally changing in time.

M. Nei: *Biological relevance of equilibrium theories*

In my opinion there has been too much emphasis on the study of equilibrium theories of selective polymorphism in the past. I note that the majority of these theories have never been really used by biologists to interpret their data from natural populations. This is because, in addition to the difficulty of estimating genotype fitnesses, the biological world is always changing so that the mode of selection for a locus or a group of loci never stays constant. I also note that there are not many cases of balanced polymorphism that were documented unequivocally. Probably the only authenticated case of a single-gene overdominance is that of sickle-cell anaemia gene, which causes resistance to malaria in heterozygous condition. The ecology of the vector mosquito Anopheles gambiae, however, suggests that the condition for this polymorphism in Africa has existed only for about 7,000 years (about 300 generations) (Livingstone, Amer. Anthrop. 60: 533, 1958). Furthermore, because of the recent improvement of public health, this gene is expected to gradually disappear from the human population.

In recent years the amino acid sequences of cytochrome c, haemoglobin α and β chains have been studied in many different organisms. Comparison of these sequences among different species indicates that gene substitution is almost always occurring at a locus and the long-term polymorphism of a pair

of specific amino acid sequences encompassing different
organisms has not been reported.

S. Karlin: Equilibrium and non-equilibrium studies

It is often claimed that equilibrium situations are unrealistic, and it is only relevant to examine transient phenomena. There certainly are population cases, (e.g., some bacterial, fungi, insect, fishes) in equilibrium (perhaps in a
statistical sense) or close to equilibria for sufficient
durations, and other population cases (especially humans)
where parameters are constantly changing. It is certainly
desirable to be able to offer a complete analysis of both the
dynamic and equilibrium behavior of the population, but this
is a prohibitive task in the present state of our science.
Equilibrium and non-equilibrium studies can only be carried
out in parallel and the analyses of each lend insights for
the other.

Knowledge of the stable equilibria configurations does give
some information pertaining to the dynamics of the system.
More specifically, from the equilibria contingencies one can,
for example, infer to some extent features and possibilities
of the underlying selection or migration structure. If the
strength (the largest eigenvalue of the approximating linear
system) and positions of the stable equilibria states are
worked out then it is possible to map out the domains of
attraction to the relevant equilibria and correspondingly
assess sensitivity to initial conditions. A description of
all the equilibria states also provides a control for any
simulation or computation schemes used in studying the transient process. By using probabilistic and stochastic process
concepts it is further possible to allow even for certain
classes of random changes in the parameters of the model,

(e.g., see TPB Dec. pp.355-405).

M. Nei: *The problem of speciation*

I agree with Nevo that the process of speciation is
not well understood at the present time and a more intensive
study has to be made. In my opinion one of the most important
questions with respect to speciation is how quickly reproductive isolation is developed between two incipient species.
In this conference I presented some of the results of our
theoretical studies on the fixation time of mutant genes
affecting reproductive isolation. They agree with the experimental result that the time after divergence between two
hybridizable species varies greatly among different pairs of
species in each of the mammalian and amphibian groups (Wilson
et al., Proc. Nat. Sci. U.S. 71: 2843, 1974). Nevertheless,
our study is dependent on a number of assumptions which must
be confirmed by experiments. Particularly, the genetic mechanism of reproductive isolation is not well understood, though
a number of authors have studied this problem. It is hoped
that in the future this problem will be studied at the molecular level as well as at the gene level. I have already discussed some possible scheme of the evolution of hybrid inviability or sterility at the molecular level (Nei, Molecular
Population Genetics and Evolution, North-Holland Pub. Co.,
Amsterdam, 1975).

B. Wallace: *Speed of reproductive isolation*

Nei asks how quickly reproductive isolation is developed
between two incipient species. Some information regarding
sexual isolation is known: K.F. Koopman, 1950, Evolution 4:
135-148; L. Ehrman, 1971, Amer. Nat. 105: 479-483; L. Ehrman,
1973, Amer. Nat. 107: 318. Many selected traits prove to be
additive on a log scale; if the selection involves an impor-

tant component of fitness (such as resistance to insecticides or toxic salts), hybrids between two dissimilarly selected populations may rather quickly exhibit inferior fitness relative to non-hybrid individuals. Thus, the essential features by which two groups are recognized as species may arise rather suddenly. This view suggests a corollary: Much of the correlation observed between genetic distance and morphological similarities of pairs of species may reflect individual correlations of each of these items with the length of time elapsed since the crucial, nearly instantaneous steps in speciation (reproductive and sexual isolation) occurred.

M. Slatkin: *Supergenes and coadaptation*

I would like to raise a different issue, one which comes from the classical evolutionary literature and which disappeared either because of lack of understanding or due to some negative selection. In the models that have been discussed there appear to be two kinds of coadaptation which are being analyzed and not recognized as such. One is the kind which involves linkage. With this model of coadaptation, the implication is that there is correlation of function and location on the chromosomes, that is, you get linked genetic complexes or what have been called super-genes and exemplified by inversion in Drosophila. Other species have also been held as examples of these supergenes. In that sort of analysis linkage theory is essential and the question seems to be under what circumstances would coadaptive gene complexes of that sort evolve. Another type of theory, which at present is totally different, concerns the existence of modifier genes where the alleles which are correlated in function are not necessarily located together, but that evolution of one locus is somehow influenced by the condition of a small or large number of other loci, and that evolution at these different loci somehow will proceed, not just in relation to external selection pressures, but in relationship to internal genetic conditions. I merely wish to point out the difference between these models, both of which at various times have been justified as relevant for coadaptive gene complexes or genomic unity.

W.F. Bodmer: *Gene duplication and coadaptation*

Concerning this question of coadaptation of gene sets and supergenes I believe one has to distinguish two situations. Many of the supergenes in higher organisms probably originated by gene duplication and, therefore, they are there together anyway, and one might expect that their evolution would be such that they would have similar functions. One has to distinguish this as an origin of a linked set very clearly from situations where quite different genes are brought together by translocation or inversion. I think in the case where the genes have an origin by duplication they are more likely to have some relationship of coadaptation and much less where they have been brought together. Actually, there are very few cases that I know of where one can be sure that such genes have been brought together. I think it has often been postulated that modifiers work better if they are closely linked with the gene they are modifying. In fact, the first examples of possible supergenes involved situations of incompatibility systems in plants where one had assumed there was modification which was probably linked with the major genes. That brings the two types of situations Slatkin referred to, close together.

B. Wallace: *Linkage and coadaptation*

Slatkin's comments raise points that were made in my own presentation but which may stand repetition. First, it should be noted that the breakage points that have given rise to naturally incurring inversions in the genus <u>Drosophila</u>, when summed over all species, are essentially randomly distributed. Second, it is known from at least two laboratory experiments that the presence of one inversion polymorphism in a population tends to exclude the incorporation of further polymorphisms; there appears to be a ceiling on the advantage of heterozygous individuals. Third, the model I described favors the accumulation of dissimilar homologous control regions within chromosomal segments that differ in gene arrangement. This accumulation would correspond to Slatkin's coadaptation that requires linkage; I would stress, however, the need to eliminate duplication-deficiency cross-over strands rather than linkage, itself. Linkage, that is, is a consequence and not a cause. Fourth, the model also suggests (1) that there exist optimal combinations of sensor sequences at different loci and (2) that no single combination is best because dissimilar sequences at homologous loci operate most efficiently. It seems to me that these latter, essentially epistatic relationships between loci correspond to Slatkin's second kind of coadaptation. Under the model which I described however, the two kinds are related: the second favors the initiation and development of the first.

D. Cohen: *Coadaptation and equilibrium*

The problems of coadaptation and of equilibrium are closely related. Strong epistasis between genes will undoubtedly slow down the rate of approach to equilibrium, and in fact where only a very unique combination of alleles is advantageous, an equilibrium state may never be reached, and we don't need either inversion segments or some other types of linkage supergenes to get strong interaction in alleles.

I would like to suggest the following distinction between two extreme types of genetic characters. To the first belong those characters which have little or no epistatic relations with other characters. These would be characters which are expressed late in the developmental sequence or those which have little functional interaction with other characters, for example, skin color. The frequency of alleles in the loci controlling such characters will change rapidly even under moderate selection pressures.

The second group of characters are those which have strong epistatic interactions with many other characters. Such epistatic interactions represent strong functional interactions during development or in the phenotypic expression. Such characters may change very little or not at all under selection because of the very low probability for the simultaneous occurrence of a favorable combination in many loci. These characters are in fact "frozen" at a non-equilibrium level.

An interesting aspect of the distinction between these two types of characters is that they tend to be determined also by their evolutionary history.

Repeatedly changing selection pressures in the past will cause repeated changes in allelic frequencies and prevent the establishment of a set of specific epistatic interactions, so that these characters will remain free to change in response to future directional selection. On the other hand, strong continuing selection for perfection of any complex system under constant conditions will favor those alleles which are able to function better because of subtle and complex interactions with alleles in other loci. Such characters will gradually become more and more "frozen" until they could no longer respond to future directional selection. A release of a "frozen" character may occur when a species suddenly finds itself in a new environment and without competition. Under such conditions, the probability of success of a new combination is greatly increased.

Functionally and developmentally complex organ systems can respond to directional selection, e.g. a change in body size or limb size and proportions. This is made possible by the control of the complex set of developmental interactions by one or a few dominant characters. For example, limb size is mainly determined by the growth of the bones. All the other tissue systems in the limb adjust themselves by some feedback mechanism to the development of the dominant character. Thus, there is no need for the highly improbable combination of co-adapted genetical changes in muscles, blood vessels, nerves, skin, etc., when a genetic change occurs which affects the size of a limb by changing the growth pattern of the bones. I believe that this type of feedback interactions during development is an essential prerequisite for the evolution of complex metazoan organisms.

A. Beiles: *Protective aspects of coadaptation*

Regarding another aspect of coadaptation, I would like to call attention to inversions, supergenes and linkage. In some organisms generation time is shorter than the time it takes for environmental conditions to change. For such situations coadaptation can be a device to avoid "quick" adaptations to temporary conditions. Thus, alleles which are suitable for different, and recurrently occurring, conditions are not lost.

For this reason one should be hesitant to generalize the conclusions drawn from short generation organism like Droso-phila. Organisms with longer generation times should be critically assessed.

M.W. Feldman: *Genic associations*

There is every reason to expect the "disequilibrium" problem to go the same route as the neutrality controversy. Small
effects are very difficult to measure, and between loci with
multiple alleles the maximum expected associations may not be
as high as for diallelic loci. In addition, for multiple
allelic loci the range of recombination values for which
association is expected may well decrease. I am not optimistic
that questions concerning selection and neutrality will be
successfully approached using intergenic associations as a
measure or tool. On the other hand, the topic still poses
intrinsic problems of great interest. What is the appropriate
two locus generalization of the infinite alleles model? Is it
universally true that to have strong associations between
loosely linked genes there must be strong epistasis? Is it
always the case that the amount of selective interaction which
can maintain an association between tightly linked genes is
less than that necessary to maintain the same association between loosely linked genes? For most of these questions we
have cases in which accepted theory is violated. It behooves
us to find for what general classes of models such violations
occur.

Concerning Slatkin's two nodes of coadaptation, both are
discussed in my paper with Krakauer. They do not seem as
different to me as they do to him. It is a problem of the fate
of a new gene on the one hand (modifier) and of a new chromosome on the other (inversion). By assuming that the modifier
gene is extremely close to the major genes, merely by renaming
the chromosomes, the two methods coincide.

A. Beiles: *Linkage disequilibrium and clines*

Previously I have suggested that in numerous situations

populations seem to be exposed to (a) short and drastic phases,
alternating with (b) long neutral periods. If this model is
true, clines and disequilibria formed by selection will consequently undergo numerous "quiet" generations. But while
linkage disequilibria will rapidly decay by recombination,
clinal differentiation will persist. Such clinal patterns,
under neutral conditions, have already been described by
Karlin and Richter-Dyn. In other words, under such dual
operation of selection and neutrality, one should expect only
rare detections of linkage disequilibria, unless protected by
inversions, etc., while clines should be found much more
frequently.

S. Karlin: *Aspects of multi-locus problems*

The analysis of general multi-locus selection patterns
remains an important open problem of population genetics. It
would be of value to obtain a finer correspondence of the
character of the stable equilibria possibilities showing its
dependence on the selection parameters, recombination rates,
mating pattern and other factors. I have discovered the
following apparently robust principle, (see TPB, June, 1975)
which may have practical consequences.

There are two types of polymorphisms occurring in multi-
loci selection models: (i) A balance of the relative genotypic selection forces pervades the evolutionary process
establishing a stable polymorphic state while the recombination
mechanism exercises minor effects. (ii) Selection forces
alone (without recombination) would lead to the elimination of
certain chromosomal types but in the presence of some recombination rates the full complement of chromosomal types are
created and sustained.

The characteristics of the polymorphic outcomes for the

circumstances of (i) and (ii) are markedly different. For
the situation of (i), usually a single, rather central, globally stable polymorphism is manifested for each set of recombination rates. Equivalently, the multi-allele one-locus
model corresponding to no recombination but subject to balancing selection essentially determines the equilibrium configuration which qualitatively persists for all values of the
recombination parameters. In the situation of (ii) one or
more stable polymorphisms are created mostly for tight linkage
where chromosomal types divide into two groups such that the
frequencies among one group are each of very small magnitude,
while those of the other group are represented with more substantial frequency. Thus, for small recombination rates where
usually a number of such polymorphism exist, all express
strong linkage disequilibrium and the actual polymorphism
achieved is sensitive to the initial population makeup. With
moderate or loose linkage these polymorphisms are in most
circumstances not maintained. The fitting and discrimination
of multi-locus data to conform to the possibilities (i) and
(ii) should be of interest and relevant to the interpretations
of some observations.

What kind of frequency data at a locus can one expect if
this locus is maintained segregating as part of a multi-locus
selection balance? In this vein, the following is a valid
result. If observations made at a single locus are statistically consistent with heterozygote advantage then the data
can well reflect a polymorphic stable equilibrium involving
several loci. Put in a contrapositive form, if the observations at a locus significantly preclude an explanation based
on overdominance then segregation at the locus in question
cannot be ascribed to a supergene complex maintaining polymorphism attributable solely to selection balance.

Much attention in multi-locus studies concerns linkage disequilibrium evaluations. A serious drawback in the analysis of the linkage disequilibrium function is that it is principally a two-loci two-allele concept. For the case of multiple alleles at two loci the proponents of this concept advocate the examination of all pairwise disequilibrium functions where in each evaluation two alleles (one at each locus) are distinguished and the other alleles are lumped creating a single alternative allele at each locus. The unresolved difficulties in statistical theory attendant to the interpretation of partial correlation coefficients are also encountered in dealing with this vector of D values. It appears that too much emphasis has been placed on D as a tool in the determination of the behavior of two-loci models. A fortiori, its weaknessess are substantially magnified when treating multi-loci-multi-allele systems.

It is challenging to find different intrinsic measures of higher order interactions among loci and alleles. There are many speculations and pronouncements relating to "association of genes", "strong vis a vis weak epistasis", "degrees of linkage interactions". These concepts are vague and often ill-defined. How does one effectively distinguish strong as against weak epistasis? What are meaningful sets of indices for evaluating and contrasting different selection regimes? More specifically, how does one order different multi-locus selection patterns which are often inherently not comparable? How do we assess the consequences of different sets of recombination frequencies especially for more than two loci? What does it mean for loci or gene frequency patterns to be "associated"? This is generally done in terms of pairwise disequilibrium functions which we have already remarked on to be unnatural.

Claims have been advanced that most genes are unlinked and, therefore, multi-locus studies should concentrate mainly on the case of loose linkage. To get a proper perspective even for the case of loose linkage it is probably essential to study simultaneously the equilibrium structure for all ranges of recombination frequencies. The issue of whether much or little linkage disequilibrium exists in nature appears to be quite unresolved and probably all possibilities are present. There are increasingly many documented examples of high disequilibrium for many inversion and specific gene combinations in Drosophila, in the HLA system in man and elsewhere. On the other hand some experimenters find no significant linkage disequilibrium. How does one decide or assess when significant linkage disequilibrium is present. The analysis via Chi squares and contingency tables is not clearcut or unequivocal.

A 6
B 7
C 8
D 9
E 0
F 1
G 2
H 3
I 4
J 5

832

QH455 .P66
Population genetics and ecology : proceedings of t

Randall Library – UNCW

NXWW

304900223186